Gamma Ray Transients
and
Related Astrophysical Phenomena
(La Jolla Institute, 1981)

AIP Conference Proceedings
Series Editor: Hugh C. Wolfe
Number 77

Gamma Ray Transients and Related Astrophysical Phenomena
(La Jolla Institute, 1981)

Editors
Richard E. Lingenfelter
Hugh S. Hudson
Diana M. Worrall
Center for Astrophysics and Space Sciences
University of California at San Diego

American Institute of Physics
New York   1982

Copying fees: The code at the bottom of the first page of each article in this volume gives the fee for each copy of the article made beyond the free copying permitted under the 1978 US Copyright Law. (See also the statement following "Copyright" below). This fee can be paid to the American Institute of Physics through the Copyright Clearance Center, Inc., Box 765, Schenectady, N.Y. 12301.

Copyright © 1982 American Institute of Physics

Individual readers of this volume and non-profit libraries, acting for them, are permitted to make fair use of the material in it, such as copying an article for use in teaching or research. Permission is granted to quote from this volume in scientific work with the customary acknowledgment of the source. To reprint a figure, table or other excerpt requires the consent of one of the original authors and notification to AIP. Republication or systematic or multiple reproduction of any material in this volume is permitted only under license from AIP. Address inquiries to Series Editor, AIP Conference Proceedings, AIP.

L.C. Catalog Card No. 81-71543
ISBN 0-88318-176-2
DOE CONF- 810871

PREFACE

The La Jolla Institute and the Center for Astrophysics and Space Sciences at the University of California, San Diego supported the workshop on Gamma Ray Transients and Related Astrophysical Phenomena, held in La Jolla on August 5-8, 1981. This workshop brought together active researchers with a wide range of observational and theoretical backgrounds to review and study the explosive phenomena responsible for the various gamma ray transients - solar flares, cosmic gamma ray bursts and longer lived gamma ray line transients. Although these transients arise from very disparate sources, they nonetheless may have in common a number of plasma processes and emission mechanisms. X-ray burster observations and theories were also reviewed with emphasis on their relationship to gamma ray bursts. The wealth of recent observational data, particularly from the SMM, HEAO, and VENERA satellites, made this workshop especially timely.

In this volume the authors of both the invited and contributed papers were given the opportunity to elaborate upon their talks to provide more detail than could be presented at the meeting. We hope that the publication of these proceedings will serve as stimulus to further research.

The workshop organizing committee consisted of Edward L. Chupp, University of New Hampshire, Thomas L. Cline, NASA Goddard Space Flight Center, Paul Gorenstein, Harvard/Smithsonian Center for Astrophysics, Richard E. Lingenfelter (Chairman), University of California, San Diego, Reuven Ramaty, NASA Goddard Space Flight Center, Gerald H. Share, NRL Space Sciences Division, Ian B. Strong, Los Alamos National Laboratory, and Stanford E. Woosley, University of California, Santa Cruz.

We are particularly grateful to Cindy Duffy, who coordinated the arrangements for the workshop and helped assemble this volume, to Adolf R. Hochstim and E. Margaret Burbidge whose support and enthusiasm made the whole show possible and to the La Jolla Institute and the Center the Astrophysics and Space Sciences, for their sponsorship.

The Workshop and these Proceedings were financed by independent research funds of the La Jolla Institute.

Richard E. Lingenfelter
Hugh S. Hudson
Diana M. Worrall

La Jolla, California

TABLE OF CONTENTS

I. GAMMA-RAY TRANSIENTS

   A. GENERAL CHARACTERISTICS

GAMMA-RAY BURST SYSTEMATICS
R.W. Klebesadel, E.E. Fenimore,
J.G. Laros and J. Terrell 1

A REVIEW OF THE 1979 MARCH 5 TRANSIENT
T.L. Cline 17

FREQUENCY OF GAMMA-RAY BURSTS
$> 3 \times 10^{-6} \text{erg/cm}^2$
G.H. Share, K. Wood, J. Meekins and
D.J. Yentis 35

OBSERVATIONS OF GAMMA-RAY BURSTS FROM
10 keV TO 9 MeV
G.H. Share, J.D. Kurfess, S. Dee,
E.L. Chupp, J.M. Ryan, D.J. Forrest,
J. Lanigan, E. Rieger, G. Kanbach and
C. Reppin 45

GAMMA-RAY BURST OBSERVATIONS BY THE HEAO-3
HIGH RESOLUTION GAMMA-RAY SPECTROMETER
W.A. Wheaton, J.C. Ling, W.A. Mahoney,
G.R. Riegler and A.S. Jacobson 55

OBSERVATIONS OF A GAMMA-RAY BURST AND OTHER
SOURCES WITH A LARGE-AREA, BALLOON-BORNE
DETECTOR
R.B. Wilson, G.J. Fishman and
C.A. Meegan 67

A SEARCH FOR ASSOCIATIONS OF RADIO PULSES
AND GAMMA-RAY BURSTS
P. Inzani, G. Sironi, N. Mandolesi and
G. Morigi 79

   B. POSITIONS AND DISTRIBUTION

GAMMA-RAY BURST POSITIONS
K. Hurley 85

STATISTICAL STUDY OF GAMMA-RAY BURST
SOURCE LOCATIONS
G. Pizzichini 101

THE OBSERVATIONAL CONSEQUENCES OF AN
INTRINSIC BURST LUMINOSITY DISTRIBUTION
    M.C. Jennings 107

DERIVATION OF THE LUMINOSITY OF A GAMMA-
RAY BURST
    J.C. Higdon 115

C. ENERGY SPECTRA

GAMMA-RAY BURST SPECTRA
    B.J. Teegarden 123

10 JUNE 1974 TRANSIENT
    J.C. Ling, W.A. Mahoney, J.B. Willett and
    A.S. Jacobson 143

TIME VARIATIONS OF AN ABSORPTION FEATURE IN
THE SPECTRUM OF THE GAMMA-RAY BURST ON 1980
APRIL 19
    B.R. Dennis, K.J. Frost, A.L. Kiplinger,
    L.E. Orwig, U. Desai and T.L. Cline . . .153

SEARCH FOR TIME VARIATIONS IN 511 keV FLUX BY
ISEE-3 GAMMA-RAY SPECTROMETER
    J.P. Norris, T.L. Cline and
    B.J. Teegarden 163

D. EMISSION PROCESSES

RELATIVISTIC PLASMAS
    R.J. Gould 169

THE ROLE OF CYCLOTRON RESONANCE PROCESSES IN
THE SPECTRA OF ACCRETING NEUTRON STARS
    J. Trümper 179

THE LOW ENERGY SPECTRA OF GAMMA-RAY BURSTS
    R.W. Bussard and F.K. Lamb 189

ON THE INTERPRETATION OF GAMMA-RAY BURST
CONTINUA AND POSSIBLE CYCLOTRON ABSORPTION
LINES
    E.E. Fenimore, J.G. Laros, R.W. Klebesadel,
    R.E. Stockdale and S. Kane 201

GAMMA-RAY LINES FROM SOLAR FLARES AND COSMIC
TRANSIENTS
    R. Ramaty, R.E. Lingenfelter and
    B. Kozlovsky 211

ON THE THEORY OF GAMMA-RAY AMPLIFICATION THROUGH
STIMULATED ANNIHILATION RADIATION (GRASAR)
    R. Ramaty, J.M. McKinley and F.C. Jones ..231

EMISSION MECHANISMS OF THE MARCH 5th 1979
GAMMA-RAY TRANSIENT
    E.P.T. Liang 241

E. SOURCE MODELS

SURFACE AND MAGNETOSPHERIC PHYSICS OF NEUTRON
STARS AND GAMMA RAY BURSTS
    D.Q. Lamb 249

THE THERMONUCLEAR MODEL FOR GAMMA-RAY BURSTS
    S.E. Woosley 273

GAMMA-RAY BURSTS FROM YOUNG NEUTRON STARS
    K. Brecher 293

A TWO-DIMENSIONAL MODEL FOR GAMMA-RAY BURSTS
    B.A. Fryxell and S.E. Woosley 299

GAMMA BURSTS FROM NEUTRON STARS AND STELLAR
FLARES
    S.A. Colgate 309

PRECURSORS TO GAMMA-RAY BURSTS IN THE
ASTEROID IMPACT SCENARIO
    D. Van Buren 311

PROTON DECELERATION IN A NEUTRON STAR
ATMOSPHERE
    J.G. Kirk and D.J. Galloway 315

II. X-RAY BURSTS

SOME TOPICS ON THE X-RAY PULSARS AND THE X-RAY
BURSTS OBSERVED BY THE SATELLITE HAKUCKO
    M. Oda 319

ASSOCIATION OF RECURRENT SOFT X-RAY TRANSIENTS
WITH X-RAY BURST SOURCES
    T. Murakami 339

NUCLEAR FLASH MODELS FOR X-RAY BURST SOURCES
    P.C. Joss 349

A NUMERICAL SIMULATION OF THE MAGNETOSPHERIC
GATE MODEL FOR THE X-RAY BURSTERS
    S. Starrfield, S. Kenyon, J.W. Truran and
    W.M. Sparks 351

THERMONUCLEAR RUNAWAYS IN THICK HYDROGEN RICH
ENVELOPES OF NEUTRON STARS
    S. Starrfield, S. Kenyon, J.W. Truran
    and W.M. Sparks 355

III. SOLAR TRANSIENTS

SOLAR ENERGETIC PHOTON TRANSIENTS (50 keV-
100 MeV)
    E.L. Chupp 363

HIGH-ENERGY OBSERVATIONS OF STELLAR FLARES:
COMPARISON WITH THE SUN
    H.S. Hudson 383

SOLAR HARD X-RAY IMAGES OBSERVED BY ASTRO-A
    K. Ohki, S. Tsuneta, N. Nitta, T. Takakura,
    K. Makishima, T. Murakami, Y. Ogawara and
    M. Oda 395

EVIDENCE FOR DELAYED SECOND-PHASE ACCELERATION
IN SOLAR FLARES
    J.B. Willett, J.C. Ling, W.A. Mahoney,
    G.R. Riegler and A.S. Jacobson 401

SECOND-PHASE ACCELERATION VERSUS SECOND-STEP
ACCELERATION IN SOLAR FLARES
    T. Bai 409

SOLAR FLARE ENERGETICS
    R.P. Lin 419

THE IMPULSIVE FLUX TRANSFER SOLAR FLARE MODEL
    P.J. Baum and A. Bratenahl 433

IV. INSTRUMENTAL CONCEPTS

THE BURST AND TRANSIENT SOURCE EXPERIMENT FOR
THE GAMMA-RAY OBSERVATORY
    G.J. Fishman, C.A. Meegan, T.A. Parnell
    and R.B. Wilson 443

MODULATED MULTIPLE SLIT CAMERA FOR IMPROVED
LOCALIZATION OF GAMMA-RAY BURSTS
    P. Gorenstein 453

A THIRD-GENERATION SMALL SPECTROSCOPY
EXPERIMENT FOR HARD TRANSIENT EVENTS
    R.W. Klebesadel, W.D. Evans, J.G. Laros,
    G.H. Nakano, D.W. Datlowe, W.L. Imhof and
    H.S. Hudson 469

PROPOSED HARD X-RAY IMAGING AND GAMMA-RAY
BURST STUDIES FOR XTE
    J.E. Grindlay and S.S. Murray 477

ION CHAMBER GAMMA BURST DETECTOR
    S.A. Colgate 489

V.   PANEL SUMMARY AND RECOMMENDATION 497

Sec. I A Gamma Ray Transients: General Characteristics

GAMMA-RAY BURST SYSTEMATICS

R. W. Klebesadel, E. E. Fenimore, J. G. Laros, J. Terrell
University of California, Los Alamos National Laboratory
Los Alamos, New Mexico 87545

ABSTRACT

The systematics of gamma-ray bursts are considered. Categorization within this class of events appears reasonable based upon both temporal and spectral features. The gamma ray bursts can be categorized as bursts like the 1979 March 5 event, bursts like the 20 minute transient observed by Jacobsen, and classical (that is, normal) bursts. Further, the classical bursts can be subdivided into bursts which are very brief, bursts with a doublet or quasi-periodic time structure, or bursts which are long and irregular. There appears to be a correlation between the occurrence of emission lines and the doublet or quasi-periodic class and also between the occurrence of absorption lines and the long and irregular class.

INTRODUCTION

Only recently has enough data been collected to attempt a taxonomy for gamma ray bursts. Figure 1 illustrates the relationship of gamma-ray bursts to other phenomena within the field of high-energy astrophysics, studied through x- and gamma-ray astronomy. It must be stressed that this classification relates specifically to the phenomenology rather than source objects. As observations improved (both in quality and quantity) it was resolved that many X-ray sources exhibited multiple characteristics. Steady-state x-ray sources include all those which have appeared to be permanent on the time-scale of the history of the observations within this energy (slightly over 10 years). These include many variable sources, both periodic and aperiodic. In fact, variability at x-ray energies appears to be almost universal rather than exceptional.

The transient sources are those which have appeared or intensified from a level undetectable at the current observational sensitivities, then subsided once again below detectable levels. These include two rather arbitrarily defined categories. The long-term transients include those which persist for weeks or months. The first of these to be characterized by definitive observations was Cen X-2, which first exhibited a nova-like outburst in 1967[1]. Conversely, fast transients including both X-ray bursters and gamma-ray bursters are those which persist for times ranging from a fraction of a second to several minutes. The temporal behavior of X- and gamma-ray bursters is not markedly different; however, they are readily differentiable on the basis of their spectral characteristics.

Gamma-ray bursts were the first of the fast transients to be recognized as a distinct astrophysical phenomenon[2]. Temporal variability on very short time scales was a characteristic which served to identify gamma-ray bursts from the backgrounds generated by

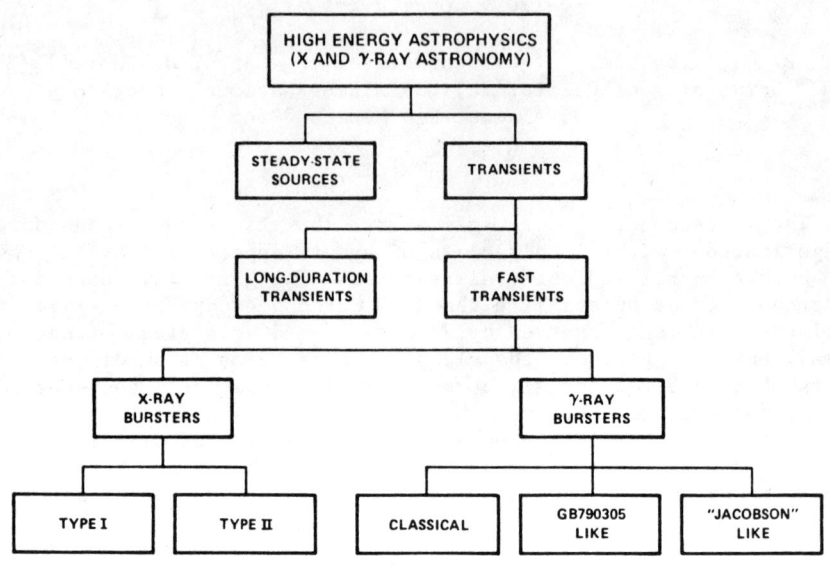

Fig. 1. The relationship of gamma-ray bursts to other phenomena in the X-ray and gamma-ray regime.

Table 1
Temporal Characteristics of Classical Gamma-Ray Bursts

Parameter	Min.	Max.
Rise time	<10 ms	1 s
FWHM of peaks	20 ms	2 s
Total Duration	0.5 s	70 s

local particles and solar flares. The gamma-ray bursts observed to date fall into three categories (see fig. 1): the "classical" gamma-ray bursts, events similar to GB790305, and longer duration events such as was observed by Jacobson et al.[3] GB790305 differs from the classical GRB's in many ways[4] but in particular was much softer. The source apparently also produced bursts recurring over an interval of several weeks[5]. There is probably at least one other source (designated B1900+14) of bursts with similar temporal and spectral characteristics and recurrent nature.[6] The "Jacobson" event was a 20 min long transient exhibiting evidence of a number of line features as observed on 1974 June 10 by the JPL balloon-borne, high-resolution gamma-ray spectrometer[3]. However, it is presently a single member of an arbitrarily-drawn category. It is probable that it is unique only because of the limited observations with such instrumentation. The remainder of this paper will be concerned with classical gamma-ray bursts and the other classes noted above will not be discussed further.

THE CLASSICAL GAMMA-RAY BURSTS

The early observations of classical gamma-ray bursts did not reveal a common pattern or groups of patterns in the temporal structure (although some similarities in structure may have been masked by the nonlinear time binning which characterized the early Vela data). The range of values of several parameters defining the

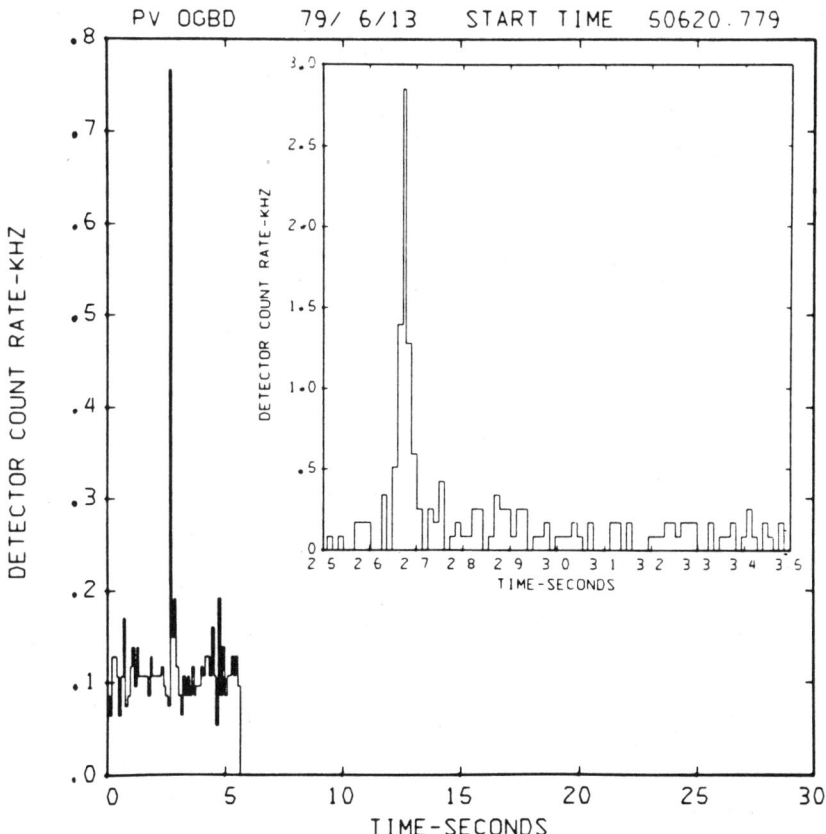

Fig. 2. Event GB790613, among the briefest gamma-ray bursts observed to date. Inset is shown the full temporal resolution available from the Pioneer Venus OGBD event record.

temporal characteristics of classical bursts are given in Table 1. Generally, these characteristics vary throughout the quoted ranges without apparent grouping.

Figure 2 illustrates the temporal structure of GB790613 which represents the fastest-rising, narrowest, and briefest of the gamma-ray bursts observed to date. The signal rises within less than 10 ms, has a width on the order of 20 ms FWHM, and (even allowing that the hinted at but unconvincing structure following the narrow spike is real) falls to undetectable levels within less than a half second. GB791116 (Fig. 3) shows the slowest-rising and longest duration single spike, without significant temporal structure within that spike. There is a small precursor which triggered the recording system (trigger time is at 3 s relative to the start of the record).

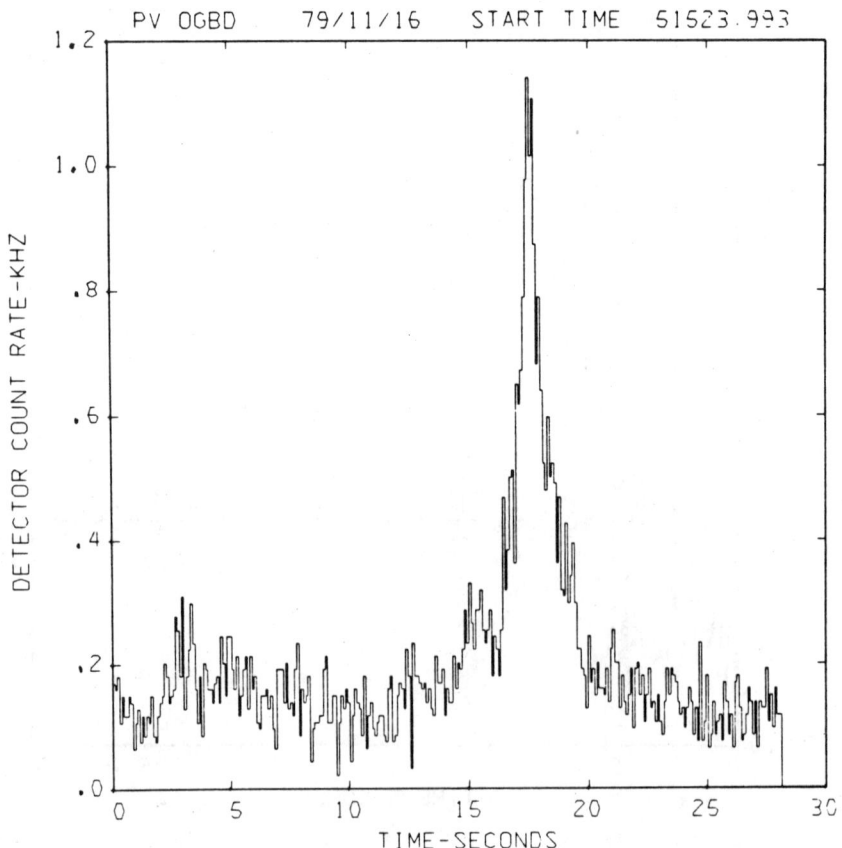

Fig. 3. Event GB791116, which includes the slowest-rising, longest-duration single spike observed to date.

The major response, however, rises within about 1 s and has a duration of about 2 s FWHM.

Many gamma-ray burst events consist of a number of well-defined, individual spikes. Often these individual spikes are distinctly separated, without evidence of net flux detectable above background in the interval between them. Such an event is exemplified by GB800709, the record of which is shown in Fig. 4. This event also spans the greatest duration, as observed at energies >100 keV, of those observed to date. This behavior very much limits the physical processes which may be responsible for these events, excluding those which would experience full completion of a cycle once initiated and not be expected to repeat on a short time scale (excepting beaming effects which would be strongly periodic).

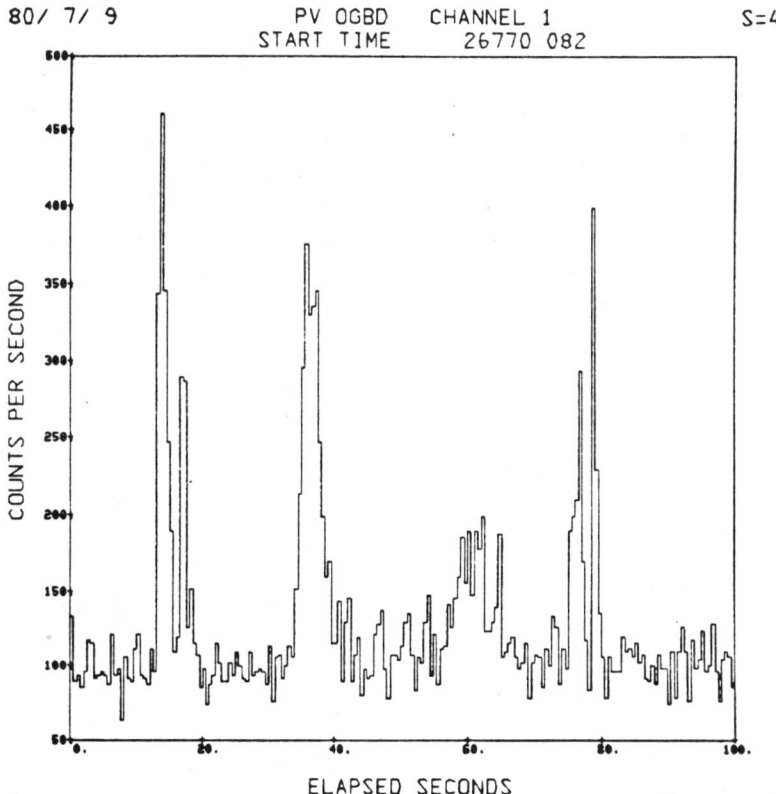

Fig. 4. Event GB800709. The longest duration (at E > 100 keV) event observed to date. This event also exemplifies the quasi-periodic events.

In contrast to the above variety of temporal behavior observed for classical events, the spectral characteristics from the early observations appeared to be very similar. The bursts completely dominated the background, over their brief duration, at energies ≳ 100 keV, in instruments with essentially omnidirectional sensitivity. Early studies of the spectral characteristics of a major subset of known bursts implied that the spectral distributions of the emissions were all consistent with a single function, $N(E) \propto \exp(E/150 \text{ keV})$ for $E \gtrsim 100$ keV[7]. Again, this spectral signature served to identify gamma-ray bursts, but was not of value in defining classifications between members of the group. More recent reanalysis of Apollo 16 spectral data for GB720427, as shown in Fig. 5,[8] demonstrated that a function characteristic of thermal bremsstrahlung fit the observations best. That function also fits best virtually all bursts[9].

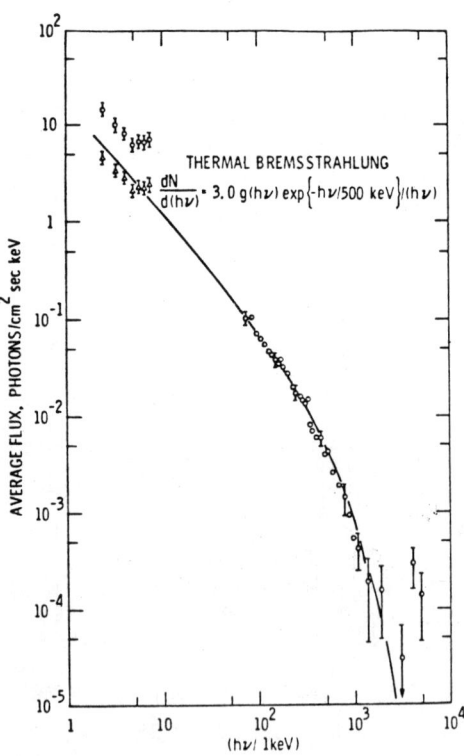

Fig. 5. The average spectrum of GB720427 as observed by Apollo 16.[8]

Similarly, the locations of the classical sources do not suggest
a systematic distribution in space. Early analyses of locations were
not sufficiently precise to identify specific source objects. The
general distribution of sources was observed to be roughly isotropic,
with no clear preference for either the ecliptic or the galactic
plane. There was, however, a slight preference for the galactic
plane at 95% confidence level[10]. A recent study including 37
locations was done with greater sensitivity[11]. These nominal
locations are shown in Fig. 6. The essentially isotropic
distribution implies that the sources lie nearby in the galaxy
(typical distance ~300 pc)[11], although a distant extragalactic origin
is also consistent. (It is interesting to note, however, that four
of these locations are rather tightly clustered in the Virgo/Coma
region.)

Precise location of sources has likewise not been particularly
helpful in revealing the nature of the source objects or a
classification among them. Except for the anomalous burst of 1979
March 5 and a distinctive triple radio source at the location of
GB781119[12], the regions defined by the precise locations were not
found to contain any striking objects either through searches of
existing catalogs or specifically directed observations. This result
is consistent with a neutron star, unaccompanied by a more visible
companion[13].

DISCUSSION OF TAXONOMY

Early observations seemed to show that classical gamma-ray
bursts had temporal characteristics which varied continuously
throughout their ranges and spectra which were very similar.
Improved observations performed by the generation of instrumentation
implemented in 1978 have greatly added to our knowledge of both the

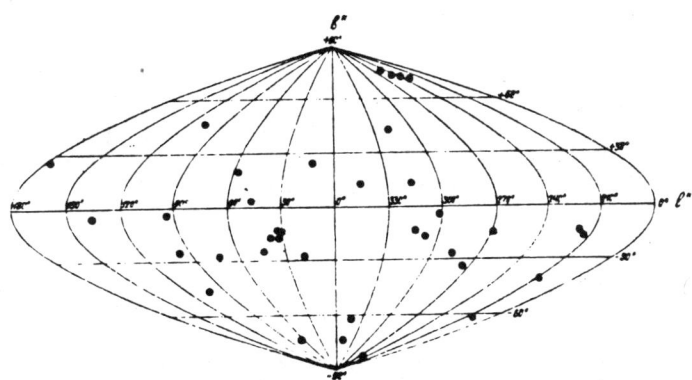

Fig. 6. The locations of 37 gamma-ray bursts, in galactic
coordinates, as determined by the KONUS experiment[11].

temporal and spectral characteristics. These larger, more detailed data bases have demonstrated systematic similarities among the bursts. In addition, some classical bursts from different locations have shown remarkably similar temporal structure. Thus there are systematic similarities which provide a framework for a system of classification.

A reanalysis of data from the scanning X-ray detectors aboard the Vela spacecraft has disclosed that two verified gamma-ray bursts were also observed by the X-ray detector because the sources fortuitously fell within the collimated field-of-view[14]. Since the X-ray detectors have an effectively more sensitive threshold for observing these events than do the omnidirectional gamma-ray

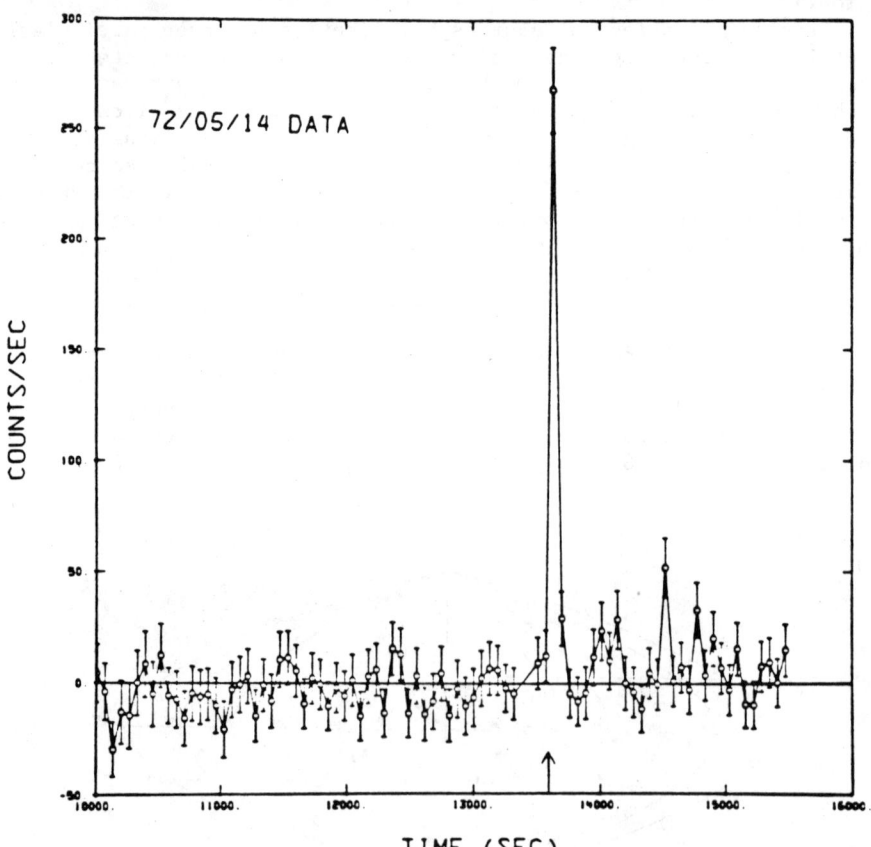

Fig. 7. The response of a scanning, collimated X-ray detector to GB720514, showing evidence of recurrence, at X-ray energies, to ~1000 s. The arrow indicates the time of gamma-ray burst onset.[14]

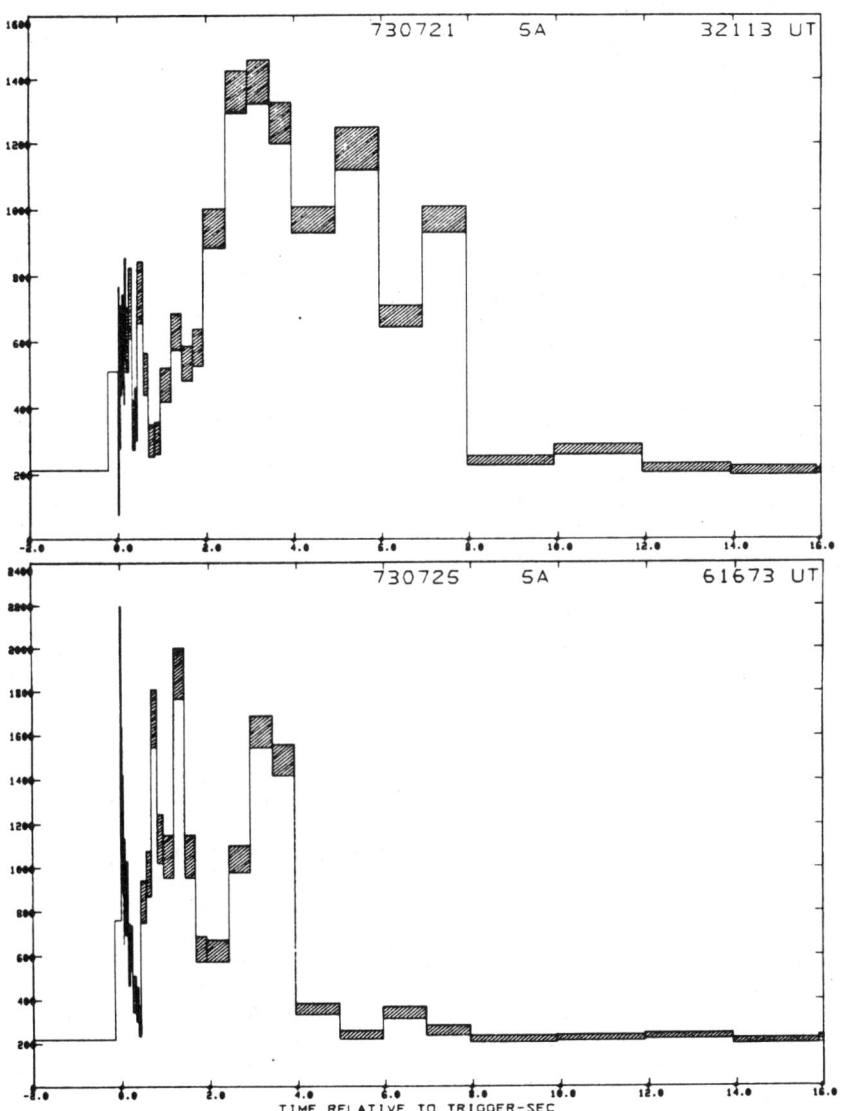

Fig. 8. Records of events GB730721 and GB730725 which show similarities in temporal structure and may represent recurrent outbursts from a single source.

detectors, the X-ray observations are well suited to searching for persistent X-ray emission and recurrences from the source. In fact, both bursts observed in this manner gave no evidence of significant persistent emission (at 3 < E < 12 keV) extending beyond the duration of the burst defined at gamma-ray energies, but did provide evidence of a short recurrence on time scales of hundreds of seconds[14]. Figure 7 shows the time history of the observations in the direction of the source of one of these bursts, GB720514. The arrow marks the occurrence of the gamma-ray burst (the duration of which is only a little longer than the width of the arrow). A recurrence (4.7σ response) is observed at about 890 seconds after the gamma-ray burst. Note also that the source did not emit persistently at X-ray energies. Although the nominal spectral indication suggests that these later bursts are softer than the initial burst, the spectral measurement is not definitive. Since both of the bursts which were observed by this instrument exhibit this long-term recurrence, it may be that this is a common characteristic of gamma-ray bursts or a characteristic that could be useful in future taxonomy.

The first evidence of a possibly recurrent source was found in the Vela data for events GB730721 and GB730725. These events occurred at an interval of only four days, provided crude locational data consistent with a single source, and showed similarities in temporal structure and total intensity. The temporal structure, as shown in Fig. 8, can best be described as emission at a period of about two seconds. Perhaps this waveform could be produced by rotational modulation of the signal, but the similarity of the two records would suggest that even the phasing of such modulation was closely reproduced in the two events.

From the Pioneer Venus data there is found an even more marked similarity in temporal structure between events GB781124 and GB800419, as shown in Fig. 9. Since the locations deduced for these two events are greatly separated, this implies that very similar conditions exist at different sources. It must be noted that the time scale of event GB781124 was compressed by a factor of 0.7 in order to achieve a satisfactory degree of agreement between these two records. Given this adjustment, a cross-correlation analysis showed that the two records are comparable nearly to the degree expected for statistically independent records of the same event. This degree of similarity between these events strongly suggests that the temporal features are imposed by the physical parameters constraining the generating process. The temporal profile is not at all suggestive of a periodicity. A number of other bursts resemble these two but these are weaker and less amenable to critical comparison.

Besides these recurrences or very detailed similarities between pairs of bursts, the classical bursts display systematic similarities which suggest subdivision into at least three different types (see Table 2). Based upon increasing complexity of temporal structure: Type I are the brief events, type II are doublet and quasi-periodic events, and type III are the long and the irregular events.

The possibility of a subclass consisting of the very brief events has been noted previously[11]. Type I gamma-ray bursts are exemplified by GB790613 (shown in Fig. 2). Even if those brief

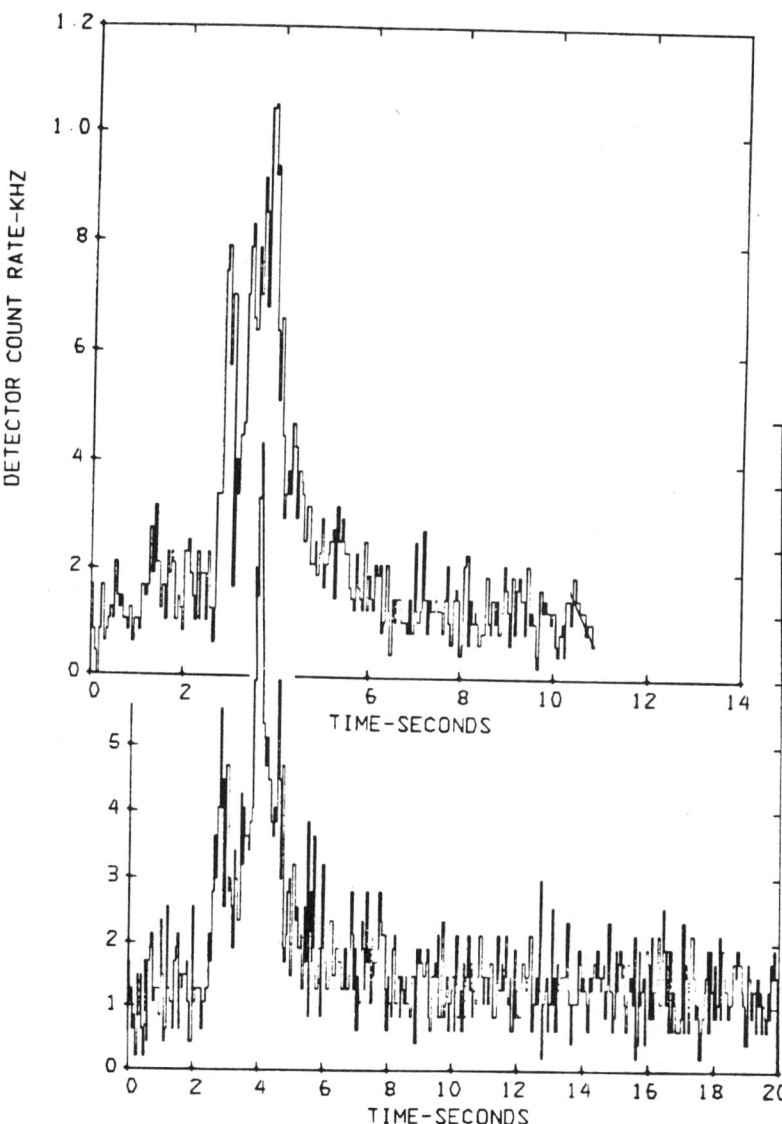

Fig. 9. Very similar response observed by the Pioneer Venus OGBD to events GB781124 (lower curve) and GB800417 (upper curve). Note that the time scale has been compressed for GB781124 relative to that for GB800417.

events (such as GB790305) exhibiting anomalously soft spectra are rejected, there remain a number of members of this group. These show very similar structure, although their very brevity precludes critical comparison because of limitations in the temporal resolution.

Another category of bursts includes those which are quasi-periodic with near zero flux between the peaks. Event GB800709, shown in Fig. 4, is an example of this type. Although a subjective impression of periodicity is conveyed, random variations and departures from a rigid period produce unconvincing results in power spectrum analysis. This class possibly includes doublet pulse events, since these may show, as in the case of GB781104[15] (illustrated in Fig. 10), a nearly periodic repetition of doublets.

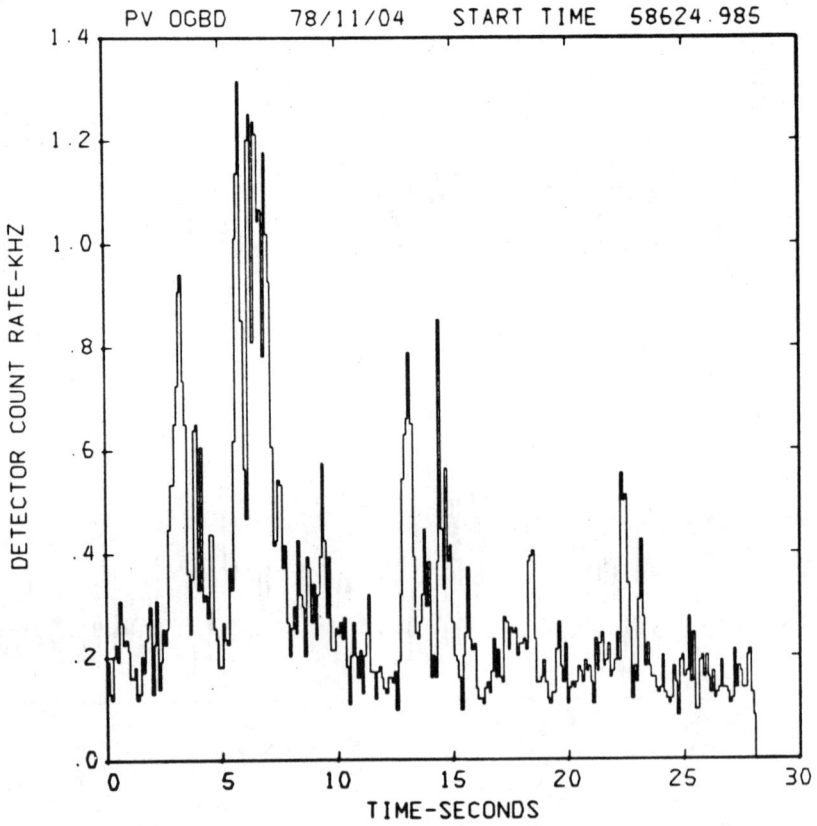

Fig. 10. Event GB781104, displaying a nearly periodic series of double pulses[15].

Finally, there is a class where the envelope varies slowly and irregularly, generally remaining rather well above background for tens of seconds. These also typically show rapid and apparently random variations within the overall envelope. Events of type III do appear to share more than a superficial resemblance. Event GB790307, shown in Fig. 11, is presented as an example of this class. GB720427, observed by Apollo 16[8], is another event sharing these temporal characteristics.

Of course, there remain a number of events which are difficult to place into any of these possible classes. Many of these are simply so weak that statistical fluctuations confuse attempts to identify a pattern in the temporal structure. Others, such as GB781119, do not fit well into any of these classes, but neither do

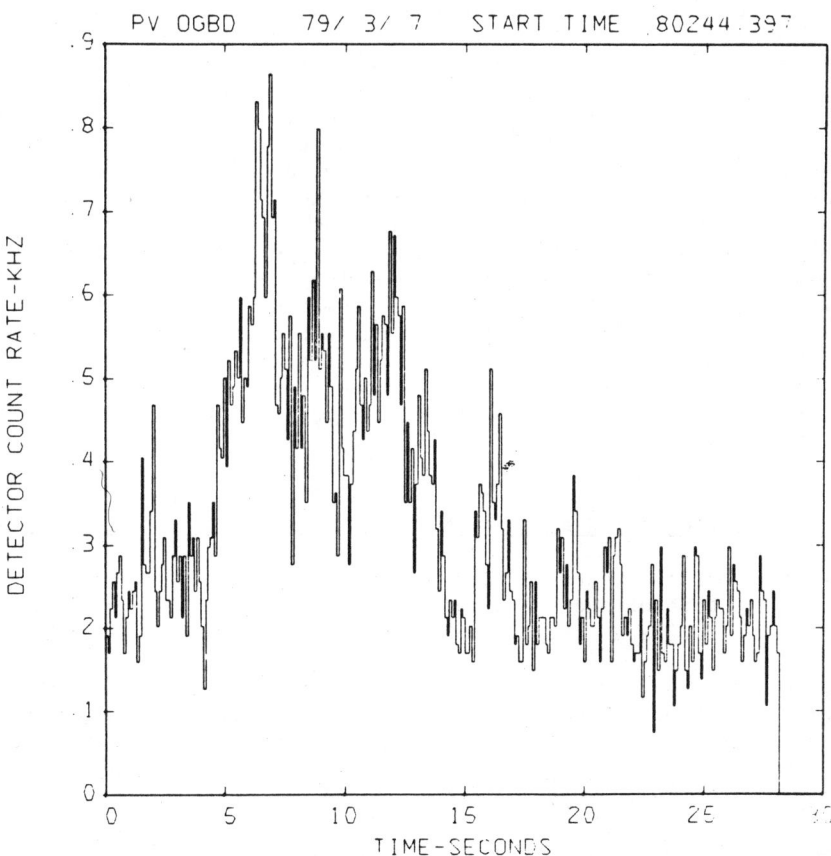

Fig. 11. Event GB790307 - an example of events showing long duration and irregular variations in intensity[15].

they seem to warrant additional classes. With the continuation of observations and the extension of the data base there will no doubt be further insight provided in this scheme of taxonomy.

Having addressed a taxonomy based upon temporal structure, it is well to consider differences in the spectral characteristics of these events as another possible means of classification. Spectral line features have been observed in some, but not all, events. Both emission and absorption features have been reported[9], occurring singly and (in one case only) jointly. These lines provide evidence that the gamma-ray bursts originate within small regions near neutron stars containing extreme magnetic fields[9]. Consideration of the temporal structure associated with the spectral features may be expected to provide further insight into the actual generating mechanisms.

The presence of low-energy absorption features is very well correlated with the type III classification. Upon closer inspection of the records of these events there seems to be a degree of similarity in even the details of temporal structure, but statistics are too poor to convince a skeptic. As has been noted previously, this temporal class includes the Apollo event. The low-energy spectrum of this event, although unfortunately not defined in the region $10 \leqslant E \leqslant 100$ keV, does show evidence of spectral structure which may well represent an absorption line.[8]

Those events giving evidence of line emission, on the other hand, correspond to the type II bursts. The fact that these events do share evidence of line emission in their spectral distributions and that some events seem to combine the doublet and the quasi-periodic characteristics does suggest that they may represent a single phenomenological type. There remain, then, three definable classifications of events within the category of "classical" gamma-ray bursts, together with some number of additional events not readily classifiable.

Table 2. Classification of Classical Gamma-Ray Bursts

Class	Temporal Characteristic	Spectral Characteristic
I	Brief events	Not defined
II	Doublet and quasi-periodic	Emission lines ~400 keV
III	Long and irregular	Absorption lines ~50 keV

It is, of course, not clear whether these types represent uniquely different source types or, rather, only manifestations of different processes occurring on the same type of source object. It may be hoped that observations will be extended and improved providing further insight into the nature of these events.

ACKNOWLEDGMENT

Work at the Los Alamos National Laboratory was performed under the auspices of the U.S. Dept. of Energy and partially supported by the National Aeronautics and Space Administration under contract 98331A.

REFERENCES

1. J. R. Harries, K. G. McCracken, R. J. Francey, and A. G. Fenton, Nat., 215, 38 (1967).
2. R. W. Klebesadel, I. B. Strong, and R. A. Olsen, Astrophys. J. Lett. 182, L85 (1973).
3. A. S. Jacobson, J. C. Ling, W. A. Mahoney, J. B. Willett, Gamma Ray Spectroscopy in Astrophysics (eds. T. L. Cline and R. Ramaty), 228, (NASA TM79619, 1978).
4. T. L. Cline, et al., Astrophys. J. Lett. 237, L1 (1981).
5. E. P. Mazets, S. V. Golenetskii, V. N. Il'inskii, R. L. Apetkar and Y. A. Guryan, Nat., 282, 587 (1979).
6. E. P. Mazets, S. V. Golenetskii, Y. A. Guryan, A. F. Doffe, Physico-Technical Institute preprint 631 (1979).
7. T. L. Cline and U. D. Desai, Astrophys. J. Lett. 196, L43 (1975).
8. D. Gilman, A. E. Metzger, R. H. Parker, L. G. Evans, and J. I. Trombka, Astrophys. J. 236, 951 (1980).
9. E. P. Mazets, S. V. Golenetskii, R. L. Apetkar, Y. A. Gurigan, and V. N. Dyinskii, Nat. 290, 378 (1981).
10. I. B. Strong, R. W. Klebesadel, W. D. Evans, Ann. NY Acad. Sci., 262, 145 (1975).
11. E. P. Mazets and S. V. Golenetskii, A. F. Ioffe Physico-Technical Institute preprint 632 (1979).
12. R. M. Hjellming and S. P. Ewald, Astrophys. J. Lett. in press (1981).
13. J. G. Laros et al., Astrophys. J. Lett. 245, L63 (1981).
14. J. Terrell, E. E. Fenimore, R. W. Klebesadel, and U. D. Desai, submitted to Astrophys. J. Lett (1981).
15. W. D. Evans, R. W. Klebesadel, J. G. Laros, and J. Terrell, Nat. 286, 784 (1980).

A REVIEW OF THE 1979 MARCH 5 TRANSIENT

T. L. Cline
Laboratory for High Energy Astrophysics
NASA/Goddard Space Flight Center, Greenbelt, MD 20771

ABSTRACT

The understanding of the 1979 March 5 event remains a problem of central importance for researchers in gamma ray transient astronomy. Efforts in the study of the observational results and in the development of interpretations regarding the nature of this event and its relationship to the variety of other gamma ray burst phenomena have continued through the past 2 years. A consensus of opinion has not yet been reached regarding its possible origin in N49 at 55 kpc distance, versus in an invisible source 3 or 4 orders of magnitude closer, although interpretations favoring N49 appear to be presently gaining momentum. This presentation outlines the existing data in a review of what remains the most singular high-energy astrophysical phenomenon of the space age.

INTRODUCTION

I assume that most of the persons interested in this subject are familiar with the March 5, 1979 gamma ray transient; since the bulk of the observational data has been published for some time this review seems at first unnecessary. However, progress has been and continues to be made in both the experimental and the interpretative reanalyses of the measurements. Thus, although the event remains an historic fact, its study is very much a current activity; in fact, the 'correct' interpretation is undoubtedly yet ahead of us. I personally have found this event to be fascinating, both because of its apparent observational rarity and because of its role as a generator of theoretical ideas and related calculations. Will we see a gamma ray transient like this again in our research lifetimes? Probably not, at least, probably not when we shall be as prepared with third-generation instrumentation as we happened to be with second generation instrumentation on March 5th of 1979. However, if detector sensitivity and resolution can be sufficiently improved, March 5-like events of much weaker visual magnitude may be observable and capable of detailed study, whether arriving from more distant stellar or from extragalactic sources. Will an entirely new modelling (such as the grasar, to be unveiled at this conference[1]) be found to be applicable? If so, some new experimental approach involving the measurement of, for example, polarization, or submillisecond oscillations, or gamma ray coherence, may provide another observational breakthough, in addition to that which gamma ray line spectroscopy now provides.

We can first review the question of whether the 1979 March 5 event is so different from other gamma ray bursts as to be termed 'unique', which I have described it to be in an earlier review.[2]

The variety of gamma ray burst data we have been shown at this conference, in fact, leads one to believe that every gamma ray burst is unique! However, the first basic difference that appears to confront us is that while the typical or so-called 'classical' gamma ray bursts seem generally to have time histories of random structure, with randomly evolving spectral features, the March 5 event appears to be especially ordered, with a singularly fast rise time, a single very intense peak, and a subsequent, well-defined

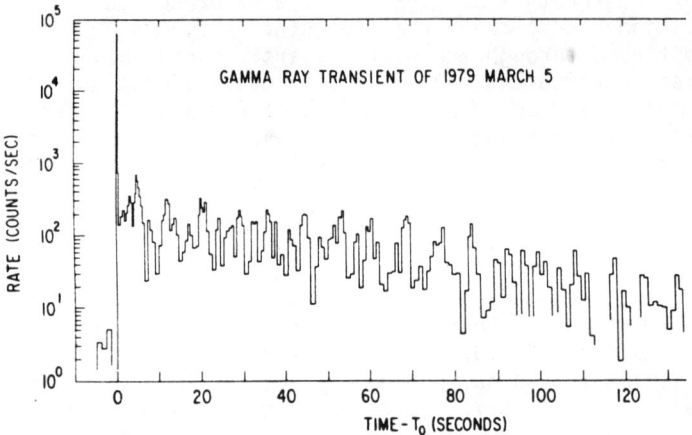

Fig. 1. Time history of the 1979 March 5 transient as observed with the ISEE-3 space probe[3].

oscillation. The other basic difference is the fact that the March 5 event is the only one with a candidate source object. Of course, its features may only seem to be atypical due to their greater resolution afforded by the great apparent brightness of the event; also, its source identification could be a cosmic accident. To 'write off' the event in this manner would be to completely ignore a possibility of great value to astrophysics, in my opinion.

The March 5 event possesses a 420-keV spectral feature[4] in common with some other classical bursts[5,6]; this, together with the regular oscillation (which is not clearly exhibited by classical events) has prompted the general belief that both originate in neutron star phenomena. However, they may of course originate by means of two or more entirely differing mechanisms. In fact, the 420-keV feature also has a novel and entirely differing implication: as we shall hear later in this conference[1], the neutron star redshift explanation is no longer the only possible interpretation of the 420 keV feature. Thus either the March 5 or the classical event connections to neutron star redshifts or both may yet be found to be incorrect or imprecise.

It is a fair statement, unfortunate though it may be, that no mathematical treatment has yet been evolved to systematically classify or categorize gamma ray bursts; thus, the probability of uniqueness of the March 5 event has not been quantitatively estimated. To pursue this, we can look at some of its features in greater detail.

OSCILLATIONS

Of the several unusual features of the March 5 event, including the rise, the intensity, the oscillations, the spectrum and the direction, the oscillations were invoked first as providing basically new information.[7] Except for a weak but suggestive ~

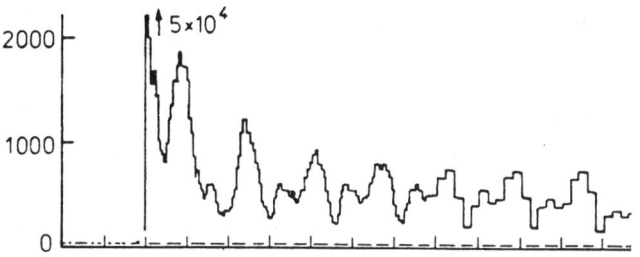

Fig. 2. (Above) Venera-11 observations of the March 5 event (horizontal scale = 5 seconds/mark), indicating the first clear evidence for periodicity in a gamma ray transient (from Mazets et al.[4]).

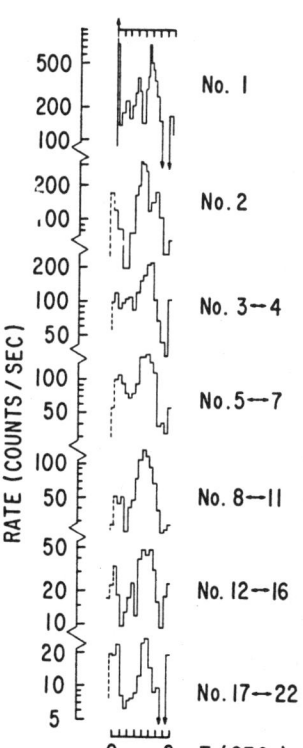

Fig. 3. (Left) ISEE-3 observations of the first 22 cycles of the March 5 event, plotted on an 8-second per period basis folded with an increasing number of cycles per plot, to compensate for the decreasing statistical validity per cycle. The large pulse and the secondary features remain constant in time, yielding an average period of 8.00 ± .05-second.[3]

4-second, ~ 5 cycle long, periodic effect in the time history of the 29 October 1977 gamma ray burst event,[8] no evidence is known for any cyclic features in classical gamma ray bursts. The clear 8-second periodicity in the March 5 event provided the first evidence of that kind for the neutron star origin model of a gamma ray transient. (Spectral evidence existed before that time for one very slow gamma ray transient[9,10] although it was not then and still is not clear what relationship that isolated phenomenon has to classical gamma ray bursts. Of course, it also is not clear what relationship the March 5 event has to classical events.) Figure 2 shows the results from the spaceprobe Venera-11 as published by Mazets and his colleagues of the Leningrad group.[6,7] Both the general 8-second repetition and the ~ 4-second modulation are clear. Similar time histories were exhibited by a companion sensor on Venera-12. The oscillations were tracked for about 3 minutes with the Goddard experiment on ISEE-3, as seen in the time history in Figure 1, but of course suffered from decreasing statistical significance as a

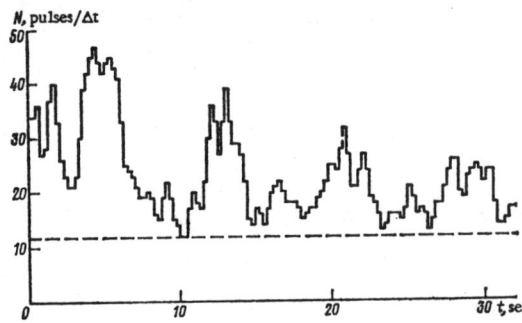

Fig. 4. The time history as observed with the Franco-Soviet instrument on Venera-12, indicating evidence for ~ 1-second features.[12]

function of time. This effect is compensated for in the presentation of Figure 3, in which an increasing number of cycles are included per plot. Here the constant phase of the compound structure is evident, providing a period of 8.00 ± 0.05 second.[3] This periodicity is clearly not a fluctuation but very definite evidence for a cyclic phenomenon; whether the 8-second feature is rotational, precessional or radial is another question. The fact that the main and secondary pulses decay differently has been invoked as evidence for a directional emission from a rotating neutron star.[11]

Weak evidence also exists for features of time structure finer than the 4-second interpulse. Figure 4 shows the Venera-12 data as published by Vedrenne and his colleagues of Toulouse and Estulin and his colleagues of Moscow.[12,13] Here, as in earlier figures, suggestive fluctuations are evident on a shorter time scale. The spectral power as a function of period derived from these results,[13] shown in Figure 5, and the power spectrum as a function of frequency of 3 minutes of Pioneer-Venus Orbiter data[14] of Evans and his colleagues at Los Alamos, shown in Figure 6, both indicate second-order structure in the ~ 0.7 to 1.1-second region. (The ISEE-3 results do show some detail but its time history may be somewhat spin-modulated, although not severely, since the source direction is only 3.5 degrees from the ISEE-3 spin axis.) Finally, one can suspect that even finer time variations could exist within the first oscillation. In Figure 7 the Leningrad Venera-11 data hint at a ~ 0.2-second fluctuation.

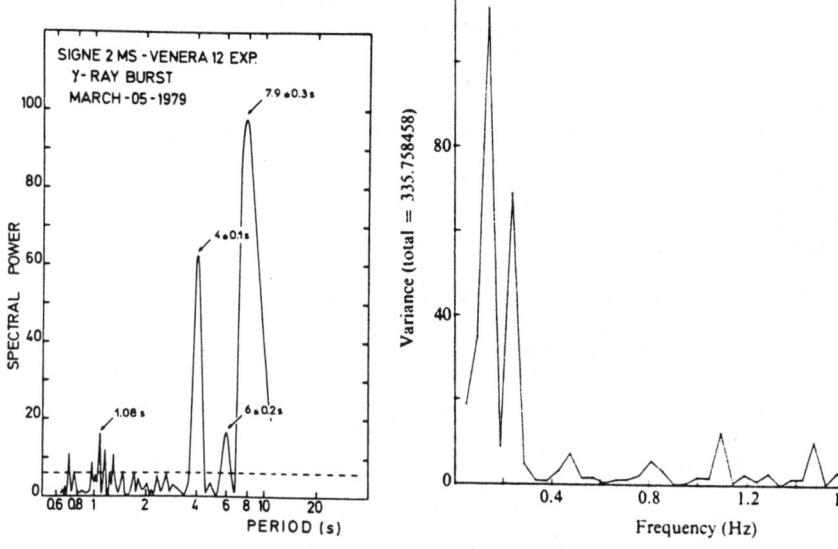

Fig. 5. Spectral power evidenced in the Franco-Soviet Venera-12 results.[13]

Fig. 6. Power spectrum from the Pioneer-Venus-Orbiter measurements.[14]

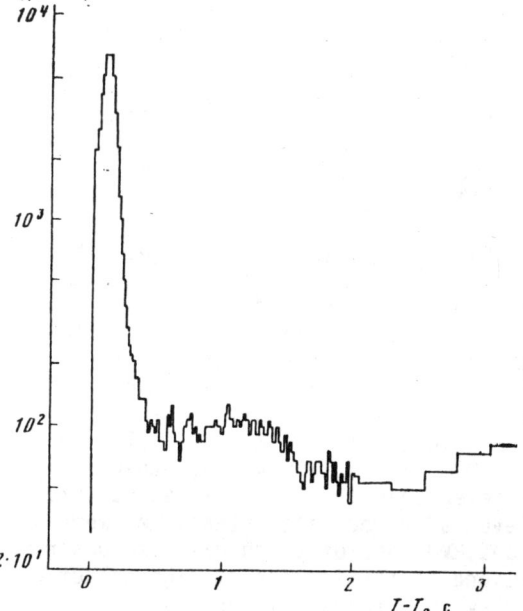

Fig. 7. Detailed Venera-11 time history of the early part of the event, including the first intense peak, its decay and the buildup of the first oscillation.[4,7]

BURST TIME HISTORY

The initial high-intensity burst of radiation in itself may be even more anomalous than the oscillating portion of the March 5 event. Its features, not shared by the general population of classical gamma ray bursts, are a fast rise, a single peak, a smooth decay and an instantaneous intensity near detector-saturation for most of the instruments (of the order of one count per 10 microseconds). In fact, the unknown spectrum of hard X-rays below detector threshold for these gamma ray detectors presents some uncertainty of possible measurement distortion; it is hoped that the effect is not major, a belief supported by the fact that the observations generally mutually agree. For example, the onset rise is faster than can be measured with every instrument, yet the peak counting rate is seen to occur usually about 20 milliseconds into the event. (Unfortunately the ISEE-3 γ-ray spectrometer failed 2 months earlier, depriving us of a comparison measurement made with germanium rather than with scintillators.) This onset consists of a sudden increase of several orders of magnitude within the resolution time of, for example, 1 millisecond in the ISEE-3 Goddard/MPI scintillator,[3] as shown in Figure 8. The exponential time constant of intensity increase is therefore less than 200 microseconds, implying a light travel distance of less than 60 km.

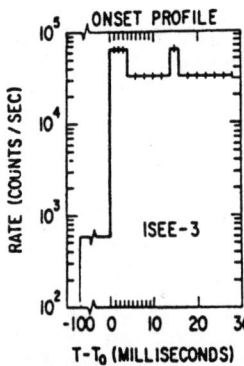

Fig. 8. (a) Onset of the high-intensity portion of the 1979 March 5 transient.[3] A time constant of less than 0.2 ms is inferred from the increase of two orders of magnitude from near background to essentially full intensity within a resolution time of 1 ms. (In this instrument the time to accumulate 64 photons is recorded to 1 ms accuracy; the first several readings are in fact 1 ms and 2 ms accumulations.) This < 1 ms full rise in the onset shape is also seen with Pioneer-Venus-Orbiter.[3,14]

No other gamma ray transient exhibits as brief a rise time, within a factor of 10 to 30 depending on the various detector limitations considered. (For example, the Goddard Helios-2 detector is instrumented with slower, 4 millisecond temporal resolution; the entirety of the fast spike as observed with that instrument[3] is shown in Figure 9. Four years of gamma ray bursts were monitored before the Helios-2 spacecraft was turned off; only the March 5 event produced a trigger response in the 4-ms circuits; all other events triggered in its 32 or 250-ms modes.) However, other events have been observed that are briefer than the 150-millisecond total duration observed here. Two events of moderate intensities were detected in 1979 with Venera and PVO, one of which was also observed with Venera-12 giving a directional determination at high galactic and celestial latitude (K. Hurley, R. Klebesadel, private communication). The question of whether such fast, single-spike events occasionally detected, including perhaps 3 more in the archived Vela data (R. Klebesadel, private communication), are merely the tips of more complex classical-event icebergs or are March 5-like events, cannot yet be answered.

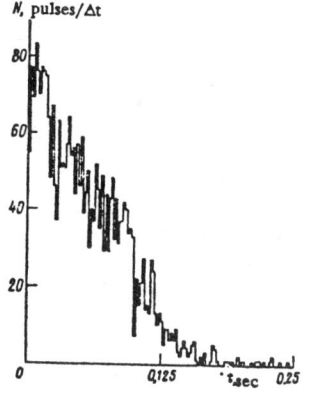

Fig. 9. The 150-millisecond wide initial burst. A transition from a slow decay of ~ 150 ms time constant to a steeper decay of ~ 35 ms time constant is seen about 100 ms into the event. These data are plotted on a ~ 4 ms per accumulation time basis.[3]

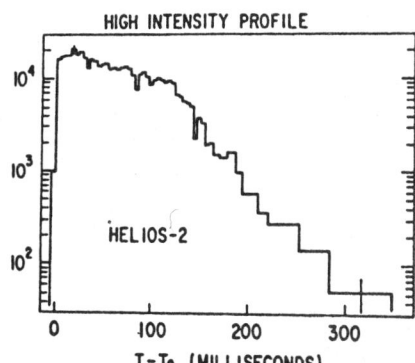

Fig. 10. The data, on a ~ 2 ms per accumulation basis, from the Franco-Soviet experiment on Venera-11, collected during the initial one-quarter second of the March 5 event.[12]

The question of very fine time variations in the March 5 event time history is interesting for several reasons. (These range from the existence of 0.1 to 100 ms oscillations, due either to an inherently fast neutron star spin period or to model-dependent transition damping vibrations,[15] to the possible existence of a very short duration nonrandomicity in the counting rate that would relate to a possible coherence, as in certain other models.[1]) Detailed studies of the first 125 milliseconds of the event, examples of which are shown in Figures 9 and 10, have been carried out using data from all available high time resolution sources, including the Einstein Observatory X-ray monitor. The Venera time histories hint at a ~ 25 msec effect with marginal statistical accuracy (K. Hurley, private communication) and the Einstein time history, of only 55 msec duration, can be interpreted to be consistent with two 27 msec features (M. Weisskopf, private communication); these departures from randomicity appear to be tantalizing but inconclusive.

INTENSITY AND SPECTRUM

The total intensity of this burst is that of a typical, strong gamma ray burst, but the instantaneous intensity of the peak is by far the greatest yet observed. Figure 11 shows a scatter plot of event total intensity versus maximum instantaneous intensity, using the sample of Helios-2 events detected from January 1976 to the

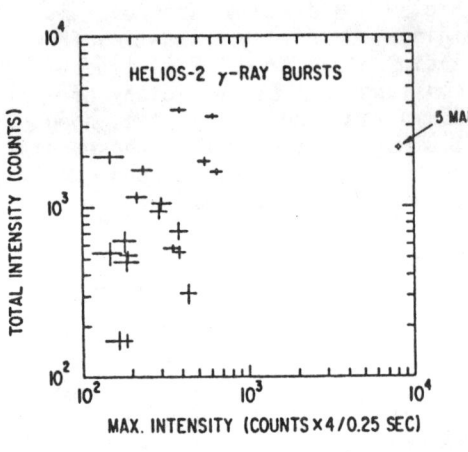

Fig. 11. Diagram of total versus maximum intensity for the gamma-ray bursts observed with Helios-2 from 1976 up to the 1979 March 5 event. The more extensive burst observations of the Vela system cannot be used for this purpose since they employ a geometrically lengthening time base whereas bursts do not always have maximum intensity near onset. If a search is made for Vela transients of March 5-like character, however, a comparison of total versus onset intensity gives a similar result.

March 5 event. It is also the case that this is the most intense event monitored with the Vela satellites for the decade beginning in 1969 (R. Klebesadel, private communication). The observed intensity is at least several times 10^{-3} erg cm^{-2} sec^{-1}, a lower limit due to the unknown fluxes below the 30 to 50-keV instrument thresholds and to the unknown effects of pulse pile-up at these energies (considering that most instruments were running near saturation). The intensity of the oscillating portion is $\sim 10^{-2}$ that of the peak; since this event is so intense relative to most gamma ray bursts, it is possible that some of the classical events of average or weak intensity could possess an analogous but unobservable periodic decay component.

The spectrum of the March 5 event is another of its anomalies. Although both this event and some classical events have the famous 420-keV feature,[4,5,6] this event has a considerably softer spectrum below $\sim$ 100 keV than do typical classical bursts. Figure 12 illustrates the spectra of the peak of the March 5 event and of the oscillating portion,[4] compared with that of a typical classical event. The 420-keV feature is clearly evident in the intensity peak (the loss of the ISEE-3 germanium spectrometer, 2 months earlier, deprived us of a knowledge of the inherent width of this peak). Besides the excess in the 30 keV region and the 420-keV feature, the other well-known aspect of this spectrum is the lack of

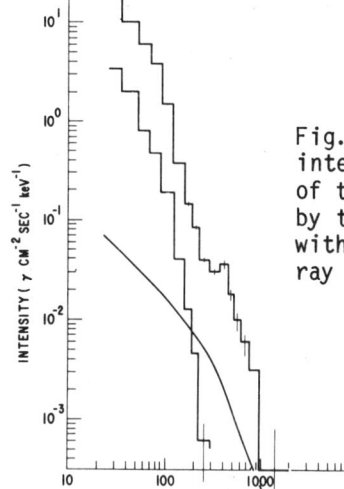

Fig. 12. The photon number spectra of the intensity peak of the March 5 event and of the oscillating portion, as observed by the Leningrad group,[4] as compared with that of a typical 'classical' gamma ray burst.

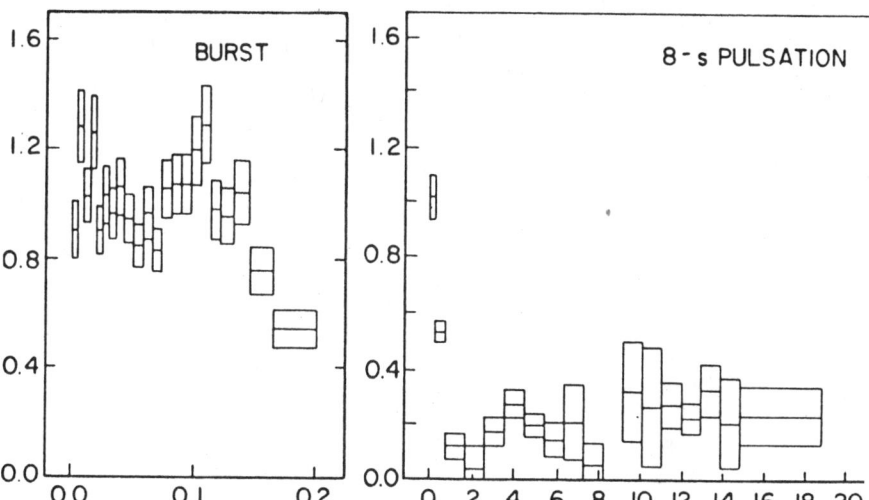

Fig. 13. Pioneer-Venus-Orbiter measurements of the spectral hardness of the March 5 event (in the above/below 100-keV region), illustrating variable behavior at the onset and shoulder of the intensity spike, and ≈ in phase with the first two oscillations.

higher energy photons, particularly above 1 MeV. This deficiency is probably due to the fact that photons above the pair production threshold energy would be, of necessity, removed by that process due to the extreme density at the source, particularly if the source is very distant.[16] The spectrum of the oscillating portion of the event is even softer than of the onset, with no detectable counts as high in energy as 400 keV. The only other published spectral observations of this event are shown in Figure 13, illustrating a spectral hardness factor (counts > 100 keV/50-100 keV) as a function of time.[11] The hardness is maximum at the onset and at the time of the change of rate of intensity decay, or shoulder, at ~ 100 msec into the event. As shown in the second illustration, the spectral hardness in the periodic portion is less than in the onset, in agreement with the Leningrad results, but is shown to vary in phase with the 8-second oscillation, with higher energy photons in greater quantity during the times of greater intensity.[11]

RECURRENT EVENTS

The March 5 event was the first gamma ray transient to exhibit recurrent events, i.e., to be associated with repeated bursts having a common source direction. Three events of very weak intensity, detected only with the Leningrad experiments on Venera-11 and -12, were observed to follow the March 5 event by delays of ~ 0.60, 29 and 50 days with intensities ~ 3, 1 and 0.5 percent that of the March 5 intensity, respectively.[15] Their time histories, shown in Figure 14, are not the same (on a reduced scale) as the March 5 profile but generally slower, although their spectra, as seen in

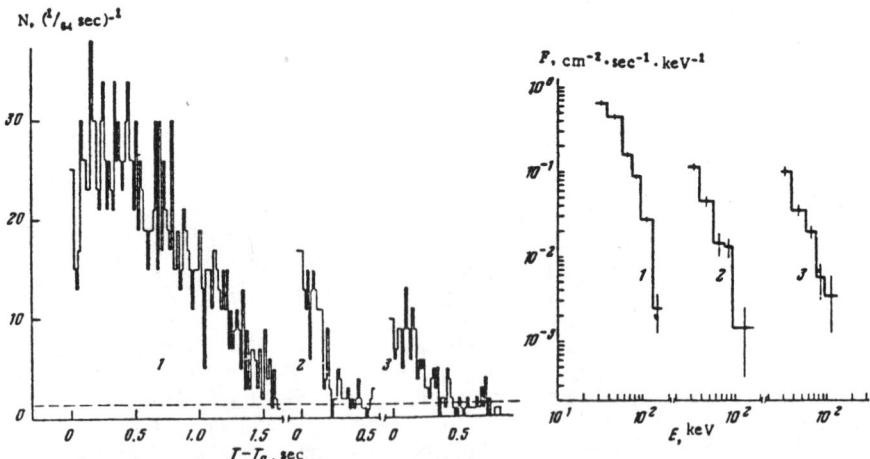

Fig. 14. Counting rate histories of the delayed March 6, April 4 and April 24 events, found to have source directions consistent with that of the March 5 event.[15]

Fig. 15. Spectra of the events of the previous figure, seen to be not unlike that of the decay portion of the March 5 event.[15]

Figure 15, are similar to that of the softer, oscillating portion although statistically limited due to their very low intensity. Their directions agree with that of the March 5 event (the Venera-11 to Venera-12 time delays provide an accurate one-dimensional source measurement in conjunction with a rough source direction estimated from the directional count rate characteristics). Given their sequential connection to each other and to the March 5 event, it is entirely reasonable to assume this to be the discovery of a common burst source emitter. Only one other series having a common source direction has been detected;[17] however, that 3-event series does not follow any intense primary or "parent" gamma ray transient. Its source direction is in the galactic plane at ~ 45 degrees galactic longitude, consistent with either a nearby or a very distant galactic source object.

DIRECTION

One remarkably fortuitous historic accident was that of the occurrence of the March 5 event during the October 1978 to December 1979 time frame. This is the interval when the interplanetary gamma ray burst network possessed its full instrument complement and was positioned in trajectories far from the Earth for their maximum interspacecraft event timing resolution. The identification of the supernova remnant N49 in the neighboring galaxy of the Large Magellanic Cloud as a possible source object was made as one of the first accomplishments of that network.[18,19] This measurement made possible the enlarged scientific controversy over the nature of the physical process involved by immediately creating the distinct possibility that the source object was at a 55-kpc distance, two to three orders of magnitude farther than would have been assumed had the directional information consisted instead of the formerly available resolution. That would have been several square degrees, providing only rough overlap of the LMC and various stellar regions in the constellation of Dorado at 30-odd degrees negative galactic latitude. Figure 16 illustrates the source location, and shows the extent of the densest optical part of N49. Another aspect of this identification is that, by contrast, high-precision directional studies of classical gamma ray bursts made with this network have provided source fields that are optically empty down to at least the 18th magnitude.[20,21] An additional factor is the fact that N49 is an X-ray emitter, strengthening the case for its identification, since one might assume there would be some X-ray/gamma-ray correlation. Finally, it is also the case that not only other precise burst source locations but previous larger source field determinations also found no correlation with either transient or steady X-ray sources or other objects that could be obvious source candidates[22,23]. Figure 17 shows the source position plotted on the X-ray contours of N49 measured (quite accidentally) within a few days of the event with the Einstein Observatory.[24] A remarkable aspect of this X-ray survey, however, is the fact that no point source was found: this does not necessary provide a weakening of the association of the event with N49 since it yields an upper limit (at ~ 10-9) to the ratio of point source X-ray strength to burst strength independent of source distance. (The X-ray measurements also included a comparison, made by chance before and after the

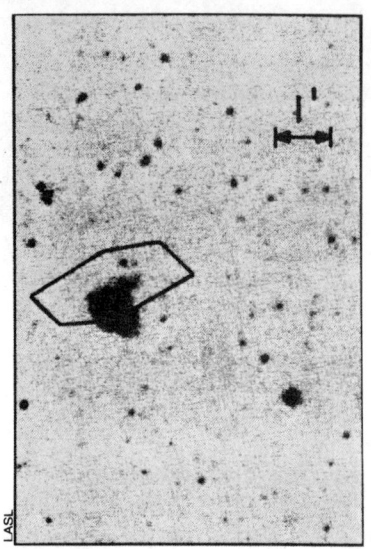

Fig. 16. Source location of the March 5 event shown on a reproduction of optical field, in which the densest portion of the N49 snr nebula is visible.[19]

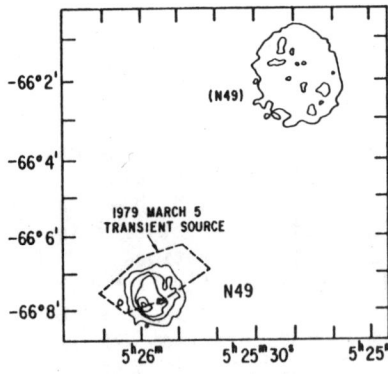

Fig. 17. Source location of the event[19] plotted on the X-ray surface brightness contour map of the N49 and (N49) region, as observed with the Einstein Observatory high-resolution imager.[24] Here the nebula is seen as 2 arc minutes wide, and the nearest neighboring nebula, (N49), is resolved at the weakest intensity level. No point X-ray source was resolved. No change in X-ray intensity above ~ 2×10^{-12} erg cm-2 s-1 was observed from shortly before to several days after the event. A comparison X-ray study with Einstein has been recently made to search for a slow, delayed effect, but the results are not available at this time.

burst, showing again an upper limit of ~ 10-9 to the intensity change ratio.) Thus, the fact that N49 is an extended object, suggesting a greater likelihood of chance association, is contrasted by the facts, supporting its identification, that it is a supernova remnant - an object associated with neutron star phenomena, and that it is an X-ray emitter. Further, the visible stars in the source field have been determined to be also at the same distance as the LMC,[25] thereby removing their distraction from this association. Of course, the argument for chance association is only a factor of two larger than if the candidate object were a point object, since the error box and N49 consist of about the same solid angle. That probability, considering the total solid angle of extended X-ray emitters and the total number of point X-ray sources (times the source box size) is between 10^{-5} and 10^{-6}, too small to be a strong argument against the identification. The basic argument against N49 is of course its distance and the $\sim 10^{45}$ erg sec^{-1} emission implied if the radiation in this event is isotropic.

DISCUSSION

The observations of the March 5 event, although inconclusive regarding the question as to whether this event is so different (that it must have been produced by an entirely different mechanism), clearly provides support for the hypothesis that its nature and possible origin at N49 are phenomena so unusual that they deserve critical study *per se* (rendering the question as to whether its understanding will enlighten us as to the nature of 'classical' gamma ray bursts, or vice versa, as secondary). A variety of early papers concerning this event tended to dismiss the N49 association as chance, avoiding the very real problems in the theoretical contortions necessary to get around the photon self-absorption inherent in fitting a physically possible mechanism to the 10^{45} erg sec^{-1} luminosity required by an isotropic N49 emission.[4,24,26,27,28] It had been known for some time that gamma ray burst intensities implied a several hundred pc source distance if treated at face value, given that the spectra entended beyond the pair production threshold;[16,29] recent updates show that, given nearby classical burst source distances,[30] the March 5 event source distance would be accordingly only ~ 2 pc, keeping the optically thin Bremsstrahlung treatment.[31] Creative treatments of the March 5 issue, taking the N49 association as a serious and possibly fruitful possibility, have only recently started to gain momentum.

The first detailed calculations of the March 5 event spectrum, treating the process of pair production, annihilation and scattering with an optically thin synchrotron mechanism (in the intense magnetic field of a neutron star requiring a ~ 20 percent gravitational redshift), fit the observations surprisingly well. Figure 18 shows the calculations of Ramaty and his coworkers, as outlined in a November 1979 gamma ray burst conference.[32] The low energy excess emission, the position and width of the line and the general shape of the continuum are all seen to be duplicated, in essence. The more worrisome problem of the source mechanism was first addressed in a sequel to this work, in which the same group showed that an earlier model of the storage of energy in neutron star vibrations, involving the release of (undetectable) gravitational radiation, provided a good fit of the vibrational damping time to the 150 msec width of the initial burst (see Figure 19). This model provided the necessary storage mechanism that made the 55-kpc distance physically

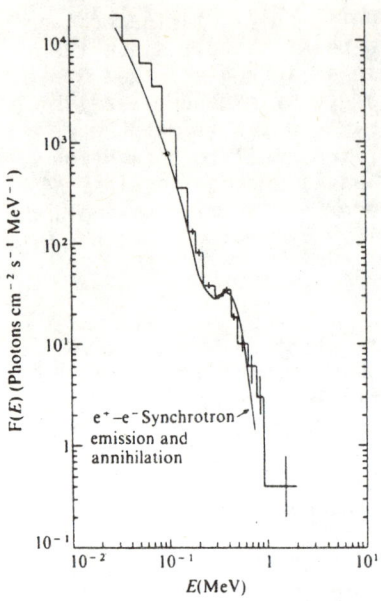
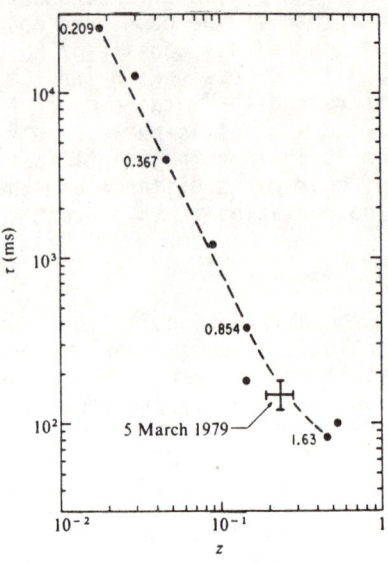

Fig. 18. The calculated spectrum of the intense portion of the March 5 event, based on an annihilation-creation model incorporating synchrotron losses and a redshift effect in a neutron star environment,[32] compared with the observations.[4]

Fig. 19. The quadrupole gravitational radiation damping time versus gravitational redshift for neutron stars, with the dashed line connecting calculated values having the same equation of state, compared with the March 5 event.[15]

possible, and linked the internal, gravitational process to the external, magnetospheric, observable phenomenon. More recently, the Ramaty model has been extended by Liang to include the higher energy effects of the inverse Compton process.[33] These calculations not only fit the observed spectrum a little better, but also provide a derivation, from first principles, of the luminosity at ~ 10^{44} erg sec^{-1}, giving an order-of-magnitude fit to the N49 source. I do not illustrate it here since it is presented in this Conference.[33] Given that these calculational exercises provide consistency, or better, with the N49 source association, we are freed from the necessity of assuming that it must be a chance association. Final proof of which view may be correct is quite possibly a long way off, however.

An ongoing reanalysis of the March 5 event directional data has provided a more precise source location;[34] this new error box is inside N49 although not at its center (see Figure 20). The 0.1 arc min^2 size of this source field is only about 5 percent that of the original conservative treatment of essentially the same data, making searches for point objects possibly a lot easier, but reducing the

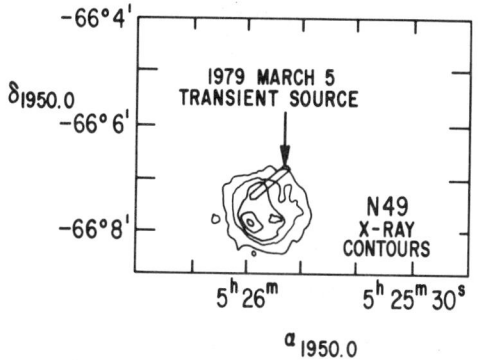

Fig. 20. The revised 1979 March 5 event source location, entirely consistent with N49 but in an eccentric location, 15 to 55 arc seconds from its center.[34] Given an unknown radial component, a possible location of the source may even be at the edge of the shell. This is the most precisely determined gamma ray source location in existence.

probability of chance identification with N49 by only a factor of two (since the N49 solid angle remains 2 arc min^2). This result would imply a motion of the neutron star from the center of the remnant, assuming N49 to be source, of $\sim 900 \pm 400$ km sec^{-1}, if the age of the remnant is 10,000 years. This is probably not an entirely unreasonable value. Of course the neutron star involved in the March 5 emission may not be the parent neutron star of the N49 snr itself, given the snr and neutron star density in the LMC.

What then is the relationship of this event to the bulk of gamma ray bursts - why is there a gap in the size spectrum? No one would extend the N49 association to the point of attributing the March 5 event, as the brightest of all bursts, to an LMC origin because that galaxy is the nearest of the external galaxies that, in turn, produce the weaker, isotropically distributed, classical bursts. This size spectrum difficulty is avoided in one way by assuming that March 5-like events are rare, either intrinsically rare, as are supernovae themselves (perhaps for reasons that are related), or apparently rare, because of a beamed emission anisotropy (or a mixture of both); in this case the detection of one event in a decade from the neighboring galaxy LMC is not really less likely than from our own galaxy.[35] (The nearby origin model has no corresponding way out of the size spectrum difficulty, other than to invoke a simple accident of proximity, like the accident of coincidence with the direction of N49 itself.) Continuing this speculative notion, a possible link between the March 5 event and the other bursts is the series of three weak, delayed events, compared with the existence of another series of three small common-origin events that does not itself follow any identifiable large event. The source direction of that series, as noted above, is in the galactic disk at ~ 45 degrees galactic longitude.[17] That direction is consistent with a galactic source distance of up to perhaps 20 kpc, half the distance to the LMC, putting those events into a model of consistency with an N49 origin for the delayed March 6--April series, except that any primary or 'parent' event in this case was invisible, due perhaps to an anisotropy of emission. Are, in fact, the larger 'classical' gamma ray bursts also 'secondary, delayed' emissions, but with repetition times outside the presently measurable several-year extents, perhaps linked to their greater intensity, relative to the March 6--April series? This would imply the existence of many apparently invisible March 5-like events, i.e., of the requirement of a narrow solid angle of emission, perhaps several hundred square degrees. Yet some link surely exists, defining the occurrence rate of the various kinds of transient events from each neutron star source. Finally, one can question the basic assumption of the neutron star gamma ray transient source concept, accounting for the 420 keV lines as evidencing a ~ 20 percent redshift, gravitationally required for a 1.4 M_0 neutron star, of the 511-keV line. The stimulated emission concept is shown in this conference to produce a 420-keV feature without the presence of a gravitational redshift.[1] Clearly a great deal of research is needed to investigate the emission characteristics implied, such as the amount of coherent beaming, that may clarify, eliminate or unify these ideas in terms of the relationship to gamma ray phenomenology of the 1979 March 5 event.

I wish to point out that I have not treated some very recently published theoretical studies of the March 5 event, but I do wish to refer them to your attention.[36,37,38,39,40]

Acknowledgements

I wish to thank J. Newby for assistance in the preparation of this manuscript.

REFERENCES

1. R. Ramaty, J. McKinley and F. C. Jones, Proceedings of this Symposium
2. T. L. Cline, Comments Ap. 9, 3 (1980).
3. T. L. Cline et al., Ap. J. (Letters) 237, L1 (1980).
4. E. P. Mazets, S. V. Golenetskii, V. N. Il'inskii, R. L. Aptekar' and Yu. A. Gur'yan, Nature 282, 587 (1979).
5. B. J. Teegarden and T. L. Cline, Ap. J. (Letters) 236, L67 (1980).
6. E. P. Mazets, S. V. Golenetskii, R. L. Aptekar', Yu A. Gur'yan and V. N. Il'inskii, Nature 290, 378 (1981).
7. E. P. Mazets et al., P.T.I. (Leningrad) preprints 599 and 618 (1979).
8. G. Pizzichini, Ap. Space Sci. 75, 205 (1981).
9. A. S. Jacobson, J. C. Ling, W. A. Mahoney and J. B. Willett, "Gamma Ray Spectroscopy in Astrophysics", NASA TM-79619, ed. T. L. Cline and R. Ramaty., p. 228 (1978).
10. R. E. Lingenfelter, J. C. Higdon and R. Ramaty, "Gamma Ray Spectroscopy in Astrophysics", NASA TM-79619, ed. T. L. Cline and R. Ramaty, p. 252 (1978).
11. E. E. Fenimore, W. D. Evans, R. W. Klebesadel, J. G. Laros and J. Terrell, Nature 289, 42 (1981).

12. G. Vedrenne, V. M. Zenchenko, V. G. Kurt, M. Niel, K. Hurley and I. V. Estulin, Soviet Astron. Letters 5, (6), 314 (1979).
13. S. Barat et al., Astr. Ap. 79, L24 (1979).
14. J. Terrell, W. D. Evans, R. W. Klebesadel and J. G. Laros, Nature 285, 383 (1980).
15. R. Ramaty, S. Bonnazzola, T. L. Cline, D. Kazanas, P. Mészáros and R. E. Lingenfelter, Nature 287, 122 (1980).
16. W. K. H. Schmidt, Nature 271, 525 (1978).
17. E. P. Mazets and S. V. Golenetskii, P.T.I. (Leningrad) Preprint 632 (1979).
18. W. D. Evans, R. Klebesadel, J. Laros, T. Cline, U. Desai, B. Teegarden and G Pizzichini, I.A.U. Circular # 3356, May 11 (1979).
19. W. D. Evans et al., Ap. J. (Letters) 237, L7 (1980).
20. J. G. Laros et al., Ap. J. (Letters) 245, L63 (1981).
21. T. L. Cline et al., Ap. J. (Letters) 246, L133 (1981).
22. T. L. Cline et al., Ap. J. (Letters) 229, L47 (1979).
23. T. L. Cline et al., Ap. J. (Letters) 232, L1 (1979).
24. D. S. Helfand and K. S. Long, Nature 282, 589 (1979).
25. G. J. Fishman, J. G. Duthie and R. J. Dufour, 1981, Ap. Space Sci. 75, 135 (1981).
26. F. A. Aharonian and L. M. Ozernoy, Astron. Tsirkulyar # 1072, 1 (1979).
27. G. S. Bisnovaty-Kogan and V. M. Chechetkin, S.R.I. (Moscow) preprint # 561 (1980).
28. I. I. Kumkova and I. G. Mitrofanov, 1980, Pis'ma Astr. Zh., 6, (4), 213 (1980).
29. G. Cavallo and M. J. Rees, Mon. Not. R. Astron. Soc., 183, 359 (1978).
30. D. Gilman, A. E. Metzger, R. H. Parker, L. G. Evans and J. I. Trombka, Ap. J. 236, 951 (1980).
31. M. Ruderman, Erice reprint (1981).
32. R. Ramaty, R. E. Lingenfelter and R. W. Bussard, Ap. Space Sci., 75, 193 (1981).
33. E. P. T. Liang, Nature 292, 319 (1981).
34. T. L. Cline et al., in preparation.
35. T. L. Cline, U. D. Desai and B. J. Teegarden, Ap. Space Sci 75, 93 (1981).
36. J. I. Katz, preprint (1981).
37. R. Hoshi, C.F.A. preprint 1495 (1981).
38. Qu Qin-yne, Wang De-Yu, Xu Ao-Ao, and Li Ze-qing, Proceedings of the 17th Cosmic Ray Conference, Paris, 1981, Paper XG 2.2-4 (1981).
39. A. K. Drukier, Proc. 17th Cosmic Ray Conference, Paris 1981, Paper XG 2.2-11 (1981).
40. T. M. K. Marar, A. K. Jain, D. P. Sharma, K. Kasturirangan and U. R. Rao, Proc. 17th Cosmic Ray Conference, Paris 1981, Paper XG 2.2-3 (1981).

FREQUENCY OF GAMMA-RAY BURSTS >3 x 10^{-6} ERG/CM^2

G.H. Share, K. Wood, J. Meekins, and D.J. Yentis
E.O. Hulburt Center for Space Research
Naval Research Laboratory, Washington, D.C. 20375

ABSTRACT

The Large Area Sky Survey (LASS) instrument on HEAO-1 has confirmed the detection of 12 of 18 γ-ray bursts with intensities >2.5 x 10^{-6} erg/cm^2 observed between August 1977 and January 1979. The LASS measurements are all consistent with the reported source locations of the bursts based on occultation by the Earth. Six new, and as yet unconfirmed, bursts have been detected between October 1977 and January 1978. Our observations provide new measurements of the rate of γ-ray bursts: 125 ± 50/yr, S >3 x 10^{-6} erg/cm^2; 50 ± 20/yr, S >10^{-5} erg/cm^2. These and other measurements establish a Log N - Log S relationship for bursts which is consistent with a spherical distribution of sources within our Galaxy which have a broad range (~10^3) of intrinsic luminosities[5].

INTRODUCTION

Recent experimental and theoretical work indicates that most of the sources of cosmic gamma ray bursts lie within our Galaxy. This conclusion has been reached by a variety of approaches. These include energetic considerations[1], the number vs. intensity distribution (Log N - Log S)[2-5], detectability into the MeV range[6], spectral and spatial observations[7,8], and observed periodicities[9].

In this paper we present results obtained with the Large Area Sky Survey (LASS) instrument on the HEAO-1 satellite which provide additional support for the galactic origin of bursts based on the Log N - Log S distribution. Although LASS was designed to survey the sky in the 0.5 - 25 keV X-ray band, its large cross section has made it a sensitive detector of γ-ray bursts.

γ-ray bursts were not likely to have been detected in the aperture of this instrument. Only about one burst >5 x 10^{-6} erg/cm^2 is expected to be detected in the aperture in 100 years. Instead the bursts were detected after many, if not all, of the primary photons were scattered one or more times in the proportional counters and surrounding material of the satellite. This made the instrument a roughly omnidirectional detector without degrading its temporal response to the bursts. The UHURU X-ray detectors earlier observed γ-ray bursts in much the same manner[10] but with reduced sensitivity owing to their smaller cross section.

HEAO-1 was in operation from August 1977 until January 1979. During this time period, the LASS instrument detected at least twelve confirmed γ-ray bursts with intensities >3 x 10^{-6} erg/cm^2 and identified six new, as yet unconfirmed, bursts.

0094-243X/82/770035-10$3.00 American Institute of Physics

INSTRUMENTATION

The full complement of LASS detectors consisted of 7 proportional counters, each filled with a Xe-CH_4 gas mixture. These detectors are shown schematically in Figure 1. Six of the counters

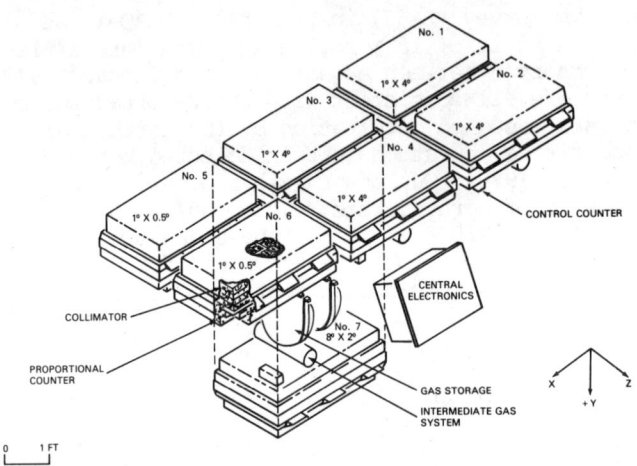

Fig. 1 Layout of the seven LASS proportional counters on the HEAO-1 satellite.

were on one side of the HEAO-1 satellite; each had an effective area to X-rays of ~1600 cm^2. The seventh counter was on the opposite side of the satellite along with other experiments and had an effective area of ~1800 cm^2. HEAO-1 was in a circular orbit at an altitude of ~450 km and an inclination of 23°. The satellite rotated with a period of ~35 min with its spin axis pointed toward the sun.

We have utilized data from the beginning of October 1977 until the end of January 1978, during which time LASS was in a low-gain mode, covering the 0.5 to 100 keV range. Data were primarily obtained from the three proportional counters designated as 3, 4 and 5 in Figure 1. Detectors 3 and 4 had 1° x 4° FWHM apertures and detector 5 had a 1° x 1/2° FWHM aperture for X-rays <20 keV. After January 1978 data were primarily obtained from detectors 3 and 5.

Sixteen spectral channels covering the energy range from 0.5 to 100 keV sampled energy losses separately in detector 5 and the sum of detectors 3 and 4 during each 0.32 s. At certain times the instrument was placed in a high-time resolution mode in which data from a given counter was accumulated at 5 ms resolution. During these intervals spectral information was only available every 0.64 s.

Owing to the fact that the bursts were detected after degradation of the incident photon beam in the satellite and detector materials, it is difficult to directly estimate the instrument's energy range and efficiency. A comparison of the LASS response to a solar flare with that of the ISEE-3 solar hard X-ray detector[11] indicates that LASS is predominantly sensitive to incident photons in the 75-200 keV range when the source is not within the instrument's aperture. The LASS efficiency for detecting bursts was estimated by comparing its response to four γ-ray bursts whose intensities were measured by Vela[12]. On the average we find that one count with an energy loss >20 keV in a given LASS detector is equivalent to 6×10^{-8} erg/cm^2 as measured by Vela. The uncertainty is ~50%. This factor was used to compute the intensity of three bursts detected by both LASS and the KONUS experiment[13] on Venera 11 and 12 in September 1978. We obtained agreement in intensity within ~50% with the measured KONUS intensities >150 keV, and somewhat worse agreement for intensities measured >30 keV. For purposes of estimating γ-ray burst intensities from LASS we therefore adopt the normalization factor of $(6 \pm 3) \times 10^{-8}$ erg/cm^2-count in each detector.

Shown in Figure 2 is a comparison of the instrument's response to a composite of three γ-ray bursts (29 October and 10 November 1977, and 21 May 1978) and to the hard X-ray source Cygnus X-1 in its low activity state. The peak in the energy-loss spectrum occurs near 60 keV for γ-ray bursts, but it is near 6 keV for X-ray sources. This spectral difference has enabled us to make a sensitive search for weak γ-ray bursts without significant contamination from soft X-ray sources passing through the aperture of the instrument. We will describe this search in the next section.

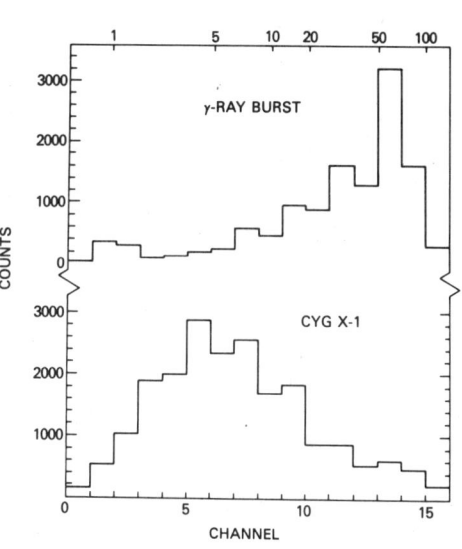

Fig. 2 Measured energy-loss responses of the LASS detectors to a composite of three γ-ray bursts and to the X-ray source Cygnus X-1.

OBSERVATIONS

a) Previously Detected Bursts

In Table 1 we list the γ-ray bursts having intensities $>2 \times 10^{-6}$ erg/cm^2 (>150 keV) reported up until the end of November 1978 for which LASS data is available. These observations were made by the Vela[12], Helios[14], Meteor[15] and Venera[13] satellites. The times listed were those obtained from LASS wherever possible,

TABLE 1

γ-RAY BURSTS OBSERVED DURING HEAO-1 LIFETIME

DATE		TIME U.T.			LASS OBSERVATION DETECTED	NADIR RA	DEC
		HR	M	S			
1977 OCT.	20a	07	54	55	yes	239.7°	+13.3°
OCT.	29a	11	41	20	yes	35.6°	+ 4.0°
NOV.	10a	17	12	45	yes	12.3°	+22.3°
NOV.	12b	05	57	05	yes	249.8°	-13.6°
1978 MAY	8a	00	34	25	no	16.9°*	-18.0°*
MAY	8a	20	14	37	no	270.9°	-12.0°
MAY	19a	7	21	49	no	41.8°	19.5°
MAY	21a	21	53	47	yes	210.3°	-21.1°
SEPT.	14a,c	16	42	12	yes	57.9°	-18.7°
SEPT.	18a,c	19	51	08	yes	195.3°	15.8°
SEPT.	21a,c	03	56	00	yes	348.0°	-12.2°
SEPT.	30c	06	47	29	yes	57.1°	- 4.7°
OCT.	6c	10	59	50	no	182.0°*	9.0°*
OCT.	6c	14	24	56	yes	256.6°	-19.2°
OCT.	25c	23	53	02	yes	220.0°	- 6.5°
NOV.	19a,c	09	26	56	yes	192.1°	15.0°
NOV.	21a,c	01	33	57	no	259.0°	-15.0°
NOV.	24a,c	03	55	30	no	108.3°	21.6°

a - From Vela and Helios, K. Klebasadel and T. Cline (private communication)
b - From Meteor (ref. 15).
c - From Venera (ref. 13).
* - Uncertain due to unknown time of arrival of wavefront at Earth.

otherwise they were obtained from the references cited. The last two columns give the position of the nadir (epoch 1950) as viewed from HEAO at the time of the bursts. Occultation by the Earth and the LASS response can be used to assist in the localization of the sources of the bursts.

Twelve of the eighteen bursts were detected by LASS. The weakest event that was detected occurred on 12 October 1978 and had an integrated intensity of 2.6×10^{-6} erg/cm^2 above 150 keV[13] (4×10^{-6} erg/cm^2 >30 keV). We shall briefly describe our observation of these bursts. The time history of the 20 October burst is shown in Figure 3. A data drop-out occurred in

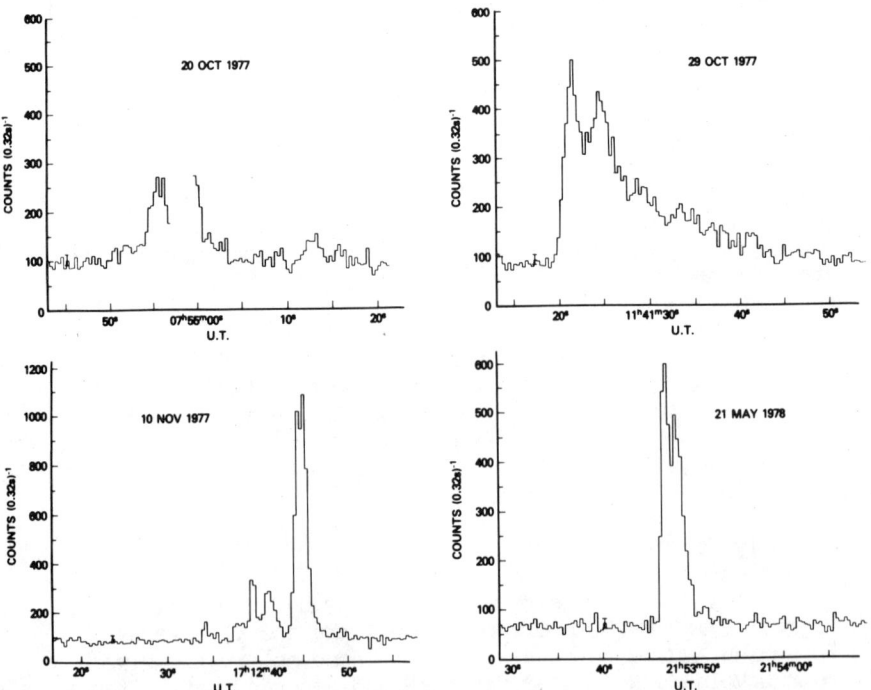

Fig. 3 Time histories of four γ-ray bursts observed by LASS.

the middle of the burst. Use of Earth occultation excludes ~50% of the allowable source region determined from Prognoz 6[16]. More detailed temporal and spectral data are available from the γ-ray spectrometer on HEAO-1[17].

The 29 October time history is also shown in Figure 3. It was the most intense burst observed. Several secondary peaks are visible superimposed on an exponential decay. A detailed analysis of this event[9] indicates strong evidence for periodic emission with a period of 4.2 ± 0.25 s. This is the second burst source to exhibit periodic emission[18]. Other indications of quasi-periodic behaviour on similar timescales has also been discussed previously[19,20]. Earth occultation does not help to reduce the uncertainty in source position[16].

The narrow burst of 10 November shown in Figure 3 is preceded by weak activity that was not previously observed by other instruments[17]. No new positional information was obtained from occultation. The remaining burst plotted in Figure 3 occurred on 21 May 1978 and had a double pulse profile characteristic of many bursts.

LASS has confirmed the detection of the weak burst ($\sim 4 \times 10^{-6}$ erg/cm^2) on 12 November 1977 by Meteor[15]. Its time history is shown in Figure 4. Statistics are limited but it appears to have a rather complicated time structure.

The time histories of the remaining bursts are not presented, but are similar to those shown in the first catalog of γ-ray bursts observed by the KONUS experiment on Venera[21].

We have utilized the celestial positions of nadir with respect to the HEAO-1 satellite in an attempt to reduce the uncertainties in the source locations of the events cataloged by the KONUS experiment and which occurred from September through November 1978[21]. LASS detected all

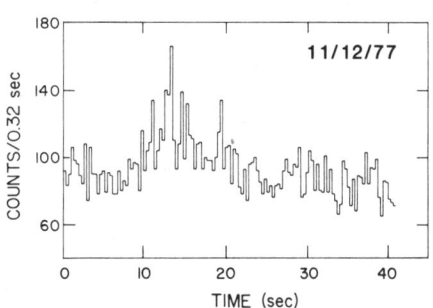

Fig. 4 Time history of the weak γ-ray burst discovered by Meteor[15].

the bursts which had source positions unocculted by the Earth. It failed to detect any bursts whose quoted positions were occulted. The LASS observations, unfortunately, did not resolve any of the ambiguouties. The fact that all of the detectable events were observed indicates that LASS had a relatively omnidirectional response to the bursts.

b) Detection of New Bursts

We have used the HEAO-1 LASS data base from 1 October 1977 to 25 January 1978 to search for additional weak γ-ray bursts.

Three detectors were operating satisfactorily during this time period and have been used in the search. They were detectors 3, 4 and 5 (see Figure 1). The search began using data from the 1° x 1/2° counter (detector 5) in order to reduce spurious triggers due to soft X-ray sources passing through the field of view of the instrument.

The method involved first fitting the counting rate >20 keV with a linear function over a time interval of 40.96 s (i.e. 128 bins of 0.32 s each). Either a 3.5σ fluctuation above the fit in any 0.32 s bin or a value of χ^2 >155 (~10% probability) relative to the best fit straight line over the full 40.96 s interval prompted further study. In the next phase all bins exceeding the fit by 2σ were summed and the ratio of low energy (E <20 keV) to high energy (E >20 keV) counts above the fitted value was calculated. If the ratio was <0.75, it was likely that the detectors were not responding to an X-ray source passing through the field of view (see Figure 2). If this energy criterion for the 1° x 1/2° detector was met, the entire procedure was repeated for the summed data from detectors 3 and 4. If the data from these modules passed the same tests, the full 40.96 s of data were plotted for visual inspection. Spurious events and statistical fluctuations were then screened out and any residual response to X-ray sources could be identified. For events passing these tests, a check was made to determine whether they were produced by solar flares or geomagnetic disturbances.

As a result of this study, six new γ-ray bursts have been detected. Information on these tentative identifications are given in Table 2. The weakest burst recorded had an integrated intensity of 3×10^{-6} erg/cm^2. These integrated intensities were estimated by comparing the LASS response to γ-ray bursts detected by Vela and Venera as discussed earlier. The events are displayed in Figure 5 at a resolution of 0.32 s. The 4 October, 24 November, and 9 December 1977 events exhibit multipulse structure (it is not clear whether the rise occurring in the latter part of the 5 January 1978 event is associated with the burst).

TABLE 2

TENTATIVE γ-RAY BURST IDENTIFICATIONS MADE WITH HEAO-1 LASS

DATE	TIME U.T. HR M S	INTEGRATED INTENSITY ERG/CM^2	NADIR RA	DEC
1977 OCT. 4	23 21 10	~9.0×10^{-6}	175.6°	-11.6°
OCT. 7	03 26 23	~3.0×10^{-6}	1.0°	+ 3.7°
NOV. 24	17 24 16	~1.4×10^{-5}	171.6°	- 9.2°
DEC. 9	01 58 40	~5.0×10^{-6}	122.0°	12.7°
1978 JAN. 5	21 12 46	~9.0×10^{-6}	29.6°	13.9°
JAN. 11	18 21 36	~1.5×10^{-5}	1.5°	8.6°

Based on these six new bursts and the three previously identified by other satellites (the burst detected by Meteor is not included because it was not found by our search routine) during the October 1977 to January 1978 time period, we have been able to derive information on the number vs. intensity distribution of bursts. In order to do this we first need to determine the efficiency with which bursts are detected by our automated search routine. This efficiency is dependent on burst width and structure as well as on integrated intensities. We used a Monte Carlo routine for simulating counts from both the background and burst.

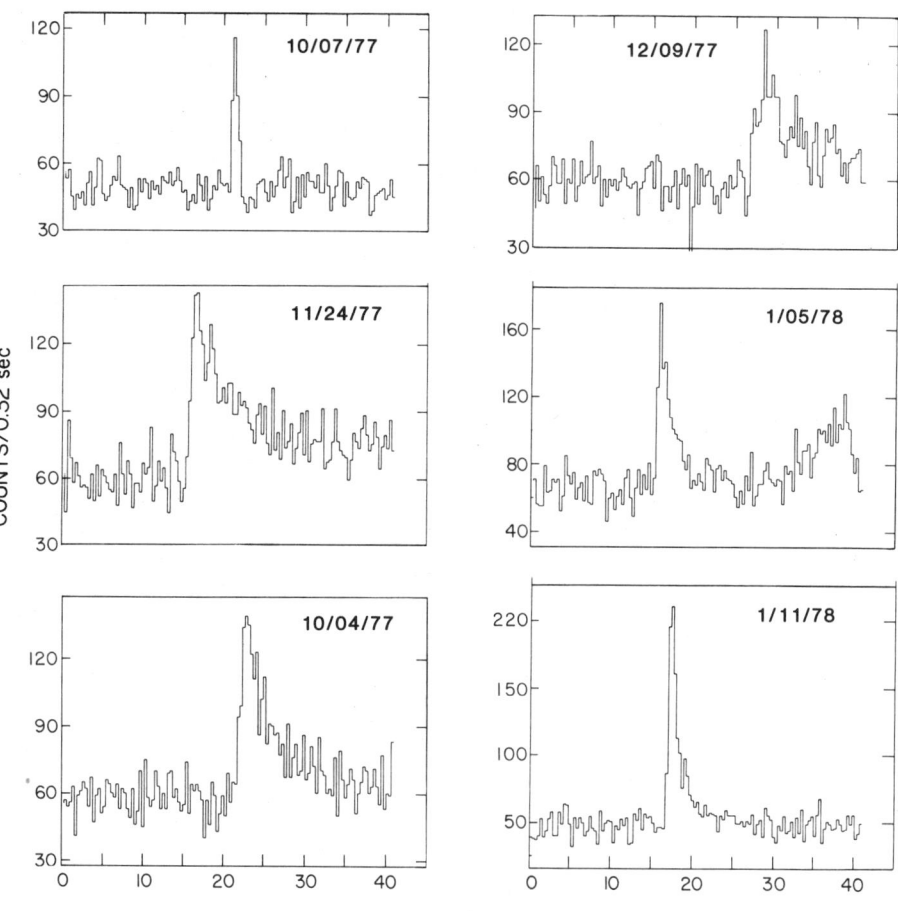

Fig. 5 Time histories of the six bursts discovered by the LASS experiment on HEAO-1.

The burst counts were randomly distributed on top of the background and followed a general triangular shape of varying duration. The simulated burst data were then analyzed using the identical search routines utilized to discover the six new bursts. For bursts with a full width duration of ~7 sec (a typical average value for gamma ray bursts) we find that the search is 100% effective for integrated intensities $>10^{-5}$ erg/cm^2. The efficiency drops to near zero $\leq 10^{-6}$ erg/cm^2. We note here that many bursts have sharper features which improve our sensitivity for detecting bursts longer than the 7 s used in our model. The effective observation time was 0.12 yr, which includes the effect of Earth occultation. We have therefore plotted two points on the Log N - Log S plot given in Figure 6. The first point includes all bursts $>10^{-5}$ erg/cm^2 and required no correction for efficiency of detection; the point at 3×10^{-6} erg/cm^2 included all the bursts above

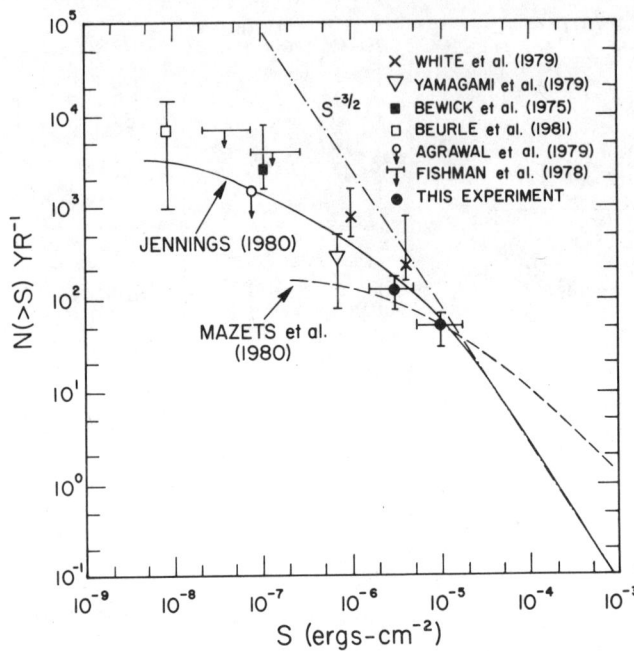

Fig. 6 Measurements of the Log N - Log S distribution of γ-ray bursts compared with the model of Jennings[5]. Other observations: White[25], Yamagami[26], Bewick[27], Beurle[28], Agrawal[29], Fishman[30]. The results from the Venera satellite, Mazets et al.[8], have not been corrected for instrumental threshold effects; see text.

that energy corrected for the calculated efficiency of detection. Our uncertainty in estimating the intensity of the burst is shown by the horizontal bar.

DISCUSSION

The current situation with respect to the Log N - Log S distribution of γ-ray bursts is summarized in Figure 6. As discussed in the previous section the values obtained from this experiment include the six new bursts. To date there has been no confirmation of these events. The Prognoz 6 satellite, which was operational during this time period, did not detect any of the events[22]. The intensity of these bursts was near the limiting sensitivity of the Prognoz detectors[23]. In addition these bursts may have relatively soft spectra and would not have been detectable had Prognoz been operating in its 300-1460 keV range. The measured values are 50 ± 25 yr^{-1} for $S > 10^{-5}$ erg/cm^2 and 125 ± 40 yr^{-1} for $S > 3 \times 10^{-6}$ erg/cm^2.

Figure 6 also gives pertinent observations made with other experiments. We have not plotted the early distributions obtained from Vela[24] and IMP 7[25] for clarity; they agree well with the plotted model of Jennings[5] down to $\sim 7 \times 10^{-5}$ and $\sim 4 \times 10^{-5}$ erg/cm^2, respectively, before flattening due to instrumental effects. The most extensive survey of bursts has been made by the KONUS experiment on Venera[8]. The derived distribution

from this experiment is approximated by the dashed line. This
distribution has not been corrected for instrumental threshold
effects which become important below 5×10^{-5} erg/cm^2. The authors
estimate that the true rate near 3×10^{-7} erg/cm^2 may be as much as
a factor of 5-10 higher. These corrections must be kept in mind
while comparing their results with other measurements. Some
questions have been raised[5] by the disagreement of the Venera
results with other measurements at high S values. However,
our results are consistent with theirs in the $10^{-6} - 10^{-5}$
erg/cm^2 range. The Leningrad group has raised questions about
the detailed utilization of such Log N - Log S plots for determining the spatial distribution of burst sources[8]. But, they
still conclude that their results are clearly different from an
$S^{-3/2}$ distribution which is characteristic of an isotropic
distribution of sources of equal intrinsic luminosity.

The other measurements shown in the Figure were obtained
with large area balloon-borne detectors. They achieve sensitivities orders of magnitudes below most satellite systems and
also show the divergence from the $S^{-3/2}$ distribution. This
divergence implies that the burst sources are galactic in origin.
It might therefore be expected that their spatial distribution
should exhibit a strong correlation with galactic latitude and
longitude[2-4]. This is not the case[8], however. Recently
Jennings[5] has shown that a galactic model which allows for a
distribution of intrinsic luminosities of the burst sources is
consistent with both the observed Log N - Log S distribution, and
the galactic latitude and longitude distributions. He finds
that luminosity distributions which are moderately peaked toward
lower values and which have a broad dynamic range $>10^2$ will
dominate over geometry effects in certain regions of the Log N -
Log S distribution. Shown in the Figure is his calculated Log N
- Log S function for an eccentrically viewed galactic distribution
of burst sources with a power-law distribution of intrinsic
luminosities having a dynamic range of 10^3. The intensity range (S)
over which our measurements have been made are in the domain where
the softness of the luminosity function, rather than the geometry,
dominates the Log N - Log S distribution.

ACKNOWLEDGMENTS

We wish to thank T. Chubb and H. Friedman for their advice and
support of this work. This research was supported by NASA under
grant number H-87023A. P. Reed and W. Brizzi assisted us with the
preparation of the manuscript.

REFERENCES

1. M. Ruderman, Annals N.Y. Acad. Sci. 262, 164 (1975).
2. G.J. Fishman, Ap. J. 233, 851 (1979).
3. M. Yoshimori, Aust. J. Phys. 31, 189 (1978).
4. M.C. Jennings and R.S. White, Ap. J. 238, 110 (1980).
5. M.C. Jennings, Ap. J. in press (1981).

6. W.K.H. Schmidt, Nature 271, 525 (1978).
7. E.P. Mazets et al., Nature 290, 378 (1981).
8. E.P. Mazets et al., Sov. Astron. Lett. 6, 318 (1980).
9. K. Wood et al., Ap. J. in press (1981).
10. D. Koch et al., Proc. Los Alamos Conf. on Transient Cosmic γ & X-Ray Sources (1973).
11. S. Kane, private communication (1979).
12. R. Klebesadel, private communication (1979).
13. E.P. Mazets et al., Sov. Astron. Lett. 5, 87 (1979).
14. T. Cline and U. Desai, private communication (1979).
15. E.P. Mazets, Sov. Astron. Lett. 4, 188 (1978).
16. A.V. Kuznetsov et al., Sov. Astron. Lett. 5, 147 (1979).
17. F.K. Knight et al., Ap. and Space Sci. 75, 21 (1981).
18. T.L. Cline et al., Ap. J. 237, L1 (1980).
19. U.D. Desai, Ap. and Space Sci. 75, 15 (1981).
20. G. Pizzichini, Ap. and Space Sci. 75, 205 (1981).
21. E.P. Mazets et al., A.F. Ioffe Physico-Technical Institute Preprint No. 618, Leningrad (1979).
22. K. Hurley, private communication (1979).
23. C. Barat et al., Ap. and Space Sci 75, 83 (1981).
24. I.B. Strong and R.W. Klebesadel, Nature 251, 396 (1974).
25. T.L. Cline and U.D. Desai, Ap. and Space Science 42, 17 (1976).
26. R.S. White et al., Proc. 16th Int. Cosmic Ray Conf. 12, 41 (1979).
27. T. Yamagami et al., Proc. 16th Int. Cosmic Ray Conf. 1, 223 (1979).
28. A. Bewick et al., Nature 258, 686 (1975).
29. K. Beurle et al., preprint Imperial College, London, (1981).
30. P.C. Agrawal et al., in X-Ray Astronomy (COSPAR), Pergamon Press, Oxford, 515 (1979).
31. G.J. Fishman et al., Ap.J. (Letters) 223, L13 (1978).

OBSERVATION OF GAMMA-RAY BURSTS FROM 10 keV TO 9 MeV

G.H. Share, J.D. Kurfess, S. Dee*
E.O. Hulburt Center for Space Research
Naval Research Laboratory, Washington, D.C. 20375

E.L. Chupp, J.M. Ryan, D.J. Forrest, J. Lanigan
University of New Hampshire, Durham, N.H. 03824

E. Rieger, G. Kanbach, C. Reppin
Max Planck Institute, Garching b. Munchen, West Germany

ABSTRACT

The γ-ray spectrometer on the Solar Maximum Mission (SMM) satellite has detected at least 14 confirmed cosmic γ-ray bursts during its first 14 months in orbit. Individual peaks observed in some of the bursts differ significantly in hardness from one another. Similarity of the time profiles in different energy bands suggest that photons spanning two decades in energy are produced by the same mechanism. All of the bursts were detected to energies $\gtrsim 1$ MeV. Detection of photons into the MeV region can be used to set limits on the distance to γ-ray burst sources[17]. Spectra of four events falling within the instrument's aperture are presented. Two appear to be well fit by single power laws, but the indices are strikingly different. The other two require either two power laws or an exponential function. A thermal bremsstrahlung fit would require temperatures $\gtrsim 1$ MeV. No clear evidence has been found for the existence of narrow line features in any of the bursts. No evidence for broadened line features is seen in any of the four spectra presented.

INTRODUCTION

Spatial, temporal and spectral characteristics of γ-ray bursts provide the means of establishing their origin. With the possible exception of the 5 March 1979 event[1], compelling evidence for a galactic origin of the bursts [2,3] has been established based on their number-intensity and spatial distributions. The bursts are observed to exhibit a variety of temporal characteristics which have been summarized by Mazets and Golenetskii[4]. The March 5 event is also unusual in this context because it was the first burst to be observed exhibiting a clear periodicity[5]. One other burst has also been found to exhibit periodicity[6]. Recent spectral observations made from the Venera spacecraft have shown that many of the bursts exhibit absorption and emission features in the hard X-ray range (see also the paper by B. Dennis et al. in these proceedings) and emission features at higher energies[7]. Supporting evidence for emission features at the higher energies has come from measurements made from ISEE-3[8]. Association

---

* Participant in the NRL - U.S. Naval Academy Ensign Program.
0094-243X/82/770045-09$3.00 American Institute of Physics

of these features with hard X-ray cyclotron phenomena and red-shifted γ-ray emissions support the suggestions that the γ-ray bursts are associated with neutron stars[7].

With the launch and successful operation of the Solar Maximum Mission (SMM) satellite, a new source of information on the spectral characteristics of γ-ray bursts is now available. SMM includes a NaI spectrometer with an effective area of 150 cm^2 at 500 keV, good energy resolution (7% at 662 keV) and a large field of view (~2π)[9]. Its primary purpose is the investigation of solar γ-ray emission. Results from these solar investigations have been reported elsewhere and are summarized by E.L. Chupp in these Proceedings.

In this paper we present time histories and count-rate spectra of some of the γ-ray bursts detected by this instrument from 20 February 1980 through May 1981. Work is continuing on the detailed analysis of these events. The data in this preliminary form allow us to draw conclusions about the spectral variability within the bursts, similarity of hard X-ray and MeV time profiles, the presence of any high-energy cut-offs, the presence of narrow line features and the spectral shapes of the bursts.

OBSERVATIONS

The γ-ray bursts have been identified using three different methods. An automated search covering the 350-800 keV energy range has been conducted at NRL as part of the normal production processing of the data. These data are accumulated every 16.38 s. In order to reduce effects due to background changes, the search is performed over specified sections of each orbit rather than the

Table 1

Bursts Detected by the SMM γ-Ray Spectrometer

Date M/D/Yr	Time U.T.	E_{max} MeV	Other Detections
03/07/80	05:09	4	P.V., HXRBS
03/24/80	23:57	4	P.V.
04/19/80	01:19	6	HXRBS, P.V., ISEE
04/21/80	03:08	2	HXRBS, P.V.
05/12/80	23:32	2	P.V.
05/24/80	15:54	3	ISEE, P.V., HXRBS
06/02/80	13:20	5	ISEE, P.V., HXRBS
06/23/80	01:29	4	--
07/06/80	08:44	1	P.V.
07/09/80	07:24	7	P.V.
08/12/80	18:55	3	HXRBS, P.V.
08/15/80	18:21	8	P.V., HXRBS
09/19/80	19:21	5	--
09/20/80	14:11	5	--
10/16/80	06:03	5	HXRBS, P.V.
11/19/80	17:06	8	P.V., HXRBS
12/20/80	18:30	6	P.V.,
03/01/81	12:35	7	--
03/04/81	08:50	1	--

P.V.   Pioneer-Venus (R. Klebesadel, priv. comm.)
HXRBS SMM (B. Dennis, priv. comm.)
ISEE   ISEE 3 (U. Desai, priv. comm.)

entire orbit. A quadratic function is fit to the data and fluctuations both on 16.38 s and longer timescales are investigated. Events detected in this manner are then visually screened using microfilm plots derived from the production analysis. The second method, employed at the Max Planck Institute, involves visual screening of the variation in rates observed in the 300-350 keV region. The third source of information is from other experiments such as the γ-ray detectors on Pioneer-Venus[10], ISEE-3[11], and the Hard X-ray Burst Spectrometer on SMM[12]. We estimate that the SMM γ-ray spectrometer is sensitive to γ-ray bursts with integrated intensities above 100 keV of $\gtrsim 5 \times 10^{-6}$ erg/cm^2.

Listed in Table 1 are the 19 bursts observed by SMM. None of the events are associated with solar flares observed at optical wavelengths. Included are the dates and times of the bursts, the approximate maximum energy at which a flux was detected, and other instruments which detected the same burst. All of the events were detected into the MeV range. The five events currently lacking independent confirmation should be treated with caution. After correcting for effective exposure time, the rate of detection of bursts by SMM appears to be consistent with other measurements $\gtrsim 10^{-5}$ erg/cm^2 (>100 keV) (see reference 2 and the results from HEAO-1 discussed in these Proceedings by G. Share et al.)

Plotted in Figures 1 and 2 are the time histories of sixteen of these events recorded in two energy bands: 25-140 keV and a 50 keV window near 300 keV. The 25-140 keV data are obtained from one of two 8 cm^2 auxiliary X-ray detectors included as part of the γ-ray spectrometer experiment. These detectors have a time resolution of 1 s. The data in the 50 keV window are derived from a special channel of the γ-ray spectrometer which has a time resolution of 64 ms (these data are plotted at 1 s resolution). The weak response for some of the events in these energy bands may have been due to absorption in the satellite materials prior to detection. Both sets of data were prescaled prior to being transmitted from the satellite; this accounts for the apparent non-statistical variations at low intensities.

The time histories exhibit the different characteristic profiles found for γ-ray bursts[4]. Durations vary from ~1 s to ~1 min. Structure at even finer time resolution has been observed as well. For example the 19 April 1980 event (Fig. 1c) is found to exhibit at least two pulses, each lasting ~250 ms (see ref. 13). Changes in spectral hardness are observed within the events. For example: the first peak of the 7 March 1980 event (Fig. 1a) is considerably harder than the second peak and extends up to 4 MeV; the middle peak of the 21 April 1980 event (Fig. 1d) is significantly harder than the 1st and 3rd peaks; and the early part of the 15 August 1980 event (Fig. 2b) is considerably harder than the latter part.

Four of the bursts were detectable at 2 s resolution in a 4-6.4 MeV window in the spectrometer. A comparison of these time profiles with those obtained in the hard X-ray range (>25 keV) indicates that the durations of individual pulses are similar over two decades in energy. This similarity is illustrated for the 19 April 1980 burst in Figure 1 of an earlier report on the SMM data[13].

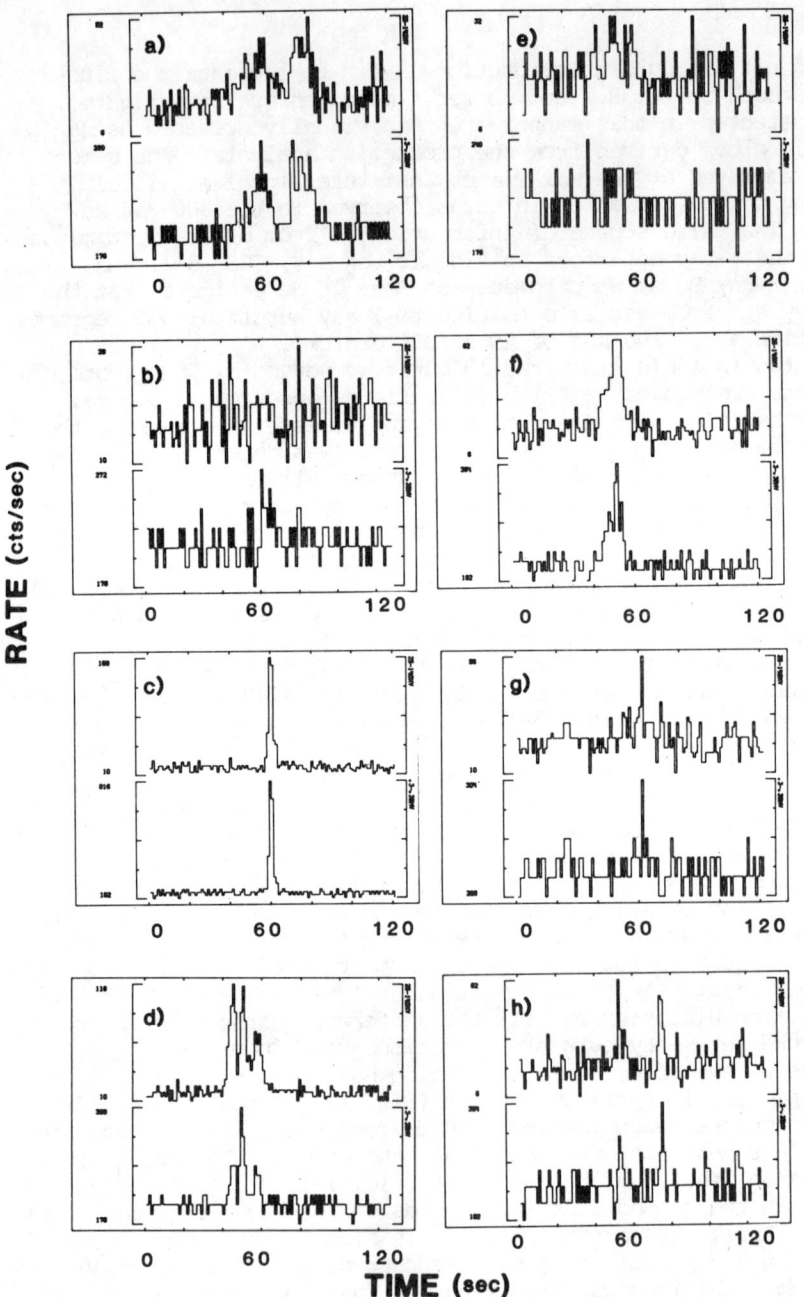

Fig. 1 Time histories of γ-ray bursts. 25-140 keV (top) and 50 keV range near 300 keV (bottom) plotted at 1 s resolution.
a) 03/07/80; b) 03/24/80; c) 04/19/80; d) 04/21/80; e) 05/12/80; f) 05/24/80; g) 06/02/80; h) 07/09/80.

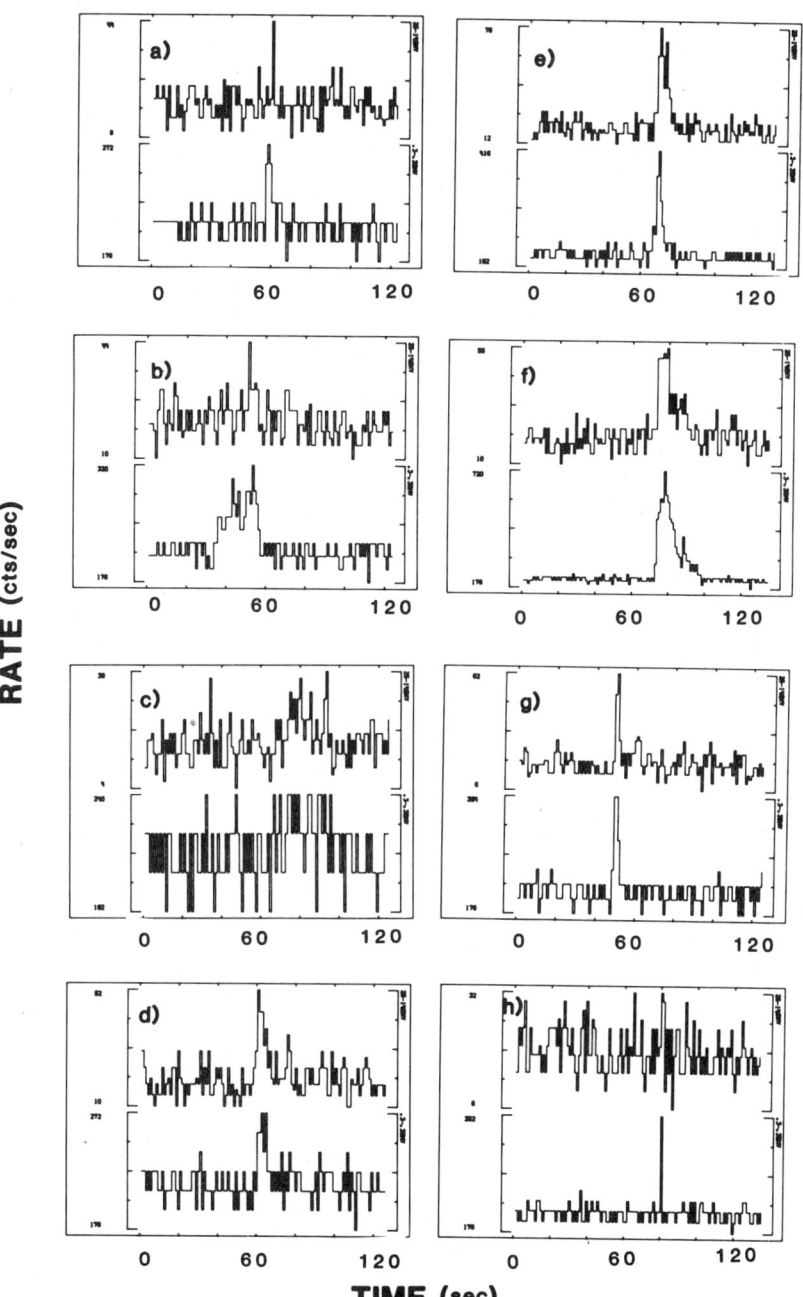

Fig. 2 Time histories of γ-ray bursts; 25-140 keV (top) and 50 keV range near 300 keV (bottom) plotted at 1 s resolution.
a) 08/12/80; b) 08/15/80; c) 09/19/80; d) 10/16/80; e) 11/19/80; f) 12/20/80; g) 03/01/81; h) 03/04/81.

We have obtained time-integrated count-rate spectra of the bursts listed in Table 1. These spectra were corrected for background utilizing data from 3-5 min intervals either immediately prior to or following the events. Spectral channels were summed together in order to obtain sufficient statistics. We note from Table 1 that all the bursts were detected above ~1 MeV.

The shape of a measured spectrum is dependent on the angle of incidence of the radiation. Spectra from bursts incident from the rear will exhibit significant absorption effects due to scattering in the satellite. The incident angle cannot be unambiguously inferred utilizing data from SMM alone. Fortunately,

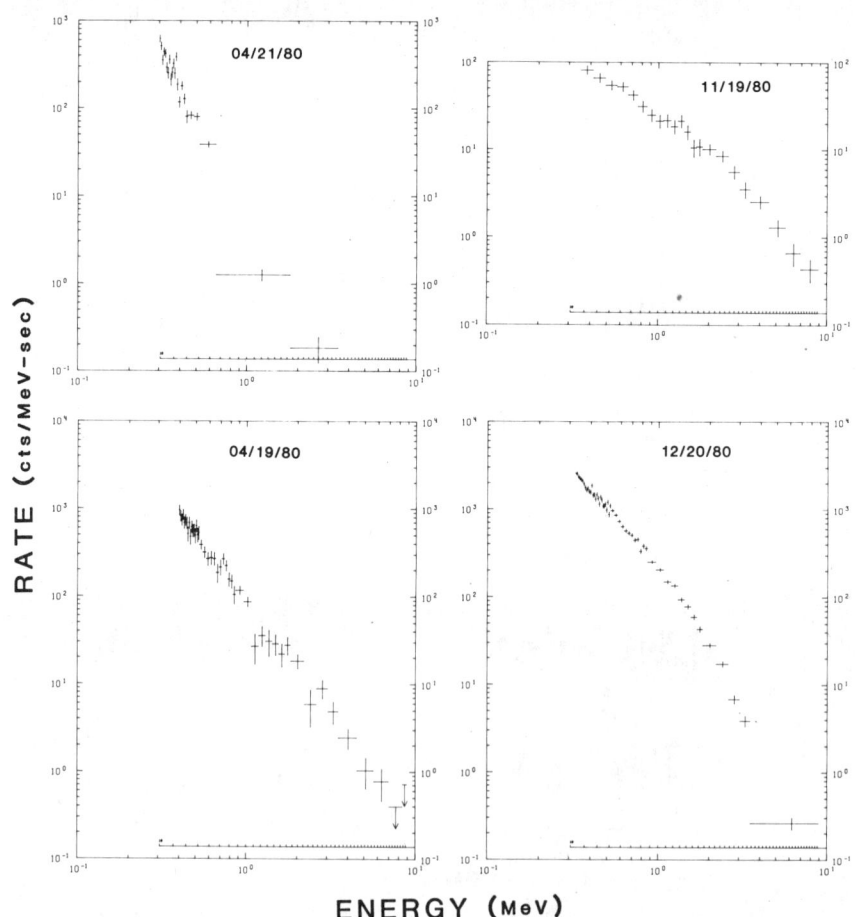

Fig. 3 Time averaged count-rate spectra of four γ-ray bursts plotted from 0.3-10 MeV. Spectra are corrected for background but not for angle of incidence. <u>The response function of the detector has not been unfolded.</u>

long baseline timing measurements are available from the γ-ray detector on the Pioneer-Venus spacecraft[10]. Some of these measurements have been communicated to us by R. Klebesadel and J. Laros prior to publication. A comparison of the times of detection of the bursts at SMM and Pioneer-Venus have allowed us to determine which of the events have probable source locations within the aperture of the spectrometer. These timing measurements exclude the Sun as a possible source for all but the 19 April event[13].

Four of the events have source locations not likely to have been obscured by the satellite. The time-integrated spectra of these bursts are plotted in Figure 3. These spectra have not been corrected for incident angle and instrument response function, so care must be taken in their interpretation. The 19 and 21 April 1980 spectra are describable by power-law functions in energy (dN/dE $\propto E^{-\alpha}$) with spectral indices (α) of ~2.1 and ~4.5, respectively. On the other hand, the 19 November and 20 December 1980 spectra appear to be best described by either two power-law functions with a break above 1 MeV, or by an exponential function. Detailed analysis of these spectra is in progress. It is of interest to note that there is no evidence for a sharp high-energy cut-off in any of these bursts.

There is a suggestion of a narrow line feature in the 19 April spectrum near 730 keV; however, it is of marginal significance $\sim 2\sigma$[13]. Only one other event exhibited a suggestion of a narrow line feature. That event occurred on 7 March 1980. At present, we do not believe that either of these features provide clear evidence for the detection of narrow line γ-ray emission during bursts. More detailed analysis is in progress.

DISCUSSION

The spectral variability observed in broad energy bands in three of the events observed by SMM is characteristic of γ-ray bursts[14,16]. The erratic nature of this variability in bursts has not been explained. Because of the 16.38 s accumulation time for spectra in the SMM instrument, variability studies at high spectral resolution will only be possible for events lasting 30 s or longer.

The similarity of the time profiles observed at 2 s resolution in the 25-140 keV and 4-6.4 MeV bands indicates that photons spanning about two decades in energy are produced by the same mechanism in regions separated by $\lesssim 2$ light seconds.

As we discussed in our earlier report on these data[13], emission in the 10-15 keV region was observed for ~30 s prior to and ~3 min following the 2 s burst on 19 April 1980. Whether this low-energy emission is associated with the burst or whether it is of solar origin is uncertain at present. If this low-energy radiation was associated with the γ-ray burst, it would indicate that a second, perhaps thermal, emission mechanism accompanies some bursts. The field-of-view of the X-ray detectors is partially blocked by an aluminum plate. By comparing the aperture with the possible source positions we may be able to unambiguously determine the source of this low-energy emission.

Spectra of bursts observed by the KONUS experiment on Venera appear to be well fitted by thermal bremsstrahlung functions with temperatures ranging from 50 to 1000 keV[7,14]. Recent analysis of the Apollo γ-ray burst[18] indicates that the data <1 MeV is well fit by a 500 keV thermal bremsstrahlung spectrum. However, an additional high-energy tail is necessary to fit the data above 1 MeV. The spectra (uncorrected for instrument response) of the four bursts discussed in this paper appear to be of two types. The first type is well fit by a power law up to the highest energy observable (e.g. 6 MeV for the 19 April event). The second type requires either two power laws or an exponential (possibly thermal bremsstrahlung) function to fit the data. Temperatures >1 MeV would be required to fit the data in the latter type. The SMM spectra do not appear to be consistent with a lower temperature thermal spectrum plus a high-energy tail, as found for the Apollo event. Detailed fits to the data will be done after the spectra are corrected for instrument response.

The fact that all of the bursts are observed into the MeV region by the SMM spectrometer can be used to set a limit on the typical distance of γ-ray burst sources. Schmidt[17] has shown that isotropically emitting intense burst sources become optically thick to their own radiation in the MeV region due to photon-photon pair production. He then estimated that burst sources with measured intensities $>10^{-4}$ erg/cm^2 must be closer than $\sim$2 kpc from Earth if emission >1 MeV is observed. Using his model and the fact that the bursts observed by SMM typically reached $\sim$5 MeV and had intensities $>10^{-5}$ erg/cm^2, we estimate that most burst sources above this intensity lie within $\sim$3 kpc of the Earth, assuming that their emission is isotropic.

The steepening observed >1 MeV in the 19 November and 20 December 1980 bursts may be intrinsic to the emission mechanism of the source. An alternative interpretation is that the steepening is caused by opacity in the vicinity of the source due to photon-photon pair production, as we discussed above.

We have found no clear evidence for narrow line emission in any of the 19 bursts detected. Two of the bursts exhibit narrow line features at the 2-3σ level, but at the present stage of analysis neither is convincing.

The detection of broadened emission lines in the 400 keV region have been reported by the Leningrad group[7]. These features have been interpreted as being due to shifted and broadened electron-positron annihilation lines produced near the surface of neutron stars[7,19]. None of the four bursts discussed in detail in this paper show evidence for such broadened features near 400 keV. Spectral analysis of the other bursts detected by SMM may be compromised by possible distortion due to passage of the incident radiation through the satellite. Further discussion of these events must await more detailed analysis.

ACKNOWLEDGMENTS

We wish to thank R. Morin and M. Strickman for assisting in the development of the software used in this work. P. Reed assisted in the preparation of this manuscript. This research has been supported by NASA under contracts NAS5-2376 (UNH) and S.70926A (NRL), and in the Federal Republic of Germany by BMFT contract 0101K 017-ZA/WS/WRK 0275:4.

REFERENCES

1. W.D. Evans et al., Ap. J. Lett. 237, L7 (1980).
2. M.C. Jennings, Ap. J. (in press), (1981).
3. E.P. Mazets et al., Sov Astron. Lett. 6, 318 (1980).
4. E.P. Mazets and S.V. Golenetskii, Ap. and Space Sci. 75, 47 (1981).
5. T.L. Cline et al., Ap. J. Lett. 237, L1 (1980).
6. K. Wood et al., Ap. J. (in press), (1981).
7. E.P. Mazets et al., Nature 290, 378 (1981).
8. B.J. Teegarden and T.L. Cline, Ap. J. Lett. 236, L67 (1980).
9. D.J. Forrest et al., Solar Physics 65, 15 (1980).
10. W.D. Evans et al., Science 205, 119 (1979).
11. W.D. Evans et al., Nature 286, 784 (1980).
12. L.E. Orwig et al., Solar Physics 65, 25 (1980).
13. G.H. Share et al., Proc. 17th Int. Cosmic Ray Conf. (Late Vol.), (1981).
14. E.P. Mazets et al., Preprint no. 719 A.F.Ioffe Physical Technical Institute (1981).
15. F.K. Knight et al., Ap. and Space Sci. 75, 21.
16. W.L. Imhof et al., Ap. J. Lett. 191, L7 (1974).
17. W.K.H. Schmidt, Nature 271, 525 (1978).
18. D. Gilman et al., Ap. J. 236, 951 (1980).
19. R. Ramaty and P. Meszaros, NASA preprint 82124 (1981).

GAMMA-RAY BURST OBSERVATIONS
BY THE HEAO-3
HIGH-RESOLUTION GAMMA-RAY SPECTROMETER

Wm. A. Wheaton
James C. Ling
William A. Mahoney
Guenter R. Riegler
Allan S. Jacobson
Jet Propulsion Laboratory
California Institute of Technology
Pasadena, California 91109

ABSTRACT

Observations of cosmic gamma-ray bursts with the JPL High-Resolution Gamma-Ray Spectrometer on HEAO-3 are discussed. Two bursts seen on 1979 November 16 are of particular interest. The first event occurred at 14:16:41 UT and lasted for eight seconds. This event was detected only by the instrument's five CsI shield segments. The second event occurred 61 sec later, at 14:17:42 UT, and lasted 18 sec. This event was clearly detected by the high-resolution germanium detectors. Because of the close temporal coincidence of the two events, we consider the possibility that both originated from one celestial source, and that the scanning motion of HEAO-3 (nominal spin period 20 min) was such as to exclude the first from the (approximately 30° FWHM) field of view of the germanium detectors and include the second. Directional information from the relative response of the shield pieces and Earth occulation constraint are all consistent with this interpretation and high-resolution spectral data from the second burst are discussed in this light.

Recent unpublished analyses of long-baseline timing data from interplanetary spacecraft show that the response in the HEAO high-resolution detectors was probably Earth albedo scattering, which can be an important systematic effect in low-Earth orbit.

INTRODUCTION

Observations of the spectra of gamma-ray bursts have recently provided important clues to their origin and nature[1,2]. Spectral features interpreted as redshifted 511 keV emission and as cyclotron absorption and emission have greatly strengthed the idea that bursts originate near magnetized neutron stars. High-resolution gamma-ray burst observations may potentially provide a powerful tool for diagnosing conditions near the surfaces of neutron stars but those to date have mostly been made with scintillators[1] or moderate-sized solid state detectors[2].

The JPL High-Resolution Gamma-Ray Spectrometer on HEAO-3 operated from 23 September 1979 until cryogen exhaustion on 1 June

0094-243X/82/770055-11$3.00 Copyright 1982 American Institute of Physics

1980. The instrument's CsI anti-coincidence shield continued to function until 1981 May 28, during which time it was a sensitive monitor for detecting gamma-ray bursts in all directions not blocked by the Earth or spacecraft structure. Here we describe the capabilities of the HEAO-3 instrument as a gamma-burst detector, and discuss one of the most interesting bursts observed, that on 1979 November 16, which produced a substantial response in the high-resolution germanium detectors. We also discuss Earth albedo scattering by gamma-ray bursts as an important systematic effect in low-Earth orbit experiments which may not have received adequate attention heretofore.

INSTRUMENT AND OBSERVATIONS

The JPL High-Resolution Spectrometer has been described in detail by Mahoney et al[3]. It consisted of a cluster of four high-purity germanium detectors collimated to a 30° FWHM field of view by a large CsI shield in electronic anti-coincidence (Figure 1).

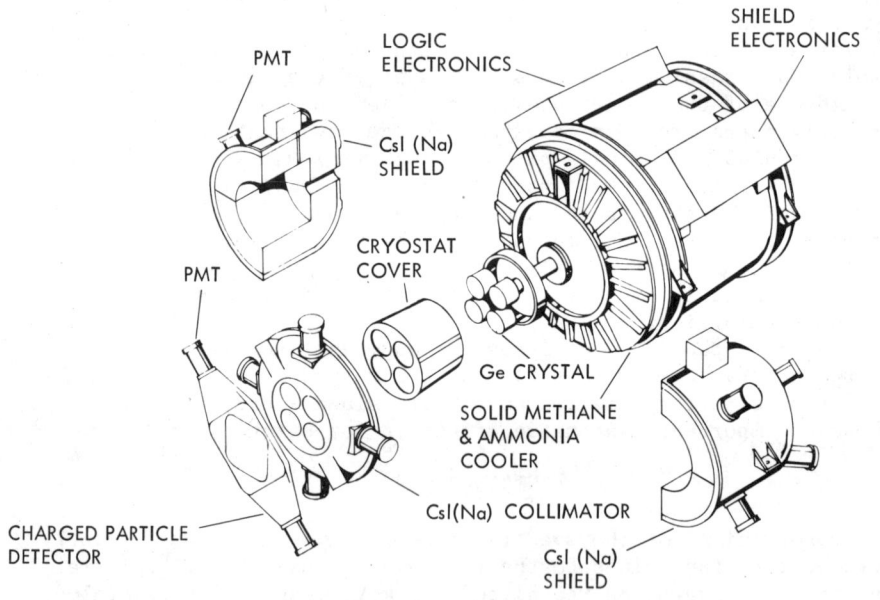

Figure 1. JPL Gamma-Ray Spectrometer

The shield was in five independent pieces, a disk-like collimator and four segments shielding the sides and rear. Events with an energy above 80 keV were accumulated separately in each shield piece and read out every 1.28 sec. Counts in a window centered on 511 keV and events above 3.8 MeV were also accumulated and read out every 10.24 sec for each shield piece. The five shield pieces together formed a gamma-burst detector with roughly isotropic response, and approximately 1000 cm² effective area in directions not blocked by the Earth or spacecraft structure. Transient counting rate spikes, mostly 1.28 sec or less in length, were frequently seen in the CsI shields (cf. Figure 2). These spikes, similar to ones reported previously [4,5], are apparently due to long-lived

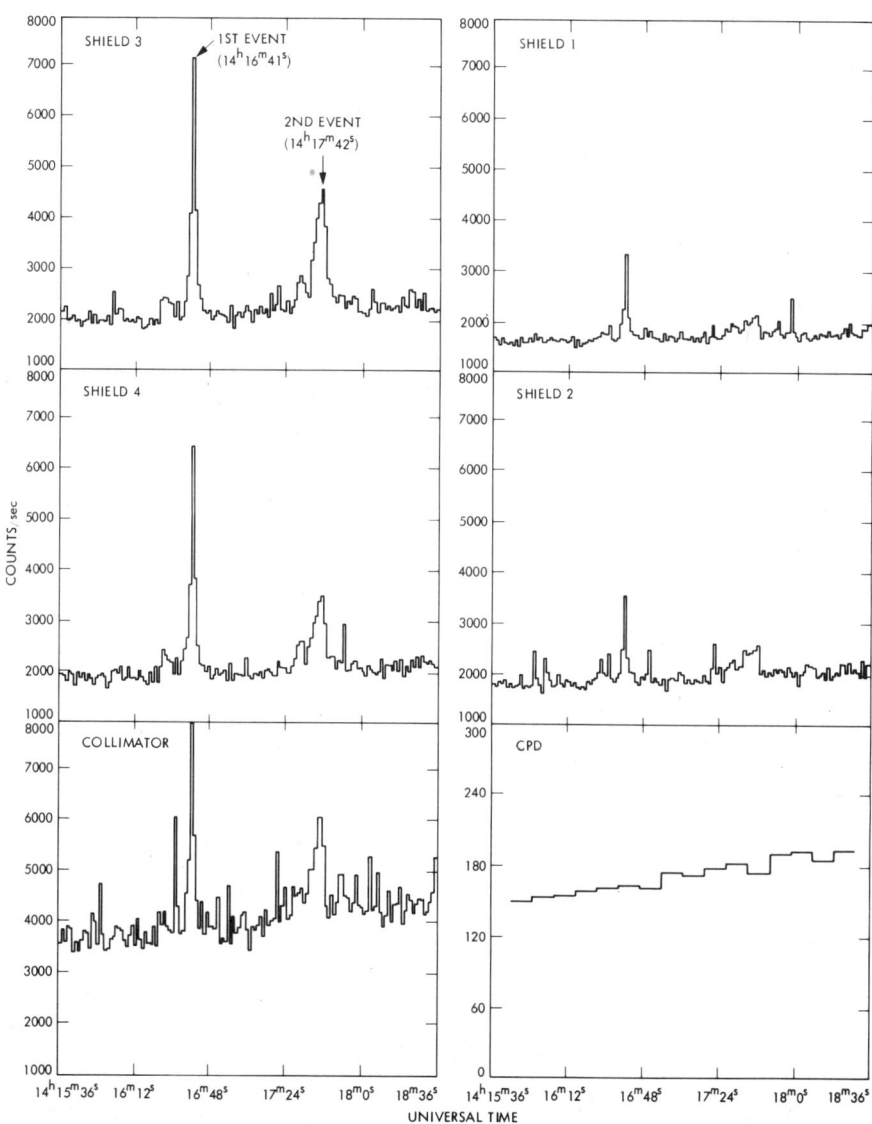

Figure 2. Gamma-ray burst in shields and collimator.

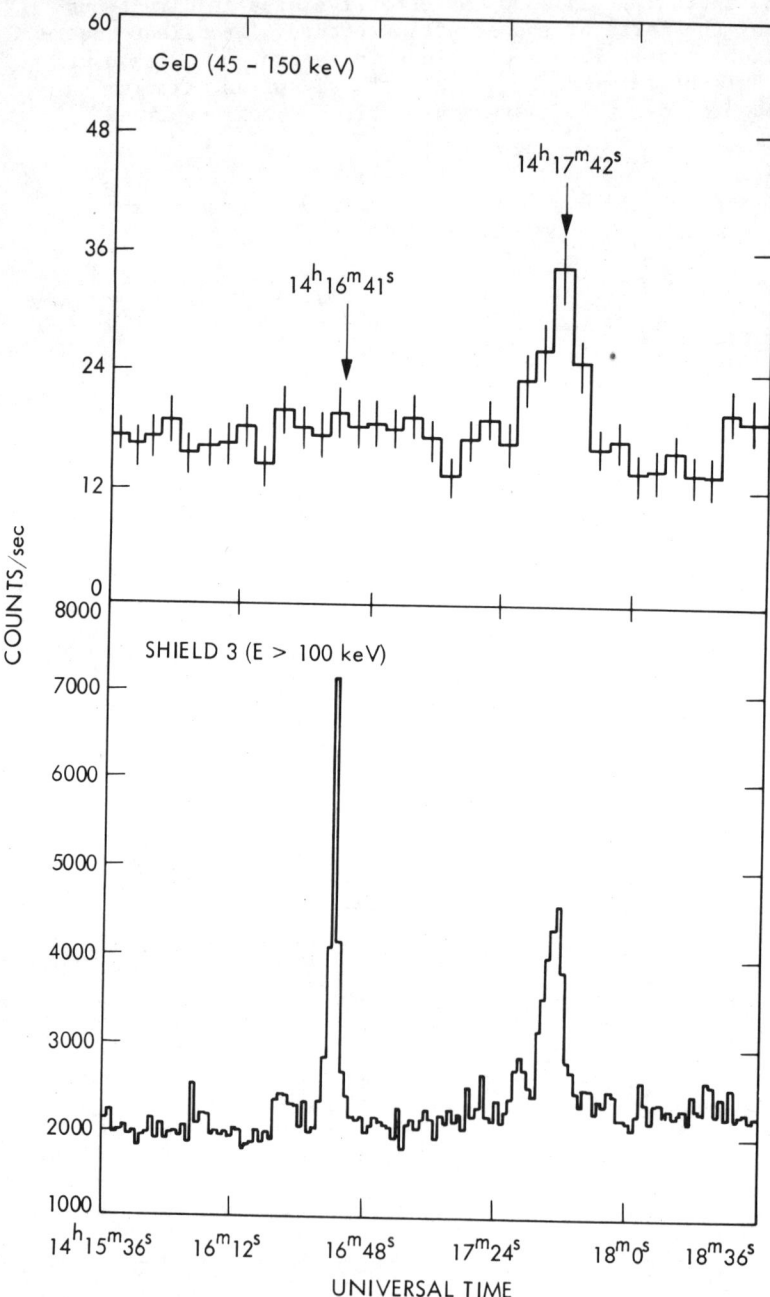

Figure 3. Shield 3 and high-resolution germanium detector responses compared.

phosphorescent states excited in the crystals by large energy-loss cosmic ray events; they do not appear to have a cosmic origin. They increased the noise level in the shields to about three times the value expected from Poisson statistics, so that the sensitivity for burst detection was about 1×10^{-6} ergs/cm^2.

The four germanium detectors (GeD's) had a total volume of 400 cm^3. Events from 50 keV to 10 MeV were telemetered event-by-event up to a maximum rate of 16 events per sec for each of the four GeDs. The initial energy resolution of the GeDs ranged from 2.5 keV (FWHM) at low energies to 7 keV at 10 MeV.

HEAO-3 normally operated in such a way as to keep the spacecraft spin axis nearly in the solar direction. During each spin, nominally 20 minutes, the instrument look axis scanned the great circle on the sky perpendicular to the Sun. During the first six months of the mission, at least four gamma-ray bursts were observed in coincidence with other spacecraft; these were the 16 November 1979 bursts, and bursts at 15:57 UT on 9 December 1979, 07:29 UT on 13 February 1980, and 01:20 UT on 19 April 1980. A description of the other bursts and of the observations after cryogen exhaustion is in preparation.

RESULTS

Figure 2 shows the 16 November 1979 burst as it appeared in shield pieces 1-4 and in the collimator. The differing responses are due to the differing sizes and orientations of the shield components and to their shadowing by each other and by the spacecraft. Shield segments one and two were always oriented towards the sun; the weakness of their response essentially rules out a solar origin for either event. Besides the double burst structure, note also the precursors to each large burst and the different widths of the two main events. There is a suggestion that the second event was accompanied by a broad hump or plateau of emission which continued for about 30 sec at a flux level an order of magnitude less than the peak in the second burst, but because of the normal variability observed in the shield rates this cannot be established for certain form HEAO-3 data alone. While it is possible that the two bursts were from physically unrelated sources which occurred by chance at nearly the same time, the similar relative responses of the five shield pieces for both strongly suggests that both originated in the same object. Figure 3 shows the response in the GeDs compared to shield 3, and Figure 4 shows the geometrical configuration. The full response curves of the detectors have about twice the radius of the 50% response curves shown. The Earth's horizon was in the field of the second burst, but was strongly attenuated by the collimator at the time of the first. We have considered two possible alternative explanations for the apparent absence of the first event from the GeD data in Figure 3. The first, the direct hypothesis, is that the spin of the spacecraft carried the source into the field of view by the time of the second burst. The second, the scattered hypothesis, is that the burst direction lay far from the field of

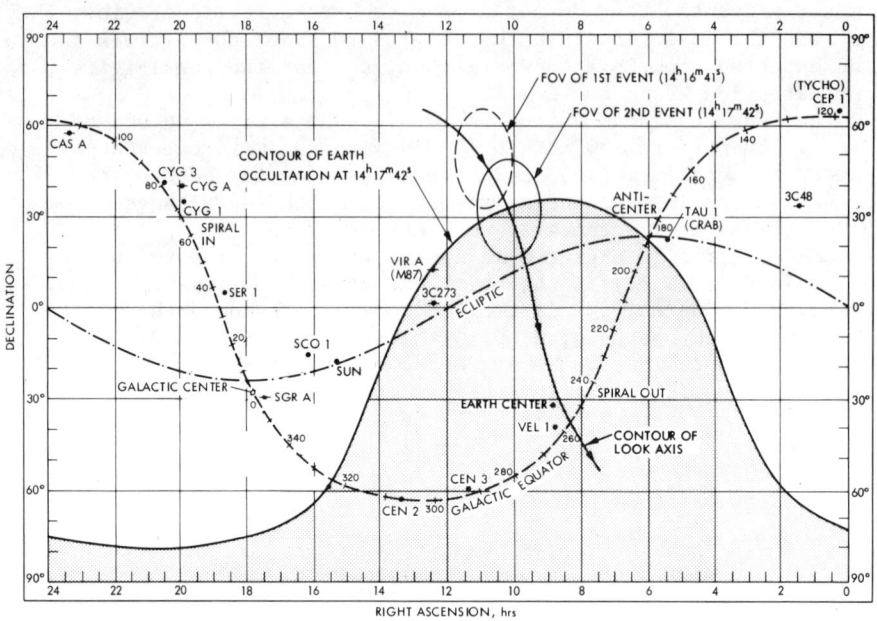

Figure 4. Detector fields of view (50% response) at burst times.

view and that the GeDs responded to photons Compton scattered from the Earth's atmosphere into the aperture.

Together with the direct hypothesis, the data give enough information to determine an approximate source position on the sky and also provide some spectral information. The direct hypothesis requires that the field of view at the time of the second event include the burst position, thus forcing it to lie above the horizon, within about 30 degrees of the scan circle, but not in the region viewed by the instrument at the time of the first burst. Comparison of the GeD and shield counting rates for the second burst requires the position to lie near the edge of the field of view at the time of the second event. The data from shields 3 and 4 and the collimator show that the burst flux was more than 3 photons per cm^2 per sec. above 80 keV, whereas the GeDs show only about 10 cts/s excess over background in the same energy range. Considering the total GeD effective area of 60-75 cm^2 at this energy shows that the source location must have been such as to reduce the collimator response to less than 10% of its value on-axis, corresponding to a source position at least 25 degrees off-axis. At such a large angle the two shield segments on the side away from the source would be substantially shadowed by the collimator, giving the very much reduced response observed for shields 1 and 2 compared to shields 3 and 4. Considering all these factors requires the source to lie at an R. A. of 8 to 10 hr, and a declination of +45 to + 55 degrees. Since this position lies on nearly the opposite side of the Earth from the spacecraft, atmospheric scattering should be only very minor, confirming the consistency of the overall picture.

Based on the direct hypothesis, the spectrum of the burst as determined from the GeDs can be analyzed. Because the efficiency is strongly energy-dependent in the wings of the collimator response and because the position of the source is not well determined, the data are primarily useful for the detection of narrow spectral features. Figure 5 shows the low energy spectrum observed in the GeDs for the second event. The excess around 56 keV is not statistically

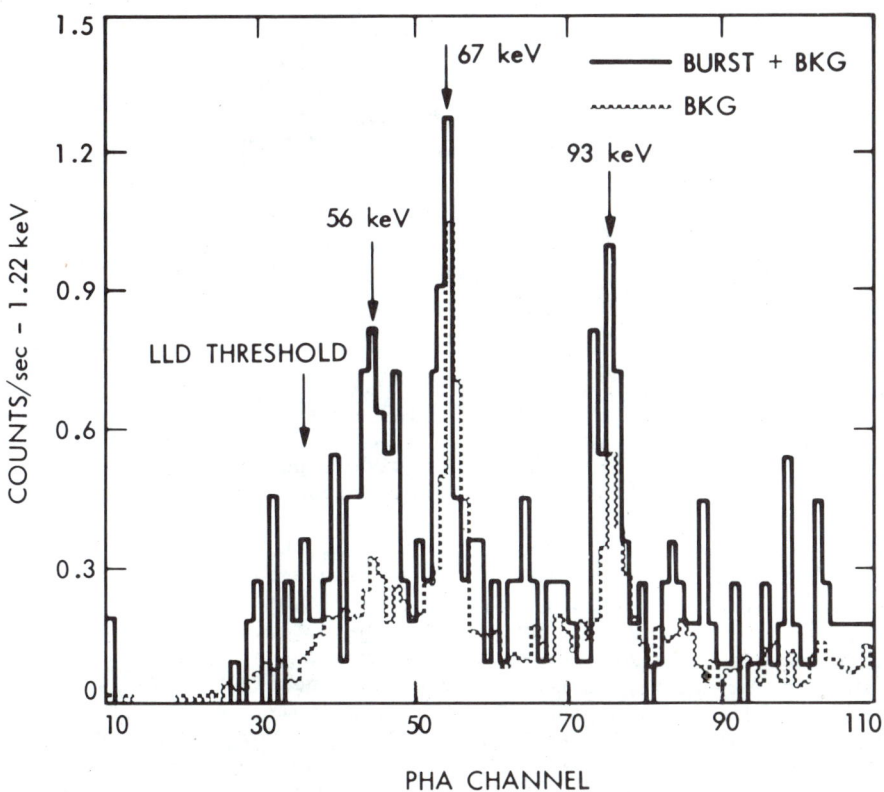

Figure 5. Low energy burst and background spectra, for the second event. PHA channel width is 1.22 keV.

significant. Figure 6 shows this more clearly; in it we compare the data to a simple model consisting of the background (measured before and after the burst) plus a featureless gamma-ray burst source spectrum input, as modified by the rapidly changing efficiency just above the LLD. The χ^2 is 8.72 for 8 degrees of freedom. The excess of 11 counts over the expectation from the smooth model

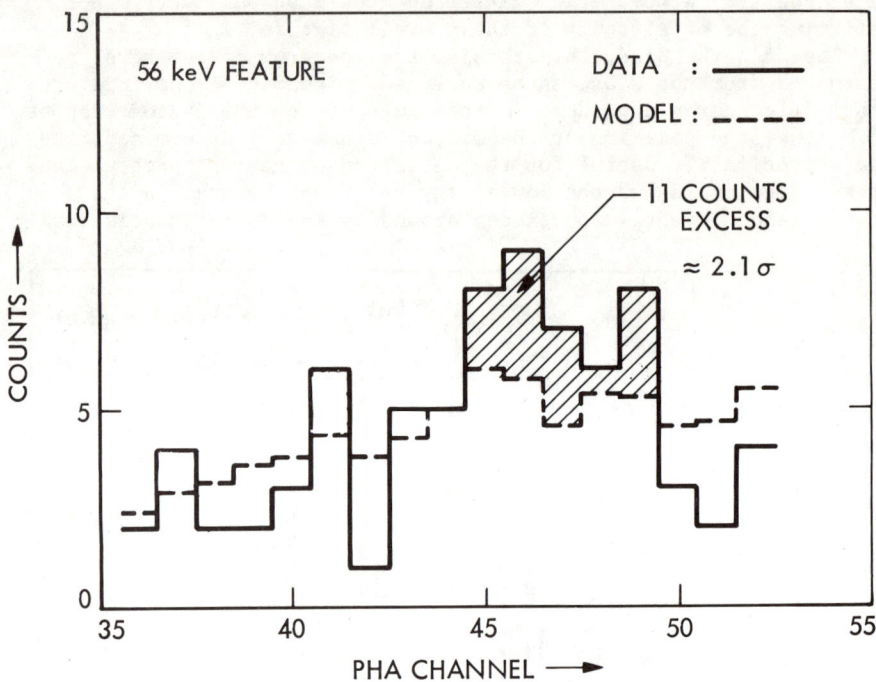

Figure 6. Fit of a featureless model spectrum to the data near 56 keV, on an expanded scale.

of 25 counts is only 2.1 sigma, and was found while searching several hundred channels. The excess in Figure 5 near 93 keV is less convincing still, and it appears at the same energy as a strong background line. Figure 7 shows the net GeD spectrum at higher energies. If 400 keV is now considered the correct energy at which to expect redshifted annihilation lines in bursts[1,2], the excess in this region may be taken more seriously. An analysis similar to that described above gives $\chi^2 = 16.4$ for eight degrees of freedom; the excess of 8 counts over the expected 10 counts in the line region is 2.5 sigma. The flux of $(0.30 \pm 0.07)/f$ photons/ cm^2 is uncertain because of the unknown collimator response f $(0.0 < f < 1.0)$, due to the uncertainty in source position; using the 25 degree off axis limit estimated above together with the aperture response at 400 keV indicates only that $f < 0.3$, corresponding to a total burst flux of over 1.0 photons/cm^2 in the 400 keV feature.

According to the alternative, or scattered, hypothesis, the response in the GeDs to the second event was due to Compton scattering of burst photons into the instrument aperture by the Earth's atmosphere. Such scattering would both smear out any narrow spectral features and shift their energies, so the effect on the results

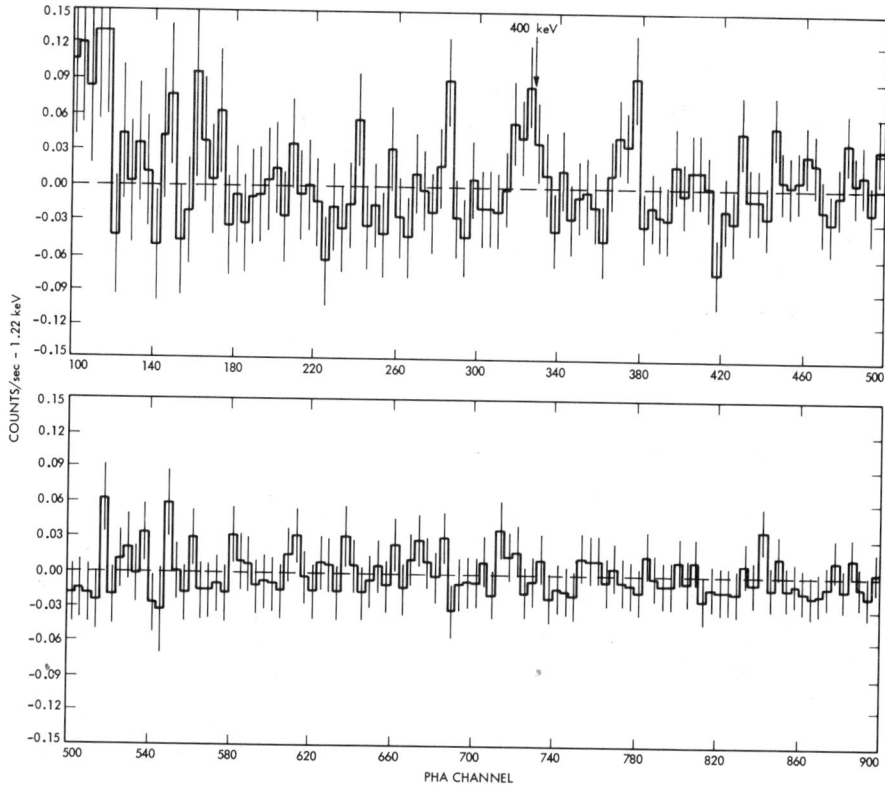

Figure 7. Background-subtracted spectrum of the second burst from 122 to 1098 keV.

obtained above based on the direct hypothesis would be quite important. After the direct analysis was substantially completed we learned from J. Laros and T. Cline that data from the interplanetary burst detection network indicate a position totally at variance with that derived above from the direct hypothesis. Although the interpretation was complicated by a roughly collinear configuration of spacecraft which occurred at the time, the true position of the source lies somewhere in the region between R.A. = 0 hr, Dec = -15 degrees, and R.A. = 1 hr, Dec = -30 degrees. The interplanetary data do essentially confirm the hypothesis of a single origin for both bursts, as the 61 sec separation is observed by all spacecraft independent of their position.

DISCUSSION

While multiple-peak structure in gamma-bursts is common, the occurrence of such widely separated peaks with at most rather minor intervening activity is unusual. According to Klebesadel et al.[6], the second event is the slowest-rising, longest-duration single spike yet seen. The total duration of > 80 sec also makes it one of the very longest (above 100 keV) bursts observed to date, a distinction it shares with GB800709[6], which it resembles in general character. If gamma-ray bursts are due to non-stationary accretion onto neutron stars[7,8], it seems unlikely that the characteristic multiple peak structure is caused by fragmentation of the captured body during the initial infall. For example, for a straight-in trajectory onto a solar-mass neutron star to reproduce the 61 sec separation observed here, the second body would have been 1.4×10^5 km distant at the time of the impact of the first, and it is difficult to understand how the two could ever have been connected. Infall with impact parameter of the order 10^4 km should not essentially alter this consideration. Disk or magnetospheric time scales seem more promising; or the 61 sec may reflect some time internal to the star itself.

Assuming the preliminary position from the interplanetary burst network is correct, one may calculate the incident photon energy corresponding to a given observed feature in the approximation that the range of scattering angles allowed by the collimator is not too large. For a scattering angle of 130 degrees, the excesses observed in Figure 5 at 56 and 93 keV correspond to energies of 68 and 132 keV before the scattering; the 400 keV feature has no corresponding energy for this angle.

This observation underscores the importance of Earth albedo effects for the analysis of gamma-burst data from detectors in low Earth orbit. Intuitively it is not very surprising that scattering is significant. Above 10 keV the Compton scattering probability in air is much greater than the absorption probability. Below about 100 keV, in the Thompson regime, the energy change of the scattered photon is small. But if we imagine a limit in which both absorption and energy loss were negligible, then the atmosphere would become a perfect diffuse reflector, and one would expect that, as for an ideal specular reflector, the reflected flux would be essentially the same as the direct flux. Experiments which operate in low-Earth orbit can be affected in at least three ways: first, as here, the albedo flux can confuse methods of position determination which rely on the relative response of detectors with differing geometries; second, as here, spectra will be degraded in resolution and shifted in energy, especially if the detectors view a wide solid angle and hence a wide range of scattering angles; and third, because of path-length differences, time structure on scales below a few milliseconds is liable to be washed out to some degree.

The arguments of the previous section show that the direct interpretation, although presumeably incorrect, is consistent with these data. Although the result could be sharpened somewhat by a much more elaborate analysis, including especially more consideration of the effect of spacecraft structure and instrument housing on the shield response, we believe it likely that the direct hypothesis would still remain viable based on HEAO-3 data alone. While the JPL instrument was not designed for this type of observation, strong atmospheric or spacecraft scattering may make the inversion of counting rate data to obtain a unique source position a badly conditioned problem for instruments of this type, in the sense that many widely-differing solutions may fit the data almost equally well. If so this issue may need to be considered early in the development of such experiments so that an effective solution can be reached.

ACKNOWLEDGEMENTS

We are indebted to John Laros and Tom Cline for reporting their unpublished position of the 16 November 1979 burst W. A. W. was supported under a National Research Council Resident Research Associateship.

This research was performed at the Jet Propulsion Laboratory under NASA contract NAS7-100.

REFERENCES

1. Mazets, E. P., Golenetskii, S. V., Aptekar, R. L., Gur'yan, Y. A., and Il'inskii, V. N. Nature 290, 378-382 (1981).
2. Teegarden, B. J., and Cline, T. L., Ap. J. (Letters), 236, L67 (1980).
3. Mahoney, W. A., Ling, J. C., Jacobson, A. S. and Tapphorn, R. M., Nucl. Instr. and Meth., 178, 363 (1980).
4. Fishman, G. J. and Austin, R. W. Nucl. Instr. and Meth., 140, 193-196 (1975).
5. Fishman, G. J., Meegan, C. A., Watts, J. W., and Derrickson, J. H., Ap. J. (Letters), 223, L13-L15 (1978).
6. Klebesadel, R. W., Fenimore, E. E., Laros, J. G., and Terrell, J., "Gamma-Ray Burst Systematics", proceedings this conference (1981).
7. Harwit, M., and Salpeter, E. E., Ap. J. (Letters), 186, L37-L39 (1973).
8. Mazets, E. P., Golenetskii, S. V., Ilyinskii, V. N., Guryan, Y. A.., Aptekar, R. L., Panov, V. N., Sokolov, I. A., Sokolova, Z. Y., and Kharitonova, T. V., preprint, submitted for publication to Astrophys. and Space Sci., May 1981.

OBSERVATIONS OF A GAMMA-RAY BURST AND OTHER SOURCES WITH A LARGE-AREA, BALLOON-BORNE DETECTOR

R. B. Wilson
Physics Department
The University of Alabama in Huntsville
Huntsville, AL 35899

and

G. J. Fishman and C. A. Meegan
Space Sciences Laboratory
NASA/Marshall Space Flight Center
Marshall Space Flight Center, AL 35812

ABSTRACT

Observations of a weak cosmic gamma ray burst of integrated intensity 2×10^{-6} erg cm^{-2}, two solar flare events, and pulsed emission profiles of A0535+26 and NP0532 are reported for several energy intervals in the energy range from 45 to 520 keV. The measurements were made a with a NaI(Tℓ) detector array flown on a balloon to 4 g cm^{-2} residual atmosphere from Palestine, Texas, on October 6-8, 1980, for 28 hours. The detector is a prototype of the Burst and Transient Source Experiment (BATSE) to be flown on the Gamma-Ray Observatory (GRO).

INTRODUCTION

Recent Soviet burst observations from the Venera spacecraft,[1] indicating that the frequency of bursts with intensity greater than 10^{-7} erg cm^{-2} is 100-400 yr^{-1}, are encouraging for the continuing study of transient events by balloon-borne systems. Typical large-area (∼1 m^2) detectors have sensitivities in the range 10^{-7} to 10^{-8} erg cm^{-2}, depending on burst duration, spectrum, and time structure. This is 10-100 times the sensitivity of operational satellite-borne instruments, allowing the observation of more distant and/or intrinsically weaker bursts. The anticipated development in the near future of balloon systems capable of carrying ∼1000 kg detectors for long-duration balloon flights (> 10 days) will allow more detailed measurements of fine temporal structure and spectral changes of the less frequent, stronger bursts.

During the past 5 years, the NASA/Marshall Space Flight Center has developed and flown large-area NaI (Tℓ) detectors for gamma-ray burst observations.[2] The earlier flights provided upper limits for the frequency of weak bursts,[3] which led to models indicating the Galactic origin of the typical gamma-ray burst sources.[4,5] This paper describes the observation of a gamma-ray burst by the most recent balloon experiment and briefly describes other sources also observed during the same flight.

0094-243X/82/770067-12$3.00 Copyright 1982 American Institute of Physics

EXPERIMENT

The instrument is an array of four independent detector modules, each consisting of a 50.8 cm x 1.27 cm disc of NaI(Tℓ) viewed by three 12.7-cm photomultiplier tubes (PMT's) inside a conical diffuse reflecting light box, as shown in Figure 1. A separate light collector box, observed by two 12.7-cm PMT's, contains a 55.2-cm disc of 0.64-cm-thick NE102 plastic scintillator, primarily used to anticoincidence charged particle events traversing the module. The energy resolution of the detector has been measured to be 0.40 FWHM at 88 keV.

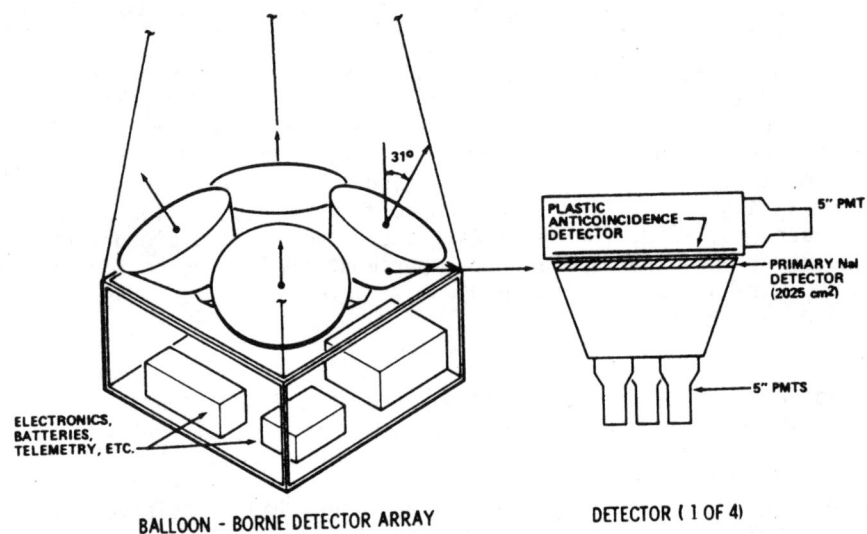

Figure 1 - Diagram of the four-detector array and a cross-section of one of the detectors.

The modules are oriented with the normal to the detector faces at 31 ± 1 degrees to the zenith and spaced at 90 degrees in azimuth (Figure 1). The rotation angle of the platform about the zenith direction is measured using three orthogonal magnetometers. To allow uniform coverage of the sky and more easily detect steady emission from point sources, a small motor was used to rotate the platform continuously at approximately 4.8 min per revolution.

The primary telemetered data consist of 36 ms time integrations for eight energy intervals spanning 45-180 keV and eight intervals spanning 180-520 keV for each of the four modules. In addition, 7.2-s time integrations for 128 energy channels quasi-logarithmically span the energy range from 45 keV to 12 MeV for one module (multiplexed), and 7.2-s time integrations for eight energy channels and 128 phase bins for the sum of all four modules. The last data type is used for observations of fast pulsars (periods under 1 s) and is described in more detail in a later section. A fraction (0.23) of the data is available as 4.5-ms time integrations for the sum of all four modules for four energy channels between 45 and 520 keV. The interval transmitted can be determined by an on-board microprocessor in response to recognition of a large burst, allowing fine time resolution as warranted by sufficient counting statistics. It was not triggered during the flight described in this paper.

Two methods of in-flight energy calibration were used: (1) a 10 µC Ba^{133} source, normally shielded by a lead/tin enclosure, could be exposed and retracted by command, producing a peak in the spectrum at 350 keV, or (2) the anticoincidence system could be reversed, allowing the observation of charged particle throughpeaks (or muons at sea level).

The detector array was flown on a balloon launched from Palestine, Texas, on October 6, 1980, reaching a float altitude of 4.5 g cm^{-2} at 0230 UTC October 7, where it remained within ± 0.5 g cm^{-2} until termination at 0530 UTC October 8. Usable data were obtained throughout until 0400 UTC October 8, except for a 1-hour interval centered on 1200 UTC October 7 during the transition from Palestine to a downrange telemetry station.

RESULTS

A. Cosmic Gamma-Ray Burst

The 25 hours of float data were searched visually using a 0.72-s integration time for coincident features in two or more modules in energy intervals of approximately 45-75, 75-180, and 180-520 keV. Two events were detected in all four modules, with an additional event seen in only two. Two of the events (including the event not seen by all four detectors) are known to be of solar origin and are discussed in the next section. An event at 13:43 UTC October 7, 1980, (Figure 2) is detected in all three energy intervals. This event is seen well above background by all four detectors. No other events, except the solar events described later, were seen down to 20 percent of the intensity of this event, at which point statistical fluctuations dominate. Thus we are clearly in a regime where the log N-log S distribution for bursts is relatively flat.

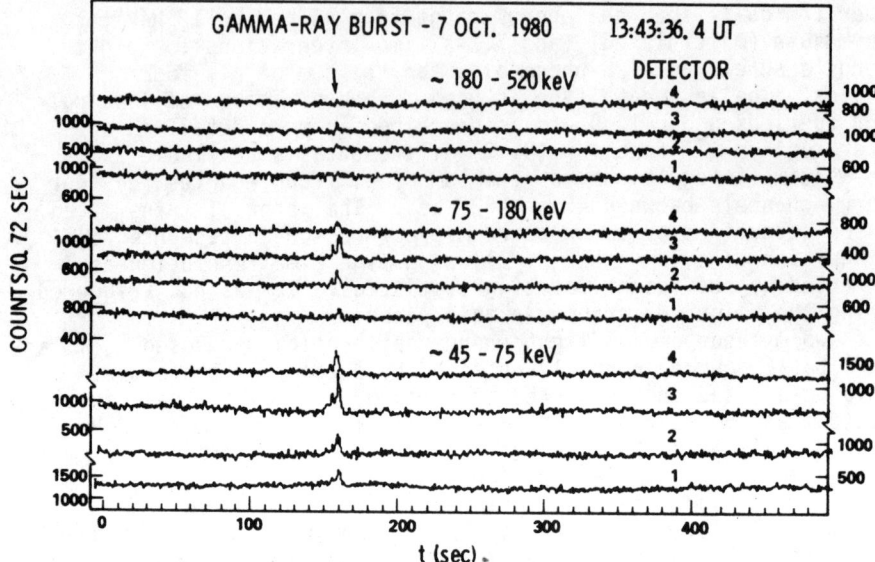

Figure 2 - Counting rates of four detectors in three energy ranges. Time of the gamma-ray burst is indicated.

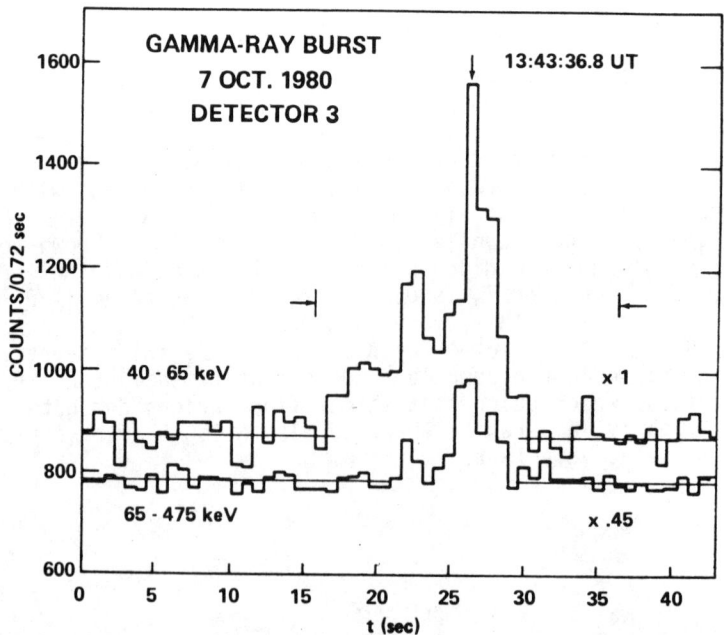

Figure 3 - Expanded plot during the time of the burst shown for detector 3, the strongest responding detector.

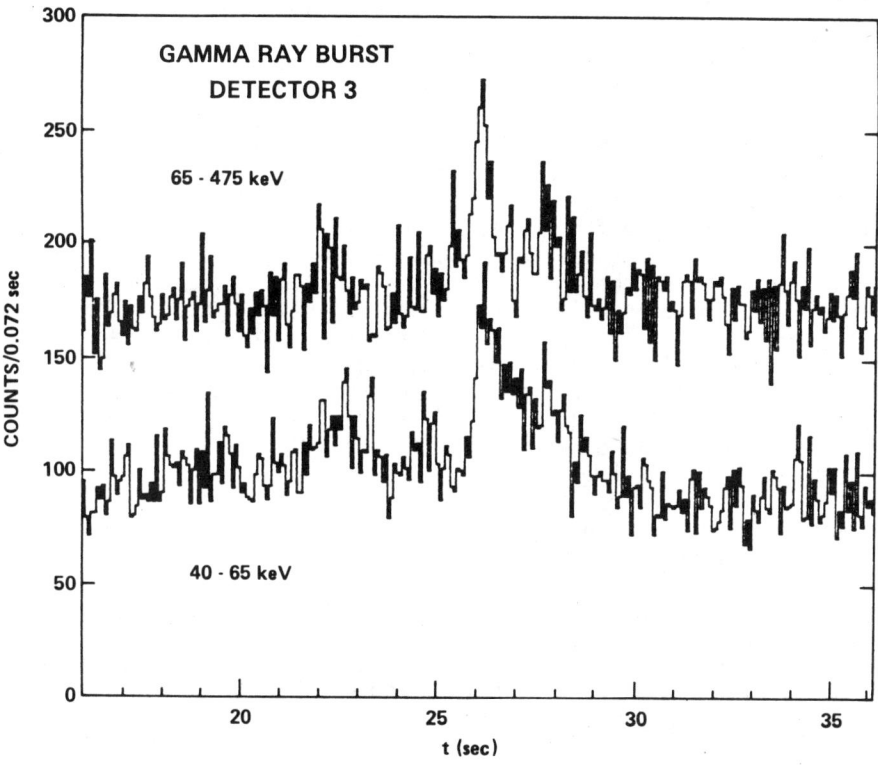

Figure 4 - High time resolution plot (72 msec) of the central burst region, indicated by arrows in Figure 3. Note that the two energy bands are reversed from Figure 3.

Figure 3 shows the 0.72-s resolution data of Figure 2, detector no. 3, expanded near the time of the burst. This detector was most nearly facing the burst ($\sim$20 degrees off axis). Figure 4 shows the central 20 s of the burst with 0.072-s resolution. A rapid-rise feature is seen in this figure at 26 s, with a rise time of $\sim$0.2 s in both energy intervals. However, the trailing edge from this feature is much longer at the lower energies, $\sim$ 2 s compared to $\sim$0.3 s at higher energies. The burst also appears to begin earlier at the lower energies, by almost 4 s, although this may be due to the higher signal-to-noise ratio in this energy interval.

The arrival direction of the burst can be measured by using the ratio of source counts in a given energy interval between the four independent modules. An energy calibration was performed for each module, using the on-board X-ray source, charged particle through-peak and an atmospheric background feature at $\sim$800 keV. It was found that variations in gain and threshold of 10 to 20 percent existed between detectors. Correcting the burst data for the differences in detector response, and using an assumed burst spectrum of $E^{-1.5}$, the four detector counting rates were consistent with a point source. Data used to derive the location were integrated over the entire burst, in the energy range 65 to 165 keV. The burst location derived was relatively insensitive to the assumed spectrum of the burst and the energy interval used. The most probable location of the burst in the local frame is: azimuth = 109 degrees east of north; zenith angle = 48 degrees, with an estimated error radius of 10 degrees. This transforms into a celestial position of R.A. = $12^h 55^m$, Dec. = $+9°$, also with an error radius of 10 degrees. The estimated error radius is based primarily on an estimate of systematic effects such as differences in detector responses and accuracy of calibration, scattering in the overlying atmosphere, and the assumed burst spectrum. These sources of error are greater than that of counting statistics. Using the preceding source location, the size of the burst can be calculated using the measured aspect angle. The detected, integrated burst size from 50 to 210 keV is 4×10^{-7} erg cm^{-2}. Assuming an $E^{-1.5}$ spectrum above 150 keV, and correcting for atmospheric absorption at 4 g cm^{-2}, the estimated size of the burst ($E \geq 50$) at the top of the atmosphere is 2×10^{-6} erg cm^{-2}.

The derived location is only 15 degrees from the Sun, just 1.5 error radii away. Although this is not sufficient to eliminate the Sun as the burst source, we believe that it is highly unlikely due to: (1) the very hard spectrum of the burst, (2) the typical gamma-ray burst time fluctuations and duration, and (3) the lack of GEOS solar X-ray flare activity at the time of the burst. (No data are available from the SMM spacecraft since it was in the nighttime portion of its orbit (L. Orwig, priv. comm).

The preceding burst location transforms into the following galactic coordinates: $\ell^{II} = 305°$, $b^{II} = +71°$, with a 10-degree error radius. The fact that a relatively weak gamma ray burst is seen at a high galactic latitude is surprising. At this intensity, the log N-log S curve for bursts is decisively bent over from the -3/2 curve (Ref. 1, 4, 5, and as noted above) and a galactic plane concentration should be observed. We note, however, that very few bursts at this strength have been located.[1] We may conclude one of the following: (1) this was a relatively rare, intrinsically weak, nearby burst (e.g., d = 65 pc, I = 10^{36} ergs) or (2) the burst source is far above the galactic plane with an intrinsic burst energy consistent with that derived from log N-log S measurements (e.g., I = 10^{39}-10^{40} ergs, d = 2-6kpc). This would also be a rare event, since this distance is many scale heights above the galactic plane for most galactic populations.[4]

B. Solar Flares

An event at 17:43 October 7, 1980, (Figure 5) is seen in all four modules but is much softer than the background (and the gamma-ray burst described previously), so that it is not observed above 75 keV. It was also observed by detectors on the Solar Maximum Mission (SMM) spacecraft (G. Share, L. Orwig, priv. comm.).

A second solar event also confirmed by SMM (G. Share priv. comm.) at 14:12 UTC October 7 occurred 3.5 hours before meridian passage, with the zenith angle of the Sun at 64 degrees. Due to the low intensity and soft spectrum, it was only seen by the two detectors having the lowest energy thresholds.

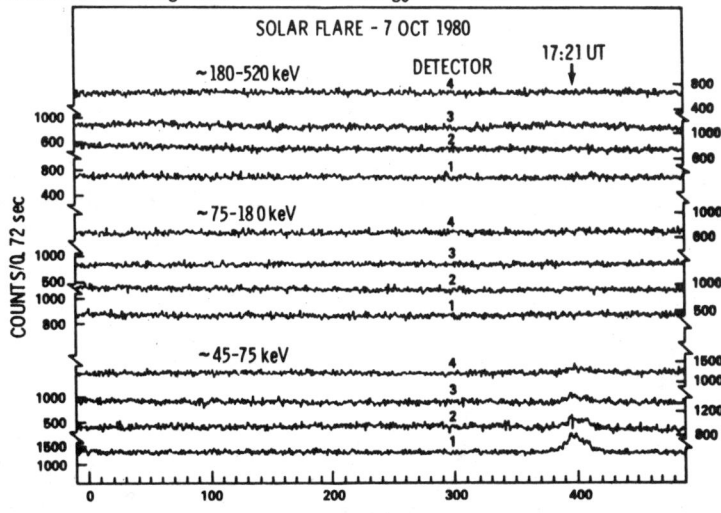

Figure 5 - Solar flare observed by all four detectors and confirmed by instruments on the Solar Maximum Mission (SMM). The scales are similar to that of Figure 2.

C. NP0532

Periodic hard X-ray sources such as NP0532 can be observed by the uncollimated detector array if the pulse period is known. In order to efficiently process periodic (pulsar) data at high counting rates, data were multiscaled onboard at the approximate period of the source under study for 7.2 s in each of eight energy intervals, using 128 phase bins. The folding period was obtained by counting a 2 MHz clock derived from an oscillator stable to 1 ms per day. Pulsar data were accumulated at the periods of NP0532 (the Crab pulsar) and PSR0833-45 (the Vela pulsar). The latter data have not yet been analyzed. A deviation from the instantaneously correct folding period (due to frequency offsets, Doppler shifts and intrinsic period change of the source) of up to 1.2 μs (in the case of NP0532) can then be corrected in subsequent data analysis. However, it was found that for approximately 3 hours either side of NP0532 transit the deviations in phase experienced were much smaller. This allowed accumulations of 300 s with less than a one-bin shift of a constant phase point from consecutive accumulations. The resulting preliminary pulse profile for NP0532 is shown in Figure 6. Some degradation in phase resolution is known to exist in this profile which is expected to improve slightly by processing the profile data on shorter time intervals.

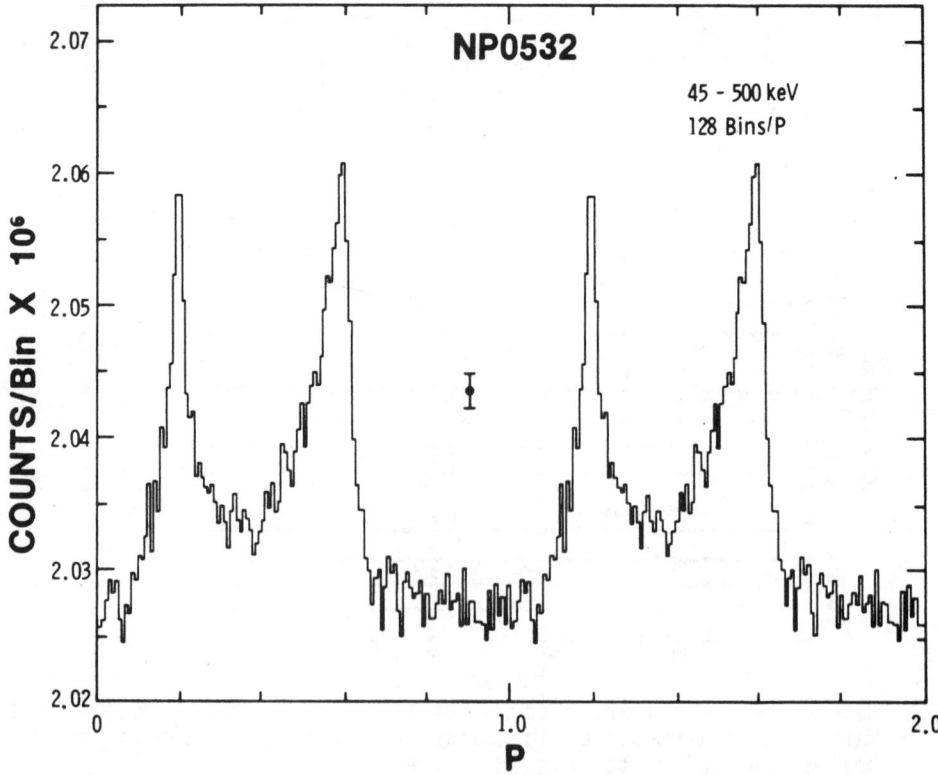

Figure 6 - Preliminary pulse profile of the Crab pulsar, NP0532.

The pulse profile is nearly identical to that measured 10 years earlier by Kurfess[6] in the energy range 100 to 400 keV. However, the present data show somewhat faster rise and fall times for both the main and interpulse than that measured by Strickman et al.,[7] in the energy range 20 to 250 keV. The ratio of main to interpulse flux is very similar in all three measurements. The FWHM of the main pulse is similar to that measured at lower energies by Kestenbaum, et al.,[8] but the trailing edge of the interpulse obtained here is sharper.

D. A0535+26

It was noted after the balloon flight that the transient, hard X-ray source A0535+26 was in an active state at the time of our flight.[9] A period search from 100 to 106 s in 45-75 keV for 7 hours centered on meridian passage was performed, with the results of the null hypothesis χ^2 test shown in Figure 7. The period is measured to be 103.68 ± 0.01 s; the smoothness of this curve and the large chi-squared value indicates the high signal-to-background ratio of the observation. The resulting pulse profile (Figure 8) is substantially different from that obtained during previous outbursts at lower energies[10,11]. The profile is, however, similar to that observed by Fishman and Watts,[12] but with a much larger main peak relative to the narrow dip just before it. The data in different energy ranges show some variation of pulse shape, with the flat interval in phase 0 to 0.35 showing excess emission at higher energies.

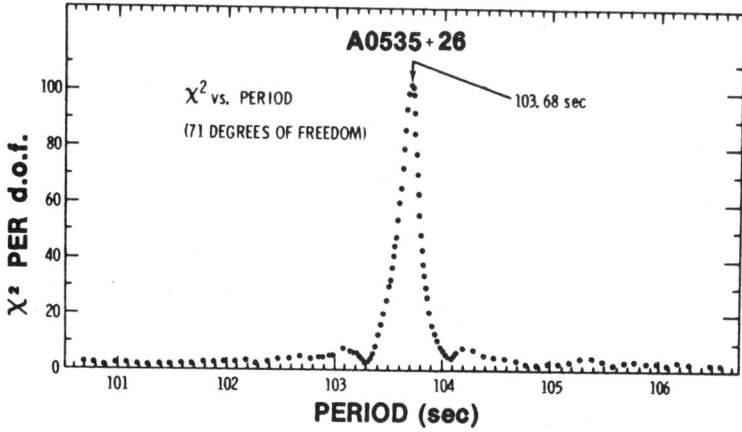

Figure 7 - Chi-squared vs. no. period plot near the period of the transient hard X-ray source A0535+26. The structure in the wings of the peak is due to interference between different phases of the pulse profile.

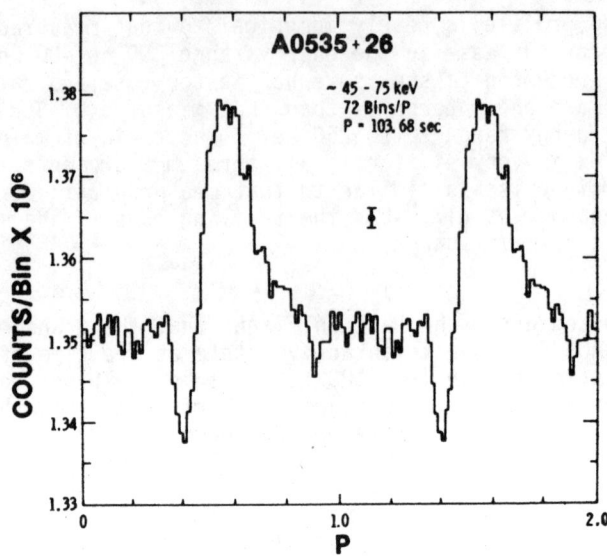

Figure 8 - Average pulse profile of A0535+26. Two cycles are shown using the same data.

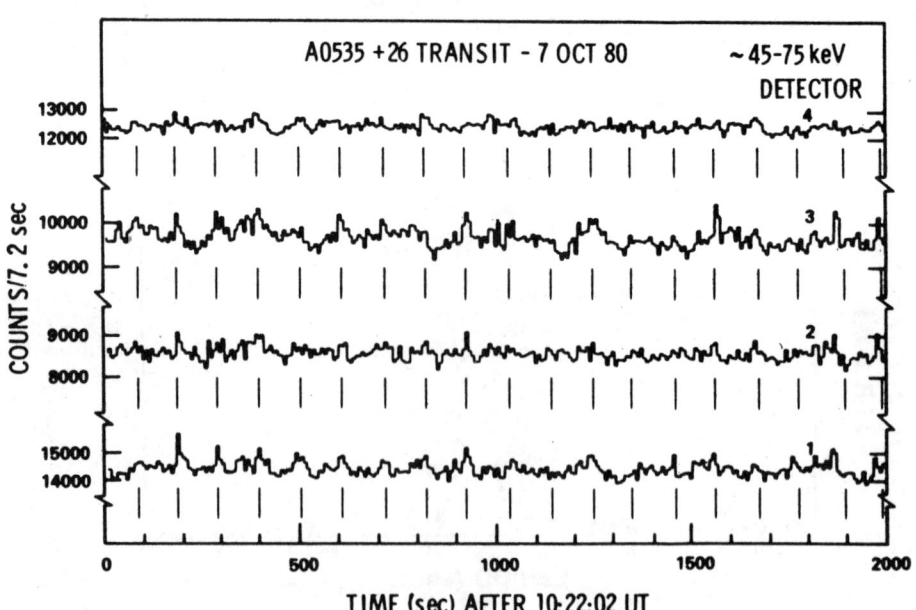

Figure 9 - Raw counting rate data near time of meridian transit of A0535+26. Vertical marks indicate the times of the peak phase of the source.

The intensity of the source is sufficient to detect individual pulses in the raw data from the uncollimated detectors. Shown in Figure 9 is an interval of 2000 s of raw data in 7.2-s bins near the time of meridian passage. Individual pulses from the source can be seen, with large variations in flux from pulse to pulse evident.

FUTURE WORK

This paper briefly describes preliminary results from data obtained on a recent balloon flight of a large-area detector array. Data analysis now in progress is expected to lead to a better-determined burst location and a burst spectrum. Additional spectral data will also be derived for NP0532, A0535+26 and possibly other periodic sources. These results will be presented in future publications.

A continuing balloon flight program is planned with approximately one flight per year. These flights will be used for burst and periodic source studies, as described in this paper, and also as developmental tests of components, software and techniques to be used with the Burst and Transient Source Experiment (BATSE) for the Gamma-Ray Observatory (GRO).[13] It is hoped that future flights will be of increased duration, with an ultimate goal of several-week-long, globe-circling flights in the Southern Hemisphere.

ACKNOWLEDGMENTS

This work was partially supported by a NAS/NRC Research Associateship for one of the authors (RBW) and by NASA contract NAS8-34137. We are grateful to R. W. Austin, F. Berry, S. Dothard W. Parker, W. H. Hammon, J. Smith, W. J. Selig, and J. W. Watts for their technical support and the personnel of the National Scientific Balloon Facility, Palestine, Texas.

REFERENCES

1. E. P. Mazets and S. V. Golenetskii, Astrophys. Space Sci. 75, 47 (1981).
2. G. J. Fishman, Astrophys. Space Sci., 75, 125 (1981).
3. G. J. Fishman, C. A. Meegan, J. W. Watts and J. H. Derrickson, Ap. J. 223, L13 (1978).
4. G. J. Fishman, Ap. J. 233, 851 (1979).
5. M. C. Jennings and R. S. White, Ap. J., 238, 110 (1980).
6. J. D. Kurfess, Ap. J. 168, 639 (1971).
7. M. S. Strickman, W. N. Johnson and J. D. Kurfess, Ap. J., 230, L15 (1979).
8. H. L. Kestenbaum, W. Ku, R. Novick and R. S. Wolff, Ap. J., 203, L57 (1976).
9. M. Oda et al., IAU Circ. No. 3525 and 3527 (1980).

10. G. R. Ricker et al., Ap. J. 204, L115 (1976)
11. H. Bradt et al., Ap. J. 204, L67 (1976).
12. G. J. Fishman and J. W. Watts, Ap. J., 212, 211 (1977).
13. G. J. Fishman, C. A. Meegan, T. A. Parnell and R. B. Wilson, these proceedings.

SEARCH FOR ASSOCIATIONS OF RADIO PULSES AND GAMMA RAY BURSTS

P. Inzani, G. Sironi
Istituto IFC/CNR and Istituto di Scienze Fisiche
dell' Universita' - Milano, Italy

N. Mandolesi, G. Morigi
Istituto TESRE/CNR - Bologna, Italy

ABSTRACT

Continuous radio records obtained between July 1976 and May 1979 by automatic radiometers operating at 151 and 408 MHz from Medicina, (Italy), have been scanned for radio pulses associated with gamma ray bursts. Sixty five gamma events have been examined. For none of them a definite association with a particular radio pulse detected within ± 10 min from the burst onset has been possible. However, a statistical analysis of the delays between each gamma event and the nearest radio pulse suggests to a 89% confidence level that about 20% of the bursts are associated with a weak radio precursor with flux density $\geq 10^{-13}$ erg/sec cm^2 MHz.

INTRODUCTION

In 1976 two laboratories, IFC/CNR-Milano and TESRE/CNR-Bologna, set up a collaboration which, between July 1976 and May 1979, ran an automatic station located in Medicina, (Italy), and continuously recorded the level of celestial noise at 151 and 408 MHz looking for fast transient phenomena. The system[1] had a sensitivity of $\sim 10^{-13}$ erg/cm^2 emitted in one second through a one MHz bandwith around the observing frequency, (see Table I). Compared with $\sim 10^{-5}$ erg/cm^2 in few seconds from a typical gamma burst, such a value shows that we could detect events whose radio luminosity was a very small fraction of the typical luminosity of X and gamma events. As shown by previous searches (for a review, see ref. 1 and 2), radio bursts, if they do exist, are very rare; even under the best conditions the observer has to look for them in a background of unwanted events produced by human and atmospheric electromagnetic activity. Between July 1976 and May 1979, the Medicina station stored 13,000 hours of analog data and 1900 hours of digital data. Analyzed for the presence of pulses of celestial origin, they set a limit of 10^{-5} sec^{-1} to the rate of radio bursts with a flux $\geq 10^{-13}$ erg/sec cm^2 MHz in the UHF an VHF bands, a rate similar to that previously found by other experimenters (for a review. see ref. 1). Rarer events can be extracted from the background noise when an external source of information provides a time interval in which to look.

In the following, we present results of an 'a posteriori' search for radio pulses associated with gamma ray bursts.

RESULTS

Through a comparison of the available lists of gamma bursts, (see ref. 3 and references therein), and our radio data, sixty-five events which occurred when the radio equipment was in regular operation were selected,

0094-243X/82/770079-06$3.00 Copyright 1982 American Institute of Physics

TABLE I
The Medicina Station (lat=44.5°N, long=0.77^{h}E)

Ant. HPBW	12° x 73°		20° x 20°		60° x 40°	
Ant. Type	(1)		(2)		(3)	
Freq. (MHz)	151	408	151	408	151	408
S_{min} (3σ) (10^{-22} W/m^2 Hz)	1.0	0.3	0.4	0.1	2.7	0.7

(1) transit system at dec=-28°; (2) tracking system on Cygnus; (3) transit system at dec=+45°.

(see Table II). First of all, we reduced the onset times of the gamma bursts at the detector, t_d, to t_o, the time at the Earth, by means of published data (4) and/or

$$t_o - t_d = (R/c) \cos(\delta_1) \cos(\delta_2) \cos(\alpha_1 - \alpha_2) + \sin(\delta_1) \sin(\delta_2) \qquad (1)$$

where R is the earth-detector distance and (α_1, δ_1), (α_2, δ_2) the celestial coordinates of the burst source and detector. Because of the extension of the error boxes around the burst sources, the uncertainties on $t_o - t_d$ are generally very high; therefore, whenever possible we used data from near earth satellites. When the source position was unknown, only bursts seen by detectors on board satellites at a maximum distance from the earth of one light minute were accepted. In Table II, beside t_o, we list for each event (i) the time difference d between the nearest suspected radio burst (SRB) within ±10 minutes from t_o, (ii) the visibility V of the gamma source by our aerials, and (iii) the antenna type (see Table I).

We then examined the radio records around t_o and for each frequency prepared a list of radio pulses which: i) occurred within ±1 hours from t_o, ii) had a rise time faster than 1 min, iii) had a duration between 1 sec and 1 min, iv) exceeded the mean noise level by at least 4σ, where σ is the rms value of the fluctuations of the radio noise. The two lists were combined to form a list of SRBs, i.e., radio events which occurred at both frequencies within ±T sec from t_o and had time histories not in conflict with the sign of the delay one expects to be produced by dispersion in the insterstellar medium. Assuming a Poisson distribution of the SRB's the probability of the random association with a gamma burst within ±T sec from t_o is

$$P = 1 - \exp(-T/R)$$

where R, (sec^{-1}) is the rate of SRB's. Ten minutes is the maximum delay T_m between gamma and radio events, set by the mean value of R, we accepted

TABLE II
List of Events

Date	t	d	V	A	Date	t	d	V	A
08/16/76	58522	62	0	1	11/12/78	36490	210	?	3
12/20/76	63512	--	0	1	11/15/78	76045	-5	0	3
05/18/77	14169	--	?	2	11/15/78	80052	--	0	3
09/05/77*	36000	20	0	1	11/19/78	34016	-126	0	3
09/26/77	68760	-235	?	1	11/21/78	5735	410	?	3
10/24/77	18300	--	?	1	11/23/78	58078	507	0	3
11/05/77	6120	--	?	1	11/24/78	14130	--	>0	3
11/10/77	61964	--	0	1	12/04/78*	48009	-9	>0	3
11/13/77	7020	--	?	1	12/13/78*	85980	80	0	3
12/16/77	67920	--	?	1	12/18/78*	85718	86	0	3
01/08/78*	25848	172	0	3	12/29/78	43611	-193	>0	3
01/11/78	38460	90	?	3	01/01/79	563	-93	>0	3
01/13/78	67380	--	?	3	01/02/79	64569	-99	0	3
01/21/78	0	--	?	3	01/07/79	20155	-255	>0	3
05/08/78	72877	--	?	3	01/16/79	32411	589	>0	3
05/19/78	26509	-109	?	3	02/13/79	46993	-226	>0	3
05/21/78	78827	--	?	3	02/15/79	62461	--	0	3
09/14/78	42171	219	>0	3	02/18/79	1993	-273	0	3
09/14/78	60128	--	?	3	03/31/79	76164	-84	0	3
09/16/78	36206	-26	?	3	04/02/79	5671	-36	>0	3
09/18/78	71373	-163	?	3	04/04/79	2655	172	0	3
09/21/78	14156	--	?	3	04/05/79	73431	-171	?	3
09/21/78	73324	106	?	3	04/05/79	75446	-266	?	3
09/23/78*	86100	--	0	3	04/08/79	0	150	?	3
09/30/78	24451	-71	>0	3	04/11/79	82513	467	?	3
10/11/78	39153	-293	?	3	04/12/79	79208	298	0	3
10/12/78	4597	203	0	3	04/14/79*	53143	-273	0	3
10/12/78	62036	-406	>0	3	04/18/79	27665	-165	>0	3
10/22/78	84272	--	?	3	04/19/79	58306	149	0	3
10/25/78	85989	--	>0	3	04/27/79*	24595	-5	>0	3
10/26/78	29062	-392	>0	3	04/30/79	72847	140	0	3
11/02/78	45249	-329	>0	3	05/04/79	31466	40	>0	3
11/04/78	58668	-253	0	3					

* = solar event; V = source visibility by the antenna; A = antenna type; d = delay between radio and gamma pulse (sec); t = time of the gamma burst (sec).

TABLE III
Event Classification

Visibility	>0		0		?	
Type	no SRB	SRB	no SRB	SRB	no SRB	SRB
Solar	-	4	1	4	-	-
Cosmic	2	13	4	12	12	13

DISCUSSION

Forty-four of the sixty-five events of Table II have an SRB within ± 10 min from t_o. The great majority of the SRB's however, are local origin disturbances randomly associated with the gamma event. The distribution of the events versus the Medicina solar time shows in fact that the SRB's are more common at daytime. It also appears that the level of radio noise had a constant increase over the years, (in fact, the number of events without SRB's drastically fell between 1976 and 1979).

We cannot, however, extract from Table II the genuine gamma radio associations. In fact, i) the dispersion between the two radio signals which form an SRB is generally small, (<10 sec), and never sufficient to eliminate the possibility of atmospheric or man-made disturbances; ii) the luminosity in the radio band, evaluated for those events whose position in the sky is known, is a negligible fraction of the luminosity at gamma energies and does not put in evidence any peculiar event; iii) the panorama of the time profiles of the SRB's is undifferentiated and does not suggest the existence of various classes of events. The only exception is perhaps the August 16, 1976 event. An association with radio pulses detected about one minute after the gamma event was suggested in 1977[5]. Subsequently published x-data from a balloon experiment made the association doubtful.[6,] We cannot, however, hope to improve the situation because to the best of our knowledge, all the available data have been published.

We are left with the statistical analysis of the sample of 65 bursts.

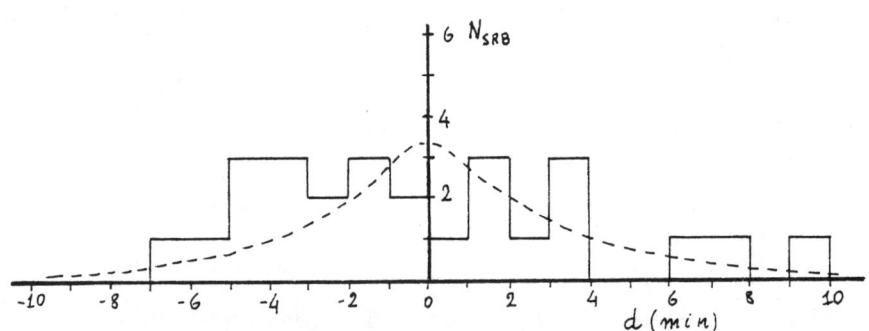

Fig. 1 - A distribution of SRB's versus the delay between radio and gamma burst

To extract astrophysical information, the events have been divided in various categories (Table III): i) 9 bursts are solar origin events. They are marked by an asterisk in Table II and have been used to check the performance of the radio equipment; ii) for 31 events the position of the gamma source is known and 15 of them were within the field of view of our aerials. Because 2 of the 15 did not show any associated SRB, $F = 2/15 = 13\%$ is a lower limit to the fraction of gamma bursts whose pulsed radioemission within ± 10 min from t_o is below 10^{-13} erg/sec cm^2 MHz, the limit set by the receiver's sensitivity. Cleanest data, i.e., the data recorded at least five hours before or after the local noon, give $F \geq 33\%$; iii) 12 of the 16 bursts out of the antennas field of view have an associated SRB, therefore P = (0.75 ± 0.30) is the probability of random gamma-radio associations; iv) of the 56 non-solar bursts, 38 are SRB. Combined with P they give the fraction G = (0.2 ± 0.3) of bursts whose radioemission is greater than $\sim 10^{-13}$ erg/sec cm^2 MHz.

G can also be estimated if one plots the 26 SRB's of Table II with positive or unknown visibility versus the delay d between the SRB nearest to a gamma burst and the gamma burst itself, (see Figure 1). Were all the SRB's due to local noise, a symmetric distribution would be expected. We observe on the contrary an excess of four events where the SRB precedes the gamma burst. Because the expected number of non-solar gamma bursts within the aerial field of view is 23 (see Table III) the excess is equivalent to G = (0.2 ± 0.1), a fraction different from zero. The probability that the excess of 4 is just a fluctuation in a symmetric distribution is 11%, therefore the confidence level of the estimated value of G is 89%.

CONCLUSIONS

Radio records collected between July 1976 and May 1979, matched to existing lists of gamma ray bursts, show that at least 80% of gamma bursts emit at radio frequencies, in the UHF and VHF band, less than 10^{-13} erg/sec cm^2 MHz. For the remaining 20% there is, however, a weak indication, (89% confidence level), that the gamma bursts have a radio precursor of intensity $\geq 10^{-13}$ erg/sec cm^2 MHz. Unfortunately, there are no reasonable chances that the experimental situation will improve in the near future. In fact the level of local radio noise steadily grows and as far as we know we will have to wait some years before new coordinated systems of gamma burst detectors enter in operation.

Were the present result confirmed, it would set strong constraints on the possible model of burst sources. The radio precursor would, in fact, imply that we are observing processes in which energy is gradually accumulated inside a small region. At the beginning the region is optically thin and the temperature sufficient to produce radio emission which escapes outside. As the process goes on, emission at still higher frequencies is suppressed by increasing opacity. At last the bulk of the accumulated energy is released in the gamma domain. Such a behaviour can be found in solar bursts, (see May 14, 1979 event in Table II), and could be reconciled with superflare models of gamma bursts[8]. Or one can think of an extension of the Maraschi and Cavaliere[9] model of x-ray bursters based on accretion on compact objects followed by thermonuclear explosions.

ACKNOWLEDGEMENTS

We are indebted to J. Jelley, R. Drever, H. Helmken, W. Evans, T. Cline, K. Hurley, G. Pizzichini and many other people for helpful discussions and/or providing data before publications.

Part of the present search has been financially supported by GIFCO, (Gruppo Italiano di Fisica Cosmica) of the Italian National Council of Research.

REFERENCES

1. P. Inzani, G. Sironi, P. Cazzola, S. Cortiglioni, N. Mandolesi, G. Morigi, G. G. C. Palumbo, Ap. Space Sci. $\underline{56}$, 239 (1978).
2. R. B. Partridge, in Neutron Stars, Black Holes and Binary X-ray Sources Gursky and Ruffini Ed., Page 29, (1975).
3. A. Ciapi, S. Cortiglioni, B. Falconi, E. Galli, P. Inzani, N. Mandolesi, G. Morigi, G. Sironi, submitted to Ap.J., (1981).
4. S. V. Golenetskii, Yu. A. Gurian, preprint n.702, USSR, Academy of Sciences, Leningrad 1981.
5. N. Mandolesi, G. Morigi, P. Inzani, G. Sironi, F. S. Delli Santi, F. Delpino, M. Petessi, A. Abrami, Nature $\underline{266}$, 427 (1977).
6. M. Sommer, D. Muller, Ap.J. (Letters), $\underline{222}$, L17, (1978).
7. S. Cortiglioni, N. Mandolesi, G. Morigi, A. Ciapi, P. Inzani, G. Sironi, Ap. Space Sci., $\underline{75}$, 153 (1981).
8. F. W. Stecker, K. J. Frost, Nature Phys. Sci., $\underline{245}$, 70 (1973).
9. L. Maraschi, A. Cavaliere, in Highlights of Astronomy $\underline{4}$/I, 127 (1977)

Sec. I B Gamma Ray Transient Positions and Distribution

GAMMA RAY BURST POSITIONS

K. Hurley
Centre d'Etude Spatiale des Rayonnements
(CNRS-UPS)
B.P. 4346, 31029 Toulouse Cedex, France

ABSTRACT

The international network of gamma ray burst detectors has provided redundant localizations for six gamma ray bursts with accuracies in the arcminute range. Catalog, radio, soft X-ray, and optical searches have been performed for some of these events. The results of these searches are reviewed. Although radio, X-ray, and optical candidates are found in the error boxes, no clear association between gamma ray bursts and other forms of emission has emerged to date. Optical and radio searches are continuing.

INTRODUCTION

Almost from the very discovery of gamma ray bursts, one of the major experimental goals has been to achieve localizations precise enough to allow sensitive searches for counterparts in the optical, radio, and soft X-ray ranges. Although it may be possible to localize gamma ray bursts to arcminute accuracy using a single instrument[1], the only method which has in fact been used to date is that of arrival time analysis with three or more spacecraft. In this method, the difference in arrival times at any two spacecraft is used to construct an annulus of permitted arrival directions on the celestial sphere. While simple in principle, and in fact proposed in a somewhat different context before the discovery of gamma ray bursts[2], arrival time analysis contains a number of subtleties which are not always appreciated. The first is that collecting data from many spacecraft launched and tracked by different countries is time consuming; since the data must be compared with an accuracy of several milliseconds, it is neccessary to know all the sources of systematic timing errors in the experiments and in the ground stations. (If all the spacecraft were tracked by a single ground station, then, since to first order only the difference in arrival times must be known, small systematic timing errors introduced on the ground would tend to cancel out; but this is obviously not the case when different tracking stations are involved.) Second, and this is the overwhelming advantage of the method, is that when four or more spacecraft are used, the arrival direction of a gamma ray burst is redundantly

0094-243X/82/770085-15$3.00 Copyright 1982 American Institute of Physics

determined; that is, if the three or more annuli obtained intersect over a common region, that region is neccessarily the correct one. Finally, arrival time analysis is sometimes referred to as "triangulation", but this is a misnomer; triangulation involves solving for an unknown length in a triangle when two sides and one angle are known. Unfortunately, in gamma ray burst localization, one length, that to the source, remains unknown. In theory, however, arrival time analysis could be used to triangulate the distance to the source. For example, using a network of detectors spaced 2 A.U. apart, capable of recording the arrival time of a wavefront to an accuracy much better than 1 ms., it would be possible to distinguish between a plane wavefront and a spherical one originating at a distance of less than 5 parsecs. The most accurate timing achieved to date between widely separated spacecraft, however, is in the 10 ms. range.

With the establishment of an international network of gamma ray burst detectors (Table I), it has been possible to determine redundantly the positions of six events with accuracies ranging from 0.05 arcmin^2 to 14 arcmin^2; two of these positions are inconsistent with previously published localizations using another method. Many other events have been localized non-redundantly and/or with less accuracy than this, and some three or four events are currently being studied which will give redundant localizations with accuracies in the range cited above.

This paper will review the results which have been obtained from optical, radio, catalog, and soft X-ray searches of 7, approximately arcminute size, error boxes (Table II). Three of these error boxes have appeared in the literature or will appear shortly. The other four are currently being prepared for publication in more detail than will be given here.

CATALOG SEARCHES

A catalog of over 10^5 objects was compiled some time ago by H. Helmken and co-workers at the Center for Astrophysics, using over 40 published and unpublished lists. The original catalog has been modified and added to by members of the international network (Table I) and contains, but is not limited to, positions of radio sources, pulsars, nearby stars, white dwarfs, supernova remnants, RNGC objects, Seyfert galaxies, X-ray sources, and gamma ray sources. It was through the use of this catalog that the position of the 1979 March 5 event was rapidly found to be consistent with the N49 supernova remnant[3]. Unfortunately, none of the other six events in Table II has a position consistent with a cataloged object.

RADIO SEARCHES

The position of the 1979 March 5 gamma ray burst is consistent with that of the supernova remnant N49,[4] which is known to be a radio source[5]. The 408 MHz radio emission was discovered long before the gamma ray burst, and it is not surprising to find that an SNR is also a radio emitter. More interesting, however, is the recent discovery by Hjellming and Ewald [6] that the 1978 Nov 19 error box position[7] is consistent with four radio sources. Figure 1 shows 1465 and 4885 MHz VLA maps of the region. Three of the sources may constitute a radio triplet, thus resembling the 2695 and 8085 MHz observations of Sco X-1[8]. In the case of Sco X-1, the two outer sources are roughly equidistant from the central source, which is associated with the optical and X-ray emitting object. In the 1978 Nov 19 case, however, the two outer sources, A and B in Figure 1, which are linearly polarized, have no such obvious central source; the source C in Figure 1, which is on the line joining A and B, may be the third source in the triplet, but it is not associated with any optical object down to about 21st magnitude. The association between the radio sources and the gamma ray burst is a tentative one at present, which could be strengthened by the observation of a large proper motion or short time scale fluctuations in the radio source, or by the discovery of an optical candidate[6]. Nevertheless, the discovery is sufficiently encouraging to warrant further radio searches. One such search has begun for the 1979 June 13 event using the VLA, but no results are available at this time[9].

SOFT X-RAY OBSERVATIONS

The 1979 Mar 5 region has now been observed four times by the Einstein observatory; one IPC and one HRI observation were done both before and after the gamma ray burst. The two IPC observations, which did not resolve the X-ray structure of N49, have been used to set an upper limit to the change in the X-ray flux of 0.8% before and after the event[10]. The first HRI observation revealed N49 as an extended source about 25 parsecs in diameter, and set an upper limit to the intensity of a point source superimposed on the diffuse source at 2.1×10^{-12} erg/cm^2 s in the 0.5-4.5 keV range. The upper limit to the intensity of a point source near, but not superimposed upon the diffuse source is about an order of magnitude less, but an improved analysis of the localization data for the 1979 Mar 5 event has shown that the error box is now entirely contained within the X-ray contours[11]. The resulting luminosity of any point source in the X-ray contours of N49 is less than about 4×10^{35} erg/s, or

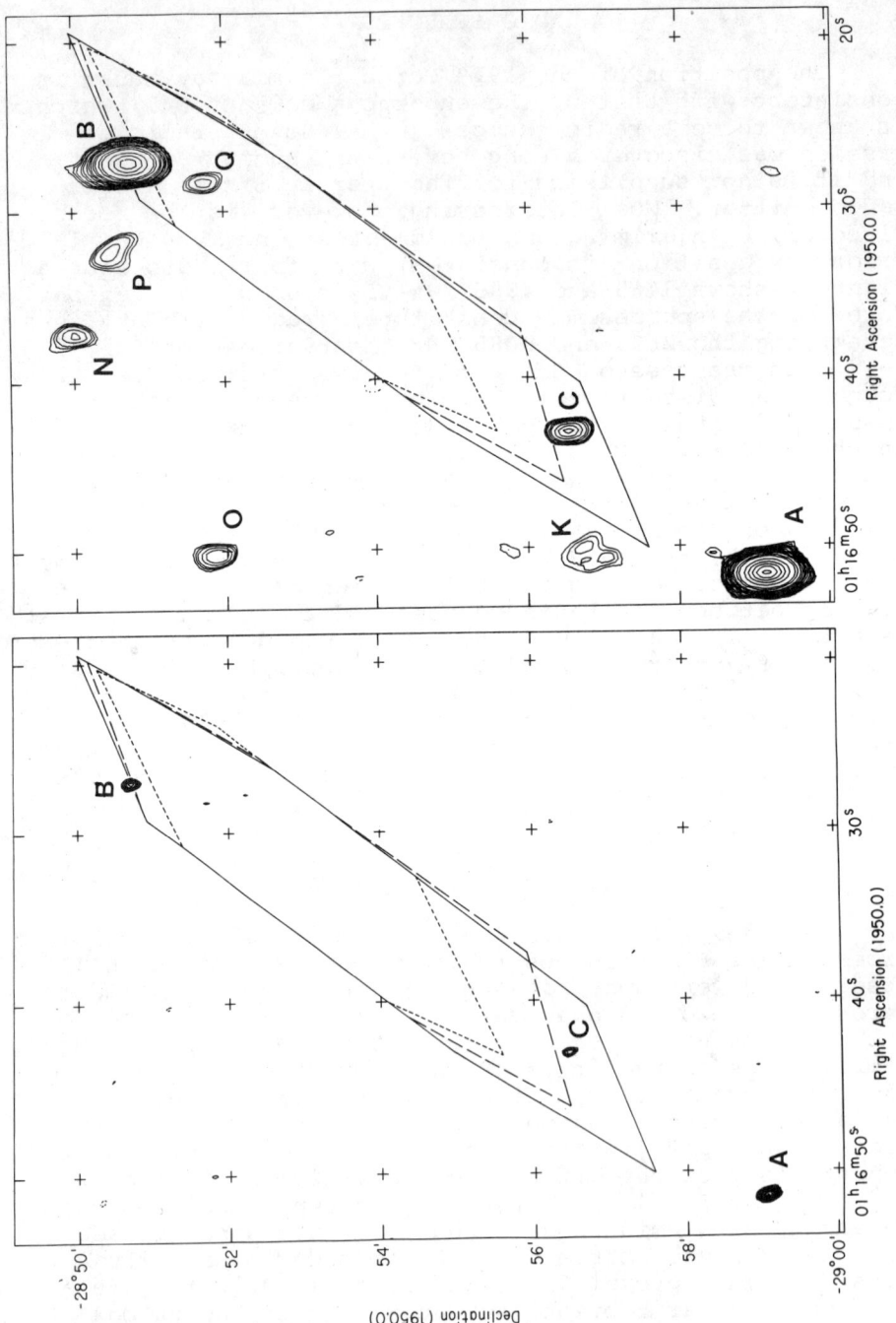

Figure 1. 4885 (l) and 1465 (r) VLA maps of the 1978 Nov 19 gamma ray burst region, from Hjellming and Ewald[6].

two orders of magnitude smaller than that of a typical pulsating binary X-ray source[10]. If the event indeed originated in N49, the energy output was about 5×10^{43} ergs if omnidirectional; thus changes in the SNR structure and consequently in the X-ray isophotes may have occurred. For this reason, a final HRI observation was requested and carried out, but the data have not yet been analyzed.

An 8000 s IPC observation was devoted to the 1978 Nov 19 error box[12]. A weak source (less than 3×10^{-13} erg/cm^2 s) was detected at the 3.4 sigma level, whose position would be consistent with either of the radio sources "Q" or "B" in Figure 1. The probability of finding a source in an error box of this size (8 arcmin^2) is estimated at about 1%. Because of this, and the fact that the source is at the limit of detection, a definite association between the source and the gamma ray burst cannot be established; however, the observation will be used later to speculate on the nature of the source.

An 8000 s IPC observation was carried out on the 1979 April 6 region; this event was a short and impulsive one, in contrast to the 1978 Nov 19 event, which was long and complex, and its total energy flux at earth was some two orders of magnitude less. The IPC observation revealed no source. _If_ the emission mechanisms were the same in the two cases, and _if_ the same absolute gamma ray luminosity was reached in both cases, then the lack of detectable soft X-ray emission for the 1979 April 6 event might be explained by its greater distance from earth. But given that the two time histories are so different, and that no clear relation between gamma ray bursts and soft X-ray emission can be established, this line of reasoning is far from certain.

At least ten error boxes are small enough to be proposed for EXOSAT observations, which could begin in 1982. Observing times well in excess of 10^4 s will probably be needed to reach sensitivities below those of Einstein. In any case, however, a concentrated effort should be made to understand the relationship between gamma ray burst sources and soft X-ray emission.

OPTICAL SEARCHES

For the seven events listed in Table II, Palomar Sky Survey, ESO Quick Blue, or SRC survey plates have been examined; the limiting magnitudes are approximately 19.5, 21.5, and 22.5, respectively. Figure 2 shows the N49 supernova remnant with a refined error box for the 1979 March 5 event[11]. The optical contours have a diameter of about 16 parsecs, or somewhat smaller than the X-ray contours. The error box does not include the center

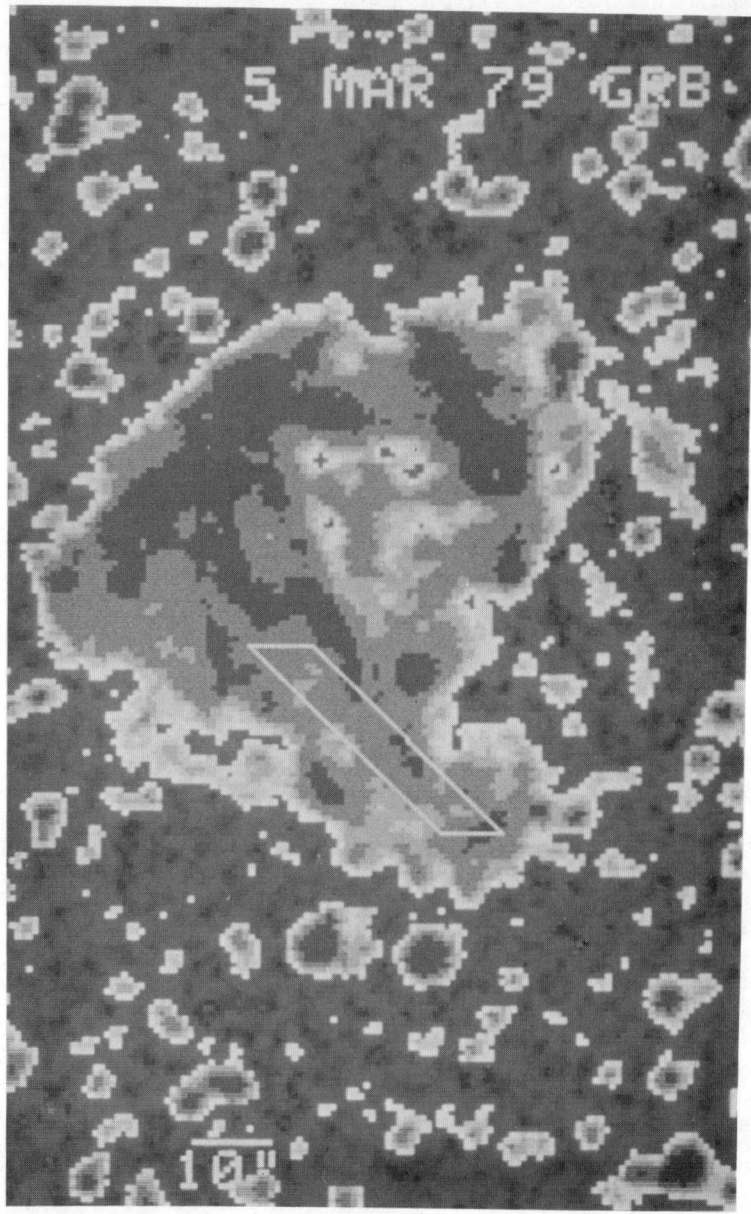

Figure 2. Optical contours of the N49 SNR from an ESO Quick Blue survey plate, and the refined 1979 March 5 error box[11]. Two or three star-like images are contained in the box.

of the optical contours, but lies along a spur, with one extremity of the error box about 3 parsecs from the center. Two or three star-like images are present in the error box, but none corresponds to any of the objects previously studied by Fishman et al.[13]. In four other cases, survey plate searches have, not surprisingly, revealed a number of objects in the error boxes. Two cases should be mentioned here. The 1978 Nov 4 error box contains an uncataloged galaxy which, in the deeper plate taken at the prime focus of the ESO 3.6 m telescope, appears to have a bright nucleus, and may be an active galaxy. The 1979 April 18 event is in a region of faint nebulosity extending from the Orion nebula, about 5° away, and identified by COS-B as a region of diffuse gamma ray emission possibly related to the CO radio emission[14]. In both cases, the association with the gamma ray burst is tantalizing, but cannot be defended from a statistical point of view: in the case of the 1978 Nov 4 event, there are many other objects in the error box, while in the case of the 1979 April 18 event, the large extent of this and other cloud complexes gives a relatively large probability of finding a chance association.

For three of the events in Table II, deep optical searches have been carried out in a systematic program of gamma ray burst error box searches involving astronomers at the Observatoire de Meudon working in collaboration with the members of the international network (Table III). An example is shown in Figure 3, a computer enhanced image of part of a plate taken at the prime focus of the ESO 3.6 m telescope[15]. For this event (1979 April 6) no object was apparent in the SRC J plate[16], while three objects can be found in the deep plate shown, which reaches about 23rd magnitude. This situation is to be expected, based on number-magnitude counts of faint objects. Furthermore, recent work on faint-object counts[17] points out a potential problem, namely, that at magnitudes fainter than 22, the density of galaxies equals or exceeds that of stars; on the 1979 April 6 plate, it has not been possible to analyze the shapes or the colors of the three images at the plate limit to establish their classification, although this is planned for future observations.

DISCUSSION

The precise localization results obtained from the international network have only begun to be exploited for optical, X-ray, and radio searches; the data obtained to date on a still rather limited sample do not yet present any coherent picture. From the results of the deep optical searches, it may be said that no gamma ray burst error box is truly empty. Thus it is unneccessary to in-

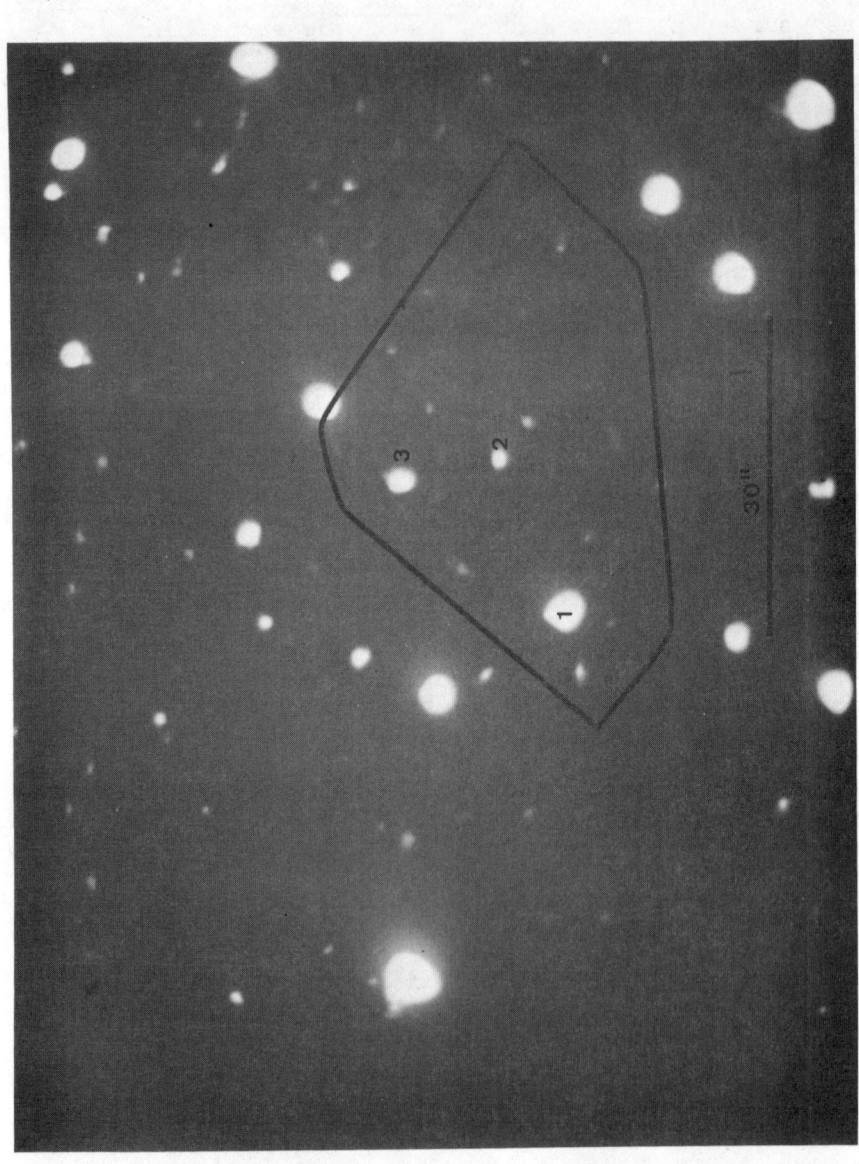

Figure 3. A computer enhanced version of an ESO 3.6 m plate of the 1979 April 6 gamma ray burst position15. 3 objects are visible in the error box.

voke invisible objects, such as lone neutron stars, as
gamma ray burst sources, although the observational data
clearly cannot be used to rule this out. If gamma ray
burst generation involves a compact object such as a
neutron star, whether alone or in a binary system, it
may be possible to detect the thermal radiation from the
surface of such a star in the soft X-ray range[18]. It is
possible that the soft X-ray source in the 1978 Nov 19
error box represents such a detection, although a very
wide range of neutron star temperatures and distances
would be allowed: for example, an object with 10 km
radius would produce the observed flux if it had a temperature of 10^6 °K and were located at about 4 kpc, or
if it had a temperature of 300000°K and were located at
100 pc[18].

It would be difficult, although perhaps not impossible, to detect the surface radiation of a lone neutron
star optically. If such an object radiates as a black
body (and there are reasons to suspect that it may not[18,19]) then it might be visible with a blue magnitude of
about 25, assuming again a temperature of 10^6 °K, a radius
of 10 km, and a distance of 100 pc. At this limiting
magnitude, a plate may be expected to have some 10 objects per square arcminute[17], however, and only the smallest error boxes could be searched. The advantages of the
Space Telescope in this sort of search would be substantial.

For two of the redundantly localized events in Table
II, an inconsistency has been found between the positions
obtained from the international network and those given
by Mazets et al.[20,21]. For the 1978 Nov 4 event, the limit of the Mazets et al.[20] error circle, derived using
the cosine law detector response, is at least 1° from the
redundantly determined position. The 1979 Jan 13 position given by Mazets et al.[21] was obtained by arrival
time analysis, yielding an annulus of position, and by
the cosine law detector response, which limits the localization to part of the annulus. While the annulus includes the redundantly determined position of the international network, the part of the annulus isolated by
the cosine law detector response is at least 5° away from
that position. These two discrepancies suggest that
shadowing or backscattering effects in the Venera spacecraft may be playing an important role in the detector
response in some cases.

CONCLUSION

The precise localization of gamma ray bursts will
continue for the next few years at about the same level
of accuracy using Pioneer Venus Orbiter, ISEE 3, Prognoz
9, Venera 13 and 14, and possibly other near earth space-

craft. Optical, radio, and soft X-ray searches will be performed whenever possible, in order to build a large enough data base to detect some sort of recurrent pattern. The need for smaller error boxes is now becoming evident; gamma ray burst experiments aboard the two International Solar Polar Mission spacecraft will provide them.

ACKNOWLEDGEMENTS

I am indebted to the members of the international collaboration for the use of their data, especially C. Barat, T. Cline, U. Desai, I.V. Estulin, W.D. Evans, R. Klebesadel, J. Laros, and M. Niel. The optical searches were carried out with the collaboration of C. Chevalier, S.I. Ilovaisky, and Ch. Motch. Many ESO staff members have been extremely helpful, particularly S. D'Odorico, P. Grosbøl, H.-E. Schuster, and R. West. This work was supported by CNES Contract 81-212.

TABLE I. INTERNATIONAL NETWORK OF GAMMA RAY BURST DETECTORS

INSTITUTES	SPACECRAFT	ORBIT	OPERATING DATES	INSTRUMENT DESCRIPTION REFERENCE
GODDARD SPACE FLIGHT CENTER	HELIOS 2 ISEE 3	HELIOCENTRIC LAGRANGE POINT	1/76 - 12/79 9/78 - PRESENT	22 23
LOS ALAMOS NATIONAL LABORATORY	VELA PIONEER VENUS ISEE 3	GEOCENTRIC VENUSCENTRIC LAGRANGE POINT	1967 - PRESENT 5/78 - PRESENT 9/78 - PRESENT	24 25 24
CENTRE D'ETUDE SPATIALE DES RAYONNEMENTS/ INSTITUTE FOR SPACE RESEARCH (MOSCOW)	PROGNOZ 6,7 VENERA 11,12 PROGNOZ 9 VENERA 13,14	GEOCENTRIC HELIOCENTRIC GEOCENTRIC HELIOCENTRIC	9/77-2/78, 10/78-6/79 9/78 - 4/80 START 1982 - END 1981 -	26,27 27
GODDARD SPACE FLIGHT CENTER/ MAX-PLANCK INSTITUTE FÜR ASTROPHYSIK/ SPACE RESEARCH INSTITUTE, UTRECHT/ CENTRE D'ETUDE SPATIALE DES RAYONNEMENTS	INTERNATIONAL SOLAR POLAR MISSION	HELIOCENTRIC	1986 -	

TABLE II. RESULTS OF SEARCHES FOR GAMMA RAY BURST COUNTERPARTS

EVENT	ERROR BOX AREA, ARCMIN^2	SURVEY PLATE SEARCH	DEEP OPTICAL OBSERVATION	RADIO OBSERVATION	EINSTEIN OBSERVATION	CATALOG SEARCH
1978 NOV 4	14	50 OBJECTS	50 OBJECTS	- - -	- - -	NO OBJECTS
1978 NOV 19	8	SEVERAL OBJECTS[6,7]	SEVERAL OBJECTS[12,13]	SEVERAL SOURCES[6]	WEAK SOURCE[12]	NO OBJECTS
1979 JAN 13	5	34 OBJECTS	46 OBJECTS	- - -	- - -	NO OBJECTS
1979 MAR 5	0.05	SEVERAL STARS (FIG.2)	- - -	RADIO SOURCE[5]	DIFFUSE SOURCE[10]	N49
1979 APR 6	0.25	NO OBJECTS[16]	3 OBJECTS[15]	- - -	NO SOURCE	NO OBJECTS
1979 APR 18	2	5 OBJECTS	- - -	- - -	- - -	NO OBJECTS
1979 JUN 13*	1	NO OBJECTS	- - -	IN PROGRESS	- - -	NO OBJECTS

* Not yet redundantly localized

TABLE III. GAMMA RAY BURST POSITION OPTICAL SEARCH PROGRAM*

OBSERVING PERIOD	EVENT	OBSERVATORY	INSTRUMENT	COMMENTS
1980 APR 16-17	1979 JAN 13	ESO	3.6 M PRIME FOCUS	4 PLATES
1980 JUN 20-21	1978 NOV 4	ESO	3.6 M PRIME FOCUS	2 PLATES SHORT EXPOSURES
1980 SEP 5-6	1979 APR 6	ESO	3.6 M PRIME FOCUS	2 PLATES
1981 JUN 1-2	1978 NOV 4	ESO	3.6 M PRIME FOCUS, MACMULLAN CAMERA	2 EXPOSURES, BAD WEATHER
1981 SEP 20-22	1978 NOV 19 1979 APR 6	ESO	3.6 M PRIME FOCUS	
1981 OCT	1979 APR 18	OHP	1.93 M CASSEGRAIN FOCUS LALLEMAND CAMERA	
1981 NOV 24-28	1978 NOV 19 1979 MAR 5 1979 APR 6	ESO	1.54 M CASSEGRAIN FOCUS ESO CCD CAMERA	
1982 JAN	1979 APR 18	CFH	3.6 M PRIME FOCUS MEUDON ELECTRONIC CAMERA	
1982 FEB 16-21	1979 MAR 5 1979 APR 18	ESO	3.6 M PRIME FOCUS MEUDON CCD CAMERA	
1982 MAR	1979 JUN 13	OHP	1.93 M CASSEGRAIN FOCUS LALLEMAND CAMERA	

*Carried out by C. Chevalier, S.I. Ilovaisky, and Ch. Motch, Observatoire de Meudon

REFERENCES

1. Helmken, H., and Gorenstein, P., Proceedings of the 12th ESLAB Symposium, Frascati, Italy,1977, ESA SP-124, 353
2. Giacconi, R., Ap. J. Lett., $\underline{173}$, L79, 1972
3. I.A.U. Circular 3356, 1979
4. Evans, W.D., Klebesadel, R.W., Laros, J.G., Cline, T.L. Desai, U.D., Hurley, K., Niel, M., Vedrenne, G., Vedrenne, G., Estulin, I.V., Kuznetsov, A.V., and Zenchenko, V.M., Ap. J. Lett., $\underline{237}$, L7, 1980
5. Mathewson, D.W., and Clarke, J.N., Ap. J., $\underline{180}$, 725, 1973
6. Hjellming, R.M., and Ewald, S.P., Ap. J. Lett., $\underline{246}$, L137, 1981
7. Cline, T., Desai, U., Pizzichini, G., Teegarden, B., Evans, D., Klebesadel, R., Laros, J., Barat, C., Hurley, K., Niel, M., Vedrenne, G., Estulin, I., Mersov, G., Zenchenko, V., and Kurt, V., Ap. J. Lett., $\underline{246}$, L133, 1981
8. Wade, C.M., and Hjellming, R.M., Ap. J., $\underline{170}$, 523, 1971
9. Hjellming, R.M., 1981, private communication
10. Helfand, D.J., and Long, K.S., Nature, $\underline{282}$, 589, 1979
11. Cline, T., Desai,U., Teegarden, B., Evans,D., Klebesadel,R., Laros,J., Barat,C., Hurley,K., Niel, M., Vedrenne, G., Estulin, I., Kurt,V., Mersov,G., Zenchenko, V., Weisskopf, M., and Grindlay, J., preprint, 1981
12. Pizzichini, G., Danziger, J., Grosbøl, P., Tarenghi, M., Cline, T., Desai, U., Mushotzsky, R., Teegarden, B., Evans, W., Klebesadel, R., Laros, J., Barat, C., Hurley, K., Niel, M., Vedrenne, G., Estulin, I., Mersov, G., Zenchenko, V., and Kurt, V., Paper presented at the 15th ESLAB Symposium on X-Ray Astronomy, Amsterdam, Holland, 1981
13. Fishman, G., Duthie, J., and Dufour, R., Ap. and Space Sci., $\underline{75}$, 135, 1981
14. Caraveo, P.A., Barbareschi, L., Bennett, K., Bignami, G., Hermsen, W., Kanbach, G., Lebrun, F., Masnou, J., Mayer-Hasselwander, H., Sacco, B., Strong, A., and Wills, R., Paper XG4.3-1, 17th ICRC, Paris, France, July, 1981
15. Chevalier, C., Ilovaisky, S., Motch, Ch., Barat, C., Hurley, K., Niel, M., Vedrenne, G., Laros, J., Evans, W., Fenimore, E., Klebesadel, R., Estulin, I., and Zenchenko, V., Accepted for publication, Astron. and Astrop. Lett., 1981
16. Laros, J., Evans, D., Fenimore, E., Klebesadel, R., Barat, C., Hurley, K., Niel, M., Vedrenne, G.,Estulin, I., Zenchenko, V., and Mersov, G., Ap. J. Lett., (in press), 1981

17. Tyson, J., and Jarvis, J., Ap. J. Lett., **230**, L153, 1979
18. Helfand, D., Chanan, G., and Novick, R., Nature, **283**, 337, 1980
19. Seward, F., HEAO Science Symposium, NASA CP-2113, 368, 1979
20. Mazets, E., Golenetsky, S., Ilinsky, V., Panov, V., Aptekar, R., Gurjan, Yu., Sokolov, I., Sokolova, Z., and Kharitonova, T., Ioffe Physico-Technical Institute Preprint 618, 1979
21. Mazets, E., Golenetsky, S., Ilinsky, V., Panov, V., Aptekar, R., Gurjan, Yu., Sokolov, I., Sokolova, Z., and Kharitonova, T., Ioffe Physico-Technical Institute Preprint 637, 1979
22. Cline, T., Desai, U., and Teegarden, B., Ap. and Space Sci., **75**, 93, 1981
23. Teegarden, B., and Cline, T., Ap. and Space Sci., **75**, 181, 1981
24. Klebesadel, R., Evans, D., and Laros, J., Ap. and Space Sci., **75**, 5, 1981
25. Evans, D., Fenimore, E., Klebesadel, R., Laros, J., and Terrell, N., Ap. and Space Sci., **75**, 35, 1981
26. Chambon, G., Hurley, K., Niel, M., Vedrenne, G., Zenchenko, V., Kuznetsov, A., and Estulin, I., Space Sci. Inst., **5**, 73, 1979
27. Barat, C., Chambon, G., Hurley, K., Niel, M., Vedrenne, G., Estulin, I., Kuznetsov, A., and Zenchenko, V., Space Sci. Inst., (in press), 1981

STATISTICAL STUDY OF GAMMA-RAY BURST SOURCE LOCATIONS

Graziella Pizzichini
Istituto TESRE/CNR
Via Castagnoli 1, 40127 Bologna Italy

ABSTRACT

Some tests have been applied to a catalog of 65 Gamma-Ray Burst source locations, in order to find if their distribution in galactic coordinates should be considered isotropic. The statistical significance of the overlapping of some error boxes is also discussed.

INTRODUCTION

Many gamma-ray burst source locations are now available, but the sources of the events have not yet been identified. The only possible exception is the supernova remnant N49 in the Large Magellanic Cloud, which could be associated with the unusual event of March 5, 1979[1]

It is known that the distribution of source locations on the celestial sphere does not show any "outstanding" properties, such as a concentration on the galactic plane or towards the galactic center; therefore, no definitive answers should be expected from a statistical study of source locations. In the present work such a study has, nonetheless, been attempted in order to find possible clues in favor of or against a galactic origin of the events. We have also investigated the possibility that some events, whose locations have overlapping error boxes, are due to the same source. Two sources which produced several bursts are listed by Mazets and Golenetskii[2]: the source of the March 5, 1979 event, with a total of 4 events, and B1900+14, with 3 events.

PROCEDURE ADOPTED

In order to avoid the selection effects due to the different sensitivities of different detectors, we have used only the catalogs of events detected by the KONUS experiment on Venera-11 and -12[2,3,4]. The catalogs give locations and error boxes for 60 sources, corresponding to 65 events, since, as we already mentioned, two of the sources produced 4 and 3 bursts, respectively. The tests, almost always a χ^2 test, have been applied both to "source" counts and to "event" counts. The results are given in Tables I-VI

CONCLUSIONS

In many cases the χ^2 obtained is high, corresponding to a low probability. This is due essentially to the excess of events with

Table I Distribution Above and Below the Galactic Plane

ℓ^{II}_{min}	ℓ^{II}_{max}	b^{II}_{min}	b^{II}_{max}	No. of	No. of
degrees		degrees		events	sources
0	360	0	90	22	20
0	360	-90	0	43	40
	Total			65	60
	χ^2			6.8	6.7
	$P_1(\chi^2)$			0.010	0.010

Note. If we compute $\sum_{r=22}^{65} = \frac{65!}{r!\,(65-r)!\,2^{65}}$, which is the probability of finding at least 22 events in one hemisphere, if the distribution is isotropic, we find that this probability of 0.0065.

Table II Distribution in Galactic Latitude

ℓ^{II}_{min}	ℓ^{II}_{max}	b^{II}_{min}	b^{II}_{max}	No. of	No. of
degrees		degrees		events	sources
0	360	-90	-50	6	6
0	360	-50	-30	13	10
0	360	-30	-10	19	19
0	360	-10	10	11	9
0	360	10	30	6	6
0	360	30	50	5	5
0	360	50	90	5	5
	Total			65	60
	χ^2			13.6	12.7
	$P_6(\chi^2)$			0.04	0.05

Table III Distribution in Galactic Longitude of Sources with $|b^{II}|<60°$

ℓ^{II}min ℓ^{II}max degrees	b^{II}min b^{II}max degrees	No. of events	No. of sources
-45 45	-60 60	11	11
45 135	-60 60	15	13
135 225	-60 60	11	11
225 315	-60 60	18	15
Total		55	50
χ^2		2.53	0.2
$P_3(\chi^2)$		0.5	0.98

Table IV Distribution in Galactic Longitude Above and Below the Galactic Plane with $|b^{II}|<60°$

ℓ^{II}min ℓ^{II}max degrees	b^{II}min b^{II}max degrees	No. of events	No. of sources
-45 +45	0 60	5	5
45 135	0 60	8	6
135 225	0 60	2	2
225 315	0 60	3	3
-45 45	-60 0	6	6
45 135	-60 0	7	7
135 225	-60 0	9	9
225 315	-60 0	15	12
Total		55	50
χ^2		16.7	11.4
$P_7(\chi^2)$		0.02	0.08

Table V Distribution in Galactic Longitude Above and Below the Galactic Plane for all Latitudes

ℓ^{II}_{min}	ℓ^{II}_{max}	b^{II}_{min}	b^{II}_{max}	No. of	No. of
degrees		degrees		events	sources
-45	45	0	30	5	5
45	135	0	30	8	6
135	225	0	30	2	6
2225	315	0	30	7	7
-45	45	-90	0	9	9
45	135	-90	0	8	8
135	225	-90	0	11	11
225	315	-90	0	15	12
Total				65	60
χ^2				12.9	11.4
$P_7(\chi^2)$				0.04	0.08

Table VI Locations Grouped Into 6 Equal Area Regions with Centers at the Positrons Indicated

Center of Region		No. of Events	No. of Sources
ℓ^{II}	b^{II}		
degrees			
0	0	10	10
180	0	9	9
90	0	13	11
270	0	16	13
0	+90	7	7
0	-90	10	10
Total		65	60
χ^2		4.7	2.0
$P_5(\chi^2)$		0.46	0.8

$b^{II} \simeq -15°$, as already noted by Vedrenne[5], and $\ell" \simeq 270°$. "Source" counts give consistently smaller χ^2 than "event" counts, in agreement with the fact that repetition of bursts from the same source should produce a less isotropic distribution of "events".

This result favors a nearby galactic origin for most gamma-ray bursts but does not exclude the possibility that some of them, for example the March 5, 1979 event, may be of a different nature.

Several events have source locations with overlapping error boxes, but the probability that this will happen for 65 events with error boxes of several square degrees each is extremely high. There is one case of 3 error boxes, those of the events of November 24, 1978, January 1, 1979, and January 16, 1979, nested inside the 16° x 16° error box of the May 2, 1979 event. The probability that 3 or more locations out of 64 will fall into this error box is 14 percent. For the events of December 29, 1978, and February 15, 1979, the probability that at least one more location will fall into the larger error box is 3 percent; therefore, the probability that these events were not produced by the same source is small but not negligible.

The author thanks Dr. U. D. Desai for suggesting this work.

REFERENCES

1. W. D. Evans et al., 1980, Ap. J. (Letters) 237, L7.
2. E. P. Mazets and S. V. Golenetskii, 1979, Preprint Akademia Nauk, USSR n. 632.
3. E. P. Mazets et al., 1980, Preprint Akademia Nauk, USSR n. 662.
4. E. P. Mazets et al., 1981, Preprint Akademia Nauk, USSR n. 712.
5. G. Vedrenne, 1981, Phil. Trans. R. Soc. Lond. A 301, 645.

THE OBSERVATIONAL CONSEQUENCES OF AN INTRINSIC BURST LUMINOSITY DISTRIBUTION

Mark C. Jennings
Institute of Geophysics and Planetary Physics
University of California, Riverside, Ca. 92521

ABSTRACT

The Galactic monoluminostiy gamma burst models of Jennings and White[1] are generalized to include intrinsic burst luminosity distributions. LogN-LogS is substantially affected by distributions moderately peaked toward low luminosity and of dynamic range $\gtrsim 10^{-2}$. Spherical or "Halo" distributions, which are unacceptable for monoluminosity sources, are, along with Disk distributions, compatible with LogN-LogS observations upon introduction of an intrinsic variation in burst luminosity. Comparing models and data indicates maximum burst luminosities of $4 \times 10^{41} \lesssim L_2 \lesssim 3 \times 10^{43}$ ergs and burst rate densities of $1 \times 10^{-10} \lesssim \eta_o \lesssim 2 \times 10^{-10} pc^{-3}-yr^{-1}$ for Halo models, and $4 \times 10^{39} \lesssim L_2 \lesssim 8 \times 10^{40}$ ergs and $1 \times 10^{-7} \lesssim \eta_o \lesssim 2 \times 10^{-7} pc^{-3}-yr^{-1}$ for disk models.

The generalized models are used to predict source angular distributions versus LogS. It is shown that neither Disk nor Halo geometries can reconcile the present LogN-LogS and angular position observations. The significance of this is discussed.

INTRODUCTION

Assuming monoluminosity--"standard candle"--bursts Jennings and White[1] (JW) have shown that spherical or "Halo" source distributions, Disk distributions of small scale height, i.e., $\beta \gtrsim 200pc$, and distributions with strong central concentration, i.e., radial scale length, $\rho_o \gtrsim 4-5kpc$, are unable to describe the LogN-LogS observations.

However, Disk models of larger scale height do reproduce the observations. Although differing on details, Fishman[2] independently arrived at similar conclusions regarding Disk models.

A fundamental prediction of Disk models is source concentration to the Galactic plane. Yet data on burst positions by Mazets and Golenetskii[3], Mazets et al.[4], Cline et al.[5], and Hurley[6] and more recently by Mazets et al.[7] have failed to reveal planeward concentration. This problem, and the observation of bursts of differing fluence from the same apparent source (Mazets and Golenetskii[3]), have indicated the need to both relax the monoluminosity requirement and clarify the relation between LogN-LogS and angular distribution.

GALACTIC DISTRIBUTION MODELS
LogN(>S) vs LogS

The JW model is generalized by introducing a burst rate density luminosity spectrum of the form:

0094-243X/82/770107-07$3.00 Copyright 1982 American Institute of Physics

$$\mathcal{N}(L) = \frac{\alpha \mathcal{N}_0}{L_2^\alpha (1-\zeta^\alpha)} L^{\alpha-1} \text{ pc}^{-3}\text{-yr}^{-1}\text{-erg}^{-1}, \quad L_1 \leq L \leq L_2, \quad \zeta \equiv L_1/L_2 \quad (1)$$

L is the integrated burst luminosity. α determines the distribution shape and ζ controls its dynamic range.

While intrinsic luminosity distributions affect Disk model LogN-LogS, the basic conclusions are the same as those for monoluminosity models, i.e., the data are adequately reproduced by distributions of larger scale height. Because of this the Disk LogN-LogS results will not be discussed in detail here but will be presented elsewhere.

It is in the case of Halo distributions that an intrinsic luminosity distribution has its most profound effect. Figure 1 presents the behavior of LogN-LogS as a function of distribution shape α. For convenience fluence is expressed in units of a maximum luminosity burst at the near edge of the Galaxy, i.e.:

$$\tilde{S} \equiv S/(\frac{9L_2}{4\pi R^2}), \quad R \equiv \text{Galaxy Radius} \quad (2)$$

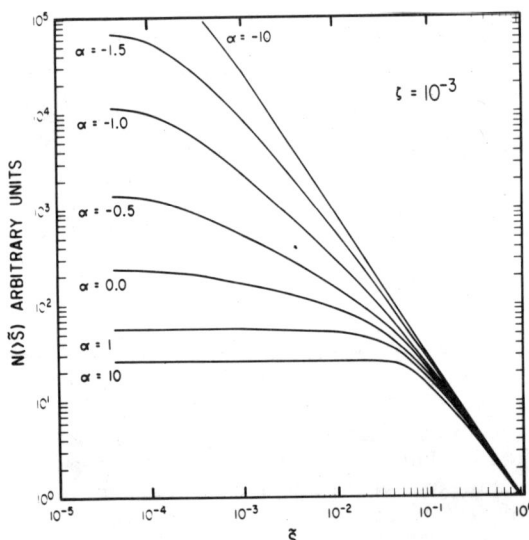

In these units an intrinsic luminosity distribution manifests itself only when $\tilde{S}<1$. For $\tilde{S}>1$ dLogN/dLog$\tilde{S}$=-3/2 independent of the luminosity distribution. Most α values, i.e., ≳-1.5 or ≳0.0 produce number spectra equivalent to monoluminosity models of respective luminosity L_1 or L_2 (cf. (JW) Fig. 5). However, over the limited range -1.5≳α≳0.0 LogN-Log$\tilde{S}$ becomes sensitive to the luminosity distribution, exhibiting a slope approximating α for ζ≳$\tilde{S}$ ≳0.04.

Fig. 1. The effect on LogN-LogS of varying the intrinsic luminosity distribution shape.

The effect of varying the distribution's dynamic range is shown in Figure 2. Here one sees that ζ controls the minimum observable fluence and the $\tilde{S}$ interval over which dLogN/dLog$\tilde{S}$∼α. The influence of α and ζ on LogN vs Log$\tilde{S}$ arises from the interplay of the luminosity distribution and underlying geometry. This interaction may be briefly summarized. As $\tilde{S}$ decreases below one, LogN-Log$\tilde{S}$ enters an interval, (0.04≲$\tilde{S}$<1), of declining geometrical contribution and downward divergence from $\tilde{S}^{-3/2}$. This is followed by a luminosity distribution dominated interval, (ζ≲$\tilde{S}$≲.04), of slope ∼α,

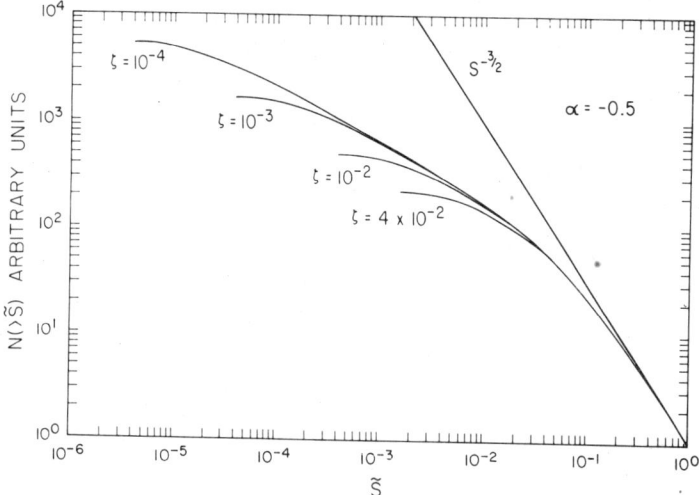

Fig. 2. The effect of LogN-LogS of varying the intrinsic luminosity distribution dynamic range.

and finally by an interval $(.04\zeta \leq \tilde{S} \leq \zeta)$, of geometry contribution resurgence where the diminishing availability of additional bursts is unable to sustain LogN growth.

From Figures 1 and 2 it is concluded that intrinsic luminosity distributions of sufficient dynamic range, i.e., $\zeta \lesssim 10^{-2}$, and appropriate shape ($-1.5 \lesssim \alpha \lesssim 0.0$) can significantly alter the form of Halo model LogN-LogS from that produced by monoluminosity bursts. A similar conclusion holds for Disk models. Further, it is noted that all acceptable distributions are "soft", i.e., peaked toward low luminosity. The form assumed for the luminosity spectrum (equation (1)) is quite general, yet there is no a priori evidence excluding other possible forms. However, it can be shown that the range and softness characteristics are a fundamental requirement of any potential distribution. The above conclusions differ markedly from those of Yoshimori[8], who concluded that LogN-LogS is unaffected by luminosity distributions, and Fishman's[2] findings that soft power law distributions are unacceptable and that if a distribution exists the majority of bursts have a characteristic energy near the mean of a narrow distribution or near the high energy cutoff of a hard distribution. The reasons for these discrepancies have been identified but a full discussion is beyond the scope of this paper.

The softness of acceptable luminosity distributions has interesting observational consequences. It explains why burst repetition --undetected by early, lower sensitivity experiments--has been reported in higher sensitivity, long time-base observations (Mazets and Golenetskii)[3]. New importance is also placed on low S observations since these are not simply geometrically diminished standard events but comprise the bulk of the phenomenon.

Halo models have been compared with the data using the technique described in JW[1]. This procedure produces two conserved quantities,

$$\mathcal{N}_o L_2^{-\alpha} = \text{const.}, \text{ and } \mathcal{N}_o L_2^{-3/2} = \text{const.}, \tag{3}$$

which may be used to uniquely determine $\mathcal{N}_o$ and L_2 as a function of α and the LogN-LogS observations. For Halo models ζ is in most cases determined by the need to reproduce the faintest observed bursts and/ or satisfy certain observational LogN upper limits. The results of the model-data comparison are presented in Table I and Figure 3.

Table I Summary of Halo Model Parameters

α	ζ	L_2(ergs)	$\mathcal{N}_o \text{pc}^{-3}\text{-yr}^{-1}$	$S(\tilde{S}=1)\text{erg-cm}^{-2}$
-1.5	2×10^{-4}	3×10^{43}	1.1×10^{-10}	10^{-2}
-1.0	10^{-3}	2×10^{42}	1.4×10^{-10}	6.3×10^{-4}
-0.5	10^{-3}	4×10^{41}	2.3×10^{-10}	1.2×10^{-4}
0.0	-------------INFERIOR FIT--------------------------			

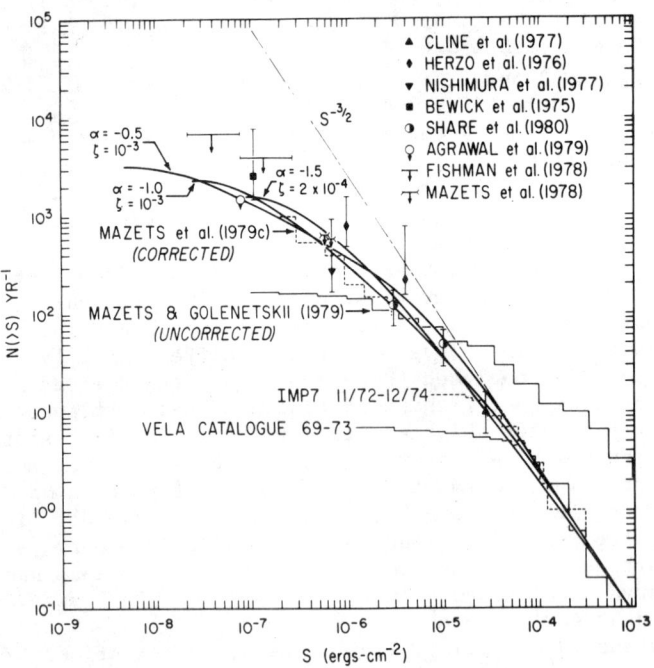

Fig. 3. Comparison of adopted intrinsic luminosity distribution models with LogN-LogS observations.

THE ANGULAR POSITION OF BURSTS

In addition to explaining LogN-LogS, an acceptable model must reproduce the angular distribution data which has been rapidly increasing in quantity and quoted accuracy due to the higher sensitivity experiments of the interplanetary network. For a Galactic distribution of sources one expects an approximately isotropic distribution when $S > \frac{9L_2}{4\pi R^2}$ for Halo populations, and $S > \frac{L_2}{4\pi \beta^2}$ for Disk populations. At lower S anisotropy should appear, with latitude (b^{II}) being the most sensitive coordinate for Disk populations and longitude (ℓ^{II}) being most sensitive for the Halo. In view of this the generalized models have been used to determine $<\sin|b^{II}|>$ vs S for Disk models and the ratio of bursts in the anti-center hemisphere (AC) to the number of bursts in the center hemisphere (C) vs S for Halo models. This permits direct comparison of the LogN-LogS data with the properties of the angular distribution.

The results of these calculations are shown in Figure 4, which juxtaposes LogN-LogS with the Disk and Halo anisotropy indicators as a function of S. The line at $<\sin|b^{II}|> = .465$ in Figure 4b represents Mazets et al.'s[7] compilation of 69 events. The 1σ error bar assumes a conservative ±10° b^{II} error per position, substantially larger than the 3-5° total error estimated by Mazets and Golenetskii[3]. The line at $N(>S)_{AC}/N(>S)_C = 1.035$ in Figure 4c is the anti-center/center ratio for Mazets et al's[7] data. Ignoring events from Hurley[6], which are not a homogeneous body of data, Mazets et al's[7] compilation contains 59 Venera -11 and -12 positions obtained during 385 days. The most favorable view of these events assumes they are a complete sample of the interval with the largest possible limiting S. $N(>S)=56$ yr^{-1} corresponds to a limiting S of 8×10^{-6} erg-cm^{-2} which is indicated by the verticle line. Figures 4b and c indicate that significant anisotropy should be seen at this S--a fact not evidenced by the observations. If the 59 events are an incomplete sample to their limiting S--as they clearly are--the discrepancy is exacerbated, since on average fainter events show more anisotropy. The prediction of detectable anisotropy appears to present a fundamental barrier to the reconciliation of the present LogN-LogS and positional data if LogN-LogS reflects Galactic geometry. Specifically excluded are suggestions (Mazets et al.[7], JW[1]) that the angular data presents a view of only nearby members, i.e., $r<\beta$, of a Disk population. This interpretation is viable only if the angular position events populate the $S^{-3/2}$ domain of LogN-LogS; however, by $S=8\times 10^{-6}$ erg-cm^{-2} the observed relation has begun to diverge downward.

Several ways of reconciling the data suggest themselves. The observations may contain systematic errors. It is noted that the angular results depend critically on Venera -11 and -12 data. This is troublesome in view of the continuing high S disagreement between the Venera LogN-LogS relation and that of other observers. Independent corroboration of the Venera results is clearly needed. The concept of LogN-LogS as a reflection of Galactic geometry may be

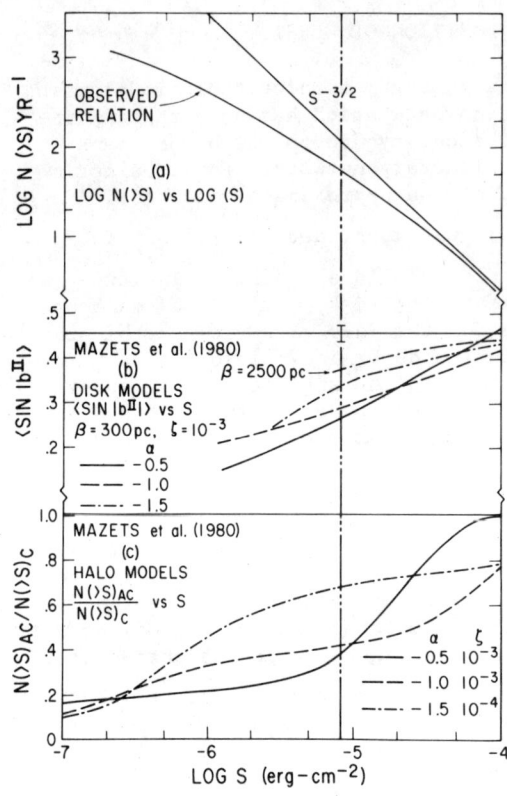

Fig. 4. Comparison of (a) LogN-LogS and anisotropy indicators, (b) $\langle\sin|b^{II}|\rangle$ vs S for Disk models, and, (c) $N(>S)_{AC}|N(>S)_C$ vs S for Halo models.

incorrect. The LogN-LogS turnover could result from the Hubble flow[9] if sources are distributed on a Universal scale, or could arise from nearby sources having an appropriate decrease in density with heliocentric distance. A near intergalactic explanation is ruled out unless sources are anticorrelated with visible mass concentrations (JW[1]). Finally, LogN-LogS may contain little or no geometrical information. For example, it may directly describe the intrinsic burst luminosity distribution, or would be compatible with a relatively thin heliocentric shell of sources with a soft burst luminosity spectrum.

Although interpretations of varying plausibility are available, verification of the current observational results would provide invaluable guidance on the distance scale of gamma-ray bursts.

REFERENCES

1. M. C. Jennings and R. S. White, Astrophys. J. <u>233</u>, 110 (1980), (JW).
2. G. J. Fishman, Astrophys. J. <u>233</u>, 851 (1979).
3. E. P. Mazets and S. V. Golenetskii, Ioffe preprint <u>632</u> (1979).
4. E. P. Mazets, S. V. Golenetskii, V. N. Ilinskii, V. N. Panov, R. L. Aptekar, Yu. A. Guryan, I. A. Sokolov, Z. Ya. Sokolova, and T. V. Kharitonova, Ioffe preprint <u>610</u> (1979).
5. T. L. Cline, U. D. Desai, G. Pizzichini, A. Spizzichino, J. H. Trainor, R. W. Klebesadel, and H. Helmken, Astrophys. J. Lett. <u>232</u>, L1 (1979).
6. K. Hurley, Toulouse Symp. on Gamma-Ray Sources (1979).
7. E. P. Mazets, S. N. Golenetskii, R. L. Aptekar, Yu. A. Guryan, and V. N. Ilinskii, Ioffe preprint, <u>686</u> (1980).

8. M. Yoshimori, Aust. J. Phys. 31, 189 (1978).
9. M. Ruderman, Ann. N.Y. Acad. Sci. 262, 164 (1975).

DERIVATION OF THE LUMINOSITY OF A GAMMA RAY BURST

J. C. Higdon
Steward Observatory, Univ. of Arizona, Tucson AZ 85721

ABSTRACT

The intrinsic luminosity, erg/s, in a gamma ray burst is calculated using two assumptions: first, the power is independent of the burst duration, and, second, the spatial distribution of the bursters is similar to the spatial distribution of the pulsars. It is found that the experimentally determined distribution of burst durations is significantly biased by the observational selection effects. It is shown that a typical luminosity of 5×10^{35} erg/s can adequately reproduce the burst (log N, log S) curve, an isotropic latitude distribution, and the experimentally determined distribution of durations.

INTRODUCTION

The observation of absorption features at 30 to 70 keV and broad emission lines at ~400 keV in the spectra of gamma ray bursters is strong evidence that these bursters are neutron stars.[1] If the bursters are neutron stars, their spatial distribution can be assumed to follow the distribution of radio pulsars, which are tracers[2] of the young ($<10^8$ yr) neutron star population.

The power of the bursters is assumed to be independent of duration primarily because, with a data base of $\approx 10^2$ bursts, it is statistically meaningless to separate the bursts into classes. The use of duration in the derivation of the luminosity of a burster follows from Mazets et al.[3], who suggested that the bulk of the range of fluence of $\sim 10^{-7}$ to 10^{-3} erg/cm^2 results primarily from variation in burst duration, and thus the spread in burst luminosity, erg/s, is much smaller than the range of fluence.

OBSERVATIONAL SELECTION EFFECTS

Consideration of the experimental selection effects is essential to the derivation of the intrinsic luminosity of the bursts. Studies of the experimental selection effects must be done because the observations of the Konus experiments, aboard Venera 11 and Venera 12[4] are the primary source of the burst positions in the sky and of the duration distribution in this work. At medium fluences $\approx 10^{-5}$ erg/cm^2 and low fluences $\sim 10^{-6}$ erg/cm^2 the Konus experiment undercounts burst events.[3] Figure 1 shows the (log N, log S) curve derived by the author from the Konus observations[4] with well-determined positions during September 1978 to February 1980. Multiple reoccurrence events were not considered. The experimentally determined curve is significantly flatter than S^{-1}, between 10^{-5}

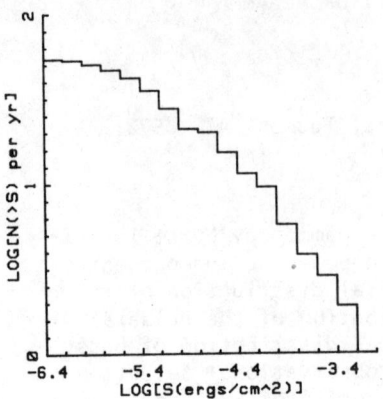

Fig. 1. The cumulative (log N, log S) distribution determined for the bursts with well-determined positions from the observations of the Konus experiment.[4]

and 5×10^{-7} erg/cm^2 the (log N, log S) curve of Figure 1 has a slope of ≈ 0.3

Following Mazets et al.,[3,5] the effect of experimental selection is considered here by comparison of the experimentally determined (log N, log S) curve, i.e., Figure 1, with the numerically calculated model for (log N, log S). The difference between these curves, as a function of fluence, is considered to be the probability for the Konus experiment to measure a position of a burst, or its duration.

The experimentally determined distribution for the burst duration is also biased by the selection effects. Since the undercounting of the Konus experiment depends on the fluence, erg/cm^2, of the burst, the shorter bursts are more likely to be undercounted. The actual duration distribution will contain a greater fraction of short duration bursts when compared to the measured distribution. The measured duration distribution will be considered as a group of coupled variables which the numerical model must reproduce; the actual distribution will be treated as a group of free parameters. To decrease the number of free parameters in the model, I will model only bursts with durations up to 70s in 5s intervals. Eighty-three of the 85 bursts with tabulated durations possess durations shorter than 70s.[4]

GALACTIC NEUTRON STAR DISTRIBUTION

The galactic neutron star distribution calculation employs the galactic distribution of radio pulsars by Manchester.[6] Figure 2 shows the radial distribution[6] of pulsars as a function of galactic radius, assuming that the Earth is 10 kpc from the galactic center; the vertical lines represent the statistical error bars of Manchester's fit. This analysis assumed that the pulsar distribution has cylindrical symmetry, i.e., there are no spiral arms. Manchester[6] found that the pulsar distribution perpendicular to the galactic plane can be adequately modeled by an exponential distribution with a scale height of 0.35 kpc. As can be seen from Figure 2, the pulsar distribution within a kpc of the Earth can be considered uniform with respect to the galactic radius.

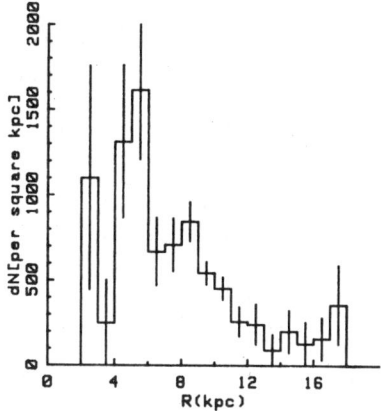

Fig. 2. The pulsar radial distribution determined by Manchester.[6] The vertical lines represent the statistical error bars of this model fit.

NUMERICAL CALCULATIONS AND COMPARISON WITH OBSERVATIONS

The numerical calculations are found from a Monte Carlo approach. This method allows comparison between fluctuations in the modeled distributions produced by the dispersion of the ensemble of the separate test runs, and the fluctuations in the observed distributions produced by small number counting statistics. The burst power, L_o, and the burst duration distribution are considered to be the parameters to be found by comparison with the observations.

The spatial distribution of bursters is assumed to be the same as that of the radio pulsars. The cumulative probability distribution of the bursters as a function of galactic radius was derived from Figure 2. The cumulative height distribution was based on $\exp(-|z|/z_o)$ with $z_o = .35$ kpc. The probability distribution of the galactic angle coordinate was considered to be uniform between 0° and 360°. A variety of different duration distributions were tested.

The lack of any significant anisotropy in the arrival directions of the observed bursts[3] limited the intrinsic power of the models to be less than $\simeq 10^{36}$ erg/s. The calculation with $L_o = 10^{36}$ erg/s produced a modeled cumulative latitude distribution with a mode of 22°; the observed latitude distribution possesses a mode of 30°. A lower limit on the burst power can be found from the shape of the (logN, log S) curve. If the burst power is significantly less than 2×10^{35} erg/s, the slope for fluences greater than 4×10^{-7} erg/cm^2 is -1.5. Prior analyses[7] of burst statistics found that in the region of $\sim 10^{-6}$ erg/cm^2 the slope of (log N, log S) is smaller than -1.5.

The following presentation of modeled curves is based on the intrinsic burst power of 5×10^{35} erg/s. With $S_{min} = 4 \times 10^{-7}$ erg/cm^2, the threshold of the Konus experiment, a maximum distance of 857 pc is found for $T_{max} = 70$s. Since the pulsar distribution can adequately be modeled as being uniform within a kpc of the Earth, the calculations for $L_o - 5 \times 10^{35}$ erg/s are based on an Earth-centered coordinate system. This system allows some important computational simplifications.

Figure 3 shows the modeled (log N, log S) curve based on $L_o = 5 \times 10^{35}$ erg/s. This calculation is based on an ensemble of 15 runs of 5×10^4 test bursts. The vertical lines represent the dispersion among the 15 runs. The open rectangles represent the error bars of the observations. The fluence, S, erg/cm^2, and cumulative counts, N(>s) yr^{-1}, of the observations shown in Figure 3 are listed in Table 1. A total of 97911 test bursts contribute to Figure 3. Although the dispersion of the ensemble smooths out the detailed shape of the (log N, log S) curve, as S increases from the threshold of 4×10^{-7} erg/cm^2 to 4×10^{-4} erg/cm^2, the slope of the curve decreases from -1.3 to -1.5.

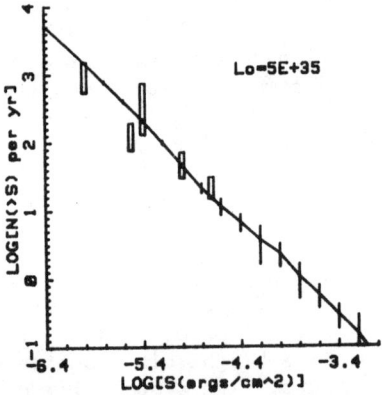

Fig. 3. The modeled (log N, log S) distribution calculated for a luminosity of 5×10^{35} erg/s and the duration distribution shown in Fig. 4. The vertical lines represent the dispersion of an ensemble of 15 runs of 5×10^4 test bursts. The rectangles represent the statistical error bars of the observations listed in Table 1.

TABLE 1

S(erg/cm²)	N(>S) yr⁻¹	reference
1.0E-06	850(+650,-300)	White et al.[8]
3.0E-06	112(+78,-35)	Share et al.[9]
4.0E-06	230(+500,-100)	White et al.[8]
1.0E-05	50(+20,-20)	Share et al.[9]
2.0E-05	22.5(+7.5,-7.5)	Fishman[7]

Following Mazets et al.,[3,5] the difference between the modeled (log N, log S), Figure 3, and the experimentally determined (log N, log S), Figure 1, is assumed to result from the selection effects. This correction factor is applied to Figure 3 before the modeled duration and latitude distributions are found. The correction factor is treated as a probability dependent on the fluence, S. This probability distribution dominates the determination of the modeled latitude and duration distributions; therefore, only 2455 of 97911 test bursts of Figure 3 contribute to the derivations of the latitude and duration distributions.

Figure 4 shows the modeled interstellar distribution as a dashed line, and the expected experimental distribution as a solid line.

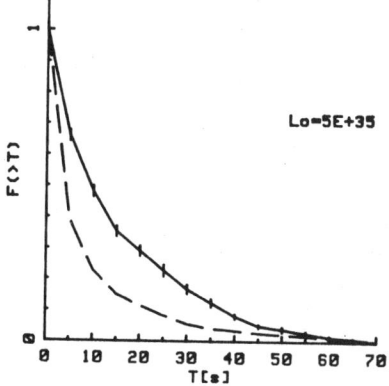

Fig. 4. The actual duration distribution is shown as a dotted line, and the modeled experimental distribution as a solid line. The vertical lines represent the dispersion of the ensemble of 15 runs.

Figure 5 shows the expected experimental distribution and the experimental distribution based on the tabulated durations of the Konus experiment,[4] which is shown as open rectangles. The vertical lines

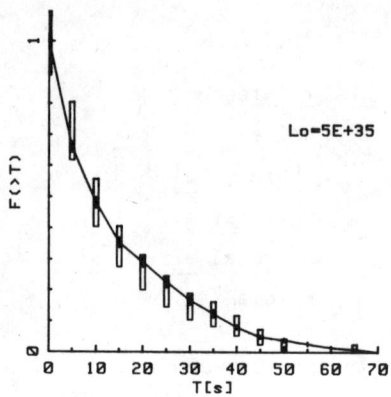

Fig. 5. Comparison of the expected duration distribution and observed distribution.[4] The vertical lines represent the dispersion of an ensemble of 15 runs. The rectangles represent the statistical bars of the observations.

represent the dispersion of the ensemble of test runs. As can be easily seen from Figure 4, the actual distribution is significantly different from the experimentally determined distribution. This difference results from the fact that the fluence thresholds bias the experiments against the detection of weak fluence bursts of short duration. This effect of dependence on the fluence, S, increases the mode of the true duration distribution of 4s to 10s for the mode of the experimentally determined distribution.[4]

In Figure 6 the modeled latitude distribution is compared with burst observations. The cumulative latitude distribution of observed bursts based on 68 events was determined from the results of the Konus experiment,[4] excluding the two locations with repetitive bursts and including the 10 locations from the compilation of Hurley.[10] The vertical lines represent the dispersion of the ensemble of 15 separate runs; the open rectangles represent the statistical error bars of the observations. The mode of the modeled cumulative distribution is at a latitude of 28°.

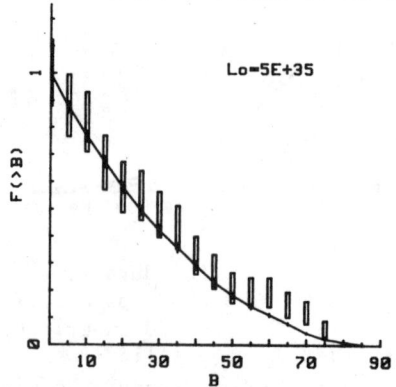

Fig. 6. Comparison of modeled latitude and observed distribution.[4,10] The vertical lines represent the dispersion of an ensemble of 15 runs.

As can be seen, the model and the observations agree approximately. The observations are higher than the model distribution and an isotropic distribution between 60° to 70°. The observed distribution, $N(>60°)$, is based on only 10 bursts; the author assumes that this excess results from a statistical fluctuaion. The average fluence of the test bursts is 1.56×10^{-5} erg/cm^2; this figure compares well with $<S>$ of 4.6×10^{-5} erg/cm^2 for 58 bursts with well-determined locations in the Konus experiment.[4]

CONCLUSION

Within the framework of the two assumptions made here, the work presented suggests that it is plausible to represent the intrinsic burst luminosity of a gamma ray burst by a single variable, and that the bulk of the range of observed fluence is produced by the duration distribution. The detailed reproduction of the observations depends sensitively on how the observational selection effects of the Konus experiment are derived. Because the author does not understand the selection effects of the Konus experiment, he suggests that the introduction of more parameters, e.g., a distribution of intrinsic power, is not warranted to improve the agreement between his model and the observations.

A perhaps surprising result of the above calculation is the large difference between the measured duration distribution and the actual distribution. This suggests that the observations are significantly biased against recording short duration bursts. As long as the burst power is not inversely correlated with the duration, the difference between the modeled and the measured durations is insensitive to the source spatial distribution.

The work of the author was supported in part by NASA under Grant NSG-7101.

REFERENCES

1. E. P. Mazets, S. V. Golenetskii, R. L. Aptekar, Yu. A. Guryan, and V. N. Illinskii, Nature, 290, 378 (1981).
2. R. N. Manchester, and J. H. Taylor, Pulsars (W. H. Freeman, San Francisco, 1977), p. 162.
3. E. P. Mazets, S. V. Golenetskii, R. L. Aptekar, Yu. A. Guryan, and V. N. Illinski, Sov. Astron. Lett., 6, 318 (1980).
4. E. P. Mazets, S. V. Golenetskii, V. N. Illinskii, V. N. Panov, R. L. Aptekar, Yu. A. Guryan, A. V. D. Dyachkov, M. P. Proskura, I. A. Sokolov, Z. Ya Sokolova, N. G. Khavenson, and T. V. Kharitonova, Preprint Fiz. Tech. Inst. Ioffe Akad. Nauk SSSR No. 662 (1980); E. P. Mazets et al., Preprint Fiz. Tech. Inst. Ioffe Akad. Nauk SSSR No. 712 (1980); E. P. Mazets and S. V. Golenetskii, Astro. Space Sci., 75, 47 (1981)
5. E. P. Mazets, S. V. Golenetskii, R. L. Aptekar, V. N. Illinskii, and V. N. Panov, Sov. Astron. Lett., 4, 188 (1979).
6. R. N. Manchester, Aus. J. Phys., 32, 1 (1979.

7. G. J. Fishman, Ap. J., 238, 851 (1979).
8. R. S. White, J. L. Long, M. C. Jennings, and B. Dayton, 16th Cosmic Ray Conference, Kyoto, Japan, vol. 12, 41 (1979).
9. G. H. Share, D. J. Yentis, W. D. Evans, K. Wood, and J. Meekins, Bull. Am. Phys. Soc., 25, 527 (1980).
10. K. Hurley, Non-Solar Gamma Rays (Proc. COSPAR Meeting, Bangalore, Adv. Space Exploration 7 (Pergamon, Oxford, 1980), p. 123.

Sec. I C Gamma Ray Transient Energy Spectra

GAMMA-RAY BURST SPECTRA

B. J. Teegarden
NASA/Goddard Space Flight Center
Greenbelt, MD 20771

ABSTRACT

A review of recent results in gamma-ray burst spectroscopy is given. Particular attention is paid to the recent discovery of emission and absorption features in the burst spectra[10,11,13,14]. These lines represent the strongest evidence to date that gamma-ray bursts originate on or near neutron stars. Line parameters give information on the temperature, magnetic field and possibly the gravitational potential of the neutron star. The behavior of the continuum spectrum is also discussed. A remarkably good fit to nearly all bursts is obtained with a thermal-bremsstrahlung-like continuum. Significant evolution is observed of both the continuum and line features within most events.

INTRODUCTION

This paper is intended as a review of the measurements to date of the spectral characteristics of gamma-ray bursts (GRB). It will also cover to a certain extent the theoretical interpretation of these results where it relates in a direct way to the understanding of the observations. Within recent years GRB spectral measurements have begun to reveal a very rich phenomenology and to make a major contribution to the understanding of the origin and physics of GRB's.

Early measurements[1-8] were made with instruments designed for some other purpose than the detection of gamma-ray bursts. With the launching of instruments specifically designed to study the spectral characteristics of GRB's[9-14] a great variety of new observations have significantly extended our knowledge of the spectral behavior of GRB's. In particular the work of Mazets and his colleagues at the A.F. Ioffe Physical-Technical Institute[9-12], Leningrad, represents the most comprehensive study of GRB spectra presently available. Their data have among other things revealed the repeated presence of absorption and emission lines in GRB spectra. "Cyclotron" absorption features are very probably present in a major fraction of the events measured[10]. In addition, lines at 400-450 keV are seen in a smaller number of events that are probably from redshifted annihilation radiation[10]. Taken together these results are strong, if not compelling, evidence for the neutron star origin of GRB's.

In this paper the early spectral results are first briefly reviewed. Then the behavior of the GRB continuum spectrum, its form, variability, and evolution, is treated. Finally the absorption and emission features, their relationship to other spectral parameters and physical interpretation are discussed.

HISTORICAL OVERVIEW

Early results on GRB spectra[1-8,15] generally suffered from a number of problems; e.g., poor statistics due to small detector area, inadequate time resolution and variable intensity modulation due to spacecraft rotation. The first published GRB spectra[1] were derived from a small instrument on the Imp-6 spacecraft intended to measure solar flare X-rays and interplanetary positrons. Spectra from six events were reported covering the energy range from 0.1-1.2 MeV. The spectral form for all events was consistent with a simple exponential form $\exp(-E/E_0)$ with E_0 = 150 keV. In a subsequent paper[2] using data from the Imp-7 spacecraft they presented further evidence that event-integrated GRB spectra could be represented by a single functional form. This time the original 150 keV exponential was modified by the addition of a power-law high energy tail ($E^{-2.5}$). The Imp-7 spectra of Cline and Desai[2] are reproduced in Figure 1. Other contemporaneous results[3,4,6,15] were consistent

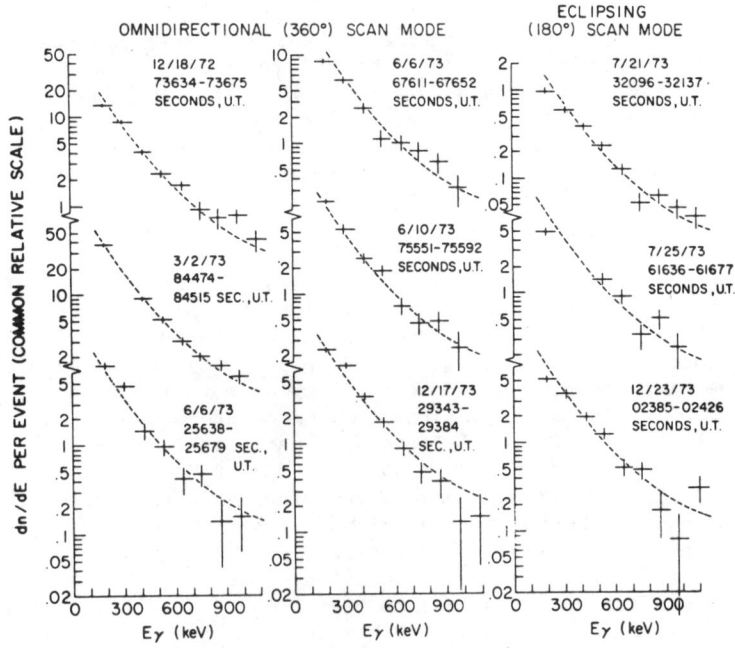

Figure 1. GRB spectra from the Imp-7 experiment of Cline and co-workers[2]. A single functional form consisting of a 150 keV exponential with a power-law tail appears to fit all the events.

with this picture. We shall see later, however, that this simple
picture was, in fact, not correct and that spectral variability from
event-to-event and evolution within individual events are, in fact,
typical characteristics of GRB's.

Of the early GRB spectral results, the most precisely
determined spectrum was that of the event of 1972 April 27 as
measured by the γ- and X-ray spectrometers on Apollo 16[6]. This
spectrum is shown in Figure 2. A combination of two detectors, (1)
an X-ray proportional counter and (2) an NaI γ-ray spectrometer,

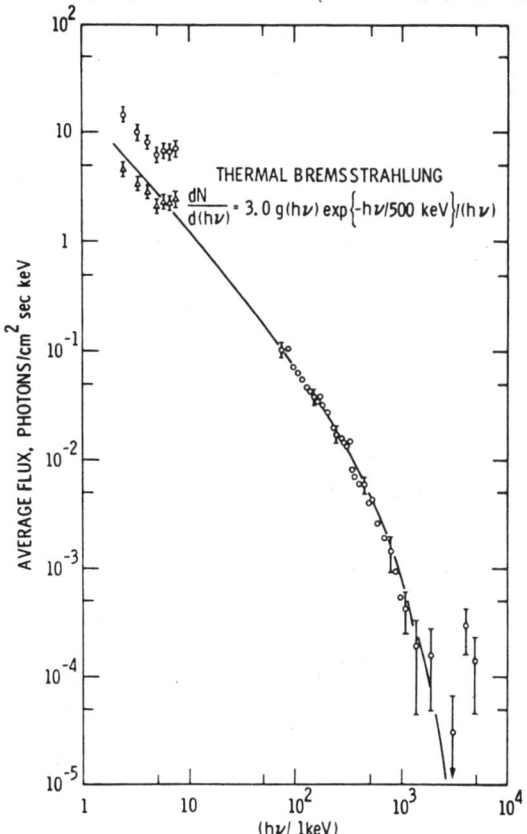

Figure 2. Spectrum of the Apollo 16 GRB[16] averaged over the entire event.

gave the widest energy coverage (2 keV to 5 MeV) of any GRB thus far
recorded. An unfortunate gap between the coverage of the two
instruments in the 10-60 keV range may have precluded them from
making the first discovery of cyclotron lines in GRB's. In a
subsequent paper Gilman et al.[16] showed that the Apollo 16 data
could be fit remarkably well by a thermal-bremsstrahlung spectrum
with a temperature of 500 keV. It should be noted that this fit to
the data was quite good over a range of three decades in energy.
The authors naturally assumed that the burst source was, in fact, a
hot optically thin plasma. To maintain the observed spectral shape,
the source must be optically thin to Compton scattering. In

addition, it is possible to place a limit on the size of the
emitting region the Apollo 16 event based on the most rapid time
variations observed during the event. Variations as fast as 110
msec were seen, which imply a source size $\lesssim$ 3000 km. These
assumptions taken together allow one to place a limit on the
distance to the source of $\lesssim$ 50 pc. This early result did, in fact,
establish a correct spectral form for GRB's (i.e., a thermal-
bremsstrahlung-like spectrum). The interpretation, however, in the
light of more recent data[10,11] is almost certainly not correct.
Later results will show that the emitting region is probably not
optically thin and dominated by free-free emission.

THE CONTINUUM SPECTRUM

The KONUS experiment of E. P. Mazets and coworkers at the A. F.
Ioffe Institute, Leningrad has yielded by far the most comprehensive
set of data on the spectra of gamma-ray bursts[9,10,11]. Their
relatively simple instrumentation consists of six NaI scintillators
each of ~ 50 cm^2 area arranged to provide complete sky coverage.
The spectrum is measured over the 30 keV-1 MeV interval in 16
channels. Identical instruments were carried on board the Venera-11
and -12 spacecraft. The fortunate combination of all-sky coverage,
absence of earth occultation and magnetospheric background effects
and a 1 1/2-year active lifetime has produced a data sample of ~ 150
GRB's. The spectrum of the 1970 April 19 GRB as measured by the
KONUS experiment[10] is reproduced in Figure 3. As was true with the

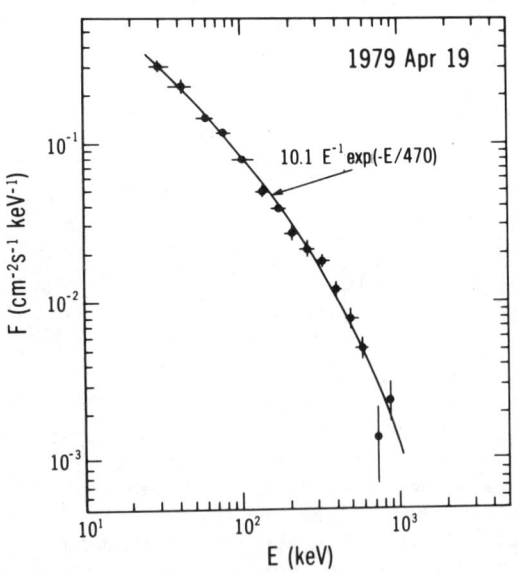

Figure 3. Spectrum of the 1979 April 19 GRB from the KONUS experiment at the maximum of the event.

Apollo 16 event, this GRB is well fit by a thermal bremsstrahlung-
like spectrum. In fact, the radiation temperature of 470 keV is
close to that of Gilman et al.[16]. However, it should be pointed out
that Gilman et al.[16] incorporated a Gaunt factor into their fit and

Mazets et al.[10] did not. This leads to somewhat higher radiation temperatures than otherwise. Of the published spectra of Mazets and coworkers, nearly all can be fit with a spectral form $dN/dE \propto \frac{1}{E} \exp(-E/E_o)$. The observed range of radiation temperatures E_o is quite broad--from $\sim$ 30 keV to more than 1 MeV. That a single spectral form provides an excellent fit to the great majority of events over such a broad range of radiation temperature is a remarkable and significant feature of GRB's and one that GRB models must now incorporate. In Figure 4 the distribution of the number of events

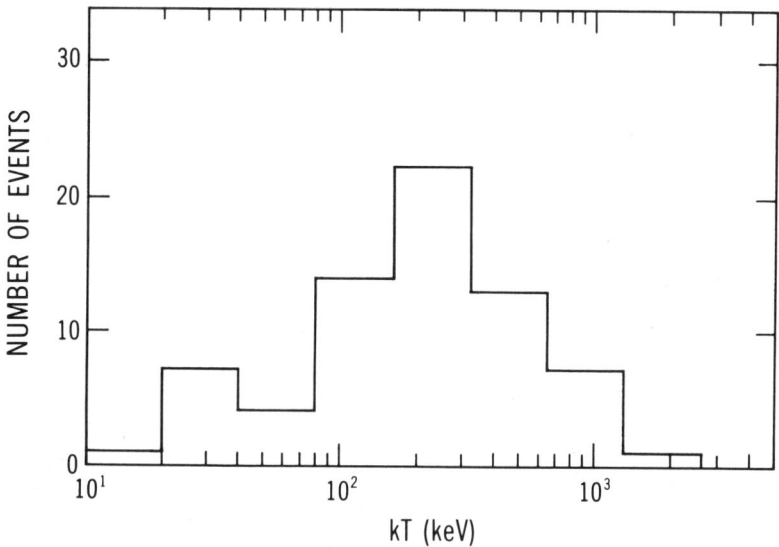

Figure 4. Number of events vs. maximum radiation temperature[11].

versus maximum radiation temperature is plotted[11]. The distribution is broad with a maximum at 200-300 keV.

Recently the first results from the γ-ray spectrometer on the Solar Maximum Mission (SMM) have been reported by Share et al. 1981[17]. In slightly more than a year the instrument has detected 17 events in the energy range 300 keV to 8 MeV. The large volume of NaI detectors make this the most sensitive instrument for GRB spectroscopy thus far flown. Unfortunately however, the low energy threshold is at $\sim$ 300 keV. They find in at least one case (1980 April 19) that the spectrum is consistent with an $E^{-2.5}$ power law in the .3-8 MeV interval. Since their energy window is different from that of Mazets et al.[10,11], it is not clear that this conflicts with the Mazets observations. It may be, however, that they have found evidence for a high energy tail that departs from the thermal bremsstrahlung-like spectrum.

Significant evolution of the spectral shape is also observed during GRB's. In fact, this appears to be more the rule than the exception. An example of such evolution is the event of 1978 November 24 plotted in Figure 5[11]. Two contiguous intervals are

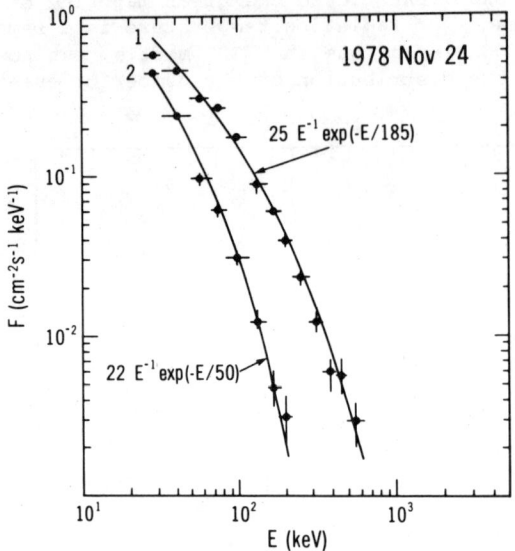

Figure 5. Spectra of the 1978 Nov 24 GRB[11] from the first and second 4-sec. intervals.

shown that are each 4 sec. duration. The earlier interval, which includes the onset of the event, has a radiation temperature of 185 keV whereas the later interval has a best-fit value of only 50 keV, quite a significant difference given the statistical validity of the data. This decrease in radiation temperature or "cooling" turns out to be a typical feature of most, but not all GRB's[10]. In Figure 6 a number of examples of the spectral evolution of GRB's are given[11]. For each event the time histories of the count rate and the derived radiation temperature are compared. The time resolution of the latter is limited by the 4 sec spectrum accumulation interval of the KONUS experiment. It is quite apparent that significant evolution of the spectrum is present in these events. It is, in fact, the prevalent behavior of most GRB's[10]. Observers should be cautioned against using event-averaged spectra, as has been frequently done in the past, because of the presence of this strong variability. Two kinds of variability are suggested by the data plotted in Figure 6. The 1978 November 24 and 1978 November 4 bursts both display temperatures that are highest at the onset of the event and then more or less monotonically decrease throughout the remainder. This decay in temperature in the 1978 November 4 event occurs at a time when the intensity is continuing to be strong and variable. kT for this event is not well correlated with the intensity. It should also be noted that the radiation temperature of the 1978 November 4 event varies from an initial value of $\sim$ 1 MeV to a final value of only $\sim$ 100 keV. In contrast, the latter two events 1979 April 19

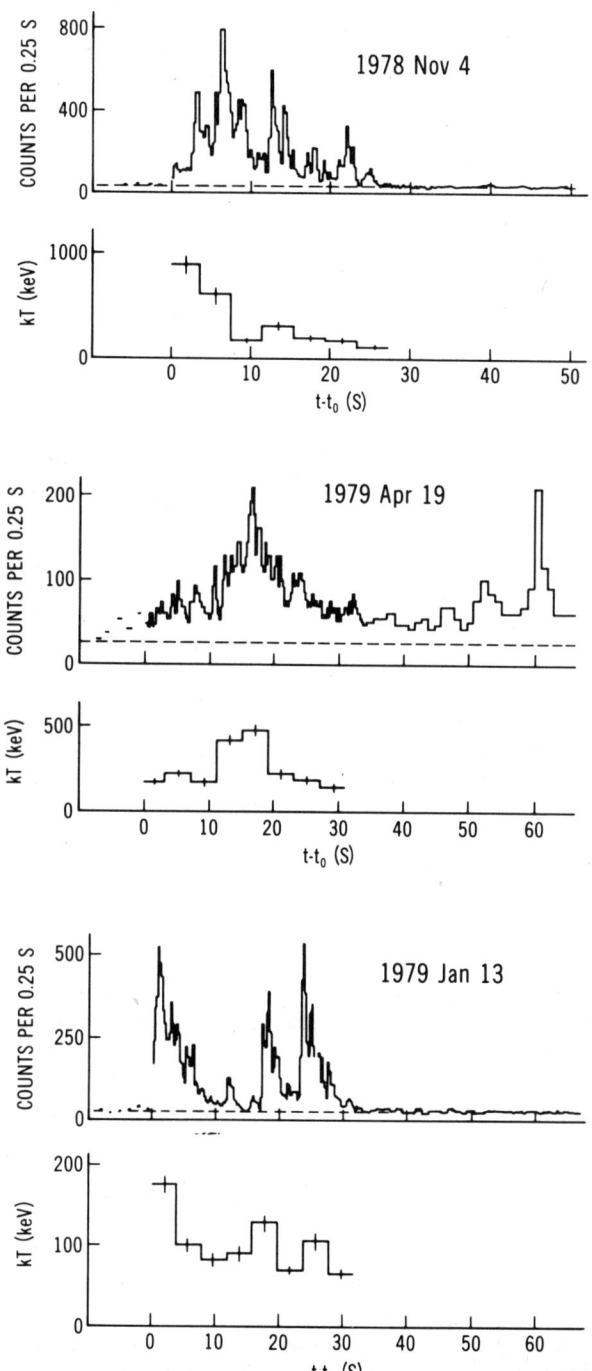

Figure 6. Time histories of the intensity and radiation temperature for 3 separate events[11].

and 1979 January 13 do appear to display a correlation between radiation temperature and intensity. The former shows a radiation temperature that peaks well into the event, and the latter shows a definite correlation of the temperature with subsidiary peaks in the count rate. It is evident that there is a rich variety of behavior present in the continuum spectra of GRB's. Some of the implications of this behavior will be discussed later in this paper.

EMISSION AND ABSORPTION LINES IN GAMMA-RAY BURST SPECTRA

a) <u>Cyclotron Emission and Absorption Features</u>

During the past year it has become apparent that emission and absorption features are common occurrences in the spectra of GRB's[10,11,13,14,18]. The most prevalent type of spectral feature that is seen is either an absorption line or broad absorption band in the 30-70 keV range[10,11]. Of the ~ 150 events recorded by the KONUS experiment, such features are seen in at least 30[10]. The proportion of events containing lines may, in fact, be even larger since the poor statistics of the smaller events may be masking the presence of these features. Two examples of "cyclotron" absorption features are given in Figure 7a,b[10]. The spectrum of the 1979 March 7 event (Figure 7a) clearly exhibits an absorption feature at ~ 45 keV having a FWHM of 15-20 keV and an equivalent width of 7 ± 0.6 keV[10]. The 1978 November 15 event[11] (Figure 7b), on the other hand displays a broad absorption band extending from ~ 100 keV down to at least the 30 keV threshold of the instrumnent. Note that there is excellent agreement between the spectra measured on the two different spacecraft. Evolution of the "cyclotron" absorption feature during an event is shown in Figure 8[10]. The first 8 sec. of the event plotted in the left panel of the figure clearly shows an absorption line at 65 keV superimposed on a thermal-bremsstrahlung-like continuum spectrum having a radiation temperature of 240 keV. The right hand panel shows the subsequent 24 sec where the line feature has completely disappeared. Furthermore, the radiation temperature has increased to 480 keV. In nearly all of the events studied by Mazets <u>et al</u>.[10,11] the absorption features are present for less than the <u>total</u> duration of the event. Furthermore, they are always strongest during the initial phase and tend to decay away more rapidly than the overall burst intensity. In general the behavior of the absorption feature is not well correlated with the behavior of the continuum spectrum of the burst[11].

At the present time, only one possible confirmation of the presence of cyclotron absorption features in GRB's is known to us. This is the event of 1980 April 19 as recorded by the Hard X-Ray Burst Spectrometer on SMM[18]. The general character of the absorption feature as well as its time variability appear to be consistent with the behavior of the events reported by Mazets <u>et al</u>.[10,11]. The issue, however, is somewhat confused by the fact that a solar origin cannot be ruled out. It should be pointed out that solar origin any of the Mazets events is ruled out since directions are determined for each event by the KONUS experiment.

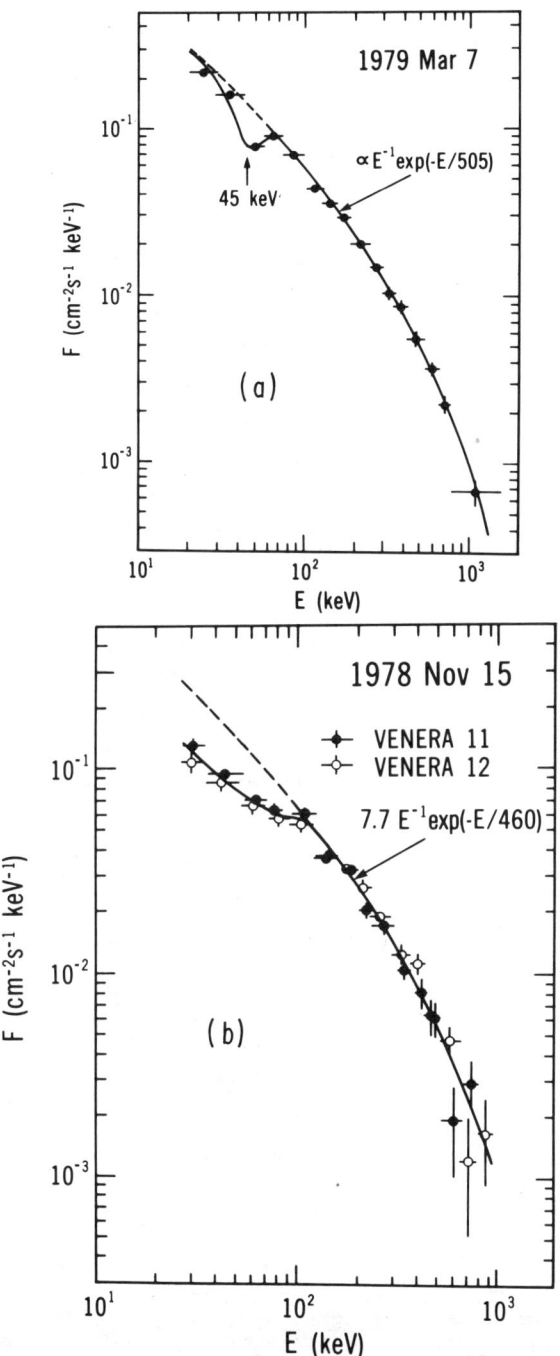

Figure 7. (a) Spectrum of the 1979 Mar 7 GRB[10] showing a "cyclotron" absorption line. (b) Spectrum of the 1978 Nov 15 GRB[11] showing a broad absorption band extending from ~ 30 to 100 keV.

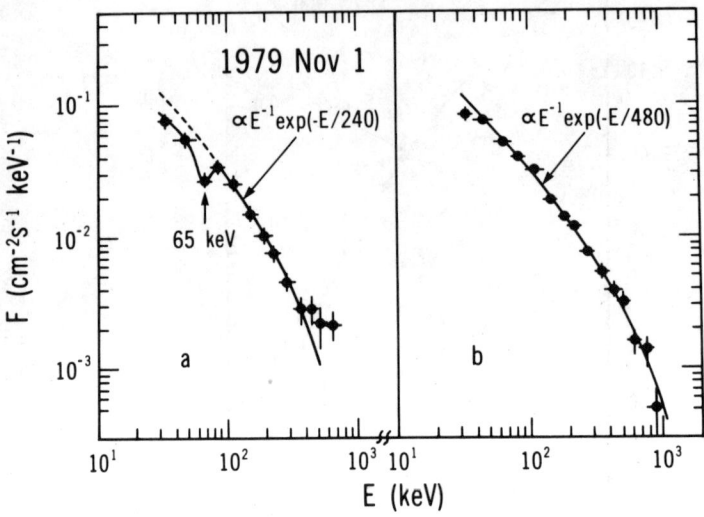

Figure 8. Spectrum from the 1979 Nov 1 GRB from the KONUS experiment. Left hand panel is the first 8 sec. of the event. Right hand panel is the subsequent 24 sec.

A very striking example of spectral evolution during a burst is given in Figure 9 where the usual "cyclotron" absorption feature is seen during the early phase of the event superimposed on a 400 keV continuum spectrum[10]. Note, however, that the lowest energy point at ~ 20 keV departs significantly from the thermal-bremsstrahlung representation, indicative of the possible presence of a separate soft component. A second spectrum is plotted that was taken during a minimum intensity period of the event occuring at an intermediate time within the burst. The two lowest energy data points have not changed. The hard 400 keV spectrum previously present above ~ 70 keV has, however, now completely disappeared. Only a very soft spectrum, which appears as a continuous extension of the unchanged low energy data points, now remains. This event seems to have two components, a soft one that remains more or less constant and a hard one that is time-variable. We shall return to this point later as a number of other observations suggest that a two-component model is appropriate.

The detailed characteristics of the "cyclotron" features appear to bear little or no relationship to the properties of the continuum spectrum. One can use the Mazets et al.[10], published data to calculate an effective "FWHM" by dividing the equivalent width by the depth of the absorption feature. For the broad band features this "FWHM" should be interpreted as a lower limit on the true width since the features generally extend below the instrumental threshold.

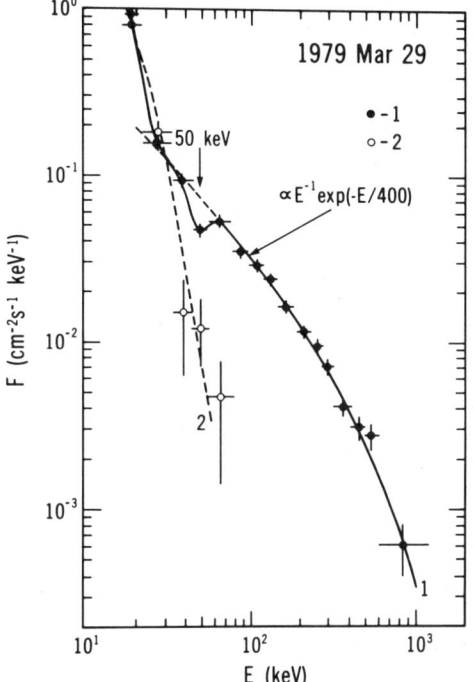

Figure 9. Spectra from the 1979 Mar 29 GRB from the KONUS experiment. Solid line is from the initial stage. Dashed line is from a deep minimum between burst peaks.

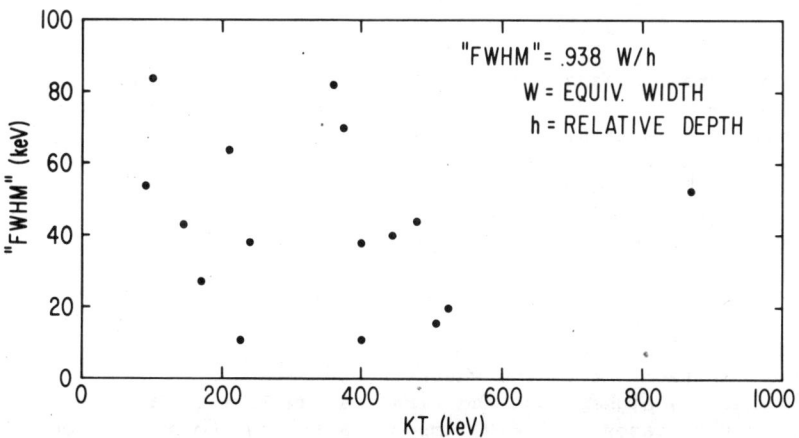

Figure 10. "FWHM" of cyclotron features vs. continuum radiation temperature for a sample of the KONUS events.

This effective "FWHM" is plotted in Figure 10 versus the continuum radiation temperature for the Mazets events where data are available[10]. There is obviously no evidence here for any correlation between the line width and the continuum temperature. This can be taken as further evidence that two different processes or regions are responsible for the line and continuum behavior.

One final example of a "cyclotron" feature is given in Figure 11[10]. This is the 1979 May 26 event which is the only published

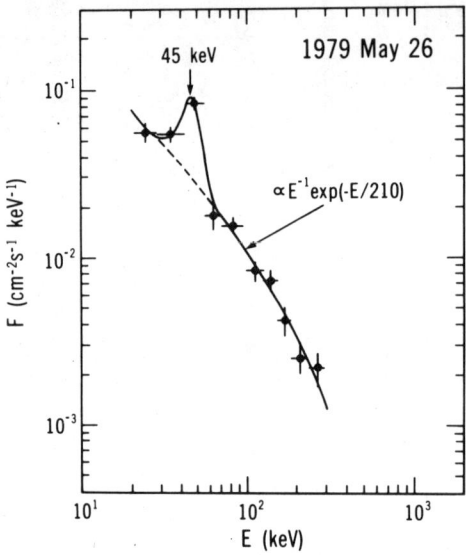

Figure 11. Spectrum of the 1979 May 26 GRB from the KONUS experiment[10]. A "cyclotron" emission feature appears at ~ 45 keV.

example of a "cyclotron" emission feature. It appears at an energy of 45 ± 5 keV and has a width of ~ 20 keV.

The persistent presence of these absorption and emission features in the 30-70 keV range of GRB spectra is strong evidence for a neutron star origin. In the intense (10^{12}-10^{13} gauss) magnetic fields that are expected to be present near the surfaces of neutron stars, the cyclotron frequency corresponds to photons in the hard X-ray range. The fact that such lines exist superimposed on a very hot continuum spectrum poses a problem. To produce a cyclotron <u>absorption</u> line requires that the electron temperature be smaller than the energy of the line (first Landau level). If the temperature were comparable to or higher than this, one would expect to see the line in emission. The one example of "cyclotron" emission (1979 May 26) may, in fact, be a case where the temperature in the emission region is higher than the first Landau level. The width of the line (in some cases $\lesssim$ 20 keV) imposes a further constraint on the temperature of the absorbing region. The Doppler broadening of the line is given by Trumper[19]:

$$\frac{\Delta E}{E} = \sqrt{8 \ln 2 \; \frac{kT}{mc^2}} \cos \theta$$

where θ is the angle between the line of sight to the observer and the magnetic field. Taking $\frac{\Delta E}{E} = 0.3$ and $\theta = 45°$ gives $kT = 16$ keV for the electron temperature. To account for these observations it appears that one must postulate that the cyclotron absorption takes place in a relatively cool (< 20 keV) layer that overlies the hot region which produces the continuum spectrum.

We return now to the question raised in the previous chapter of whether optically thin thermal bremsstrahlung can be a realistic physical description of the radiation process during a GRB. The presence of cyclotron features in the 30-70keV range implies that $10^{12}-10^{13}$ gauss fields are present in the emission region. It is then extremely unlikely that free-free transitions will be the dominant mode of radiative transport. It is much more likely in this intense field that synchrotron radiation and related processes will dominate.

b) <u>Emission Lines > 400 keV</u>

Emission lines > 400 keV in GRB spectra have been reported by Mazets <u>et al</u>.[10,11] and Teegarden and Cline[13,14]. The former, as mentioned earlier used NaI scintillators, whereas the latter employed a radiatively-cooled Germanium detector. This instrument, launched on the ISEE-3 spacecraft, had a resolution of 8 keV at the time of launch. Unfortunately, after only 4 months of operation , an electronics failure disabled that portion of the memory in which GRB spectra were stored. As a result, high-resolution Ge spectra are available for only two events.

An example of an emission line at ~ 400 keV is given in Figure 12 for the event of 1978 September 18[10]. The line appears quite pronounced and is superimposed on a 185 keV thermal-bremsstrahlung-like continuum. As was the case with the "cyclotron" lines, the 400 keV features do not appear to bear any correlation with the continuum radiation temperature. Of the 150 events recorded by the KONUS experiment, 11 display emission features in the 400-460 keV range. These features have been interpreted as redshifted annihilation radiation. Redshifts in the 15-20 percent range are expected for radiation produced near the surface of a 1 $M_\odot$ neutron star. This would shift the 511 keV annihilation line into the 400-460 keV range. There are, unfortunately, a number of factors that complicate this interpretation. First, for high electron temperatures the energy of the line will be blueshifted since all of the energy of the electrons (kinetic + rest mass) is converted into photon energy[20,21]. It is not clear, however, that electron temperatures high enough for this effect to be significant are consistent with the observations. In addition, the possibility of stimulated e^+-e^- annihilation exists if one is dealing with a cool dense electron plasma (Ramaty, private communication). It turns out

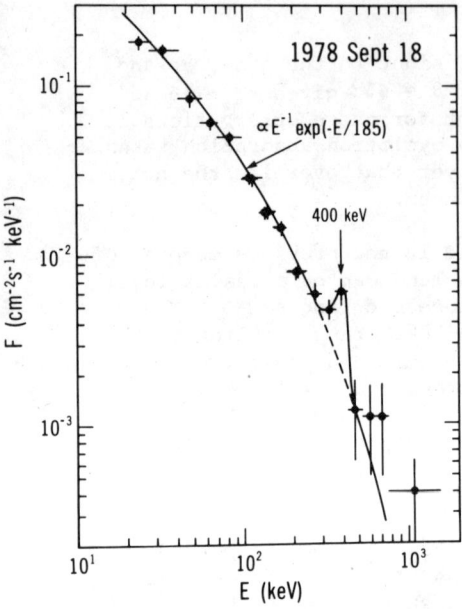

Figure 12. Spectrum of the 1978 Sept 18 GRB from the KONUS experiment.[10]

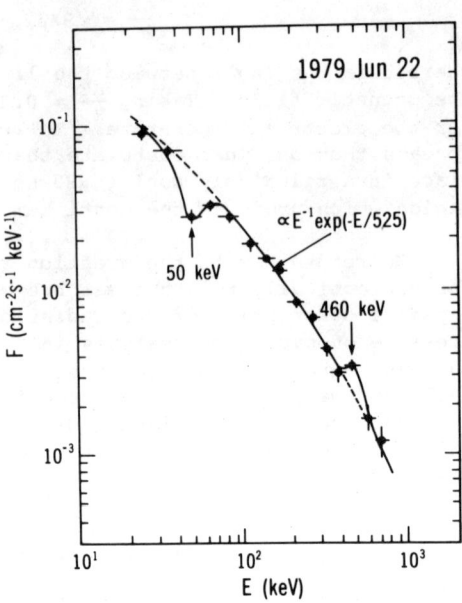

Figure 13. Spectrum of the 1979 June 22 GRB from the KONUS experiment[10]

that the energy dependence of the photon absorption coefficient is such that a peak in the 400-460 keV range can be produced.

Figure 13 is an example of an event where both a "cyclotron" absorption feature and a 460 keV emission line are observed[10] simultaneously. There is only one other such example that has been published by Mazets and coworkers[11]. This is the event of 1979 November 1 which is reproduced in Figure 14. This single figure displays much of the rich phenomenology that has manifested itself in the spectra of GRB's. The left panel of the figure shows the first 8 sec of the event where both a ~ 65 keV absorption line and a broad emission feature in the 350-650 keV interval are seen. These lines lie on top of a 280 keV continuum. A profile of this emission band with the continuum subtracted is also shown. Typically, the ~ 400 keV emission features have widths of ~ 250 keV. In the right hand panel two spectra in the later portion (respectively ~ 16 sec and ~ 24 sec after the onset) of the event are displayed. Both the absorption and emission lines have disappeared. In fact, the typical behavior for the ~ 400 keV features is appearance at the onset of the event and a lifetime shorter than that for the total intensity. Such behavior, as was seen earlier, is also commonplace with the "cyclotron" features. During the intermediate portion of the event, the radiation temperature rises to a value of 800 keV and

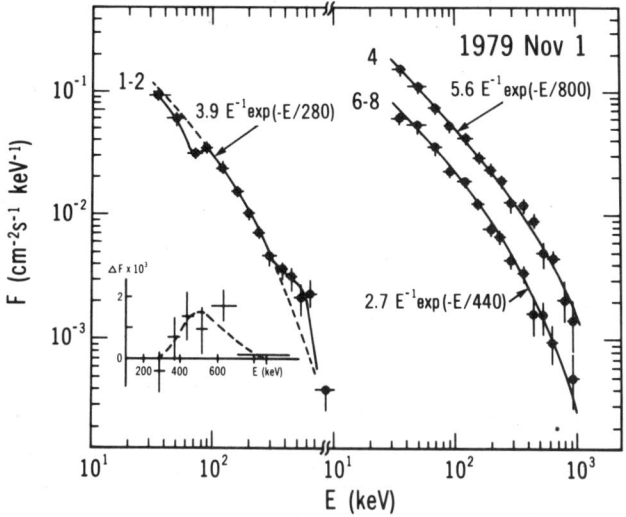

Figure 14. Spectra of the 1979 Nov 1 GRB from the KONUS experiment[11]. (1-2) first 8 sec of the event; (4) 4th 4-sec interval; (6-8) 6th, 7th and 8th 4-sec intervals.

then decays to a value of 440 keV in the latter portion. Curiously enough, the disappearance of the lines is accompanied by a heating of the continuum temperature. The general lack of correlation between the ~ 400 keV emission features and the continuum temperature can be taken as further evidence for the existence of two components in the GRB source. Furthermore, the observed broadening (~ 250 keV) of the annihilation line requires that the electron temperature be low. If the line width were entirely due to Doppler broadening, we would have $kT \cong 15$ keV. Furthermore, as pointed out by Daugherty and Bussard[22] and Mazets et al.[11], the radiation produced by pair annihilation in strong magnetic fields will be broadened by the presence of the field itself. In fact, this broadening mechanism in a field of 5×10^{12} gauss could account for the full observed width of the 400 keV lines.

Ramaty and coworkers[23,24] have proposed a model for the 1979 March 5 event where the radiation originates in a thin (~ 10^{-2} cm) surface layer that overlies a hot MeV plasma. They point out that electron cooling by synchroton radiation in a > 10^{11} gauss field will generally take place more rapidly than e^+-e^- annihilation. In their model the continuum spectrum is therefore produced by the synchroton emission from the rapidly cooling electrons and positrons which subsequently annihilate and produce observable lines whose width is characteristic of the low temperature in this thin surface layer. The 1979 March 5 event is unique in a number of ways: (1) its luminosity is brighter than any other recorded event, (2) eight-second oscillations were observed superimposed on a slowly decaying tail, (3) it has been identified with a supernova remnant, N49, in the Large Magellanic Cloud and (4) its continuum spectrum is very soft. These unique features may point to different physical conditions at the source of this event. It is nonetheless true that the Ramaty et al.[23,24] model for the 1979 March 5 event may be applicable to other GRB's.

A cool layer that is optically thin to Compton scattering appears to be required to explain the spectral behavior that has been recently observed. The fact that in many events $\gtrsim$ 1 MeV photons are present would imply that $\gamma - \gamma$ pair production in the region immediately above the absorption layer is not an important process. Schmidt[24] has used this argument to place a limit on the distances of GRB sources. He assumed an emitting region of 10^9 cm extent and obtained a distance limit of $\lesssim$ 2 kpc for events in which MeV photons are present. The weight of recent evidence is towards a much smaller size ($\sim 10^5$ cm) for the emission region which in turn may lead to a smaller distance limit for the burst sources. The problem is, however, complicated by the strong energy and angular dependence of the cross-sections[25] so that detailed calculations will be necessary to establish a true distance limit.

Figure 15 shows the spectrum of the 1978 November 19 GRB as measured by the Germanium spectrometer onboard the ISEE-3 spacecraft[13,14]. A broad marginally significant feature is present at $\sim$ 400 keV that is consistent with the spectral feature reported by Mazets et al.[10] during this same event. In addition, the ISEE-3 data show a second line at an energy of 740 keV having a width of $\sim$ 40 keV. This line is significant at the 99 percent confidence level[13]. Teegarden and Cline[13] pointed out that the energy of this line is consistent with that expected if the 847 keV first excited level of iron is redshifted by the same amount as the $\sim$ 400 keV feature. It is perhaps significant that this event has the hardest spectrum of any thus far recorded and is one of the few that apparently cannot be fit with a thermal-bremsstrahlung function. The continuum spectrum for energies $\sim$ 200 keV-2 MeV best fit with an $E^{-1.3}$ power law[13].

If the assumption of a redshifted iron line is correct, then conditions at the source are probably such that other nuclear levels can be excited. We do not a priori know the composition of the source region. If it is pure iron, then we would see only higher levels of Fe nuclei, but if it is, for example, the same as the solar system composition, then excited levels of other nuclei would probably produce significant radiation. Ramaty et al.[26] have calculated the production of gamma-rays by cosmic rays interacting with the interstellar medium. We have normalized their iron line intensity to that of our 740 keV line and then plotted the most prominent lines > 740 keV normalized by the same factor and redshifted by the same amount as the Fe line. The Ramaty et al.

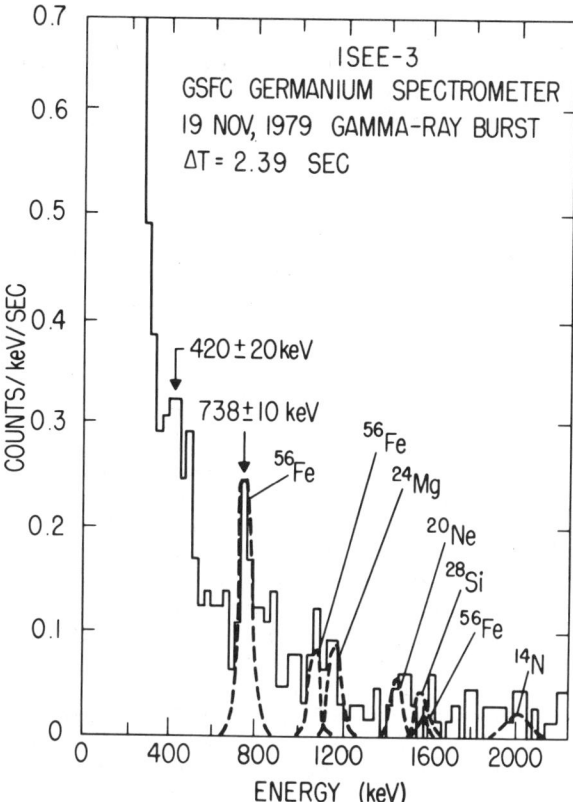

Figure 15. Spectrum of the 1978 Nov 19 event from the ISEE-3 Gamma Ray Burst Spectrometer[13,14].

calculations[26] that we used assumed an E^{-3} power law proton spectrum which may not be a good representation of the spectrum at the site of the GRB. The relative line strengths, however, are not strongly dependent on the shape of the proton spectrum but rather on the relative abundances and excitation cross-sections. It is evident in Figure 15 that these higher energy lines could make a signficant contribution to the > 1 MeV portion of the spectrum and that they might, in fact, account for the unusual hardness of the spectrum of this event.

Katz[27] has put forth an alternate explanation for the 740 keV line. He has pointed out that for sufficiently high magnetic fields ($\gtrsim 10^{13}$ gauss) single photon annihilation will become the dominant mode. In this mode a single photon with energy 1.022 MeV is emitted so that the ratio of the two lines should in this simple-minded picture be a factor of two. The actual ratio in the data are ~ 1.76, but the peaks are broad enough so that a factor of two cannot be ruled out. Ramaty et al.[28] have pointed out that a mechanism exists which can effectively shift the annihilation peaks toward lower energies and in general the one- and two-photon peaks will not be shifted by the same amount. It therefore remains unclear whether this can serve as an adequate explanation of the spectral features in the 1978 November 19 event.

SUMMARY AND CONCLUSIONS

The most important characteristics of GRB spectral behavior are summarized as follows:

1) The continuum spectrum is well represented by a thermal bremsstrahlung function with kT ranging between 30 keV and $\gtrsim$ 1 MeV. Possible evidence exists for a power-law high energy tail.

2) Significant evolution of the continuum sprectrum during the event is commonplace. The most typical behavior is a higher temperature during the early part followed by "cooling" or softening of the spectrum. There are, however, many examples of events where the temperature appears to be correlated with the intensity which may in turn possess complex structure.

3) "Cyclotron" absorption features in the 30-70 keV range are frequent, if not typical, occurrences in GRB's. At least 30 out of the 150 events recorded by the KONUS experiment display such features. Line widths vary from as small as ~ 20 keV (the probable instrumental limit) to broad absorption bands extending over more than 80 keV. "Cyclotron" features are normally present during the early part of the event and disappear before the event is finished. Little or no correlation exists between the continuum radiation temperature and the properties of the absorption lines.

4) Emission lines in the energy interval 400-460 keV are present in at least 7% of the GRB's. The features apparently are broad ($\Delta E \gtrsim 200$ keV) and uncorrelated with the continuum temperature. As with the cyclotron features they appear at the onset and disappear before the event is over.

5) A narrow (~ 40keV) line has been reported in the 1978 November 19 event at 740 keV. This event has the hardest spectrum of any thus far recorded and is one of the few that apparently cannot be fit by the thermal-brehmsstrahlung expression.

The principal conclusions that can be drawn from these observations are summarized as follows:

1) The "cyclotron" and 400-460 features can be taken as strong evidence for a neutron star origin for GRB's. The 400-460 keV lines are interpreted as redshifted annihilation radiation. The redshift is consistent with that expected from a 1-1.5 $M_\odot$

neutron star.

2) The presence of absorption features <70 keV and ~ 400 keV lines with $\Delta E/E \sim 0.5$ appears to require the existence of a cool layer overlying the hot plasma in which the continuum is produced. Rapid synchrotron cooling is a possible mechanism for the production of such a layer.

3) The presence of ~ 1 MeV photons in many events requires that the distance to the source $D \leq 1$ kpc. At larger distances γ-γ absorption would strongly suppress the high energy photons.

4) Optically thin thermal-brehmsstrahlung is probably not a viable description of the radiation mechanism at the source. In the strong magnetic fields in the source region synchrotron emission and absorption are almost certainly much stronger processes. Any theory, however, must account for the fact that the continuum spectra of nearly all GRB's are extremely well fit by a thermal-brehmsstrahlung expression over a wide range (30 keV - 2 MeV) in energy.

5) The 740 keV line could be explained either by a redshifted iron line or single-photon annihilation.

REFERENCES

1. T. L. Cline, U. D. Desai, R. W. Klebesadel and I. B. Strong, Ap. J. (Letters), <u>185</u>, L1 (1973).
2. T. L. Cline and U. D. Desai, Ap. J. (Letters), <u>196</u>, L43, (1975).
3. W. A. Wheaton, M. P. Ulmer, W. A. Baity, D. W. Datlowe, M. J. Elcan, L. E. Peterson, R. W. Klebesadel, I. B. Strong, T. L. Cline and N. D. Desai, Ap. J. (Letters), <u>185</u>, L57, (1973).
4. C. Pizzichini, G. G. C. Palumbo and A. Spizzichino, Ap. J. (Letters), <u>195</u>, L1, (1975).
5. W. L. Imhof, G. H. Nakano, R. G. Johnson, J. R. Kilner, J. B. Reagan, R. W. Klebesadel and J. B. Strong, Ap. J., <u>191</u> L7, (1974).
6. A. E. Metzger, R. H. Parker, D. Gilman, L. E. Peterson and J. I. Trombka, Ap. J. (Letters) <u>194</u>, L19, (1974).
7. S. R. Kane and K. A. Anderson, Ap. J., <u>210</u>, 875, (1976).
8. S. R. Kane and G. H. Share, Ap. J., <u>217</u>, 549, (1977).
9. E. P. Mazets and S.V. Golenetskii, Astrophys. Space Sci., <u>75</u>, 47, (1981).
10. E. P. Mazets, S. V. Golenetskii, R. L. Aptekar, Yu. A. Guryan and V. N. Ilyinskii, Nature, <u>290</u>, 378, (1981).
11. E. P. Mazets, S. V. Golenetskii, V. N. Ilyinskii, Yu. A. Guryan, R. L. Aptekar, V. N. Panov, I. A. Sokolov, Z. Ya. Sokolova and T. V. Kharitonova, A. F. Ioffe Institute preprint No. 719, Leningrad, (1981).
12. E. P. Mazets, S. V. Golenetskii, V. N. Ilyinskii, V. N. Panov,

I. A. Sokolov, Z. Ya. Soklova, N. G. Khavenson and T. V. Kharitonova, A. F. Ioffe Institute preprint No. 712, Leningrad (1981).
13. B. J. Teegarden and T. L. Cline, Ap. J. (Letters), 236, L67, (1980).
14. B.J. Teegarden and T. L. Cline, Astrophys. Space Sci., 75, 181, (1981).
15. G. G. C. Palumbo, G. Pizzichini and G. R. Vespiqnani, Ap. J. (Letters), 189, L9, (1974).
16. D. Gilman, A. E. Metzger, R. H. Parker, L. Evans and J. I. Trombka, Ap. J., 236, 951, (1980).
17. G. H. Share, M. S. Strickman, R. L. Kinzer, E. P. Chupp, D. J. Forrest, J. M. Ryan, E. Rieger, C. Reppin and G. Kanbach, to be published in the Proceedings of the 17th International Cosmic Ray Conference, Paris, France, (1981).
18. B. R. Dennis, K. J. Frost, A. L. Kiplinger, L. E. Orwig, U. D. Desai and T. L. Cline, proceeds of this conference, (1981).
19. J. Trumper, in Gamma Ray Spectroscopy in Astrophysics, NASA Technical Memorandum No. 79619, (1978).
20. F. A. Aharonian, A. M. Aloyan and R. A. Sunyaev, Yerevan Physics Institute preprint 432(39)-80, (1980).
21. R. Ramaty and P. Mészáros, Ap. J., in press, (1981).
22. J. K. Daugherty and R. W. Bussard, Ap. J., 238, 296, (1980).
23. R. Ramaty, S. Bonazzola, T. L. Cline, D. Kazanas, P. Mészáros and R. E. Lingenfelter, Nature 287, 122, (1980).
24. R. Ramaty, R. E. Lingenfelter, R. W. Bussard, Astrophys. Space Sci., 75, 193, (1981).
25. W. K. H. Schmidt, Nature 271, 525, (1978).
26. R. Ramaty, B. Kozlovsky and R. E. Lingenfelter, Ap. J. Supp., 40, 487, (1979).
27. I. Katz, preprint, (1981).
28. R. Ramaty, J. M McKinley and F. C. Jones, proceedings of this conference, (1981).

143

10 JUNE 1974 TRANSIENT

J. C. Ling, W. A. Mahoney, J. B. Willett & A. S. Jacobson
Jet Propulsion Laboratory
California Institute of Technology
Pasadena, California 91109

INTRODUCTION

We review results of a transient event (referred to in this workshop as the "Jacobson" transient[1,2]) which we observed in a balloon flight in 1974. The event, lasting for twenty minutes, is quite different in character from the types of cosmic transients reported by Teegarden[3] and Cline[4]. During this twenty-minute period, four intense gamma-ray lines at energy 413 keV, 1.79, 2.2 and 5.9 MeV appeared in the background spectrum. The widths of these lines (15 keV to 95 keV) were all broader than the instrument's resolution. No continuum emission was detected. The event observed by the main detector system was also accompanied by correlated behavior in several shield segments. In this paper, the evidences for and physical interpretations of this event will be discussed.

INSTRUMENTATION

The instrument (Figure 1) consisted of four 40 cm^3 lithium-ion drifted germanium detectors clustered in a single cryostat cooled by liquid nitrogen. Each detector had its own signal chain consisting of a bias supply, preamplifier, discriminators and a 8192-channel pulse height analyzer. The energy resolution ranged from

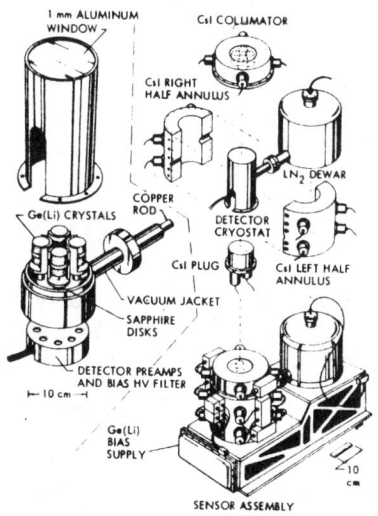
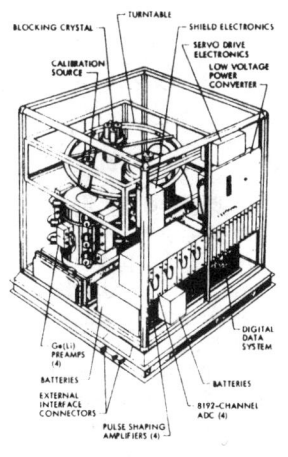

Figure 1. JPL Gamma-Ray Spectrometer   Figure 2. Balloon Gondola

0094-243X/82/770143-09$3.00 Copyright 1982 American Institute of Physics

2 keV at 100 keV to 5 keV at 6 MeV. Unfortunately, one of the detectors failed to function properly for the flight. The experiment was therefore performed with three detectors only. These detectors were also surrounded by a segmented CsI anticoincidence shield consisting of two shield halves on the side, a plug unit on the back, and a collimator on the top. The collimator had four holes aligned with the detectors which provided an angular resolution of 20° FWHM at low energy. In addition to the main detector system, there was also a 20.3 cm diameter by 10.2 cm thick NaI blocking detector mounted on a movable turntable (Figure 2) which could be positioned to cover the aperture in order to measure the true detector background. The CsI shield segments and the NaI blocking detector provided important supporting evidence for this event.

BALLOON FLIGHT AND OBSERVATIONS

The balloon was launched at 1205 UT on 10 June 1974 and reached a float altitude of 2.9 gm/cm^2 at around 1515 UT. The history of the integral count-rate in the 65-100 keV energy band measured by the three germanium detectors for the entire flight is shown in Figure 3.

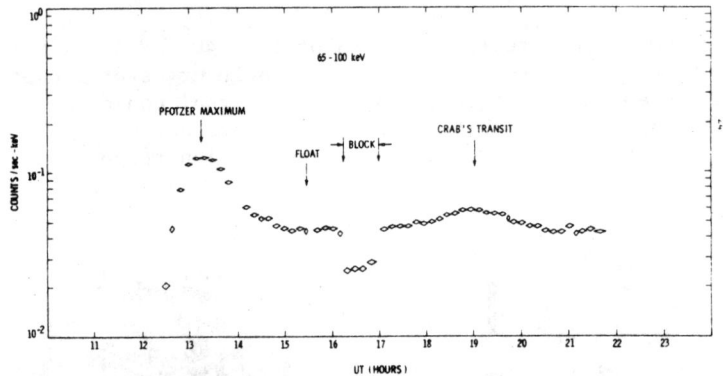

Figure 3. History of 65-100 keV count-rates.

During the first part of float, the aperture was blocked for about 45 minutes during which time the count-rate decreased by about 40%. The difference may be attributed to the downward flux consisting of atmospheric and cosmic diffuse components.

The primary objective of the experiment was to search for gamma-ray line emissions from the Crab Nebula which transited the field of view at 1900 UT. It was during this same flight that a line emission at 73 keV from the Crab Nebula[5] was first detected. The existence of such a feature has been confirmed[6-9]. The most recent result reported by the Naval Research Laboratory (NRL) group under

J. Kurfess indicates that the feature is both pulsed with the Crab's pulsar-period (similar to the cyclotron emission and absorption features observed in Her X-1[10] and 4U0115+63[11]) and "flare"-like with a time scale on the order of at least several hours, based on the JPL[5] and the Italian/Indian[6,7] observations. If this phenomenum is real, it is certainly a new type of line transient which we were also fortunate to observe on 10 June 1974. It is our expectation to attempt to verify the NRL observation with the HEAO-3 data.

A. TIME PROFILE

The twenty-minute line transient was discovered quite by accident. While studying the 6.1 MeV atmospheric gamma-ray line, we noticed that the count-rate in the adjacent energy channels around 5.9 MeV flared up for about twenty minutes near the end of the flight at 2030 UT. As we examined the data further, we discovered an even stronger feature near 1.8 MeV flared up during the same time period. Figure 4 shows the history of the count-rate in three different energy channels, each 100 keV wide, centered at 1710, 1810 and 1910 keV, respectively. At 2030 UT the rate in the 1810 keV

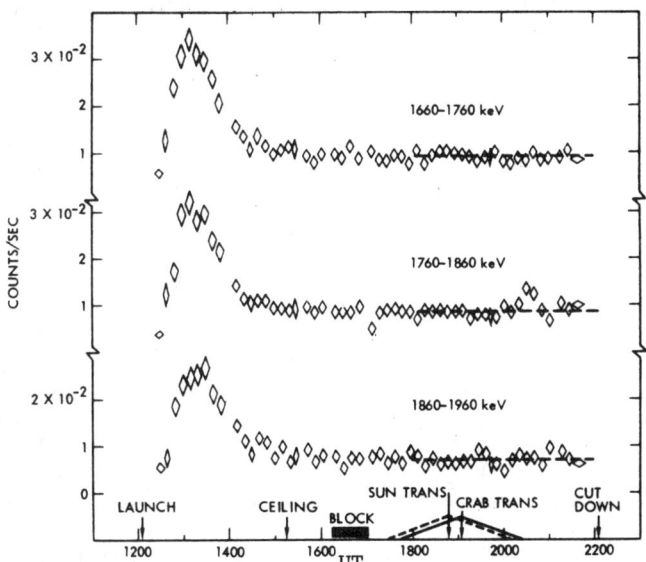

Figure 4. History of 1.8 MeV Count rates.

channel increased by about 45% over the background with no corresponding increase shown in the two adjacent channels. Since the increase only lasted for twenty minutes, the time scale was too short for a point source transit, which should have lasted for approximately two hours. If the response of the instrument was due to cosmic transient event, it should also have been detected by the shield detectors, especially the one exposed to the upper hemisphere. A comparison of the 1810 keV channel time profile with the integral rate (E > 150 keV) measured by various CsI shields (Figure 5) indicates that the collimator did indeed register a > 5σ correlated increase. In addition, each of the two shield halves indicated a

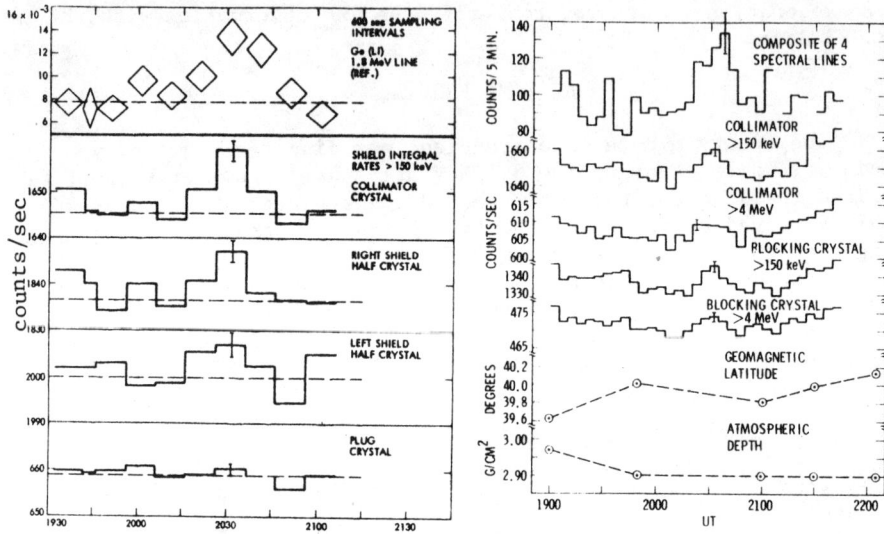

Figure 5. A comparison of the 1810 keV channel rate with the LLD rates (E > 150 keV) of various CsI shield segments.

Figure 6. A comparison of integral rate of four spectral features with various other parameters.

2.5σ increase, but no significant deviation from the background rate was shown in the plug unit in the rear. Figure 6 shows a comparison of the time profile of the integral rate of the sum of the four spectral lines with the collimator and blocking detector LLD (E > 150 keV) and ULD (E > 4 MeV) rates and with the geomagnetic latitude and altitude variations of the balloon gondola. The blocking detector LLD rate also registered a correlated > 3σ increase at the time of the transient event. In addition, both ULD rates indicated a slight increase. The transient observed by both the central and shield detectors was not correlated with effects of either geomagnetic latitude drift or changes in the balloon's altitude.

B. SPECTRUM

Figure 7 shows the energy loss spectra accumulated during the event and compared with background spectra accumulated before and after. Four line features with centroids at energy 413, 1790, 2219 and 5946 keV, respectively, were detected with > 3σ significance.

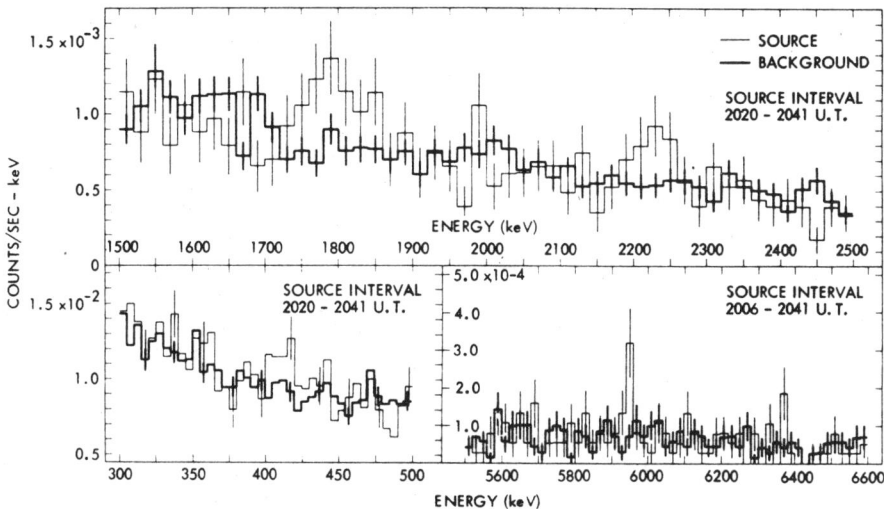

Figure 7. Energy-loss spectra accumulated during the event are compared with the background spectra accumulated before and after.

These lines had apparently different widths and possibly slightly different onset times. The strongest feature was the 1790 keV line which can be associated with the most prominent line of either Si^{28} or Mg^{26}. The 2219 keV feature is probably the well known deuterium formation line which has been observed frequently in solar flares[17]. However, there is no obvious identification of either the 413 or the 5946 keV lines. Any contenders for these features should be accompanied by other stronger lines which were not observed. A line feature about 400 keV has been observed in gamma-ray bursts[12,13] and possibly from the Crab Nebula[14]. This line is usually interpreted as the gravitational red-shift of the 511 keV annihilation line near the surface of a neutron star. If our observed 413 keV line is the 511 keV line red-shifted by $z=0.24$, then the observed 1.8 MeV line corresponds to the 2.2 MeV deuterium formation line with about the same red-shift factor. The unred-shifted component of the 511 keV line could not be observed due to the strong background component. The 5946 keV line may be associated with the two strongest Fe^{56} neutron capture lines with a red-shift factor of z=0.285.

A summary of information about these lines is shown in Table 1. The line fluxes were determined assuming that the source was along the detector axis. The location of the source is only crudely

LINE ENERGY (keV)	COUNT RATE $(SEC)^{-1}$	FLUX* $(CM^2 - SEC)^{-1}$	FWHM (keV)	TIME INTERVAL SPANNED (U.T.)
413.2 ± 1.8	$(5.5 ± 1.6) \times 10^{-2}$	$(7.0 ± 2.0) \times 10^{-3}$	15	2020 - 2041
1789.7 ± 6.0	$(6.78 ± 1.55) \times 10^{-2}$	$(3.15 ± 0.74) \times 10^{-2}$	95	2026 - 2041
2218.6 ± 6.3	$(2.67 ± 0.87) \times 10^{-2}$	$(1.51 ± 0.49) \times 10^{-2}$	70	2020 - 2041
5946.5 ± 3.7	$(7.06 ± 2.18) \times 10^{-3}$	$(1.47 ± 0.46) \times 10^{-2}$	25	2006 - 2041

*ASSUMES SOURCE ON THE DETECTOR AXIS

Table 1. Spectral line parameters.

determined by these data. Comparing the rate increase of different shield segments, and using a Monte Carlo calculation to simulate the shield response, the location of the source can be estimated and is within the field of view of the instrument. Such a calculation indicates that if the source were outside the FWHM response of the instrument, then the count-rate measured by the shield segments would be considerably higher than those observed. The energy flux in the lines is 2.9×10^{-7} ergs/cm²-sec, and the mean fluence is about 3×10^{-4} ergs/cm² over the twenty-minute duration of the event.

The portion of the sky observed by the instrument at the midpoint of the transient event is shown in Figure 8. The concentric curves are the FWHM fields of view of the instrument at energies 0.41, 1.8 and 6 MeV, respectively. The planets Mercury and Saturn were within the field of view, as were the SN-remnant IC-443, and gamma-ray source 195+5 observed by both SAS II and COS-B. The Crab Nebula and the Sun were also at the edge of the field of view. The numbered dots are stars within 6 pc from the earth. The flaring star YZ Canis Minor (dot no. 5) is about 6 pc from the earth. The largest flare previously observed from this star in the optical region had an energy release of 10^{35} ergs integrated over a four-hour interval. If these lines originated from this star, the energy release would be 10^{36} ergs, a factor of 10 higher. On 10 June 1974, four hours prior

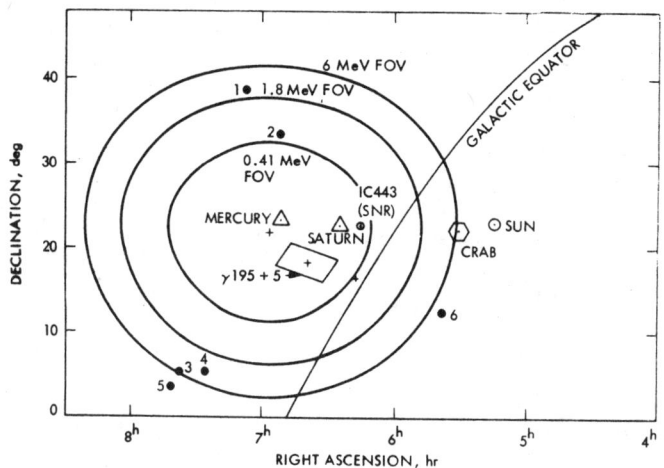

Figure 8. Field of view of 10 June 1974 Transient event.

to the JPL twenty-minute transient event, two X-ray flares[15] were observed by the Vela X-ray monitor from a source location RA=7^{hr} and DEC=8°. Unfortunately, no Vela coverage of this location was available during the period of the twenty-minute transient even though the source location was within the field of view of our instrument. Klebesadel[15] suggested that this location was also consistent with the reported position of three other gamma-ray bursts.

GAMMA-RAY TRANSIENT MODEL

A possible model for the 10 June 1974 transient was proposed by Lingenfelter, Higdon and Ramaty[16]. The model (Figure 9) required an episodic accretion of gas onto a neutron star from its companion star in a two-star system. As newly arriving gas collides with the hot gas in the accretion disk, copious neutrons and positrons will be produced. The 0.4 and 1.8 MeV lines can be identified as the gravitational red-shifted annihilation and neutron capture lines in a hydrogen atmosphere near the surface of the neutron star created by the infall gas from the accretion disk. The width of these lines suggested that the region ranged from the surface of the neutron star with z=0.28 up to an altitude corresponding z=0.2. The 5.9 MeV line may be related to the red-shifted component of the two strongest lines (7.632 & 7.646 MeV) from neutron capture iron crust of the neutron star with a surface gravitational red-shift factor of z=0.285. The unshifted 2.2 MeV photons are produced in a region well above the neutron star surface, either in the accretion disk with z=0.002 or in

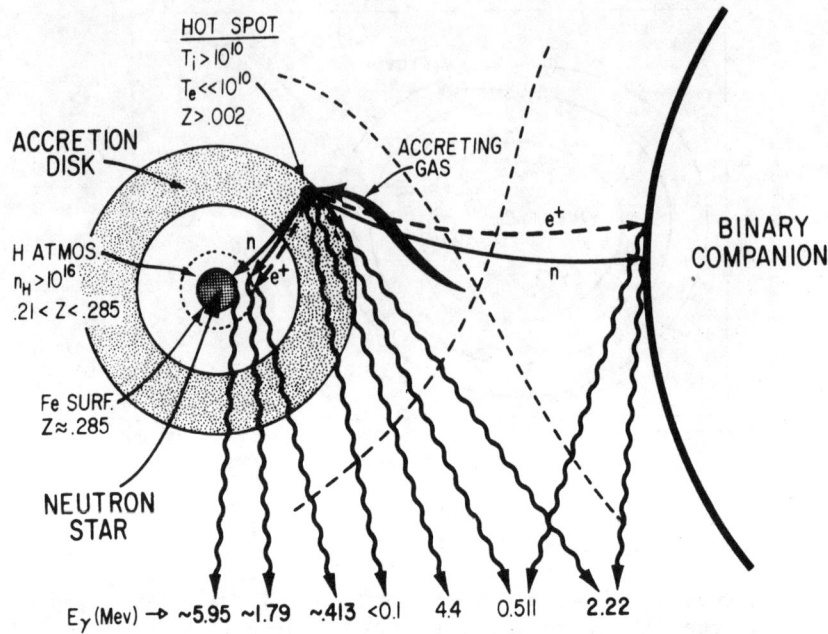

Figure 9. A schematic model for the 10 June 1974 transient event (from Lingenfelter, Higdon and Ramaty[16]).

the atmosphere of the binary companion. The unshifted 511 keV line is estimated to have comparable intensity to the unshifted 2.2 MeV line. However, the line is not detectable due to a large detector background at this energy.

CONCLUSION

The field of gamma-ray "line" astronomy has certainly made significant advances during the last decade. It is interesting to note that to date all of the confirmed line features have been shown to be varying with a wide variety of time scales ranging from seconds (the cyclotron and red-shifted annihilation lines observed in bursts[12,13] and point sources [8-11]), minutes (lines observed in solar flares[17]), hours (the 73 keV line from the Crab Nebula [5-7]) and possibly to months (the most recent result of the 511 keV annihilation line from the galactic center[18]). These are several other reported lines from the galactic center[19], galactic plane[20] and Cen A[21] which have yet to be confirmed. However, it is possible that some of them may also be time varying. The twenty-minute line is a unique phenomenon but has yet to be verified. We hope that future observations provide corroboration for this type of event.

ACKNOWDEDGEMENTS

We thank Dr. W.A.Wheaton and Ms. V.O'Brien for their helpful comments. This research was supported under NASA contract NAS7-100.

REFERENCES

1. Jacobson, A.S.1977, Invited Paper, American Physical Society Meeting, Washington,D.C., 25-28 April.
2. Jacobson, A.S., Ling, J.C., Mahoney, W.A. & Willett, J.B. 1978, GSFC Gamma Ray Spectroscopy in Astrophysics,NASA T.M.79619, 228.
3. Teegarden,B.J. 1981,this workshop.
4. Cline,T.L. 1981, this workshop.
5. Ling, J.C., Mahoney, W.A., Willett, J.B. & Jacobson, A.S. 1979, Ap. J.,231,896.
6. Ubertini,P., Bazzano,A., La Padula,C.D.,Polcaro,V.F. & Manchanda, R.K. 1980, Proc. Vth ESA-PAC Symp.,Bournmouth,U.K.
7. Manchanda,R.K., Bazzano,A., La Padula, C.D., Polcaro,V.F. & Ubertini, P. 1981, Ap. J. in press.
8. Strickman,M.S., Johnson,W.N. & Kurfess,J.D. 1981, Bull. of the 158th AAS Meeting in Calgary,Canada.
9. Johnson, W.N., Strickman, M.S. & Kurfess,J.B. 1981,this workshop.
10. Trumper,J., Pietsch, W., Reppin, C., Voges, W., Staubert, R. and Kendziurra, E., 1978, Ap. J. (Letters) 219, L105.
11. Wheaton, W.A., Doty, J.P., Primini, F.A., Cooke, B.A., Dobson, C.A., Goldman, A., Hecht, M., Hoffman, J.A., Howe,S.F., Scheepmaker, A., Tsiang, E.Y., Lewin, W.H.G., Matteson, J.L., Gruber, D.E., Baity, W.A., Rothschild, R., Knight, F.K., Nolan, P., Peterson, L.E. 1979, Nature, 282,240.
12. Teegarden, B.J., and Cline, T.L.1980, Ap.J. (Letters),236,L67.
13. Mazets, E.P., Golenetskii, S.V., Ilyinskii, V.N., Guryan,Yu.A., Aptekar, R.L., Panov, V.N., Sokolov, I.A., Sokolova, Z.Ya., and Kharitonova, T.V. 1981, preprint.
14. Leventhal, M., MacCallum, C.J. and Watts, A.C. 1977, Nature, 266,696.
15. Klebesadel, R.W. 1981, private communication.
16. Lingenfelter, R.E., Higdon, J.C. and Ramaty, R. 1978, GSFC Gamma Ray Spectroscopy in Astrophysics, NASA T.M. 79619,252.
17. Chupp, E.L., Forrest, D.J., Higbie, P.R., Suri, A.N., Tsai, C. and Dunphy, P.P. 1973, Nature, 241,333.
18. Riegler, G.R., Ling, J.C., Mahoney, W.A., Wheaton, W.A., Willett, J.B., Jacobson, A.S. and Prince, T.A. 1981,Ap.J.(Letters), in press.
19. Haymes,R.C., Walraven, G.D., Meegan, C.A., Hall, R.D., Djuth, F. T. and Shelton, D.H. 1975, Ap.J.,201,593.
20. Imhoff, W.L. and Nakano, G.H. 1977, Ap. J. 214,38.
21. Hall, R.D., Meegan, C.A., Walraven, G.D., Djuth, F.T., Shelton, D.H., and Haymes, R.C., 1976, Ap. J., 210, 631.

TIME VARIATIONS OF AN ABSORPTION FEATURE IN THE

SPECTRUM OF THE GAMMA-RAY BURST ON 1980 APRIL 19

B.R. Dennis, K.J. Frost, A.L. Kiplinger, L.E. Orwig
Laboratory for Astronomy and Solar Physics

U. Desai and T.L. Cline
Laboratory for High Energy Astrophysics
NASA/Goddard Space Flight Center, 20771

ABSTRACT

A short but intense γ-ray burst was observed on 1980 April 19 at 01^h 19^m 46^s with the Hard X-ray Burst Spectrometer on the Solar Maximum Mission. The event lasted for a total of 10 s and consisted basically of three 200 - 300 ms wide spikes with significant variability on a time scale of 20 - 40 ms. The photon number spectrum integrated over the impulsive part of the event fits a thermal bremsstrahlung function with a temperature of (330 ± 70) keV at energies from 151 to 487 keV. At lower energies the data points lie considerably below this function suggesting a broad absorption feature extending down to ≤ 28 keV, the lowest energy measured. The upper energy of this absorption feature varies between 100 and 150 keV on a time scale of ≤ 0.5 s. This event is interpreted as a typical gamma ray burst although it is still remotely possible that it is of solar origin. The spectral features and their variability are interpreted in terms of electron interactions at the cyclotron resonance frequency in magnetic fields of 10^{12} - 10^{13} gauss close to the surface of a neutron star.

INTRODUCTION

Absorption and emission features in the high energy X-ray spectra of compact astrophysical objects are believed to result from magnetic bremsstrahlung in intense fields of the order of 10^{12} - 10^{13} gauss. At these intense field strengths the plasma cyclotron frequencies are in the 10 to 100 keV energy range. Emission and absorption at such frequencies create spectral features that can be used to determine magnetic and plasma conditions in the vicinity of the compact object.

The first observation of such a feature was made by Trümper et al.[1] in the spectrum of Her X-1. This feature, now confirmed by Gruber et al.[2], can be interpreted either as a cyclotron emission line at ~ 58 keV or as an absorption dip at ~ 42 keV. Such emission and absorption features have also been detected in the spectra of some gamma-ray bursts by Mazets et al.[3,4] They observed absorption features extending below ~ 100 keV in the spectra of 20 out of ~ 150 gamma ray bursts detected on the Venera 11 and 12 space probes,

0094-243X/82/770153-10$3.00 Copyright 1982 American Institute of Physics

and they observed an emission feature at 45 keV in the spectrum of one burst. They obtained spectra every 4s throughout the events and in some bursts observed large variations in the observed features. Their observations of absorption and emission lines interpreted as cyclotron features provide the most convincing evidence to date that gamma ray bursts are produced by neutron stars.

In this letter we present high energy X-ray spectral data for a gamma-ray burst with over an order of magnitude finer time resolution than that obtained by Mazets et al.[3] The measured photon number spectrum can be interpreted as showing a broad absorption feature at energies below ~ 100 keV similar to the features reported by Mazets et al.[3,4] for other events. These observations support the idea that gamma-ray bursts are produced in the vicinity of neutron stars.

OBSERVATIONS

The gamma ray burst was detected on 1980 April 19 at 01^h 19^m 46^s with the Hard X-ray Burst Spectrometer (HXRBS) on the Solar Maximum Mission (SMM). The HXRBS is an actively shielded and collimated CsI(Na) scintillation spectrometer designed primarily to detect high energy X-rays from solar flares[5]. Although the instrument is pointed at the sun at all times, any off-axis source is detectable within a 40° FWHM circular field of view. Outside this field of view the central detector is shielded by a minimum of 1" of CsI(Na) operated in anti-coincidence.

The arrival direction of the burst was determined to be ~ 18° from the sun using the detection times at SMM, Vela 5 and Vela 6 in earth orbit, at ISEE-3 located 4.5 light seconds away in the solar direction, and at the Pioneer-Venus Orbiter[6]. There is still a remote possibility, however, that the source was of solar or near-solar origin. Nevertheless, it is clear that the source was well within the field of view of the HXRBS, and no significant spectral distortions should have occurred. We also know that, at the time of the event, the Earth was well out of the HXRBS field of view, the horizon being ⩾ 50° off axis. Thus, fluorescence from the earth's atmosphere could not have contributed to the source spectrum measured during the burst.

The time profiles of the event in several different energy bins covered by the HXRBS are shown in Figures 1 and 2. The event is typical of many gamma-ray bursts and consists basically of three sharp, 200 - 300 ms wide spikes separated by 500 - 600 ms. Significant variations in flux are evident on time scales as short as 20 - 40 ms. A gradual flux increase precedes the first spike by 1 to 2 s and a general decay is evident for several seconds after the third spike. The fluctuations in the decay phase of the event do not show significant power at any frequency from 0.5 to 10 Hz.

The time profiles in the different energy ranges plotted in Figure 2 show clearly that the spectrum changes significantly throughout the event. The first spike is barely visible above a general plateau at energies below 126 keV but it shows up clearly in the 126

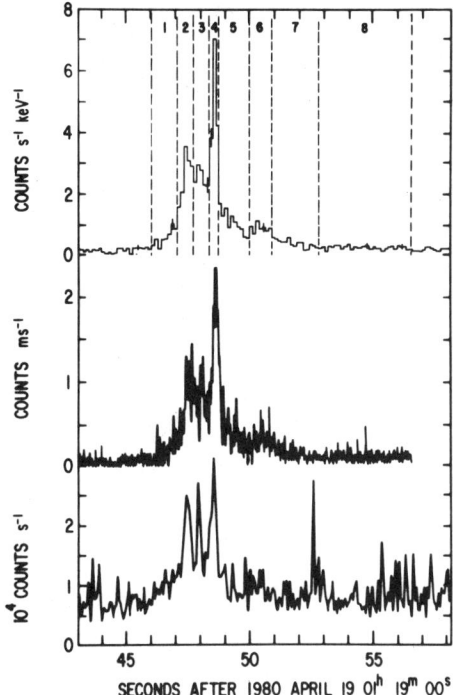

FIGURE 1 : Time profiles of the gamma-ray burst in different energy ranges and with different time resolutions. (a) Counting rate in the central detector for the energy range from 28 to 487 keV with a time resolution of 128 ms. The time intervals used for the energy spectra in Figures 3, 4 and 5 are indicated. (b) Counting rate in the central detector for the same energy range but with a time resolution of 20 ms. This data has not been corrected for instrument dead time, which changed from ~ 5% before and after the burst to ~ 15% at the time of the biggest spike in interval #4. (c) Counting rate in the CsI(Na) shield crystal for all energies $\gtrsim$ 200 keV with a time resolution of 64 ms. The spike at 01^h 19m 53s and all fluctuations below ~ 1200 counts s^{-1} are variations in the background rate.

to 151 keV and the 203 to 232 keV energy ranges and also in the shield data at $\gtrsim$200 keV shown in Figure 1. The second spike is clearly resolved only in the shield data. The third spike is also not well resolved at energies below 78 keV but shows up strongly at higher energies.

The photon number spectrum for the impulsive part of the event (intervals 2, 3 and 4 as indicated in Figure 1) is plotted in Figure 3. The spectral points were obtained iteratively by de-

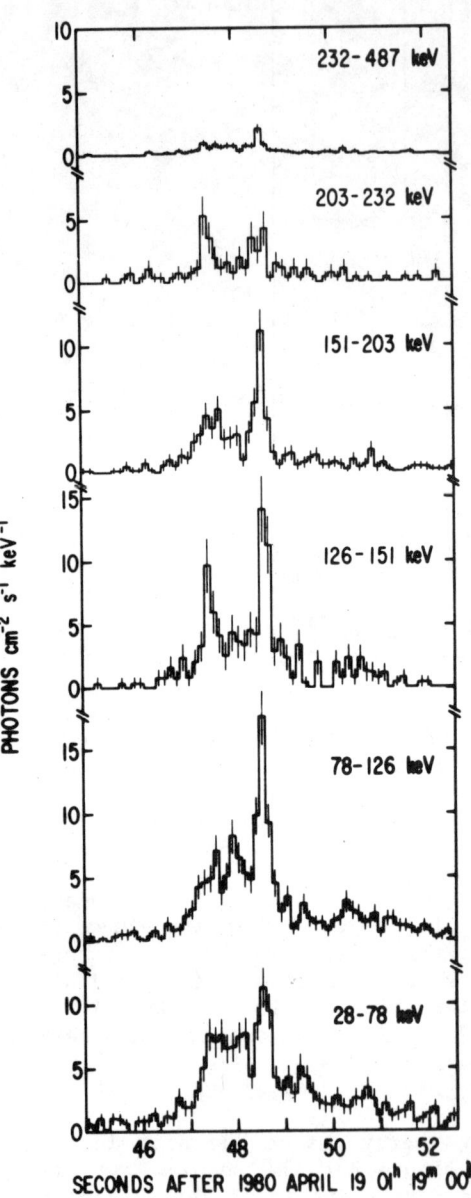

FIGURE 2 : Time profiles of the gamma-ray burst for the indicated energy ranges all with a time resolution of 128 ms.

convolving the observed 15-channel counting rate data through the instrument response matrix for an assumed incident spectrum equal to the least-squared fit exponential function indicated in Figure 3. The instrument response matrix was computed for different incident spectral shapes using the energy calibration determined for pre-launch data and in-orbit data from the the on-board Am^{241} source[5]. The computations take into account the detector efficiency and resolution, K-escape and Compton scattering using the results of pre-launch source calibrations and Monte Carlo simulations. To determine the accuracy of the calculations, comparisons were made of HXRBS spectra for several solar flares with spectra of the same flares obtained with the Gamma Ray Experiment (GRE) on SMM (Ryan, private communication), and with the hard X-ray detector on ISEE-3 (Kane, private communication). These comparisons suggest that systematic uncertainties are less than ± 30% for any of the 15 channels although all of the flare spectra are considerably steeper (E^{-4}) than the spectrum of the gamma ray burst. An additional systematic uncertainty arises from the probable 18° off axis position of the source. This results in a reduction in sensitive area by a factor of ~ 2 below the 68.6 cm^2 for an on-axis source, with the consequent increase in the calculated photon fluxes by the same factor. As indicated in Figure 3, an exponential spectrum of the form

photon flux = $A \exp(-E/E_o)$

is an acceptable fit to all the data points. Here E is the photon

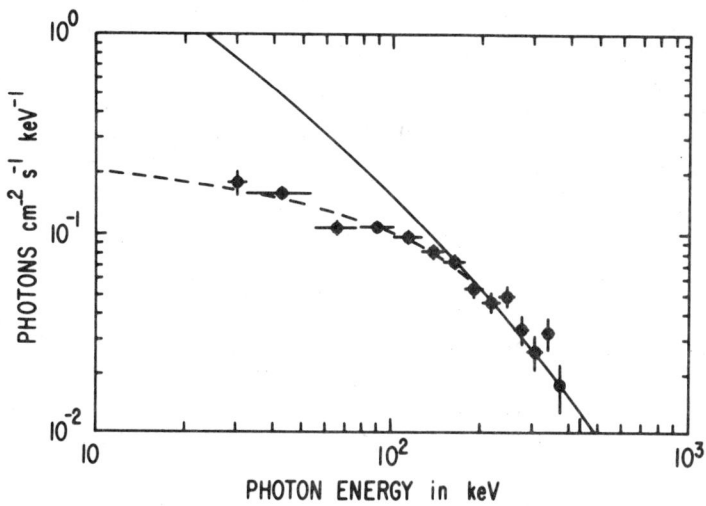

FIGURE 3 : Photon number spectrum for the intervals 2, 3 and 4 indicated in Figure 1 (impulsive part of the event). The error bars are ± 1σ based on the square-root of the observed number of counts in each energy interval. The solid curve represents the thermal bremsstrahlung function with kT = 330 keV, which gave the least-squares fit to the data points above 151 keV (channels 7-15). The broken curve represents the exponential function giving the best fit to all the data points.

energy and A and E_0 are free parameters determined in the fitting procedure. The value of E_0 obtained is 159 ± 8 keV and χ^2 is 13.5 for 13 degrees of freedom with the counting rate statistics alone used to weight the points. A double power-law function also fits the data equally well with a spectral index of -0.5 ± 0.1 at energies below 300 keV and -4.0 ± 1.0 at higher energies. The apparent dip in the spectrum in channel 3 is only significant at the 2.2σ level below the fitted exponential function. It may result from an error in the value used for the thickness of the dead layer on the CsI central crystal[7]. This dead layer, estimated to be 5 mils thick from measurements of the ratio of the 22 keV to the 88 keV photon fluxes from Cd^{109}, results in a dip in the count rate spectrum at energies above the K edges of cesium and iodine (33 and 36 keV respectively). Consequently, we would expect any inaccuracy in the assumed thickness of this dead layer to result in a dip in channel 2 although such a dip could appear in channel 3 if the incident spectrum had a steep low energy component. No such dip in channel 2 or 3 has been observed in any solar flare spectrum.

Following Mazets et al.[3] and Gilman et al.[8], we have also used the function expected for bremsstrahlung from an optically thin plasma at a temperature, T. This function is of the form

$$\text{photon flux} = B\, E^{-(1+\alpha)} \exp(-E/kT)$$

where k is Boltzman's constant, α results from the temperature-averaged Gaunt factor, and B and T are free parameters determined in the fitting procedure. The value of α changes from 0.48 to 0.148 as kT varies from 10 keV to 1 MeV [9], but for comparison with the results of Mazets et al.[3,4], we have taken α = 0 in all of our analysis. Using the correct value of α would increase the significance of the absorption feature and also increase the least-squares fit temperature. An acceptable fit of this thermal bremsstrahlung function to all 15 data points is not possible since it can never be flatter than $E^{-(1+\alpha)}$. However, an acceptable fit can be obtained to the data for channels 7 to 15 corresponding to the energy range from 151 to 487 keV and this is shown in Figure 3. The best fit value of kT is 330 ± 70 keV. At energies below 151 keV the data points lie considerably below the level of this function, at least a factor of 2 lower at 60 keV, suggesting a broad absorption feature extending down to ≤ 28 keV, the lowest energy measured. For the purposes of investigating spectral changes with time, the event was divided into the eight contiguous intervals indicated in Figure 1. The spectral data shown in Figure 4 was obtained assuming the best-fit exponential spectrum to all the data points in each interval. The thermal bremsstrahlung function was then fitted to as many of the higher energy points as possible while maintaining an acceptable value of χ^2. The resulting curves are plotted in Figure 4.

The spectra for the eight intervals show a general softening with time. The temperature obtained from the fits changes from ≳ 400 keV for interval #1 to 70 ± 20 keV for interval #8. The absorption feature appears in all intervals except #8 and possibly #5. In intervals 1, 2 and 4 the absorption feature extends from at least 150 keV down to ≤ 28 keV, the lowest energy measured. In intervals 3, 6 and 7 the absorption appears to extend only to ~ 100 keV or less. The apparent emission features at 330 keV in interval #2 and 260 keV in interval #3 are only significant at the ~ 2σ level and, in any case, they are narrower than the instrument could allow. The spectrum for interval #7 can be interpreted as containing an emission feature at 100 keV with a high temperature continuum. The lower temperature fit (kT = 50 ± 20 keV) with absorption below ~ 100 keV as indicated in Figure 4 appears more likely, however, in view of the normal "cooling" observed by Mazets et al.[4] in many gamma ray bursts, and the low temperature continuum (kT = 70 ± 20 keV) measured in interval #8.

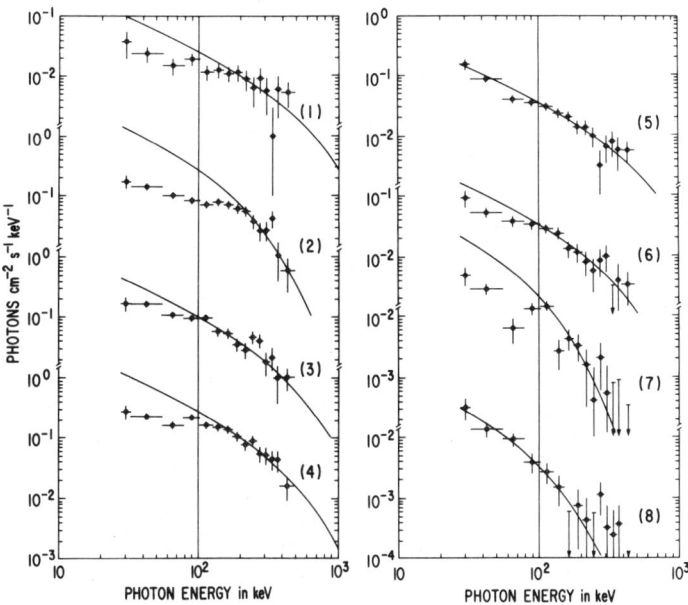

FIGURE 4: Same as Figure 3 but for the individual intervals 1-8 indicated in Figure 1. The curves represent the thermal bremsstrahlung function giving the least-squares fit to as many of the higher energy data points as possible while still obtaining an acceptable value of χ^2. The values of kT in keV for the indicated curves are 406, 229, 510 and 302 for intervals 1, 2, 3 and 4, and 375, 229, 52, and 73 for intervals 5, 6, 7, and 8 respectively.

DISCUSSION

It is now almost certain that the April 19 burst was of nonsolar origin. The timing information from five separate spacecraft gives an error box ~ 18° from the sun. The rapid flux variations, the short duration of the event, and its extremely hard spectrum are all typical of gamma ray bursts. Furthermore, the sun was very quiet at this time with no reported Hα flares. A small soft X-ray event was in progress at the C2 level as observed by the GOES 2 and GOES 3 satellites (Donnelly, private communication), and a small GRF radio burst was observed at 3750 MHz from NOAA region #2396 near the east limb at the same time (Enome, private communication). The SMM pointed instruments were observing region #2389 near the west limb and saw nothing. A coronal transient was observed off the N.W. limb by the SMM Coronagraph/Polarimeter in this time frame (Sawyer, private communication) but was probably not connected with the X-ray event. Thus, it is clear that, if the hard X-ray event was of solar or near-solar origin, it would constitute a new type of event never previously detected. Consequently, we prefer to assume that the event is a gamma ray burst and will interpret the data on that basis.

The photon number spectrum obtained for this event cannot be interpreted unambiguously. Observations of gamma rays to energies in excess of 5 MeV in this event[10] show that the simple exponential fit to the HXRBS data alone is probably not correct at higher energies. A two-power-law fit could, however, represent the data. It is unclear at present if the thermal bremsstrahlung function could also fit all the high energy data above ~ 150 keV. In any case, there are many difficulties associated with the assumption that the emission is thermal bremsstrahlung from an optically thin hot plasma[4]. Radiation other than from free-free electron transitions would predominate from the expected nonequilibrium, relativistic and strongly magnetized plasma near the surface of a neutron star. Also, this plasma, while optically thin for free-free transitions, would be optically thick for Compton scattering with consequent strong modification of the spectrum. Nevertheless, the thermal bremsstrahlung function does provide an accurate fit to the spectra of most events detected by Mazets et al.[3,4] in the energy range from 30 keV to 2 MeV and to the spectrum of the one event detected by Gilman et al.[8] from 2 keV to 2 MeV. Thus, this functional form can be considered as representing the source emission, whatever its origin. The deviations from this spectral form observed at low energies in some bursts are then naturally interpreted as resulting from absorption of this source radiation.

The observed absorption feature in the X-ray spectrum and its variation with time strongly support the model of a neutron star as the source of the gamma ray burst. The time variations in the total flux on a scale as short as 20 to 40 ms indicate that the source size is $\leq 10^9$ cm. If the burst results from accretion onto a neutron star, then the bulk of the radiation probably originates at or near the magnetic poles, relatively close to the surface. The observations show that the equivalent temperature of the source is of the order of 300 keV. As suggested by Mazets et al[3,4], the flattening of the spectrum at energies below ~ 100 keV may be produced by absorption at the cyclotron frequency by a cooler plasma overlying the source in the presence of a strong magnetic field. The fact that the absorption extends over a broad energy range down to at least 28 keV indicates that the magnetic field strength within the absorbing region must range from $<2 \times 10^{12}$ gauss to $\sim 10^{13}$ gauss. The increase in the energy of the upper edge of the absorption feature in intervals 1, 2 and 4 to $\simeq$ 150 keV presumably is a consequence of higher field strengths in the absorbing region suggesting that the source of emission at these times is closer to the surface of the neutron star.

The possibility of an emission feature appearing in interval 7 could indicate that, at this time, the source of emission had moved to a greater altitude above the neutron star, similar to the altitude of the previous absorption. The absence of any spectral feature in the spectrum for interval 8 may result from equal line emission and absorption. Alternatively, the emission region may by this time have moved to an even greater altitude with much lower magnetic fields. Detailed modeling is clearly required before any further understanding of these data can be achieved.

It is interesting to note that two similar though weaker events were detected within two days of the April 19 event. One was detected two days earlier on 1980 April 17 at 1210 UT with the X-ray detector on ISEE-3 and the other two days later on 1980 April 21 at 0308 UT with the HXRBS and the GRE on SMM[10]. The time profiles of all three events show three sharp spikes, and the energy spectra of the April 19 and 21 events are very similar. Unfortunately, neither event has been reported from more than one spacecraft and hence source triangulations have not been possible. Nevertheless, these three events are strongly suggestive of a repetitive nature of the same source.

In conclusion, if we accept the non-solar origin of this event, we have confirmed the results of Mazets et al.[3,4] on the existence of an absorption feature at X-ray energies in the spectrum of a gamma-ray burst. In addition, we have shown variability of the extent of the absorption feature on a time scale of 0.5 s during the event and detected a possible emission feature at ~ 100 keV later in the event. These results combined with the emission line feature observed by Mazets et al.[3] at 45 keV in one burst are strongly suggestive of the quantization of electron energies in intense magnetic fields in the source region. If this interpretation is correct, then the source of most gamma-ray bursts must be in the vicinity of neutron stars within our galaxy. If this event is, however, of solar or near solar origin, then it is a most unusual event and would require reinterpretation on this basis.

We acknowledge the efforts of the many people who have contributed to the success of SMM and the HXRBS. We thank the personnel of Computer Sciences Corporation led by Mr. T. O. Chewning for their programming efforts which made the analysis of the data possible. We also thank Dr. G. H. Share for many valuable discussions of the data. We are grateful to Dr. E. E. Fenimore for carrying out Monte Carlo simulations of the HXRBS, which were invaluable in determining the instrument response as a function of energy, and to Dr. S. R. Kane and Dr. J. M. Ryan for providing X-ray and gamma ray spectra for solar events measured on ISEE-3 and with the GRE on SMM.

REFERENCES

1. Trumper, J., Pietsh, W., Reppin, C., Voges, W., Staubert, R., and Kendziorra, E., Ap. J.(Letters), 219, L105 (1978).

2. Goodman, N.B., Spac. Sci. Instr., 2, 425 (1976).

3. Mazets, E.P., Golenetskii, W.V., Aptekar, R.L., Gur'yan, Yu.A., and Il'inskii, V.N., 1981a, Nature, 290, 378 (1981).

4. Mazets, E.P., Golenetskii, S.V., Ilinskii, V.N., Gur'yan, Yu.A., Aptekar, R.L., Panov, V.N., Sokolov, I.A., Sokolova, Z.Ya., and Kharitonova, T.V., preprint 719, A.F.IOFFE Physical-Technical Institute, Leningrad, submitted to Astrophysics and Sp. Sci. (1981).

5. Orwig, L.E., Frost, K.J., and Dennis, B.R., Solar Physics, 65, 25 (1980).

6. Fenimore, E.E., Laros, J.G., Klebsadal, R.W., Stockdale, R.E., and Kane, S.R., preprint, to be submitted to Ap. J. (1981).

7. Gruber et al., Ap. J. (Letters), 240, L127 (1980).

8. Gilman, D., Metzger, A.E., Parker, R.H., Evans, L.G., and Trombka, J.I., Ap. J., 236, 951 (1981).

9. Matteson, J.L., Ph.D. thesis, University of California at San Diego (1972).

10. Share, G.H., Strickman, M.S., Kinzer, R.L., Chupp, E.L., Forrest, D.J., Ryan, J.M., Reiger, E., Reppin, C. and Kanbach, G., preprint, to be published in the Late Volumes of the 17th International Cosmic Ray Conference, Paris, France (1981).

SEARCH FOR TIME VARIATIONS IN 511 KeV FLUX BY ISEE-3 GAMMA-RAY SPECTROMETER

Jay P. Norris
University of Maryland, College Park, MD 20742

Thomas L. Cline and Bonnard J. Teegarden
NASA/Goddard Space Flight Center, Greenbelt, MD 20771

ABSTRACT

The ISEE-3 gamma-ray spectrometer has provided nearly continuous monitoring of the cosmic gamma-ray background in the energy regime 125 keV to 6.5 MeV since launch of the satellite in August 1978. The data has been analyzed for possible variations in the cosmic 511 keV line flux. The detector is an unshielded, radiatively-cooled, high-purity germanium crystal with a sensitive volume 33 cm^3. At energies > 100 keV the detector has a field-of-view of $> 2\pi$ steradians which includes the galactic center. The detector resolution was degraded continuously by exposure to high-energy charged particles in the interplanetary space environment; consequently, only the first 500 day sample of data is usable. Proton and electron flux rates available from a cosmic-ray experiment aboard ISEE-3 were used to eliminate all periods during which solar energetic particles contributed significantly to our counting rate. We are able to place upper limits on the variability of the cosmic 511 keV flux of $< 2.2 \times 10^{-2}$ and $< 1.3 \times 10^{-2}$ photons cm^{-2} s^{-1} for $\simeq$ 20 day and $\simeq$ 60 day integrations, respectively, for the period October 1978 to February 1980. Comparison is made with the HEAO-3 511 keV observations of the galactic center.

INTRODUCTION

During the past decade several observations of the galactic center region have produced positive detections of emission in the vicinity of the 511 keV annihilation line [1,2,3,4,5,6,7]. The observed fluxes range from 0.8×10^{-3} up to 4.2×10^{-3} photons cm^{-2} s^{-1}. The first observations from a satellite-borne experiment appear to support the interpretation that the emission may originate from a time-varying "point" source. Since the available data have been sporadic, it would be desirable to have a continuous record of observations over an extended period. We report here upper limits to the variation of galactic 511 keV flux as observed by the ISEE-3 gamma-ray spectrometer for 500 days of nearly continuous observation.

INSTRUMENTATION

The ISEE-3 (Third International Sun-Earth Explorer) spacecraft is in a halo orbit about the Lagrangian point on the earth-sun line

0094-243X/82/770163-06$3.00 Copyright 1982 American Institute of Physics

230 earth radii (4.5 light-seconds) inward towards the sun. This location is particularly advantageous; the time-varying exposure to trapped radiation within the magnetosphere and the earth-albedo problem are completely avoided.

The spectrometer is approximately aligned with the spacecraft spin axis, that is, normal to the ecliptic plane. The instrument is recessed within the lower body of the spacecraft, preventing direct sunlight from destroying the thermal performance of two nested radiatively-cooled stages. The detector is housed in the inner stage. The outer stage is conically shaped to define a 126° (thermal) field-of-view for the inner stage. This field-of-view does not contain the sun, earth, or any significant portion of the spacecraft [8].

The detector is an unshielded p-type co-axial high-purity germanium crystal, 4 cm diam. x 3 cm depth, with an active volume of 33 cm^3. Pre-launch tests measured an energy resolution of 3-3.5 keV at 1 MeV. A higher temperature (130 K) than the predicted equilibrium temperature of 100 K for the crystal was reached when the spacecraft was placed in orbit. Consequently, the initial operating resolution of a calibration line at 570 keV from a ^{207}Bi source was ≃ 8 keV (FWHM).

OBSERVATIONS AND ANALYSIS

The experiment was intended to monitor the celestial sphere for gamma-ray bursts, providing information on any narrow spectral features within the energy range 125 keV to 6.5 MeV. In the absence of a burst trigger, a background mode operates, permitting spectral analysis of the diffuse gamma-ray background radiation. The background mode has functioned successfully since September 1978. Because of the extremely low bit rate allocated to this experiment, the instrument analyses only one photon of ≃ 550 detected in the background mode. Hence, in order to accumulate a sufficient number of counts to constitute an observation, the spectra must be accumulated for a minimum of several days.

Energy calibration was achieved by monitoring two background lines at 570 keV and 1064 keV. For the purpose of rebinning counts into fiducial channels, it was assumed that the gain was a linear function of energy. The positions and widths of the lines in accumulated spectra were fit both manually and by an automatic gaussian line fitting routine. The gain (channels/keV) decreased very nearly linearly by about 10% during the first 650 days of experiment operation. Also, significant deterioration of the energy resolution occurred: the 570 keV line broadened from the initial 8 keV FWHM to ~ 50 keV in 600 days (Figure 1). After this period, the 511 keV and 570 keV lines began to overlap significantly. The relative magnitudes of the gain and resolution changes are consistent with a decreasing charge collection efficiency, evidently caused by detector radiation damage from high-energy charged particles and fast neutrons.

Time variation in the interplanetary charged particle flux due to solar modulation and other effects will produce variations in

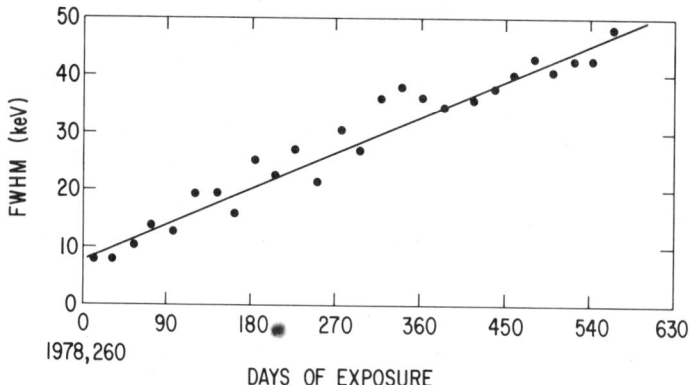

Fig. 1.- Resolution of 570 keV calibration line for 600 days from start of experiment.

the detector background that must be evaluated. Particle fluxes for several different proton energy ranges and for electrons were independently recorded by a cosmic-ray experiment aboard ISEE-3 and these data were kindly made available to us [9]. Those particles (energies ≃ 14 MeV to 18 MeV) which deposit energy in the detector within the PHA energy range originate primarily from solar flares. Therefore, to calculate the actual particle rate contribution, a canonical solar flare energetic particle spectrum proportional to $E^{-3.0}$ was assumed and the required rates extrapolated from the available data. Periods for which the calculated particle rate exceeded one-tenth of the random error for given spectral accumulation intervals were eliminated from the data base. From late 1978 to 1980 the sun was especially active. Consequently, the requirement for non-contaminated spectra resulted in eliminating up to 60% of the data.

Extraction of the astrophysical 511 keV flux from the "cleaned" data was complicated by two effects. First, a large fraction of the detected flux is instrumental in origin and its contribution is difficult to estimate. Furthermore, the capability to point the spectrometer does not exist; hence an on-source minus off-source type of observation cannot be conducted. At energies > 100 keV, the cooler stages and the spacecraft well into which the spectrometer is recessed are virtually transparent, affording the detector an effective field-of-view > 2π steradians. The galactic center region, from which a variable 511 keV flux may emanate, is constantly within the detector field-of-view. Assuming the galactic center is the most intense 511 keV source, for a galactic center observation it is sufficient to distinguish a variation in the total 511 keV flux. This subject will be addressed in the Discussion section of this paper. A second complication results from the broadening instrumental resolution due to detector exposure to particle radiation. Because of the low statistical significance of the accumulated spectra, it was not acceptable to employ a straightforward technique of fitting a gaussian line to the 511 keV feature and subtracting a continuum. Instead, to accumulate counts of ≃ 511 keV energy, an automatic "window" spectra program was employed which appropriately gain-adjusted the spectra and summed counts in four energy windows of widths 38, 50, 60, and 70 keV, centered at 511 keV. This treatment results in a systematic decrease of counts with time in a given energy window due to the broadening resolution. Therefore, a reduction technique was employed for each accumulated 511 keV window spectrum in which 1) an adjusted continuum computed from a window spectrum of width 40 keV, centered at 450 keV, was subtracted from the line plus continuum spectrum (the continuum above 511 keV was not used due to the presense of the strong calibration line at 570 keV); 2) a gaussian-shaped line which broadened with time commensurate with the observed broadening of the calibration lines was assumed for the 511 keV line, and normalized model curves were constructed which accounted for the systematic loss of counts from the energy windows; 3) the observed count rates and the models were subtracted to yield relative 511 keV count rate variations as a function of time. The count rates were then converted to fluxes by

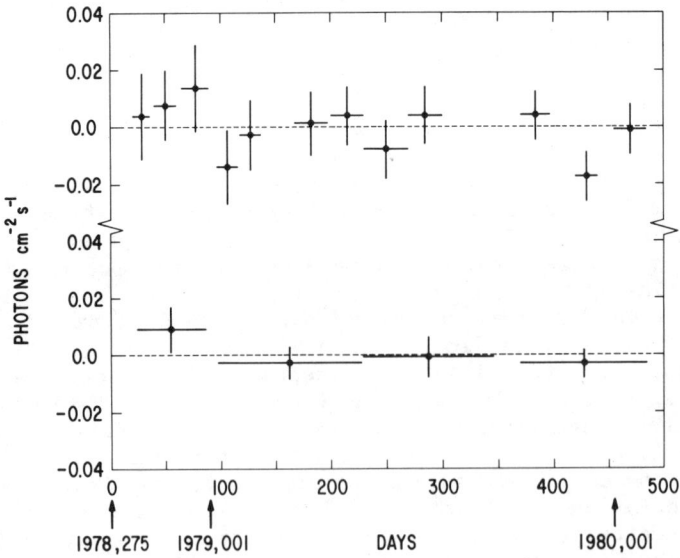

Fig. 2.- One sigma upper limits on variation of 511 keV flux. The integration intervals (live times) are approximately 20 days (top) and 60 days (bottom). The data are plotted with respect to the mean values (dotted lines).

dividing by the effective area of the detector. The wider energy windows have an inherently poorer sensitivity due to the inclusion of more background counts. For the narrowest (38 keV) window, the sensitivity decreases more rapidly with time as the 511 keV line progressively broadens, leaving less counts within the static window. The sensitivity is effected more drastically by the use of wider energy windows. The resulting fluxes for the 38 keV window are shown in Figure 2 for $\simeq$ 20 day and $\simeq$ 60 day integrations (live times). The two sigma upper limits are 2.2×10^{-2} (20 day) and 1.3×10^{-2} (60 day) photons cm^{-2} s^{-1} on the 511 keV flux variation from $> 2\pi$ steradians for the period September 1978 to February 1980. During the same period, the (> 60 MeV) particle flux determined from the ISEE-3 cosmic-ray experiment was practically constant ($\pm < 4\%$) for the intervals of retained data. The instrumental background results primarily from illumination of the spacecraft and activation of the detector crystal by this radiation. The percentage particle flux variation for any given spectral accumulation interval accounts for a maximum of less than one-fourth of the quoted limits on the 511 keV flux variation. No correlation is apparent between the (> 60 MeV) particle rate and our final results.

DISCUSSION

The galactic center observations performed with the HEAO-3 gamma-ray spectrometer support the interpretation that a time-varying source of 511 keV emission lies near the galactic center[7]. Several balloon experiments[1,2,3,4,5,6] indicate that the source flux may range from 0.8×10^{-3} up to 4.2×10^{-3} photons cm^{-2} s^{-1}. The ISEE-3 upper limits for the 511 keV flux may be examined in two ways. First, since the spectrometer field-of-view includes more than half of the galactic plane, the $\simeq$ 20 day observations exclude transient galactic sources ($\tau \geq$ one month) with fluxes $> 2.2 \times 10^{-2}$ photons cm^{-2} s^{-1} (95% confidence). If sources similar to the galactic center source were distributed uniformly thoughout the galaxy, a source within 2 - 3 kpc could have been detected by the ISEE-3 spectrometer. The lack of detection is evidence that the galactic center source is unique within our galaxy. Second, the $\simeq$ 125 day observations (60 day live time) constrain the peak galactic center source flux to $< 1.3 \times 10^{-2}$ photons cm^{-2} s^{-1} during a period of 500 days from October 1978 to February 1980.

We thank T. von Rosenvinge for providing us with particle data from an ISEE-3 cosmic-ray experiment. We also thank W. Paciesas for the automatic gaussian line fitting routine and for helpful discussions. J. P. Norris acknowledges support from a NASA Student Researchers Grant.

From a dissertation to be submitted to the Graduate School, University of Maryland, by Jay P. Norris, in partial fulfillment of the requirements for the Ph. D. degree in astronomy.

REFERENCES

1. W. N. Johnson, III, F. R. Harnden, Jr., and R. C. Haymes, Ap. J. (Letters), 172, L1 (1972)
2. W. N. Johnson, III, and R. C. Haymes, Ap. J., 184, 103 (1973).
3. R. C. Haymes, G. D. Walraven, C. A. Meegan, R. D. Hall, F. T. Djuth, and D. H. Shelton, Ap. J., 201, 593 (1975).
4. M. Leventhal, C. J. MacCallum, and P. D. Stang, Ap. J. (Letters), 225, L11 (1978).
5. M. Leventhal, C. J. MacCallum, A. F. Huters, and P. D. Stang, Ap. J., 240, 338 (1980).
6. F. Albernhe, J. F. Leborgne, G. Vedrenne, D. Boclet, P. Durouchoux, and J. M. da Costa, Astron. Astrophys., 94, 214 (1981)
7. G. R. Riegler, J. C. Ling, W. A. Mahoney, W. A. Wheaton, J. B. Willett, A. S. Jacobson, and T. A. Prince, submitted to Ap. J. (Letters).
8. B. J. Teegarden, G. Porreca, D. Stilwell, U. D. Desai, T. L. Cline, and D. Hovestadt, Gamma Ray Spectroscopy in Astrophysics, NASA TM79619, eds. T. L. Cline and R. Ramaty, 516 (1978)
9. T. von Rosenvinge, personal communication

Sec. I D Gamma Ray Transient Emission Processes

RELATIVISTIC PLASMAS*

Robert J. Gould
Physics Dept. B-019, UCSD, La Jolla, CA 92093

ABSTRACT

Recent work on the kinetic theory of a highly-relativistic electron gas and the associated photon opacity is summarized. The relaxation time for thermalization by Møller scattering is compared with the characteristic times associated with various loss processes which act to perturb the distribution away from equilibrium. It is found that, for an optically-thin plasma, bremsstrahlung dominates Møller scattering for $\Theta = kT_e/m_e c^2 \gtrsim 3.5$. Results are given for the opacity from Compton scattering, pair production in the fields of electrons and ions, inverse bremsstrahlung, and synchrotron self absorption.

INTRODUCTION

In this paper I shall give a brief review of certain properties of plasmas in which the electron gas is highly relativistic while the ions are non-relativistic. Details of the calculations leading to the results may be found in two recent papers.[1,2] In order to express the results in a more compact and transparent form, some slight changes in notation will be made, with formulas given in terms of factors that have been made dimensionless. This allows a more convenient comparison of corresponding formulas for various processes, exhibiting, in particular, the dependence on the fine structure constant (α), the total electron number density (n_e), the dimensionless electron temperature ($\Theta = kT_e/m_e c^2$), and the dimensionless electron and photon energies ($\Omega_e = E_e/m_e c^2$; $\Omega_\gamma = E_\gamma/m_e c^2$). The electron density may be made dimensionless by multiplying by Λ_e^3, where $\Lambda_e = \hbar/m_e c$ is the electron Compton wavelength; thus, we introduce $\nu_e = \Lambda_e^3 n_e$. Times may be made dimensionless by expressing them in units of $\tau_e = \Lambda_e/c = \hbar/m_e c^2$. Expressions for the photon absorption coefficient (κ), normally in cm^{-1}, become dimensionless by multiplying by Λ_e.

The existence of astrophysical plasmas of the type considered, with, say, $\Theta \sim 10-100$, has been suggested by various authors (see papers referred to in refs. 1 and 2). However, it turns out that, without specifying the detailed nature and origin of these plasmas, it is possible to draw some fairly general conclusions concerning thermalization and the maintenance of equilibrium in the distribution

---

*Research supported by NASA through grant NGR 05-005-004.

0094-243X/82/770169-09$3.00 Copyright 1981 American Institute of Physics

of electron kinetic energies. Surprisingly, little research had been carried out on this basic problem, even though a book[3] on relativistic kinetic theory has been published recently. There are still a number of problems that have not yet been worked out, although some, associated with mildly-relativistic ($\theta \sim 1$) plasmas, would be mathematically cumbersome. This review will consider only two aspects of relativistic plasmas: (i) the problem of thermalization and (ii) the calculation of the opacity. The opacity is very relevant to the thermalization problem because of the effects of, in particular, bremsstrahlung in perturbing the particle energy distribution provided the plasma is optically thin.

THERMALIZATION PROBLEMS

Møller Scattering. The cross section for elastic scattering of two relativistic electrons was first calculated by Møller. In the non-relativistic limit the process corresponds to ordinary Coulomb scattering. However, there are important and interesting differences between the relativistic and non-relativistic cases and the differences will be emphasized here as they were in refs. 1 and 2. For both relativistic and non-relativistic plasmas the particles in the high-energy tail are the last to relax by the binary collision process. Thus it is possible to have a distribution of particle energies in which the only appreciable deviation from equilibrium is at energies

$$E_e \gg \langle E_e \rangle \gg m_e c^2. \tag{1}$$

In collisions with the other electrons the particle of energy E_e will predominantly lose energy and the relaxation time can be computed from

$$\tau_M = -E_e / \langle \Delta E_1 / \Delta t \rangle. \tag{2}$$

The energy loss rate due to binary collisions is computed from

$$-\langle \Delta E_1 / \Delta t \rangle = \iint \Delta E_{lab} \, v_r \, d\sigma \, dn_2, \tag{3}$$

where ΔE_{lab} is the (lab-frame) energy exchanged from the energetic "test" electron ("1") to the plasma electron (which we can label "2" in the binary collision). When both electrons are highly relativistic the kinematic factor v_r approaches $c(1 - \cos\theta)$, where θ is the angle between their directions of motion. Note that in this limit v_r is the difference between the velocity of one electron and the projection of the other's along its direction. This is to be compared with the non-relativistic limit where v_r is simply the magnitude of the particles' relative velocity. The integrations in (3) are over the differential scattering cross section $d\sigma$ and the differential plasma density dn_2 (that is, over the distribution of plasma electron energies and directions). When the plasma distribution is thermalized, the result is, under the condition (1), which implies $\Omega_e \gg \Theta$,

$$\tau_e/\tau_M = \pi \alpha^2 (\nu_e/\Omega_e \Theta) L_M, \qquad (4)$$

where

$$L_M = \ln(\Omega_e/2\Theta) + \gamma_E - 7/8, \qquad (5)$$

γ_E being Euler's constant. The logarithm L_M is not a Coulomb logarithm. Although the result (4) is valid only when $\Omega_e/2\Theta \gg 1$, the argument of the logarithm in (5) is not large in the sense of a Coulomb logarithm. For example, for the energy loss of a highly-relativistic electron traversing a non-relativistic plasma the result would be the same as (4) with L_M replaced by[4] $4\Theta B_M$, with

$$B_M = \ln(\Omega_e^{1/2}/\Omega_p) + 9/16 - \tfrac{1}{2}\ln 2, \qquad (6)$$

in which $\Omega_p = \hbar\omega_p/m_e c^2$ is the dimensionless plasmon energy. Now generally Ω_p is extremely small and B_M, which is a coulomb-type logarithm, is of the order, say, 10 or 20. On the other hand, for $\Omega_e/2\Theta \sim 10$, $L_M \sim 1$ or 2.

The result (4) with the "Møller" logarithm (5) is thus very different from that for a non-relativistic plasma (there is an additional factor $1/4\Theta$ in the energy loss expression, but this is easy to understand). The lack of appearance of a Coulomg-type logarithm has to do with the small contribution to the energy loss rate from very small energy exchanges in binary collisions.[1] To aid in future comparisons of this result with that of other authors I have added an appendix to this review giving explicitly some of the steps of the derivation in ref. 1. It might be noted that there is, in fact, a recent paper[5] treating the problem in a somewhat different manner that has not obtained the same result, although the precise value of the authors' logarithmic factor has not been determined.

Finally, it may be of interest to know the formula for the energy loss rate of a positron traversing a relativistic electron gas when the positron energy satisfies (1). The problem differs from that for Møller scattering in that the cross section employed in (3) is now the Bhabha formula. In e^+-e^- scattering there are no exchange effects and we can consider center-of-mass scattering angles up to π (rather than $\pi/2$ for $e^\pm$-$e^\pm$ scattering) and thereby include scatterings in which the positron essentially transfers all of its energy to the plasma electron. The result[2] for Bhabha scattering is the same as (4) with L_M replaced by

$$L_{Bh} = \ln(2\Omega_+/\Theta) + \gamma_E - 8/3, \qquad (7)$$

in which Ω_+ is the dimensionless positron energy. This can also be compared with the result for a relativistic positron traversing a non-relativistic electron plasma. The result[4] would correspond to a rate with L_{Bh} replaced by $4\Theta B_{Bh}$, with [compare (6)]

$$B_{Bh} = \ln(\Omega_+^{1/2}/\Omega_p) - 11/24 + \tfrac{1}{2}\ln 2. \qquad (8)$$

<u>Perturbing Processes.</u> As opposed to Møller and Bhabha scattering, which are normal transport processes, there are several processes which act as "sink" terms in a Boltzmann equation. These processes always act to remove electrons from the tail of the distribution if they are not balanced by the associated reverse process. For example, in scattering off ions or other electrons (or positrons), relativistic electrons can emit bremsstrahlung photons. A characteristic bremsstrahlung loss time can be defined by

$$\tau_B = u_e/\varepsilon_B, \qquad (9)$$

where $u_e\ (=3n_e m_e c^2 \Theta)$ is the energy density of the electron gas and ε_B is the rate of energy emitted per unit volume. Again in terms of the time $\tau_e = \hbar/m_e c^2$ defined earlier, the result is[2]

$$\tau_e/\tau_B = 4\alpha^3 \nu_e L_B, \qquad (10)$$

where L_B is a logarithmic factor that accounts for bremsstrahlung in electron-electron and electron-ion scattering. It is also of interest to compare the Møller and bremsstrahlung times:

$$\tau_M/\tau_B = (4\alpha/\pi)(L_B/L_M)\Omega_e \Theta. \qquad (11)$$

If Ω_e is set equal to a characteristic value $\sim \Theta$ for the distribution and L_M is set equal to a characteristic value of unity, we find that $\tau_M/\tau_B > 1$ for $\Theta \gtrsim 3.5$. Thus, even though bremsstrahlung is a higher order (in α) process, it dominates Møller scattering already at moderately relativistic temperatures.

A process that is of still higher order in α is pair production in electron-electron and electron-ion collisions. The associated characteristic time can be defined as

$$\tau_{pp} = n_e/\dot{n}_{pp}, \qquad (12)$$

where $\dot{n}_{PP}$ is the rate of production of pairs per unit volume. The result is[2]

$$\tau_e/\tau_{PP} = (28\alpha^4/27\pi)\nu_e S_{PP}, \qquad (13)$$

where S_{PP} is another logarithmic factor. When compared with Møller scattering it is found that $\tau_M/\tau_{PP} > 1$ for $\Theta > 34$.

If there is a magnetic field present, the relativistic electrons will also lose energy by synchrotron radiation. Defining a loss time by

$$\tau_s = u_e/\varepsilon_s, \qquad (14)$$

where ε_s is the rate of emission of energy in synchrotron photons per unit volume, we easily obtain[2]

$$\tau_e/\tau_s = (128\pi\alpha^2/9)\nu_B \Theta^2. \qquad (15)$$

Here ν_B is a dimensionless magnetic field energy density, defined as

$$\nu_B = (\lambda_e^3/m_e c^2) u_B, \qquad (16)$$

where $u_B = \langle B^2 \rangle/8\pi$. In the presence of a photon field there would also be a loss term from Compton scattering. The corresponding time would be given by a similar expression with u_B replaced by the energy density in photons, assuming that the Thomson limit applies. Comparing the result (15) with (4) for Møller scattering, we see that if there is "equipartition" ($\nu_e \Theta \sim \nu_B$), synchrotron radiation would always dominate thermalization by Møller scattering.

<u>Interaction with Collective Modes.</u> It has been suggested that interactions with plasma wave modes can provide a thermalization mechanism for a relativistic electron gas. However, I think this is highly doubtful as a means of relaxation to equilibrium. For such interactions to drive a distribution to thermalization the wave modes must themselves be thermal in nature. No one really expects these waves, if they exist, to be thermal. Thus it seems that Møller scattering is the only established thermalization process.

OPACITY OF A RELATIVISTIC PLASMA

The problem of the opacity is of interest and is connected with that of thermalization. Here I shall summarize results of opacity calculations,[2] again expressing them in terms of dimensionless factors and in compact form. The various processes have different dependence on density, photon energy, magnetic field, etc.

Compton Scattering. For photon energy small enough that it is much less than $m_e c^2$ in the rest frame of the plasma electrons, the Compton process is in the Thomson limit. In terms of the dimensionless energy and temperature that we have introduced, this limit corresponds to $\Omega_\gamma \ll 1/\Theta$. The attenuation coefficient κ is made dimensionless by multiplying by $\lambda_e = \hbar/m_e c$. In the Thomson limit it is only a function of the dimensionless density $(\nu_e = \lambda_e^3 n_e)$ and is given by

$$\kappa_T \lambda_e = (8\pi\alpha^2/3)\nu_e \qquad (\Omega_\gamma \ll 1/\Theta). \qquad (17)$$

For more characteristic photon energies of the plasma, say, $\Omega_\gamma \sim \Theta$, we are well into the "extreme Klein-Nishina" limit of Compton scattering. In fact, this limit applies as long as $\Omega_\gamma \gg 1/\Theta$, and the result for the attenuation coefficient is

$$\kappa_{KN} \lambda_e = \tfrac{1}{2}\pi\alpha^2 L_{KN} \nu_e / \Omega_\gamma \Theta \qquad (\Omega_\gamma \gg 1/\Theta) \qquad (18)$$

where

$$L_{KN} = \ln(4\Theta\Omega_\gamma) + \tfrac{1}{2} - \gamma_E. \qquad (19)$$

Pair Production. Another process with an absorption coefficient linear in the density ν_e is pair production. This process is of higher order (in α) than Compton scattering, but it dominates the latter at sufficiently high photon energy. It is necessary to consider pair production in the fields of plasma electrons and ions. In the limit $\Omega_\gamma \gg 1$ the cross sections for the process simplify and the result can be written

$$\kappa_{PP} \lambda_e = (28\alpha^3/9) L_{PP} \nu_e, \qquad (20)$$

where L_{PP} is a sum of two logarithmic terms[2] associated with pair production in the fields of ions and electrons respectively. The pair-production opacity has only a logarithmic dependence on Ω_γ and Θ. The ratio of the coefficients for Compton scattering (in the Klein-Nishina limit) to that for pair production is

$$\frac{\kappa_{KN}}{\kappa_{PP}} = \frac{9\pi}{56\alpha} \frac{1}{\Omega_\gamma \Theta} \frac{L_{KN}}{L_{PP}}, \qquad (21)$$

and we find that pair production dominates for $(\Omega_\gamma \Theta)^{1/2} \gtrsim 10$.

Inverse Bremsstrahlung. While the Compton and pair-production opacities have a linear dependence on density, bremsstrahlung absorption yields an absorption coefficient proportional to ν_e^2. In the calculation of the absorption coefficient it is necessary to subtract the contribution from stimulated emission, yielding an effective value of κ. The result is

$$\kappa_B \Lambda_e = (8\pi^2 \alpha^3/3) \Phi \nu_e^2/\Omega_\gamma^3, \tag{22}$$

where the function Φ is

$$\Phi = (2 + w + \tfrac{3}{4}w^2)(1 - e^{-w})e^{-w} L_B, \tag{23}$$

with

$$w = \Omega_\gamma/\Theta, \tag{24}$$

and L_B is a logarithmic term.[2] The result (22) is a simplified one and includes the effects of bremsstrahlung absorption in electron-electron and electron-ion scattering. Now it seems clear that, since $\nu_e \ll 1$ for all astrophysical applications, this absorption process can never compete with Compton scattering unless $\Omega_\gamma \ll 1$.

<u>Synchrotron Self Absorption</u>. This is a purely classical process. The associated absorption coefficient is linearly proportional to ν_e times a function of the magnetic field B, the photon energy or frequency, and the temperature. The expressions given here, again quoted from ref. 2, are for a magnetic field with random orientation. Formulas in terms of simple functions can be given only in the low- and high-energy limits. In terms of a dimensionless Larmor frequency

$$\Omega_L = (e\hbar/2m_e^2 c^3) B, \tag{25}$$

we have, first in the low-energy limit,

$$\kappa_s \Lambda_e = C_s \alpha \nu_e \Omega_L^{2/3} (\Omega_\gamma \Theta)^{-5/3} \qquad (\Omega_\gamma/\Omega_L \Theta^2 \ll 1), \tag{26}$$

where

$$C_s = \frac{2^{8/3} \pi^2}{3^{7/3}} \frac{\Gamma(\tfrac{1}{3})\Gamma(\tfrac{1}{2})}{5\,\Gamma(\tfrac{5}{6})} = 4.062. \tag{27}$$

At high energies

$$\kappa_s \Lambda_e = \frac{\pi^2}{2}\left(\frac{\pi}{3}\right)^{1/2} \frac{\alpha \nu_e}{\Theta^3 \Omega_\gamma} \left(\frac{\Omega_\gamma}{\Omega_s}\right)^{1/6} \exp\left[-\left(\frac{\Omega_\gamma}{\Omega_s}\right)^{1/3}\right] \qquad (\Omega_\gamma/\Omega_L \Theta^2 \gg 1), \tag{28}$$

where

$$\Omega_s = (4/9\pi)\Omega_L \Theta^2. \tag{29}$$

The low-energy formula (26) is undoubtedly the most important one, since it applies when κ_s is largest. We might note that in this limit, when there is approximate equipartition between the energy density in the relativistic electrons and the magnetic field, κ_s is proportional to $(\nu_e/\Theta)^{4/3}$.

For low-density thermal astrophysical plasmas (if they exist) it is likely that the opacity is due to either Compton scattering or synchrotron self-absorption.

APPENDIX

Here I shall give a little more detail on the calculation of the Møller-scattering energy loss rate, giving a few intermediate steps in the derivation in ref. 1. Of particular importance is the contribution from small energy exchanges with the plasma electrons. This energy for an individual collision is $\Delta E_{lab} = E_2' - E_2$, where the prime designates the scattered electron. In the limit (1), if the "test" electron is moving along the positive x-axis, in the c.m. frame ($\bar{K}$) each electron has an energy $\bar{E}$, with

$$\bar{E}^2 = \tfrac{1}{2} E_1 E_2 (1 - \cos\theta), \tag{30}$$

where θ is the angle between their directions in the lab frame (K). However, in $\bar{K}$ the plasma electron before scattering moves essentially in the negative x-direction. The Lorentz factor γ associated with the relative direction of K and $\bar{K}$ is related to $\bar{E}$ by

$$\bar{E} \approx \gamma E_1 (1-\beta) \approx \gamma E_1 (1-\beta)\frac{1+\beta}{2} \\ \approx E_1/2\gamma, \tag{31}$$

since $\beta \to 1$. In terms of the scattering angle $\bar{\theta}$ in $\bar{K}$, E_2' is then given by

$$E_2' = \gamma \bar{E}(1 - \cos\bar{\theta}) = \tfrac{1}{2} E_1 (1 - \cos\bar{\theta}), \tag{32}$$

which is independent of θ. Then, since the scattering cross section can be written[1]

$$d\sigma \propto (\bar{X}/\bar{E}^2) d\bar{\Omega}, \tag{33}$$

where

$$\bar{X} = 4\csc^4\bar{\theta} - 2\csc^2\bar{\theta} + \tfrac{1}{4}, \tag{34}$$

we have

$$v_r \Delta E_{lab}\, d\sigma \;\propto\; \bar{X}\left[\tfrac{1}{2}E_1(1-\cos\bar{\theta}) - E_2\right] d\bar{\Omega}. \tag{35}$$

In the domain of small $\bar{\theta}$, the brackets in (35) approach

$$\tfrac{1}{4}E_1 \bar{\theta}^2 - E_2 \equiv \eta, \tag{36}$$

and

$$\bar{X} \propto (\eta + E_2)^{-2}. \tag{37}$$

Since $d\bar{\Omega} \propto \bar{\theta}\,d\bar{\theta} \propto d\eta$, the contribution from small energy exchanges is proportional to the integral

$$\int_0^{\eta_m} \frac{\eta\, d\eta}{(\eta+E_2)^2} \;\longrightarrow\; -1 + \ln\frac{\eta_m}{E_2} \tag{38}$$

for $\eta_m \gg E_2$ (which can be chosen by making the corresponding $\bar{\theta}_m$ large enough). We note that there is no contribution from the lower limit in the integral in (38). If the lower limit were set equal to a very small but finite value ϵ, the integral would be the same. This is the essential reason why it is not necessary to introduce collective excitations and why a Coulomb-type logarithm does not come in.

REFERENCES

[1] R. J. Gould, Phys. Fluids 24, 102 (1981).

[2] R. J. Gould, "Processes in Relativistic Plasmas" (submitted to Astrophys. J.).

[3] S. R. DeGroot, W. A. van Leeuwen, and Ch. G. van Weert, Relativistic Kinetic Theory (North-Holland, Amsterdam 1980).

[4] R. J. Gould, Physica 58, 379; 60, 145; 62, 555 (1972).

[5] N. E. Frankel, K. C. Hines, and R. L. Dewar, Phys. Rev. A20, 2120 (1979).

THE ROLE OF CYCLOTRON RESONANCE PROCESSES IN THE SPECTRA OF ACCRETING NEUTRON STARS

J. Trümper
Max-Planck-Institut für Physik und Astrophysik
Institut für extraterrestrische Physik
8046 Garching bei München
West-Germany

ABSTRACT

We briefly review the observational situation regarding cyclotron resonance features in the spectra of pulsating sources and then turn our attention to problems arising in the context of gamma-ray burst spectra. Finally we discuss some recent work done on the radiative transfer in homogeneous plasmas under superstrong magnetic field conditions.

INTRODUCTION

The observability of the cyclotron resonance in the X-ray spectra of accreting neutron stars rests on a number of facts:

1) In the superstrong magnetic fields threading the polar caps of neutron stars the electron cyclotron frequency

    $\hbar\omega_c$ = 11.6 keV B_{12}

    lies in the X-ray domain.

2) At typical plasma temperatures which range from ~10 keV for pulsating X-ray sources to many hundred keV for gamma-ray bursts the matter is (almost) completely ionized and electrons are the most abundant species.

3) Due to the small dimensions of the polar hot spot the magnetic field is sufficiently homogeneous, allowing the cyclotron resonance to appear as a rather narrow spectral feature.

4) For extraordinary mode photons the cyclotron absorption or scattering cross sections at the resonance are ~10^4 times larger than the Thomson cross section.

Observations of cyclotron resonance features are important because they give the most direct imformation on neutron star magnetic field strengths. Furthermore, observations of spectral and angular distributions in general, and the cyclotron resonance effects in particular, may be used as a plasma diagnostic tool giving insight into the physical conditions at the radiating hot spot on the neutron star.

PULSATING X-RAY SOURCES

The first observational evidence for cyclotron lines was obtained for the pulsating X-ray source Her X-1 in 1976[1,2], which showed a peak at 58 keV and a second possible one at 110 keV, c.f. figure 1. The existence of the first feature has been confirmed by a number of balloon and satellite experiments as summarized in figure 2. The spectrum obtained by the MPI/AIT group[3] in 1977 is shown in figure 3.

The HEAO-1 measurements of 1978 and the MPI/AIT balloon observations of 1977 had enough photon statistics to allow pulse phase spectroscopy to be made. These data revealed that the feature exists throughout the pulse of the 1.24 sec period with constant relative intensity [4,5]. The HEAO-1 data also suggest a systematic slight variation of the feature energy during the 1.24 sec pulse and show that the spectral feature is stable during the on-state of the 35-day cycle [4].

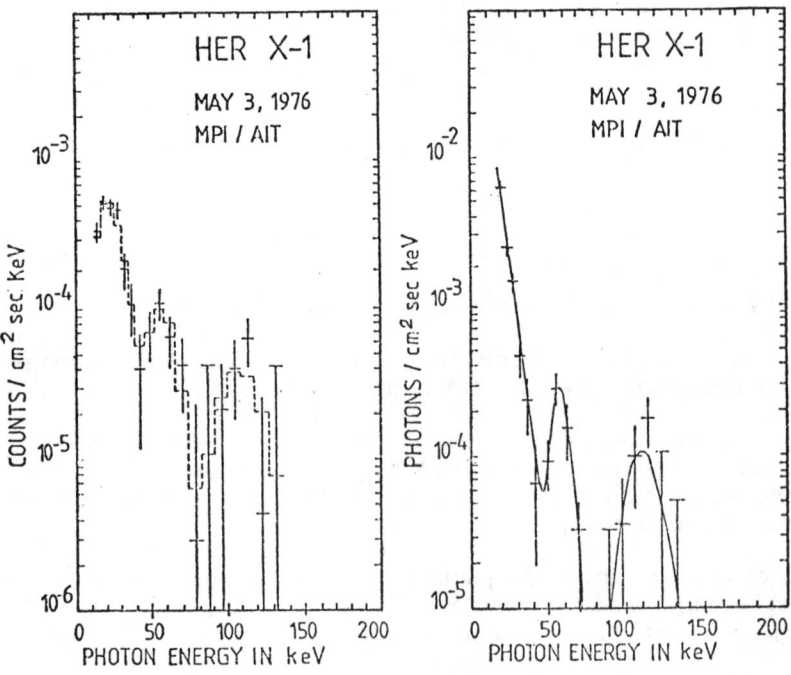

Fig. 1: Hard X-ray spectrum of Hercules X-1 obtained during the pulse phase of the 1.24 sec pulsation. The left diagram shows a raw count rate spectrum. The other diagram shows a deconvoluted spectrum, assuming spectral lines at 58 and 110 keV.

It should be noted that most of the results have been obtained with scintillation counters and proportional counters having rather low spectral resolution. Nevertheless, these data indicate that the feature is rather broad, namely 21 keV (FWHM) for an emission line or 11 keV (FWHM) for an absorption line.

This has been confirmed by measurements with a cooled Germanium spectrometer which provides much better spectral resolution but lower photon statistics [6]. The reported width is 13.5 keV (FWHM) for either an emission or an absorption line. The line energy indicated by these data is somewhat smaller than found previously, namely $43.9^{+1.2}_{-?}$ keV for an emission line. However, the difference is not very significant, and before drawing conclusions on variability one should carefully examine the detailed systematics of the different instruments.

The question whether the feature is seen in emission (at ~53 keV) or in absorption (at ~38 keV) has been of continuous interest. As already stated by Trümper et al 1978[2], the observational data do not allow to distinguish between both alternatives. This situation has

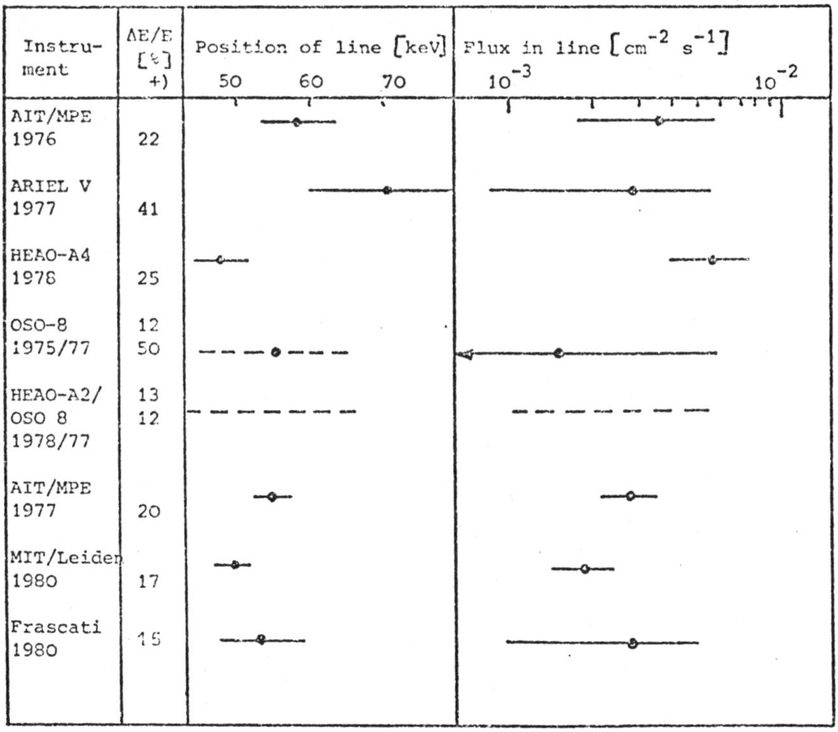

Instrument	$\Delta E/E$ [%] +)	Position of line [keV] 50 60 70	Flux in line [cm^{-2} s^{-1}] 10^{-3} 10^{-2}
AIT/MPE 1976	22		
ARIEL V 1977	41		
HEAO-A4 1978	25		
OSO-8 1975/77	12 50		
HEAO-A2/ OSO 8 1978/77	13 12		
AIT/MPE 1977	20		
MIT/Leiden 1980	17		
Frascati 1980	15		

Fig. 2: Hercules X-1: graphical representation of position and intensity of an assumed emission line according to various measurements with different instruments (from ref. 3).
+) ΔE is FWHM at 60 keV.

remained unchanged for the more recent observations. On the other hand, the deviations from a smooth continuum in the 35-60 keV region are much too sharp to be modelled by adding a second continuum. They are best fitted by a line feature, seen either in emission or absorption.

Summarizing, the existence of a first harmonic spectral feature in Her X-1 has been confirmed with rather large confidence. On the other hand, the possible second harmonic feature in the 1976 data has not been seen again. The appearance of the second harmonic may be connected with the fact that the source was particularly bright during the 1976 observation. If the higher luminosity was a result of a higher intrinsic source temperature, then the chances for detection of the second harmonic - either in emission or in absorption - may have been better. On the other hand, one cannot rule out that the second peak was just a statistical fluctuation (3.2 s.d.).

Unfortunately the number of pulsating X-ray sources showing evidence for cyclotron features has remained rather small. The only other cases are the 3.6 sec pulsar 4U 0115+63 ($E_c \sim 20$ keV, Wheaton et al. 1979[21]) which exhibits a feature looking like an absorption line, and the 7.6 sec pulsar 4U 1626-67 ($E_c \sim 20$ keV, Pravdo et al. 1980[22]). However, in the latter case the

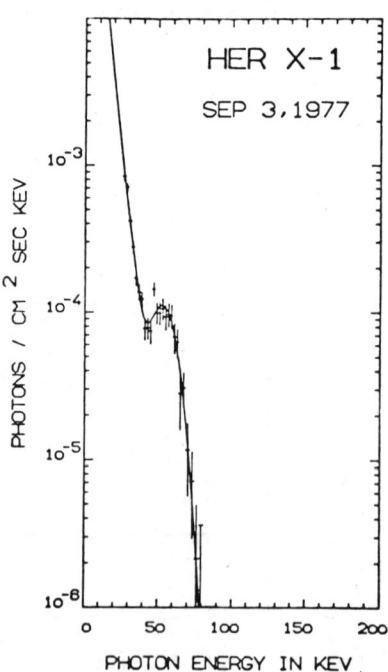

Fig. 3: Deconvoluted Her X-1 spectrum of the 1977 MPI/AIT balloon observation.

spectral feature is so broad that it also could represent a Wien peak expected from Comptonization.

Concluding this short summary we note that a number of three candidates among 19 known pulsating X-ray sources is not at all inconsistent with expectations, if one takes into account the likely dispersion of neutron magnetic fields and the existing observational constraints.

GAMMA-RAY BURSTS

Recently, cyclotron resonances have emerged in a quite different phenomenological context, namely in the spectra of gamma-ray bursts detected by Mazets and co-workers with the Venera 11 and 12 spaceprobes[23]. A large fraction ($\sim 30\%$) of these bursts show transient absorption and emission type features in the region of 30 to 70 keV. While the lower limit is coincident with the threshold energy of the instrument, there seems to be a real upper limit at 70 keV.

Of course, the most important implication of these observations is that at least a substantial portion of the gamma-ray bursters are strongly magnetized neutron stars having magnetic fields of 3 to 8 x 10^{12} Gauss. The detection of redshifted positron-electron annihilation lines in a number of events supports this identification. Since these observational data as well as corresponding theoretical models have been discussed in detail by other speakers in this meeting, I want to make only a few remarks on these interesting results.

A fraction of the events show rather narrow absorption lines with a Doppler width corresponding to ~ 10 keV, while the temperature of the emitting plasmas derived from the thermal bremsstrahlung fits is typically a few hundred keV. Therefore the source region must show extreme temperature stratifications.

There is also a remarkable case of an emission line event (26 May 1979) which shows a rather narrow width of $E \sim 15$ keV as well, while the continuum has a $kT = 210$ keV. It appears difficult to reconcile the large intensity with its small width: Even if the viewing direction were perpendicular to the magnetic field, the transverse Doppler effect would lead to a much larger width than observed.

For a substantial fraction of the absorption and emission lines the data indicate that one is dealing with just one resonance and the second harmonic is absent. For 10^{12} G fields the absorption/scattering cross section for absorption is only a factor of ~ 10 smaller at the second harmonic than that at the first harmonic. Therefore these observations require a rather narrow range of column densities of the absorbing/scattering layers in the various sources.

Finally we note that these difficulties (narrow lines,

absence of higher harmonics) could be avoided if the lines were due to Lyman α, β of Fe^{25+} (or Ni^{27+}) in a superstrong magnetic field. A comparison with recent calculations [7,8,9] shows, however, that one needs then magnetic field strengths $B > 2 \times 10^{13}$ G in order to explain the observations. This does not seem very likely.

Obviously a lot of work has to be done before we fully understand these interesting and surprising results. Of course, it also seems important to see an independent confirmation by other groups on the basis of data of equal or even better quality.

THEORETICAL CONSIDERATIONS

The classical picture of pulsating X-ray sources has been developed soon after the discovery of these exciting sources by Uhuru[10,11,12,13]. The discovery of cyclotron resonance features has given new impetus to investigations on the photon interactions with strongly magnetized plasma and the problem of radiative transfer under these conditions (see[14,15,16] and references therein). However, a solution of the general problem involving the coupling of the hydrodynamic accretion flow with the radiative heat transfer and the effects of radiation pressure has not been achieved yet in a self-consistent way.

One can expect that this problem will be of particular complexity at high accretion rates and luminosities [13]. On the other hand, in the limit of very low accretion rates, the situation could become rather simple, because the falling material in the accretion column will be transparent to the outgoing radiation, and the matter flow and radiation field decouple. One then expects a free fall of matter down to the neutron star surface. For an accretion column diameter of a few kilometers these conditions should occur at luminosities $L_x \lesssim 10^{37}$ erg/sec, for a homogeneous matter flow. If the matter comes down in clumps or filaments, the transparency of the column would be even larger. Certainly, Vela X-1 having a $L_x \sim 10^{36}$ erg/sec is a candidate source for such conditions, but they may also be present in Her X-1 ($L_x \sim 10^{37}$ erg/sec), depending on the actual values of mass, radius, column width, and "clumpiness".

The freely falling material hitting the polar cap surface will be stopped by several processes. Nuclear collisions lead to a penetration depth of ~ 50 g/cm^2 [13]. The stopping effect of Coulomb collisions between the infalling protons and ambient electrons was thought to be rather ineffective[13,17]. However, recent calculations based on Fokker-Planck coefficients for a proton undergoing small-angle Coulomb scattering show that the stopping lengths would be in the range of 1 to 20 g/cm^2, depending on the detailed conditions [18,19]. Therefore, in the case of low luminosity sources one may have a situation where a rather thin surface layer is heated by the infalling matter. The above range corresponds to a Thomson optical depth of 0.4 to 8.

The radiative heat transfer in such a hot strongly magnetized plasma sheet, allowing for Comptonization and redistribution of photons in the cyclotron line, has been treated recently by Nagel[14,15,20]. Although his assumption of a homogeneous slab - with respect to density and temperature - is certainly a simplification, the results may be relevant to the low luminosity case discussed here. Figure 4 shows the corresponding energy spectra[24] for ordinary and extraordinary photons for slabs of different optical thickness having a density of 10^{-2} g/cm^3 and a temperature of 10 keV.

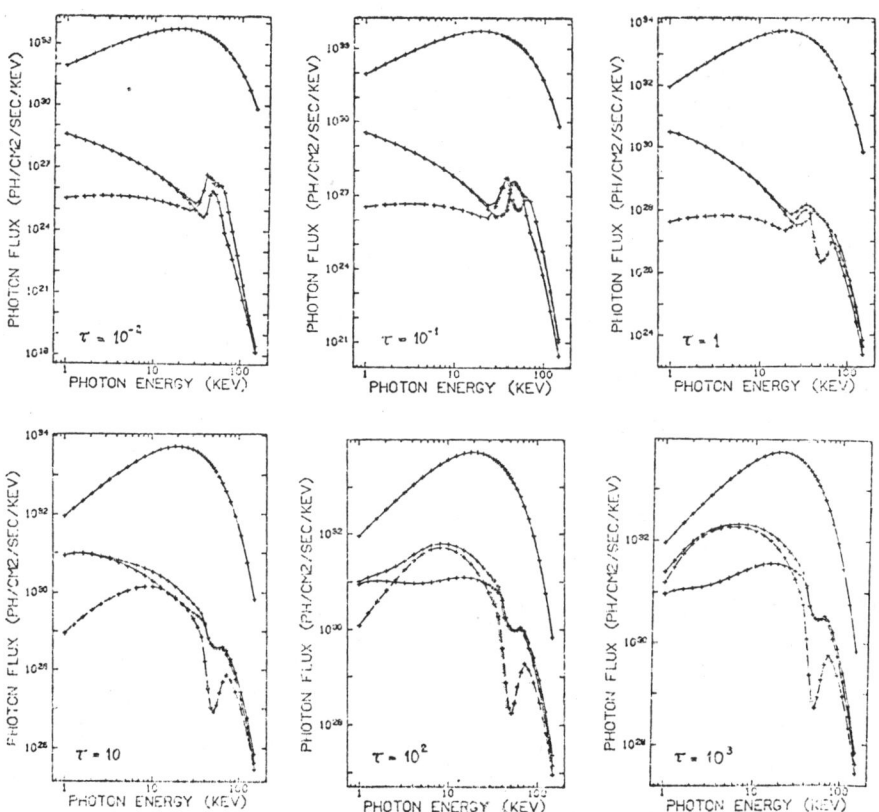

Fig. 4: Photon spectra from slabs of hot strongly magnetized plasma. Solid lines: ordinary photons, dashed lines: extraordinary photons. The uppermost curve represents the Wien spectrum. The plasma parameters are $\rho = 10^{-2}$ g/cm^3, kT = 10 keV, $\hbar\omega_c = 50$ keV. The luminosities, for an area of 1 km^2, are in order of increasing Thomson optical depths : 8×10^{30}, 6×10^{31}, 6×10^{32}, 2×10^{34}, 4×10^{35} and 10^{36} erg/sec (Ref.14,24).

As expected the spectra show an emission line for low optical depths, which becomes self-reversed at intermediate depths. In this case the extraordinary photons are trapped in the resonance and escape in the wings (see also [16]). The emission peak seen in the ordinary photons is due to the scattering from trapped extraordinary photons into ordinary ones, which then escape. This flux of ordinary photons is so strong that it fills the self-reversal dip of the extraordinary photons. At larger optical depths ($\tau \gtrsim 1$) the resonance appears in absorption. At the same time, the luminosity increases with increasing optical depth of the emitting layer and reaches $L_x \sim 10^{36}$ erg/sec for a radiating area of 1 km². Larger luminosities may be obtained if the density and/or the temperature of the radiating slab are increased. Figure 5 shows the spectrum expected for ρ = 10 g/cm³ and τ = 8. The luminosity in this case would be $\sim 10^{37}$ erg/sec for A = 3 km², sufficient to explain Her X-1.

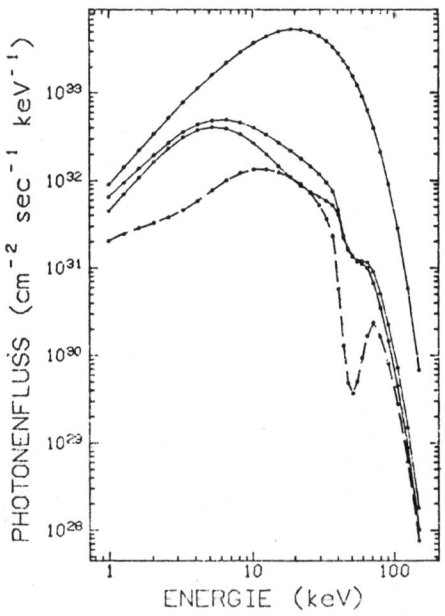

Fig. 5: Same as figure 4, for plasma parameters ρ = 10 g/cm³, kT = 10 keV, $\hbar\omega_c$ = 50 keV, τ = 8. The luminosity is 3×10^{36} erg/sec (for an area of 1 km²).

Nagel has also calculated the angular distribution of the emitted radiation[15], and showed that in some of the models the resulting pulse profiles are similar to those observed in Her X-1. In particular he finds a transition from double pulses to single pulses at an energy of

25 keV, as observed. Of course, this depends on the angles made by the rotation axis of the neutron star with its dipole axis, and with the line of sight to the earth. The best angles are 60° and 45° or vice versa.

Returning to the spectra and luminosities, Nagel's calculations nicely show that in general one expects emission lines only at rather low luminosities (c.f. figure 4). If one wants to have a high luminosity in the optically thin case one has to go to rather high densities in order to make the photon production by bremsstrahlung effective. Estimates show that in order to produce $L_x \sim 10^{37}$ erg/sec as required in Her X-1, and an emission line, one has to have $\rho \gtrsim 100$ g/cm^3. Whether such large densities of the emission region can be expected remains to be seen when models become available which couple the braking of protons and the radiative transfer problem in a self-consistent way.

REFERENCES

1. J. Trümper, W. Pietsch, C. Reppin, B. Sacco, E. Kendziorra, and R. Staubert, Ann. N.Y. Acad. Sci. 302, pp. 538-544 (1977)
2. J.Trümper, W. Pietsch,C. Reppin,W. Voges, R.Staubert, and E. Kendziorra, Astrophys. J. (Letters) 219, pp. L 105-L110 (1978)
3. R. Staubert, E. Kendziorra, W. Pietsch, R.J. Proctor, C. Reppin, H. Steinle, J. Trümper, W. Voges, Proceedings ESLAB-Symposium on X-ray Astronomy 1981
4. D.E. Gruber, J.L. Matteson, P.L. Nolan, F.K. Knight, W.A. Baity, R.E. Rothschild, L.E. Peterson, J.A. Hoffman, A. Scheepmaker, W.A. Wheaton, F.A. Primini, A.M. Levine, W.H.G. Lewin, Astrophys.J. (Letters) 240, pp.L127- L131 (1980)
5. W. Voges, W. Pietsch, C. Reppin, J. Trümper, E. Kendziorra, R. Staubert, to be published(1981)
6. J. Tueller, T. Cline, W. Paciesas, B. Teegarden, D. Boclet, P. Durouchoux, J. Hameury, and R. Haymes, to be published in Proc. 17th Int. Cosmic Ray Conf. Paris (1981
7. V.V. Burdyuzha and V.B. Pavlov-Verevkin, preprint 505, Space Research Institute Moscow (1980)
8. G. Wunner, H. Ruder, H. Herold, Astrophys. J. 247, 374 (1981)
9. H. Ruder, G. Wunner, H. Herold, J. Trümper, Phys. Rev. Letters 46, 1700 (1981)
10. J.E. Pringle, M.J. Rees, Astron. & Astrophys. 21, 1 (1972)
11. F.K. Lamb, C.J. Pethick, D. Pines, Astrophys. J. 184, 271 (1973)
12. K. Davidson, J.P. Ostriker, Astrophys. J. 179, 585 (1973)

13. M.M. Basko, R.A. Sunyaev, Astron. & Astrophys. **42**, 311 (1975)
14. W. Nagel, Astrophys. J., in press
15. W. Nagel, Astrophys. J., in press
16. I. Wassermann, E. Salpeter, Astrophys.J. **241**, 1107 (1980)
17. G.G. Pavlov, and D.G. Yakovlev, Sov. Phys. JETP **43**, 389 (1976)
18. J.G. Kirk, D.G. Galloway, preprint (1981)
19. J.G. Kirk, this conference
20. W. Nagel, PhD Thesis, University of Munich (1981)
21. W.A. Wheaton, J.P. Doty, F.A. Primini, B.A. Cooke, C.A. Dobson, A. Goldman, M. Hecht, J.A. Hoffman, S.K. Howe, A. Scheepmaker, E.Y. Tsiang, W.H.G. Lewin, J.L. Matteson, D.E. Gruber, W.A. Baity, R. Rothschild, F.K. Knight, P. Nolan, and L.E. Peterson, Nature **282**, p.p. 240-243 (1979)
22. S.H. Pravdo, N.E. White, E.A. Boldt, S.S. Holt, P.J. Serlemitsos, J.H. Swank, and A.E. Szymkowiak, Astrophys. J.**231**, pp. 912-918 (1979)
23. E.P. Mazets, S.V. Golenetskii, R.L. Aptekar, Yu.A. Gur'yan, and V.N. Il'inskii, Nature **290**, pp. 378-382 (1981)
24. W. Nagel, private communication

THE LOW ENERGY SPECTRA OF GAMMA-RAY BURSTS*

R.W. Bussard and F.K. Lamb[†]
Department of Physics
University of Illinois at Urbana-Champaign
Urbana, Illinois 61801

ABSTRACT

We examine the implications of observed γ-ray burst spectra for the physical conditions and geometries of the sources. Explanation of the continua in terms of optically-thin thermal bremsstrahlung requires a relatively large area, but a quite shallow depth. Alternatively, a spectrum similar to that observed could be produced by rapid flickering of sources with less extreme geometries, if each flicker emits a Comptonized thermal spectrum. To interpret the low energy features as cyclotron extinction, either field inhomogeneities or plasma motions are required. An alternative explanation is photoelectric absorption by heavy atoms, which requires a very high field strength, large enough to make one-photon electron positron annihilation possible. We conclude by proposing some observational tests of these possibilities.

INTRODUCTION

The γ-ray bursts have been an intriguing puzzle from their discovery to the present. Early observations showed flux variations within the bursts on time scales as short as 20 ms, and spectra averaged over the full duration of the bursts consistent with exponential photon number distributions over the energy range 100-400 keV.[1,2] Recently, Mazets et al.[3,4] have reported striking new results from observations of gamma ray bursts carried out with the KONUS instruments on board the Venera 11 and 12 spacecraft. When triggered by the arrival of a burst, these instruments measure the number of photons in the energy range from 30 keV to 2 MeV that arrive in a sequence of 4s accumulation times, for 66s following the trigger. For the first 32s, the spectrum of each 4s accumulation of photons is measured by 16-channel pulse height analysis. Mazets et al. find that the spectrum over most of this energy range is consistent with a simple analytical expression (see below). However, approximately 10% of the observed burst spectra show broad features below 80 keV where the flux is significantly less than would be predicted by extrapolating this expression to these energies. At present, the precise shapes of these features are still uncertain.[5] A few percent of the observed spectra appear to have emission features in the energy range 300-650 keV, which have been interpreted by Mazets et al. as gravitationally-redshifted electron-positron

---

*This research was supported in part by NASA grant NSG 7653 and NSF grant PHY 80-25605.
[†]Also, Department of Astronomy.

0094-243X/82/770189-12$3.00 Copyright 1982 American Institute of Physics

annihilation lines. Three of the spectra reported in reference 2 show both a depression in the spectrum below 80 keV and apparent emission in the energy range 300-650 keV.

The astrophysical sites and physical processes responsible for the production of these bursts have not yet been established, although models involving brief episodes of accretion onto neutron stars or nuclear outbursts on such stars appear quite promising.[6-9] In the present paper we use knowledge of the opacity of a hot plasma in a superstrong magnetic field and qualitative radiative transfer arguments to constrain the geometry of and physical conditions in the sources of these bursts. We have in mind sources at a distance $D \sim 100$ pc, although we retain the dependence on distance in the expressions we give. We first discuss the origin of the continuum spectrum. Next, we consider the possibility that the low-energy features are caused by cyclotron scattering in superstrong fields. In this discussion we make use of new calculations of the cross sections for resonant scattering by relativistic electrons. Finally, we consider the possibility that the low-energy depressions in the spectra are due to photoelectric absorption by heavy atoms, such as iron or nickel, in superstrong magnetic fields. A more detailed account of our results will be published elsewhere.

FORMATION OF THE CONTINUUM

The expression used by Mazets et al.[4] to characterize the continuum spectra they observed is

$$dN = AE^{-1} \exp(-E/T_o) \, dE, \qquad (1)$$

where A is the flux in photons $cm^{-2} s^{-1}$, typically $\sim 1-30$, and E and T_o are the photon energy and the characteristic temperature in keV. This corresponds to the spectrum of optically-thin thermal bremsstrahlung, if the energy dependence of the Gaunt factor, relativistic corrections, and magnetic field effects are ignored. These are, however, substantial effects for the energies and magnetic field strengths of interest.[10,11] No attempts to fit an actual bremsstrahlung emission spectrum to the observations have been reported to date. Even if the emission is due to optically-thin bremsstrahlung, the source must satisfy severe geometrical constraints in order for such a spectrum to be preserved as the radiation propagates outward through the source.

To see this, consider first a homogeneous source of projected area a and depth ℓ along the line of sight. Assuming that the source continues to be heated during the burst, an estimate of the electron density can be obtained from the emission measure. At a distance $D \sim 100$ pc, the luminosity of a typical burst is $\sim 10^{37}$ erg s^{-1} while the characteristic temperature determined by fitting expression (1) to the observed spectrum is ~ 300 keV. If the volume of the source is a fraction of that of a neutron star, say $\sim 10^{16}$ cm^3, this implies a typical electron density, $n_e \sim 10^{22}$ cm^{-3}. Alternatively, if the $\sim 10^{38}$ ergs total energy of a typical burst is originally shared equally among all the particles in the source, and if each has an initial energy ~300 keV, this implies ~10^{44} particles or a particle density ~10^{28} cm^{-3}. Since the constraints become more severe the higher the density, we shall use the lower of these two values to scale our results.

Preservation of the bremsstrahlung emission spectrum requires[12] both that the depth of the source be much less than the thermalization length at the frequency of interest, that is

$$\tau_*(\nu) \equiv \int_0^\ell (\kappa_{ff}\kappa_{es})^{1/2} dz \ll 1, \quad (2)$$

and that Comptonization in the source be unsaturated, that is

$$y \equiv \left(\frac{4k_B T_e}{m_e c^2}\right) \tau_{es}^2 = 0.67 \left(\frac{T_e}{10^9 K}\right) \tau_{es}^2 \lesssim 1. \quad (3)$$

Here κ_{ff} is the free-free absorption coefficient, κ_{es} is the electron scattering coefficient, and τ_{es} is the electron scattering optical depth. In the definition of y we have assumed $\tau_{es} > 1$ (assuming $\tau_{es} < 1$ leads to a still more severe constraint). If, as expected, the electron scattering opacity is at least comparable to that of Thompson scattering (see below), inequality (2) requires

$$\ell \ll \left(\frac{\kappa_{es}}{\kappa_{ff}}\right)^{1/2} \frac{1}{\kappa_{es}} \simeq 1.0 \times 10^6 \left(\frac{1}{Zg_\nu}\right)^{1/2} \left(\frac{h\nu}{100 \text{ keV}}\right)^{3/2}$$

$$\times \left(\frac{10^{22} \text{ cm}^{-3}}{n_e}\right)^{3/2} \left(\frac{T}{10^9 K}\right)^{1/4} \text{cm}, \quad (4)$$

where Z is the ion charge, g_ν is the Gaunt factor, and n_e is the electron density. Inequality (3) imposes a much more severe constraint, namely

$$\ell \lesssim \left(\frac{m_e c^2}{4k_B T_e}\right)^{1/2} \frac{1}{\kappa_{es}} \simeq 180 \left(\frac{10^{22} \text{ cm}^{-3}}{n_e}\right)\left(\frac{10^9 K}{T_e}\right)^{1/2} \text{cm}. \quad (5)$$

The depth limitation (5), when combined with the bremsstrahlung volume emissivity j_ν, implies that the flux from the source cannot exceed

$$F_\nu \lesssim 2.0 \times 10^4 \left(\frac{Zg_\nu n_e}{10^{22} \text{ cm}^{-3}}\right) \left(\frac{10^9 K}{T}\right) e^{-\nu/T} \text{ erg cm}^{-2} \text{ s}^{-1} \text{ Hz}^{-1} \quad (6)$$

On the other hand, the flux (1) observed at earth implies a flux

$$F_\nu \lesssim 2.0 \times 10^5 \left(\frac{D}{100 pc}\right)^2 \left(\frac{10^{10} \text{ cm}^2}{a}\right) A\, e^{-\nu/T} \text{ erg cm}^{-2} \text{ s}^{-1} \text{ Hz}^{-1} \quad (7)$$

at the source. Combining relations (6) and (7), one finds that the projected area of the source must satisfy

$$a \gtrsim 1.0 \times 10^{11} \, A \, \left(\frac{D}{100\text{pc}}\right)^2 \left(\frac{10^{22}\text{cm}^{-3}}{Zg_\nu n_e}\right) \left(\frac{T}{10^9 \text{K}}\right) \text{cm}^2. \quad (8)$$

This area, when compared with the maximum source depth allowed by equation (5), implies a source aspect ratio $\gtrsim 10^3:1$. Such a ratio could arise if the source consists of very thin sheets or filaments of plasma. Such a plasma distribution might result from accretion of plasma along magnetic flux sheets or tubes within the magnetosphere of a neutron star, or from plasma thrown up in a similar pattern from the surface of such a star by a nuclear explosion in the surface layers. However, it is hard to see why such a distribution should arise in every case, as would be necessary to account for the apparent universality of spectra of the form (1).

If the aspect ratio is not so extreme, one expects the parameter y, defined in equation (3), to exceed unity. For $y \gtrsim 10$, the spectrum that emerges from the source will be a Wien distribution, that is

$$f_\nu(t) \propto f(t) \, \nu^3 \, e^{-\nu/T(t)} \quad (9)$$

at high energies. Spectra of this type from a source that cools in a time much shorter than the detector accumulation time produce apparent spectra of the form (1), as pointed out in the context of a different model by Ramaty and Cohen.[13] For example, an instantaneous spectrum of the form (9) will produce an average count-rate spectrum of the form

$$dN \propto \left(1 + \frac{E}{T_i} + \frac{1}{2}\left(\frac{E}{T_i}\right)^2\right) E^{-1} e^{-E/T_i} dE, \quad (10)$$

if the final temperature of the source is much smaller than E. Here, T_i is the initial temperature of the Wien distribution. Like spectrum (1), this form gives $dN \propto E^{-1} dE$ at low energy and is dominated by the factor $\exp(-E/T_i)$ at energies $E \gtrsim T_i$. In this picture, each 4s accumulation samples at least several spatially or temporally separated flickerings of the source.

CYCLOTRON ABSORPTION AND SCATTERING

Let us consider now the possibility that the low energy features reported by Mazets et al. are due to electron cyclotron absorption or scattering. If the electron Landau levels are largely excited or de-excited by collisions, absorption and emission will predominate and the level populations deduced from observations will reflect the kinetic temperature of the outermost layer of the plasma. If on the other hand the levels are largely excited and de-excited by radiative processes, scattering will predominate, the level populations will reflect the brightness temperature of the radiation at the cyclotron frequency, and extinction at the cyclotron harmonics will occur even if the source is isothermal.

Denoting the ground and first excited levels by 0 and 1, respectively, the Einstein A and B coefficients for this two level system are[14]

$$A_{10} = \frac{8}{3}\pi\alpha \left(\frac{h\nu_B}{m_e c^2}\right) \nu_B = 3.9 \times 10^{15} \left(\frac{B}{10^{12}G}\right)^2 \text{ s}^{-1}. \quad (11)$$

and

$$B_{01} = B_{10} = 1.2 \times 10^7 \left(\frac{B}{10^{12}G}\right)^2 \text{ cm}^2 \text{ erg}^{-1} \text{ s}^{-1}. \quad (12)$$

Here α is the fine structure constant, and spin flip decays from the first excited level have been neglected. On using the inferred radiative flux (7), one finds that the rate of induced absorptions is

$$R_{01} = B_{01} I_{\nu_B} = 3 \times 10^{12} \, A \left(\frac{D}{100\text{pc}}\right)^2 \left(\frac{10^{10}\text{cm}^2}{a}\right) \left(\frac{B}{10^{12}G}\right)^2 e^{-\nu_B/T_o} \text{ s}^{-1} \quad (13)$$

per electron in the ground state. Here A is the flux coefficient defined in equation (1) and is not to be confused with the spontaneous emission rate (11). The rate of induced emissions per electron in the first excited level is the same. For comparison, the nonrelativistic collisional excitation rate per electron is[15]

$$R_{01}(\text{ion}) \simeq 6.3 \times 10^{-13} \, n_i \, z^2 \left(\frac{10^9 K}{T}\right)^{1/2} \left(\frac{10^{12}G}{B}\right) \ln\left(\frac{4T}{\nu_B}\right) \varepsilon_1 \text{ s}^{-1}, \quad (14)$$

where the dimensionless factor ε_1 (here ~ 0.2–1) accounts for the ability of the field to absorb transverse momentum. Thus, the rate of collisional excitations and de-excitations is less than the spontaneous decay rate if

$$n_i z^2 < 6.2 \times 10^{27} \left(\frac{T}{10^9 K}\right)^{1/2} \frac{(B/10^{12}G)^3}{\varepsilon_1 \ln(4T/\nu_B)} \text{ cm}^{-3}, \quad (15)$$

and less than the induced absorption and emission rates if

$$n_i z^2 < 4.8 \times 10^{24} A \left(\frac{D}{100\text{pc}}\right)^2 \left(\frac{10^{10}\text{cm}^2}{a}\right) \left(\frac{T}{10^9 K}\right)^{1/2} \frac{(B/10^{12}G)^3}{\varepsilon_1 \ln(4T/\nu_B)} \text{ cm}^{-3}. \quad (16)$$

In order for the low energy features to be caused by cyclotron absorption or scattering, one must have $h\nu_B \gtrsim 30$ keV, or

$B \gtrsim 3 \times 10^{12}$ G. Thus, for densities much less than 10^{26} cm^{-3}, the radiative rates are larger and the dominant process is cyclotron scattering. We shall henceforth assume that this is the case.

If we define the source brightness temperature T_ν at frequency ν implicitly by the expression $B_\nu(T_\nu) = I_\nu$, where B_ν is the Planck function, then the ratio of the population of the $(i+1)^{th}$ Landau level to the ith level is $\exp(-\nu_B/T_{\nu_B})$. On using expression (1), we find

$$\frac{n_1}{n_0} \sim 8 \times 10^{-4} \, A \, \left(\frac{D}{100\text{pc}}\right)^2 \left(\frac{10^{10} \text{cm}^2}{a}\right) \left(\frac{10^{12} G}{B}\right)^3, \quad (17)$$

corresponding to burst brightness temperatures at frequencies ~30-60 keV of ~0.15 ν_B. Thus, if the area of the source is $\gtrsim 10^6$ cm^2, the spontaneous decay rates are greater than the induced absorption and emission rates, and the populations of the excited Landau levels are small compared to the population of the ground state.

The transfer of radiation through an electron scattering atmosphere in a superstrong magnetic field has two key features that we wish to include in our treatment. First, the scattering coefficient is strongly peaked at the cyclotron frequency and its overtones as shown in Figure 1. As a result, photons with frequencies near these resonances scatter many times without traveling very far in space, whereas photons with frequencies in the nearby continuum travel much farther in the same number of scatterings. Second, Doppler shifts and electron recoils cause photons that are deep in the source to diffuse from the lines to the nearby continua and vice versa. To allow a simple treatment of these effects, we shall neglect the angular dependence of

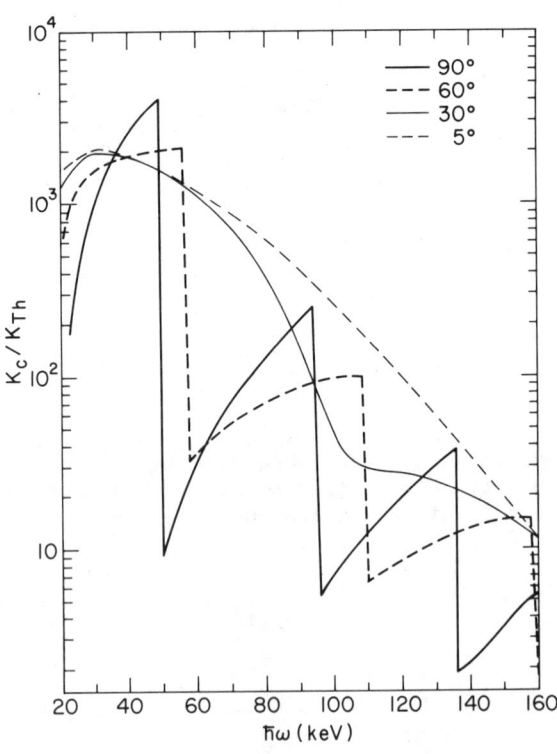

Fig. 1. The cyclotron scattering coefficient in units of that for Thompson scattering, for $B=4.4 \times 10^{12}$ G, T=300 keV, and four directions of propagation with respect to the magnetic field.

the scattering coefficient. The spectral intensity I_ν is then a function only of μ, the direction cosine of the ray with respect to the normal to the atmospheric boundary, and z, the coordinate normal to the boundary, which we take increasing inwards.

Under these conditions, the transfer of radiation is described by the equation

$$\mu \frac{dI_\nu}{dz}(\mu,z) = -\kappa_\nu I_\nu(\mu,z) + j_\nu(z), \quad (18)$$

where κ_ν is the electron scattering coefficient and

$$j_\nu(z) = \int_0^\infty d\nu' \, \kappa(\nu,\nu') \, J_{\nu'}(z) \equiv \int_0^\infty d\nu' \, \kappa(\nu,\nu') \int_{-1}^1 d\mu \, I_{\nu'}(\mu,z) \quad (19)$$

is the volume emissivity, here due entirely to scattering. The function $\kappa(\nu,\nu')$ is the product of the scattering coefficient at ν' and the probability of redistribution from ν' to ν. In order to treat the strong frequency dependence of κ_ν, we divide the frequency domain into a series of bands of width $\Delta\nu_\alpha$ over which the scattering coefficient can be assumed constant. The frequency integral in equation (19) can then be replaced by a quadrature sum, with the result

$$j_\alpha = \frac{1}{\Delta\nu_\alpha} \sum_{\alpha'} P_{\alpha\alpha'} \, \kappa_{\alpha'} \, J_{\alpha'} \, \Delta\nu_{\alpha'}. \quad (20)$$

Here $\kappa_{\alpha'}$ is the scattering coefficient in the frequency band α' and $P_{\alpha\alpha'}$ is the probability that a photon is shifted from band α' into band α in a single scattering. Now, making use of the Eddington approximation and assuming κ_α is independent of z, we can obtain a second order equation for J_α,

$$\frac{1}{3}\frac{d^2 J_\alpha}{dz^2} = \kappa_\alpha^2 \, (J_\alpha - S_\alpha), \quad (21)$$

by combining the first-order equations for the first and second moments of equation (18) in the usual way (reference 16, p. 153). Here, $S_\alpha = j_\alpha/\kappa_\alpha$ is the scattering source function.

Although the frequency domain must be divided into many bands in order to obtain accurate line profiles, an important and interesting result can be obtained simply by grouping together all the frequencies within a resonance into one band, and frequencies in the nearby continuum into another. Consider first the line at the cyclotron fundamental. Let $\Delta\nu_l$ be the bandwidth of the line, $\Delta\nu_c$ the bandwidth of frequencies in the nearby continuum from which a photon can be shifted into $\Delta\nu_l$ in a single scattering, and P_{lc} the average probability of a shift from $\Delta\nu_c$ into $\Delta\nu_l$ in a single scattering.

Then the source function in the line band is

$$S_1 = P_{11}J_{11} + \frac{1}{\kappa_1 \Delta\nu_1} P_{1c}\kappa_c J_c \Delta\nu_c, \qquad (22)$$

where the first term describes scattering within the line and the second, transfer of photons from the nearby continuum into the line. On substituting expression (22) into equation (21), one finds

$$\frac{1}{3}\frac{d^2 J_1}{dz^2} = \kappa_1^2 \lambda (J_1 - S_1^*), \qquad (23)$$

where $\lambda = 1-P_{11} = P_{c1} \ll 1$ is the probability that a photon originally in the line is shifted into the continuum in a single scattering and

$$S_1^* = \left(\frac{P_{1c}}{1-P_{11}}\right)\frac{\kappa_c \Delta\nu_c}{\kappa_1 \Delta\nu_1} J_c \qquad (24)$$

is an effective scattering source function.

Equation (23) displays the existence of a new length scale,

$$\ell_R = (\kappa_1\sqrt{\lambda})^{-1} \simeq (\kappa_c \kappa_1)^{-1/2} \simeq 0.48 \left(\frac{10^{22} \text{cm}^{-3}}{n_e}\right) \text{cm}, \qquad (25)$$

which we shall call the "redistribution length." Photons emerging from layers within the source deeper than ℓ_R are typically scattered from the line to the continuum before emerging.

Two boundary conditions are needed to specify the solution to equation (23). At the surface of the atmosphere, where the inward intensity is zero, we impose the condition $J_1(0) = \sqrt{3} H_1(0)$, the result for a gray atmosphere (reference 16, p. 55). At distances within the source that are greater than the redistribution depth but less than the thermalization depth (so that bremsstrahlung can still be neglected), the rate of diffusion of energy in frequency space into the line band balances the rate of diffusion out of this band. Thus, $J_1 \to S_1^*$ for $z \gg \ell_R$. Assuming that J_c is constant over the range of depths of interest ($z \lesssim \ell_R$, for which the continuum optical depth is small), the solution corresponding to these boundary coditions is

$$J_1(z) = \left(1 - \left(\frac{1}{1+\sqrt{\lambda}}\right) \exp(-3\sqrt{\lambda}\,\kappa_1 z)\right) S_1^* \qquad (26)$$

or, at the surface,

$$J_1(0) = \frac{\sqrt{\lambda}}{1+\sqrt{\lambda}} S_1^* \simeq \sqrt{\lambda}\, S_1^*. \qquad (27)$$

Thus, the ratio of the flux emerging at the cyclotron fundamental to that emerging in the nearby continuum is

$$r_1(0) \equiv \frac{H_1(0)}{H_c(0)} \approx \frac{P_{1c}}{\sqrt{P_{c1}}} \frac{\kappa_c \Delta\nu_c}{\kappa_1 \Delta\nu_1} . \qquad (28)$$

Now for scattering by a thermal distribution of electrons, one has $P_{c1} \sim \kappa_c/\kappa_1$ and $P_{1c} \lesssim 1$. Further, the mean frequency shift per scattering is typically about the same as the Doppler width of the line, so that $\Delta\nu_c \sim \Delta\nu_1$. This leads to the important result

$$r_1(0) \sim (\kappa_c/\kappa_1)^{1/2} . \qquad (29)$$

This result shows that redistribution between the line and nearby continuum in cyclotron scattering has an effect on the residual flux in the line that is similar to the effect of redistribution between the line core and wings in atomic scattering (compare ref. 16, p. 364).

Consider now the second and higher harmonics. Here the redistribution in frequency is predominantly from the nth harmonic into the first harmonic, since the transition to the next lower Landau level is strongly favored. For the second harmonic, the quantity analogous to P_{c1} is P_{12}, which is ~ 0.5. The redistribution length in the second harmonic is therefore not much larger than the scattering mean free path and hence λ is of order unity. Thus, the ratio of the line flux to that in the nearby continuum is

$$r_2(0) \equiv \frac{H_2(0)}{H_c(0)} \sim \frac{1}{\sqrt{P_{12}}} \frac{\kappa_c}{\kappa_2} \sim \frac{\kappa_c}{\kappa_2} . \qquad (30)$$

The quantities appearing in equations (29) and (30) may be estimated from Figure 1. Although the opacity depends strongly on the angle between the direction of propagation and the magnetic field, a photon scattering outward through the source will sample all different angles. Since the angles within 30° of $\theta = 90^0$ represent half the total solid angle, we expect such angles to be typical. At such angles the scattering coefficient has a series of line-like peaks, with $(\kappa_c/\kappa_1)^{1/2} \sim 0.1$ and $\kappa_c/\kappa_2 \sim 0.1$. Thus, according to expressions (29) and (30) the feature at the second harmonic should be as dark as that at the first. Furthermore, the branching ratios favor the transfer of radiation from higher harmonics to the first harmonic, an effect neglected in expression (29). A simple estimate suggests that the residual flux at the first harmonic could be increased by a factor $\sim$3-5 by this effect, leaving the second harmonic substantially darker. Thus, to the extent that harmonic structures are not observed, the low-energy features, if due to cyclotron scattering, must involve blends of several harmonics. Such blends would be observed if the radiating plasma moves through a range of magnetic fields during the detector accumulation interval, the field in the source is spatially inhomogeneous, or both.

PHOTOELECTRIC ABSORPTION BY HEAVY ATOMS

An alternative possibility is that the 10-50 keV absorption features reported in the spectra of γ-ray bursts are due to photoelectric absorption by heavy atoms in superstrong magnetic fields. Iron and nickel, for example, have photoelectric opacities $\sim 10^4$-10^5 times that of Thompson scattering[17] in the energy range 30-50 keV for magnetic field strengths in the range 2×10^{13}-10^{14}G.[18] The energies of the corresponding absorption edges vary approximately logarithmically with the strength of the magnetic field. Thus, even if the magnetic field in the source is quite inhomogeneous, the edge absorption will be localized in energy, producing a noticeable feature. Furthermore, photoelectric absorption in superstrong fields produces no absorption at harmonic overtones of the fundamental absorption frequency, unlike cyclotron absorption. No evidence of harmonic structure has been reported in γ-ray burst spectra.

For magnetic fields strong enough to shift the absorption edges of heavy atoms to energies above 30 keV, the properties of electron-positron annihilation radiation are quite different from those in the absence of a magnetic field.[19] In particular, for field strengths above 10^{13}G the rate of 1γ annihilation, which is forbidden in the absence of a magnetic field, is greater than that of the usual 2γ process, as shown in Figure 2. Moreover, the spectrum of the 2γ process is itself strongly modified for fields above 2×10^{13}G, as an increasing fraction of the total energy appears in one of the two photons, which therefore has an energy ~700-1000 keV (see fig. 7 of ref. 19). Thus, the presence of a 511 keV annihilation line in the spectrum of a burst implies that the field strength in the source is less than 2×10^{13}G, and hence that any absorption feature above 30 keV cannot be due to photoelectric absorption.

Although the 300-650 keV emission features that have been reported appear narrower than

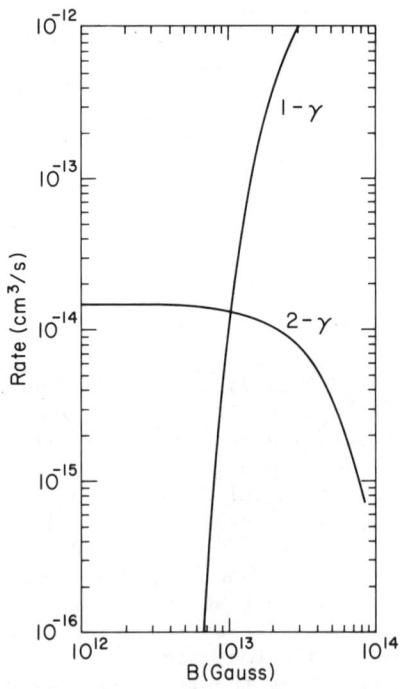

Fig. 2. Rate coefficients for nonrelativistic electron-positron annihilation into one or two photons as a function of magnetic field strength.

would be expected for annihilation at temperatures $\sim T_o$, the shapes of the features are poorly determined at present. In particular, it does not appear possible to rule out magnetic broadening and some contribution from one-photon annihilation radiation in many of these bursts. Furthermore, such features have not been detected in the great majority of the bursts that have low-energy features. One therefore cannot exclude the photoelectric interpretation on these grounds without better measurements of the high energy spectra.

Another possible difficulty with this interpretation is that magnetic fields of the strengths required would produce cyclotron extinction in the energy range from 200-1000 keV. No features have been detected in this range. A smooth continuum might still result if the scattering region samples a broad range of field strengths, but the apparent universality of spectra of the form (1) would remain a puzzle.

CONCLUDING REMARKS

Our results indicate that the continuum spectra of the γ-ray bursts could be due to optically-thin bremsstrahlung only if the source consists of very thin sheets or filaments of plasma. Such a source might result from accretion of plasma along magnetic flux sheets or tubes in the magnetosphere of a neutron star, or from plasma thrown up in a similar pattern by a nuclear explosion in the surface layers of a neutron star. The required source aspect ratio is, however, extreme ($\gtrsim 10^3$:1). Placing the burst sources at distances larger than 100 pc only makes the required aspect ratio more extreme. Another possibility, which requires a less extreme geometry, is that each burst consists of at least several flickers on time scales much shorter than the detector accumulation time. If each flicker emits a Comptonized thermal spectrum with a temperature that rises and falls, the time-averaged spectrum is very similar to that of optically thin bremsstrahlung.

We have also considered possible causes of the low-energy spectral features that have been reported. One possibility is cyclotron scattering, which requires magnetic field strengths $>10^{12}$G. Radiative transfer arguments show that if there is extinction at the first harmonic, the extinction at the second harmonic should be even greater. Thus, if the features are due to cyclotron scattering, the magnetic fields in the sources must vary by at least a factor of two, since harmonically-related features are not observed.

Another possible cause of low energy extinction is photoelectric absorption by iron or nickel. This would require substantially higher magnetic fields, $\gtrsim 2\times 10^{13}$ G. At these field strengths, electron-positron annihilation radiation will extend upwards from 511 keV to $\sim$1 MeV. A complication with this interpretation is the absence of any evidence of cyclotron extinction at 200-1000 keV.

Our results point to several key observational questions. First, is a true bremsstrahlung spectrum, including the Gaunt factor and relativistic corrections, consistent with the observed spectra? Second, is a spectrum of the form (10) consistent with the observed spectra? Third, is there any evidence that the spectra change on time scales shorter than 4s, as predicted by the flickering hypothesis? Finally, at what level of confidence can one-photon and magnetically broadened two-photon annihilation radiation be excluded for the bursts that have apparent emission features above 300 keV? Answers to these questions will provide significant constraints on theoretical models.

REFERENCES

1. Strong, I. B., Klebesadel, R. W., and Evans, W. D., in Proc. 7th Texas Symposium on Relativistic Astrophysics, Ann. NY Acad. Sci 262, 145 (1975).
2. Cline, T. L., in Proc. 7th Texas Symposium on Relativistic Astrophysics, Ann. NY Acad. Sci. 262, 159 (1975).
3. Mazets, E.P., Golenetskii, S.V., Aptekar, R.L., Guryan, Yu.A., Ilyinskii, V.N., preprint 687, A.F. Ioffe Physical-Technical Institute, Leningrad (1980).
4. Mazets, E.P., Golenetskii, S.V., Ilyinskii, V.N., Guryan,Yu.A., Aptekar, R.L., Panov, V.N., Sokolov, I.A., Sokolova, Z.Ya., and Kharitonova, T.V., preprint 719, A.F. Ioffe Physical-Technical Institute, Leningrad (1981).
5. Fenimore, E. E., Laros, J. G., Klebesadel, R. W., Stockdale, R. E., and Kane, S., preprint, LASL (1981).
6. Lamb, D.Q., Lamb, F.K., and Pines, D., Nature Phys. Sci. 246, 52 (1973).
7. Woosley, S.E., and Taam, R.E., Nature 263, 101 (1976).
8. Colgate, S.A., and Petschek, A.G., preprint, LASL (1981).
9. Woosley, S.E., and Wallace, R.K., preprint (1981).
10. Koch, H.W., and Motz, J.W., Rev. Mod. Phys. 31, 920 (1959).
11. Nagel, W., Ap. J. 236, 904 (1980).
12. Rybicki, G.B., and Lightman, A.P., Radiative Processes in Astrophysics (John Wiley and Sons, N.Y., 1979), Ch. 7.
13. Ramaty, R., and Cohen, J.M., Proceedings of the Conference on the Transient Cosmic Gamma- and X-Ray Sources (Los Alamos Scientific Laboratory, LA-5505-C, 1974), p. 146.
14. Daugherty, J.K., and Ventura, J., Phys. Rev. D 18, 1053 (1978).
15. Bussard, R.W., Ap. J. 237, 970 (1980).
16. Mihalas, D., Stellar Atmospheres (W.H. Freeman, San Francisco, 1970).
17. Wunner, G., Ruder, H., Schmitt, W., and Herold, H. 1981, preprint, submitted to M.N.R.A.S.
18. Ruder, H., Wunner, G., and Herold, H., Phys. Rev. Letters 46, 1700 (1981).
19. Daugherty, J.K. and Bussard, R.W., Ap. J. 238, 296 (1980).

ON THE INTERPRETATION OF GAMMA-RAY BURST CONTINUA AND POSSIBLE CYCLOTRON ABSORPTION LINES

E. E. Fenimore, J. G. Laros, R. W. Klebesadel, R. E. Stockdale*
University of California, Los Alamos National Laboratory
Los Alamos, New Mexico 87545
and
S. R. Kane
Space Science Laboratory, University of California, Berkeley

ABSTRACT

Most gamma-ray bursts have spectra which are fit by an optically thin thermal bremsstrahlung shape. We show that the small emitting volume implied by the observation of spectral features identified as cyclotron absorption lines from a 10^{12} Gauss field prevent thermal bremsstrahlung from providing sufficient luminosities to account for the observed fluxes unless the optical depth to Compton scattering is large. In addition, we demonstrate that complicated detector response functions can introduce artifacts which mimic absorption features below 60 keV unless the absolute gain of the detector is established and correctly applied in analysis. By comparing the continuum as observed by the Berkeley/Los Alamos ISEE-3 gamma-ray detector with the continuum observed by the Konus experiments we find a lack of agreement, suggesting that the gain of one of these experiments may be incorrect. Thus, a great deal of caution should be exercised in interpreting the apparent low energy features as cyclotron lines.

INTRODUCTION

Early observations of a few isolated gamma ray bursts showed spectra which could be characterized as a power law at low energies (<200 keV), and which fall off as an exponential at higher energies.[1-2] This shape was recognized as being due to optically thin thermal bremsstrahlung.[2] With the advent of the Konus experiment on Venera 11 and 12, bursts were routinely observed over the range ∼15 keV to ∼2 Mev. The primary spectral results from the Konus experiment are [3]: 1) virtually all of the continua can be fit by an optically thin thermal bremsstrahlung spectrum; 2) absorption features are seen in ∼25% of the events at energies between 27 and 65 Kev; and 3) emission features are observed in 10% of the spectra at energies of 400 to 460 Kev. The absorption features have been interpreted as cyclotron lines due to magnetic fields of the order of 10^{12} to 10^{13} Gauss. The emission features occur at energies expected for red-shifted 511 keV annihilation lines from a 10 Km, one solar mass object. Taken together, these results are strong evidence that gamma-ray bursts occur on highly magnetized neutron stars[3].

---

* Also at Princeton University

CONFLICT BETWEEN CYCLOTRON LINES AND THERMAL BREMSSTRAHLUNG

If thermal bremsstrahlung is assumed to be the emission mechanism for a gamma-ray burst, the observed flux can be used to determine an upper limit on the distance to the source [2]. If the burst spectrum also contains 30 to 60 keV cyclotron absorption features, the emitting volume is constrained roughly to the region of a 10^{12} Gauss magnetic field near a neutron star. The combination of these restrictions will be shown to allow the observed fluxes only if the burst sources are closer than a few parsecs.

The luminosity of a plasma with cosmic abundances emitting thermal bremsstrahlung radiation is[4]

$$L = 2.4 \times 10^{-27} <g> T^{1/2} N_e^2 V \text{ erg sec}^{-1}, \quad (1)$$

where $<g>$ is the temperature averaged gaunt factor and T is the best fit temperature. The luminosity is related to the observed flux φ by the distance D as $L = 4\pi D^2 \varphi$. An estimate for the electron density N_e is obtained by considering the optical depth of the plasma to Compton scattering

$$\tau = N_e \sigma_c \ell . \quad (2)$$

Here σ_c is half of the Thomson cross-section, and ℓ is a characteristic size of the emitting region. Since the thermal bremsstrahlung spectrum fits well over the range of energies observed, the effect of Compton scattering of the photons is probably small and τ should be near unity or less [2] (also, see below). These equations can be combined to obtain the distance to the source

$$D = \left| \frac{2.4 \times 10^{-27} <g>}{3\sigma_c^2} \right|^{1/2} \frac{T^{1/4}}{\varphi^{1/2}} \tau \ell^{1/2} \text{ cm}. \quad (3)$$

Figure 1 is intended to give various upper limits on the distance to a gamma-ray burst as a function of τ. Thus, values of T and φ were selected to allow the sources to be far away. For T we use kT = 1000 keV and the flux was set at 2×10^{-5} ergs cm^{-2} sec^{-1}. (However, since the distance is only weakly dependent on T and φ, a distance to a given burst will usually be dominated more by the uncertainties in τ and ℓ.) Another manner in which the sources can be far away is if the plasma consists of heavy ions. The solid curves in Fig. 1 are for a cosmic abundance (i.e. eq. 3), while the dashed curve assumes a pure iron plasma. Finally, an upper limit is needed on ℓ. One upper limit (3×10^4 Km) is set by the light travel time during which the intensity undergoes a significant change (0.1 sec for most bursts). A much lower upper limit is obtained if the absorption features in the 30 to 60 keV region of many gamma-ray burst spectra are, in fact, cyclotron absorption lines from a 10^{12} Gauss magnetic field. This field strength corresponds to cyclotron absorption by electrons at the first harmonic, and could only be found in the vicinity of a highly magnetized neutron star. Consider a 10 km neutron star with a 10^{13} Gauss field at its surface. Since the field falls off as the inverse cube of the radius, the absorbing region of the 10^{12} Gauss field must be contained within 20 km at the pole. We will use a spherical volume with a 20 km radius even though that ignores the fact that the emitting volume may be only above the poles. Further, we assume that the absorbing region overlies only 10% of the emitting volume. The effective radius of the emitting volume is then raised to 40 km, which serves as an extreme upper limit for ℓ if the sources contain cyclotron lines.

From Fig. 1 we see that the gamma-ray bursts can have a reasonable distance (i.e., > 50 pc) if the emitting region is as large as suggested by the light travel time and τ is close to unity. If the plasma consists of iron, then τ can be somewhat smaller. However, if the emitting region is only as large as suggested by the presence of the cyclotron lines, then most of the gamma-ray sources would have to be closer than 10 pc even if the plasma consists of pure iron. Since there are clearly not a large number of neutron stars less than 10 pc away, either the continuum is not thermal bremsstrahlung or the low energy absorption features are not cyclotron lines.

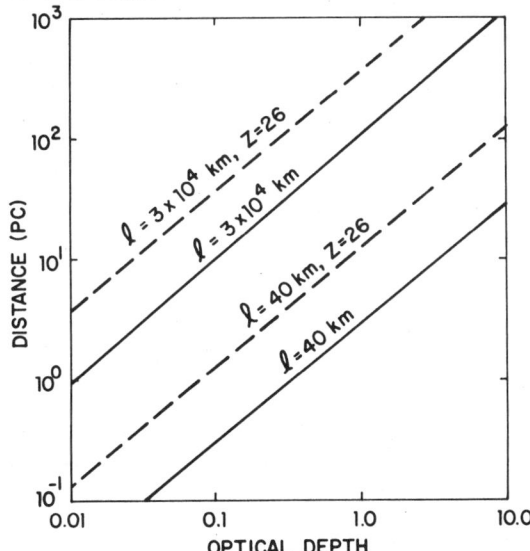

Fig. 1 Distance vs optical depth assuming conservative gamma ray burst parameters and thermal bremsstrahlung. The solid lines are for cosmic abundances and the dashed lines are for a plasma with $Z = 26$. Two representative sizes have been assumed: $\ell = ct = 3 \times 10^4$ Km is an upper limit based on light travel time and $\ell = 40$ Km is an upper limit when cyclotron lines from a 10^{12} Guass field is assumed (in which case, all bursts would have to be closer than a few parsecs).

The preceding argument required the assumption that the optical depth to Compton scattering must be near unity or less (based on the undistorted appearance of the observed thermal bremsstrahlung spectra). A perhaps even stronger restriction on τ is that τ has to be small enough so that comptonization does not destroy the cyclotron lines. To our knowledge the Comptonization of gamma-ray spectra has not been explicitly studied. For lower temperature plasmas (KT $\sim$ 10 Kev), the thermal bremsstrahlung spectra is noticeably distorted if τ is greater than unity[5] as are absorption and emission features.[6] However, at low temperatures $(1 + 4KT/mc^2)$ is the average fractional change in energy per Compton scattering[7] and, thus, an optical depth of unity might be expected to have a more severe effect on the shape of the continuum and especially on the cyclotron lines if the plasma's temperature is higher. It is not clear how to extrapolate these results for plasmas at X-ray energies to gamma-ray energies. At higher temperatures, the Compton cross-section is lower. In addition, depending on the orientation of the photons to the magnetic field, a strong magnetic field can reduce the Compton cross-section[8] and thereby increase the distance to the burst even if τ is unity or less. The large volume implied by the thermal bremsstrahlung probably requires a near spherical source rather than just the volume above a pole and therefore an angle-averaged, high-magnetic field Compton cross-section may not be much different from σ_c. Thus, there is some uncertainity in the assumption that τ is less than unity and further work on Comptonization in high temperature plasmas is needed. However, even if the optical depth to Compton scattering is as high as 10, thermal bremsstrahlung still probably conflicts with cyclotron lines because we have used extreme upper limits on all of the parameters in Fig. 1, in particular, on ℓ. The real volume implied by the cyclotron lines is probably much less than that given by $\ell=40$ km.

EFFECT OF GAIN SHIFTS ON SPECTRA

The major instruments currently observing gamma ray bursts are large field-of-view scintillators with PM tubes. The output of the PM tube is typically analyzed by an ADC to obtain the counts occuring in various spectral bins which are usually logarithmically spaced. The incident photon spectrum is then derived from the observed count spectrum utilizing the known instrument response function.

Figure 2 demonstrates the problems with scintillators operating above $\sim$30 Kev. The solid line and dashed line are two representative response functions for two incident beams of monoenergetic photons. At intermediate energies (40 to 100 Kev) the response curves have a main peak at the energy of the incident photons (e.g., E_1.) and also an escape peak at a lower energy (E_1-34 keV for scintillators containing Iodine). Incident photons at higher energies (e.g., above 400 Kev) typically have only a weak response at the incident photon energy. Instead, a

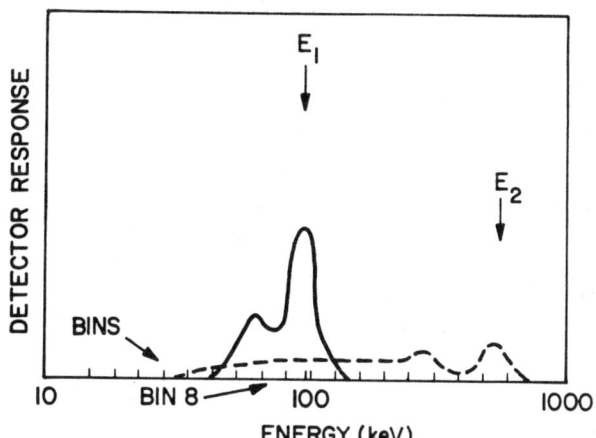

Fig. 2 Typical instrument response functions for a scintillator for two different incident monoenergetic beams.

high-energy photon often undergoes a Compton collision and is only detected as a count in some lower energy bin. Consider the analysis of a spectral bin, for example bin 8 in fig. 2. Two major effects must be accounted for. First, some fraction of the observed counts in bin 8 is actually due to incident photons at higher energies. Second, one has to correct for the efficiency of bin 8 after it has been determined how many counts in bins lower than 8 actually belong to bin 8. This redistribution of the incident photons into a count spectrum must be unfolded in some self-consistent manner.

The following steps were taken to find a self-consistent incident photons/keV spectrum. Predicted counts for various trial spectral shapes are found by folding the shape through Monte-Carlo determined detector response functions. The best fit spectrum is found by minimizing a χ^2 parameter. Note that the comparison is between observed counts and predicted counts. Thus, what is found first, is the trial function which produces predicted counts which best fit the observations. In order to display the observed counts in a photon/keV context, each bin's observed counts are scaled by the ratio of the spectral intensity (evaluted at the energy of the bin under consideration) to the predicted counts for the bin. If the resulting unfolded spectrum deviates significantly from the best fit trial spectrum, then one could be justified in suggesting spectral structure such as absorption or emission lines. The result is an incident photon spectrum which is not necessarily unique.

A potential major problem with the above technique is that there are often unknown shifts in the gain of an instrument which can cause spurious artifacts in the unfolded spectrum. For example, launch vibrations and/or thermal shocks can cause the dynodes of the PM tube to shift or the optical coupling between the PM tube and the scintillator to be degraded. Very often such disturbances can reduce the output signal intensity per keV by a factor of two. In the discussion below, the "gain" of the detector is defined to be the assumed photon energy divided by the true energy. If a gain equal to 0.5 is ignored, counts which were, in fact, caused by 100 keV photons are misinterpreted as due to 50 keV photons.

Unknown changes in the gain can cause artifacts in the unfolded spectrum. Figure 3 shows photon/keV spectra assuming two different gains from the Berkeley/Los Alamos ISEE-3 gamma ray detector for the solar event of 1980 June 7. (A solar event was used because it had excellent statistics; most of the error bars are smaller than the plot symbols.) The curve labelled G = 0.48 was analyzed with a gain (0.48 $\pm$ 0.04) derived from our on-board calibration system and comparisons to other instruments. Note that the unfolded spectrum is a typical solar gamma ray spectrum, that is a power law with an index of $\sim$2.7. There is some hint of a deviation from a power law at about 500 Kev. The curve labelled G = 0.80 was unfolded incorrectly assuming a gain of 0.8. Notice that the change in gain caused a dip in the spectrum at 40 keV (due to K-edge and K-escape effects), structure at $\sim$450 keV (due to Compton scattering effects), and a change in the slope of the continuum. Thus unknown changes in the gain can cause artifacts in the unfolded spectrum which mimic broad absorption features at low energy. We have found that low energy artifacts often occur when either CsI or NaI scintillators are used assuming an incorrect gain. In addition we have seen examples of such "lines" appearing and disappearing as a burst develops due to changing spectral shape and/or intensity.

COMPARISON OF ISEE AND KONUS

Given the possibility that gain shift can cause artifacts, it is very important to verify the cyclotron lines observed by the Konus experiment. Unfortunately, the Konus experiment is the only omnidirectional experiment that operates down to 15 Kev. However, on some occasions, gamma ray bursts occur sufficiently close to the ecliptic plane that the source falls within the field-of-view of an instrument sensitive to lower energies and collimated to view the sun or the ecliptic plane. One such burst occurred on 1978 November 4 (GB781104) and was also observed by the Berkeley/Los Alamos ISEE-3 instrument. Figure 4 presents the continuum of GB781104 as observed by Konus and our ISEE-3 instrument. The solid line labelled Konus is a 650 keV thermal bremsstrahlung spectrum plus a broad emission feature at 460 keV (taken from Mazets et al.[3]). The dotted line is the continuation of the Konus thermal bremsstrahlung spectrum at the location of the emission feature. The curve labelled ISEE is the best fit curve from the Berkeley/Los Alamos ISEE-3 gamma ray instrument along with the data points and 1 σ error bars (based solely on counting statistics). Both

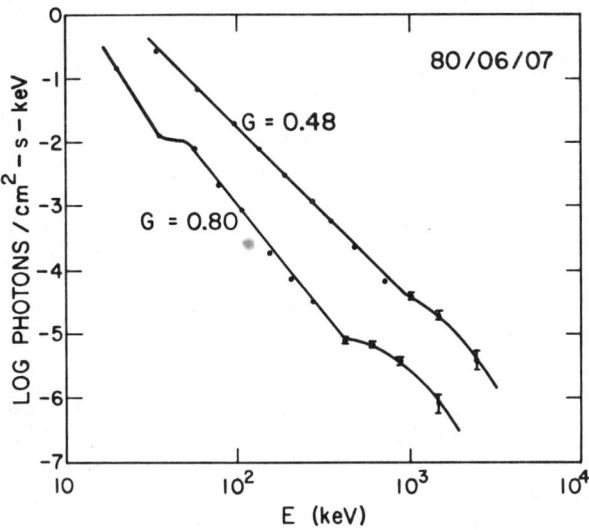

Fig. 3 A solar gamma-ray event analyzed with two different gains for the detector. A gain of 0.48 is believed to be correct and, in fact, gives a typical solar spectrum: a power law. An incorrect gain of 0.80 produces an artifact which looks like a cyclotron absorption line. The ordinate has an arbitrary scale.

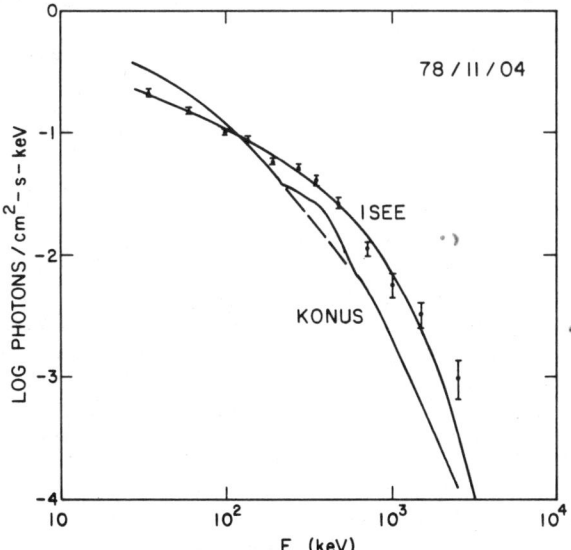

Fig. 4 A comparison of the continuum of GB781104 as observed by the Konus experiment and by the Berkeley/Los Alamos ISEE-3 gamma ray detector. The difference in slope can be explained by assuming one of the experiments was analyzed with an incorrect gain. The ordinate has an arbitrary scale.

instruments initiated the spectral observations at the same time to within ∼0.1 sec.

It is immediately obvious that the two experiments disagree on the slope of the continuum. A likely cause of this disagreement is that an incorrect gain was used in the analysis of one of the two experiments. If the ISEE gain is correct, then the gain used in the Konus experiment was too high. Note that it is high gains which produce low energy artifacts which mimic cyclotron absorption lines. Although no cyclotron lines have been suggested for GB781104, an incorrect gain for this particular burst implies that an incorrect gain might have also been used in the analysis of other Konus events, thereby leading to cyclotron lines.

The reported cyclotron lines have complex time structure and are not always at the same energy or have the same shape[3]. This is usually taken as strong evidence that the features are real and not artifacts. In addition, Dennis et al. have also reported absorption lines in GB800419 based on SMM observations[9]. Even with this confirmation, it will be important to settle the disagreement shown in fig. 4 since a gain shift can also change the apparent depth of cyclotron lines and those depths will probably be used as a diagnostic for the burst mechanism. Thus, without further information on the design, calibration, and analysis technique of the Konus experiment, it is difficult to judge how accurately the gain is known or how susceptible the unfolded spectra are to gain shifts. Until such information is available, we feel the reports of cyclotron lines should be accepted but with some reservation. We emphasize that we have not directly shown that the Konus experiment is capable of producing artifacts which mimic cyclotron absorption lines. We wish only to point out the possibility that gain shifts can produce artifacts in this type of detector and that there is an apparent disagreement between Konus and ISEE-3 which can be explained as an incorrect gain in one of those experiments.

SUMMARY

Virtually all Gamma-ray bursts have been fit with thermal bremsstrahlung. Black bodies have a completely incorrect shape and thus, one needs to explain how a gamma-ray burst on a highly magnetized neutron star produces a plasma which emits primarily by thermal bremsstrahlung. We have demonstrated in Section II that the small emitting volume implied by the observation of cyclotron lines appears to be inconsistent with the large volume required by the rather inefficient thermal bremsstrahlung mechanism. Thermal bremsstrahlung can not explain the luminosities even under the extreme conditions of a high plasma temperature, a pure iron plasma, a high surface magnetic field (10^{13} Guass), a low value for the field responsible for the absorption line (10^{12} Guass), a Compton depth of near unity despite the presence of lines in the spectra, and an emitting volume with a characteristic radius of 40 Km. However, there is the possibility that strong magnetic field effects on the Compton cross-section[8] in certain geometries might increase the allowed density within the permitted volume and therefore explain the observed fluxes. In section III, we have demonstrated that it is possible to produce artifacts in the spectrum if an incorrect gain is used. Indeed, we find evidence in GB781104 that either the Konus gain or our gain is incorrect. Thus, one possible explanation for the conflict between cyclotron lines and thermal bremsstrahlung is that the cyclotron lines are artifacts due to a gain shift. However, there has been some confirmation of cyclotron lines[9] and the lines observed by Konus have complicated temporal variability[3]. Therefore, we feel that the cyclotron lines reported from the Konus experiment should be accepted but with some reservation.

ACKNOWLEDGEMENTS

We thank A. G. Petschek for helpful discussions. The Berkeley/Los Alamos ISEE-3 gamma-ray spectrometer has been supported in part by the National Aeronautics and Space Administration under contract NAS5-25980. This work was done under the auspices of the U.S. Department of Energy.

REFERENCES

1. W. A. Wheaton, M. P. Ulmer, W. A. Baity, D. W. Datlowe, M. J. Elcan, L. E. Peterson, R. W. Kledesadel, I. B. Strong, T. L. Cline, and U. D. Desai, Astrophys. J. Lett. $\underline{185}$, L57 (1973).
2. D. Gilman, A. E. Metzger, R. H. Parker, L. G. Evans, and J. I. Trombka, Astrophys. J. $\underline{236}$, 951 (1980).
3. E. P. Mazets, S. V. Golenetskii, R. L. Aptekar, Y. A. Guryan, and V. N. Ilyinskii, Nat. $\underline{290}$, 378 (1981).
4. W. H. Tucker, Radiation Processes in Astrophysics (The MIT Press, Cambridge Mass., 1975), p. 202.
5. G. Chapline, Jr. and J. Stevens, Astrophys. J. $\underline{184}$, 1041 (1973).
6. R. R. Ross, R. Weaver and R. McCray, Astrophys. J. $\underline{219}$, 292 (1978).
7. S. A. Colgate, J. D. Colven, and A. G. Petschek, Astrophys. J. Lett. $\underline{197}$, L105 (1975).
8. V. Canuto, J. Lodenquai, and M. Ruderman, Phys. Rev. $\underline{10}$, 2303 (1971).
9. B. R. Dennis, K. J. Frost, A. L. Kiplinger, L. E. Orwig, U. D. Desai, and T. L. Cline, this volume (1981).

GAMMA RAY LINES FROM SOLAR FLARES AND COSMIC TRANSIENTS

R. Ramaty
Laboratory for High Energy Astrophysics, NASA-Goddard
Space Flight Center, Greenbelt, MD 20771

R. E. Lingenfelter*
Center for Astrophysics and Space Sciences
University of California, San Diego, La Jolla, CA 92093
and
B. Kozlovsky**
Laboratory for High Energy Astrophysics, NASA-Goddard
Space Flight Center, Greenbelt, MD 20771

ABSTRACT

Gamma ray line emission processes in solar flares and cosmic transients are reviewed and implications of recent line observations are discussed. The gamma ray line emission from solar flares results from nuclear interactions of accelerated particles with the solar atmosphere. The observed line intensities give information on the total number and spectrum of particles accelerated in the flare, on the temperature and density in the interaction regions and on the time history of the interactions. Analysis of the line observations from the June 7, 1980, flare show that the number of protons accelerated in the flare exceeded the number observed in the interplanetary medium by a factor of ~100. The bulk of the accelerated protons, therefore, remained trapped in the solar atmosphere where they produced gamma ray line emission as they slowed down.

The gamma ray emission lines observed from gamma ray transients appear to result from both the annihilation of positrons produced in photon-photon interactions and from deexcitation of nuclear levels and capture of neutrons produced in nuclear interactions. The observed line intensity provides information on the temperature, density, composition, magnetic field and redshift in the transient sources. Both the gravitational redshift and the iron enrichment implied by the gamma ray line observations strongly suggest that neutron stars are the source of the transients.

*Research supported by NASA grant NSG-7541.
**Also at University of Maryland, College Park, MD., research supported in part by NASA's Solar Terrestrial Theory Program - Permanent address: Tel Aviv University, Ramat Aviv Israel, research supported in part by the U.S.-Israel Binational Science Foundation.

INTRODUCTION

Gamma ray lines are unique tracers of the high energy processes that dominate the physics of solar flares and cosmic transients. In solar flares the lines are produced by nuclear interactions of energetic protons and nuclei with the solar atmosphere. Similar interactions in high temperature plasmas may also produce gamma ray lines in cosmic transients. In addition, the extremely high photon densities, matter densities and magnetic fields expected in the sources of these cosmic transients should lead to gamma ray line production processes that have no counterparts in solar flares. Because of the high photon densities, photon-photon collisions can become a dominant line producing mechanism via e^+-e^- pair production. The high pair densities resulting from these interactions may also lead to pair degeneracy and to possible maser action. The ultra high magnetic fields produce cyclotron emission and absorption lines, and cool and confine the e^+-e^- plasma, a necessary ingredient for producing an observable annihilation line.

The strongest line in solar flares is at 2.223 MeV from neutron capture on hydrogen in the photosphere. This result, predicted by theory and now confirmed by many observations, is consistent with line production by accelerated particle interactions. Indeed, for particles with essentially cosmic abundances and energies around a few tens of MeV/nucleon, neutron production greatly exceeds the production of any single nuclear deexcitation line. The prompt deexcitation lines, such as the 4.44 MeV line of ^{12}C, do, however, provide a direct measure of the energetic particle reaction rate. This line as well as other prompt lines have been observed from several flares, revealing the time history of the nuclear interactions.

The strongest emission line, observed in gamma ray transients, is that in the energy range from 0.40 to 0.46 MeV. It is reasonable to associate this line with e^+-e^- annihilation. The redshift from a line energy of 0.511 MeV or more could be due to the gravitational redshift of a neutron star or to maser (grasar) action discussed elsewhere in this volume. Lines from nuclear reactions also seem to be present in cosmic transients. In addition to the gravitational redshift, the principal difference between these lines and the solar lines is the drastically different particle abundance needed to explain the cosmic transient lines. In particular, an enhanced abundance of ^{56}Fe is required to account for both a nuclear deexcitation line in a gamma ray burst and a neutron capture feature in a longer duration transient. This result also supports a neutron star origin for cosmic transients.

In the first part of this paper we review the theory of solar gamma ray line production and present results of new numerical calculations based on more detailed and accurate nuclear cross sections. We illustrate the application of the theory by considering the June 7, 1980, flare for which there are reasonably detailed gamma ray data from SMM and important supporting particle data from ISEE-3, Helios-1 and IMP-8.

In the second part of the paper we discuss the physical processes responsible for gamma ray line production in cosmic transients, in particular, positron production and annihilation and nuclear line emission. We also review the possible origin of the gamma ray lines observed in a longer duration transient which is a particularly strong gamma ray line emitter.

SOLAR FLARES

Nuclear interactions of flare-accelerated protons and nuclei in the solar atmosphere produce gamma ray lines from neutron capture, positron annihilation and nuclear deexcitation. The first solar gamma ray lines at 0.511, 2.22, 4.44, and 6.13 MeV, were observed by Chupp et al[1] with a NaI detector on the OSO-7 satellite during the August 4, 1972 flare. These and other lines have been observed from several subsequent flares by detectors on the HEAO-1[2], HEAO-3[3] and SMM[4] satellites. The relative intensities of these lines were consistnet with earlier predictions.[5] A review of the observations is given by Chupp in this volume.

Neutron, Positron and Deexcitation Line Production

The strongest gamma ray line observed in nearly all of these flares is that at 2.223 MeV from neutron capture on hydrogen, $^1H(n,\gamma)^2H$. Several theoretical studies [5-10] have been made of neutron production, slowing down and capture in solar flares. Here we briefly review the interaction models used and we present results of new numerical calculations based on updated nuclear cross sections, which will be published elsewhere.

The instantaneous, or thin-target, neutron production rate is given by

$$q_n = n_H c \sum_i \sum_j a_i \int_0^\infty dE \beta \sigma_{ij}(E) N_j(E) \qquad (1)$$

where n_H is the ambient hydrogen density, a_i is the relative number of target nuclei i with respect to hydrogen, and σ_{ij} is the neutron production cross section from the interaction of target nuclei i with accelerated particles j having velocity βc, energy per nucleon E, and energy dependent number $N_j(E)$.

The time-integrated, or total thick-target, neutron yield, is given by

$$Q_n = A_o \sum_i \sum_j a_i \int_0^\infty dE\, (dx/dE)_j\, \sigma_{ij}(E) \int_E^\infty dE'\, \bar{N}_j(E') \qquad (2)$$

where A_o is Avogadro's number, $(dE/dx)_j$ is the energy loss rate per nucleon of accelerated particles j per unit path length of the ambient medium, and $\bar{N}_j(E)$ is the total number of accelerated particles j incident on the target. For more details see ref. (10).

Various forms of accelerated particle spectra were used in previous treatments[10-13] of neutron and gamma ray line production in energetic particle reactions. Two such forms, which also give[14] good fits to the particle data in interplanetary space, are exponentials in rigidity

$$N_j(E) \text{ or } \bar{N}_j(E) = A_j \exp(-P_j/P_o) \, dP_j/dE \qquad (3)$$

and Bessel functions,

$$N_j(E) \text{ or } \bar{N}_j(E) = A_j K_2\left[2(3p/m_p c\alpha T)^{1/2}\right]. \qquad (4)$$

Here A_j is proportional to the abundance of the energetic particles j, $P_j \equiv (A/Z)_j p$ is particle rigidity, $p = \sqrt{E(E + 2m_p c^2)}$ is particle momentum per nucleon, m_p is the proton mass, K_2 is the modified Bessel function of order 2, and P_o (measured in MV) and αT (nondimensional) characterize the spectrum of the energetic particles.

Equation (4) was shown[13] to be the particle number spectrum that results from stochastic Fermi acceleration with no energy losses, an energy and charge independent acceleration efficiency α and an escape time T. Furthermore, for the June 7, 1980, flare which is the first SMM event with reasonably complete gamma ray line and interplanetary particle data, this Bessel function spectrum gives a very good fit to both the proton and alpha particle spectra with essentially the same αT.

The solar atmospheric abundances[15] and the energetic particle abundances, used in the present calculations, are listed in Table 1. The particle abundances, which show a substantial enrichment of heavy nuclei, are known (e.g. ref. 16) to vary from flare to flare.

The cross sections for neutron production in energetic particle reactions have been summarized in ref. (10). We have recently updated these cross sections and added many new reactions that involve all the isotopes listed in Table 1. These new cross sections, to be published elsewhere, are used in the present calculations.

The magnitude, spectral form and charge dependence of the energy loss rate $(dE/dx)_j$ have important consequences on the nuclear yields in thick targets.

The present calculations have been carried out for slowing down in a neutral medium. The expression

$$(dE/dx)_j = (Z^2_{eff}/A)_j \, 630 E^{-0.8} \, (\text{MeV/nucleon})(\text{g/cm}^2)^{-1} \qquad (5)$$

where

$$Z_{eff} = Z\left[1-\exp(-137\beta/Z^{2/3})\right], \qquad (6)$$

gives a good fit to the tabulated[17] values of dE/dx for charged particles slowing down in neutral H. In an ionized medium the energy losses are higher[13] and hence the gamma ray yield per unit energy deposited is lower. We limit the present treatment, however, to a neutral medium.

TABLE 1 Abundances

Isotope	Ambient Particles	Energetic Particles
^{1}H	1.	1.
^{4}He	0.07	0.15
^{12}C	4.15×10^{-4}	1.07×10^{-3}
^{13}C	4.64×10^{-6}	1.28×10^{-5}
^{14}N	9.0×10^{-5}	2.14×10^{-4}
^{15}N	3.46×10^{-7}	8.57×10^{-7}
^{16}O	6.92×10^{-4}	2.14×10^{-3}
^{18}O	1.38×10^{-6}	4.28×10^{-6}
^{20}Ne	9.0×10^{-5}	2.14×10^{-4}
^{22}Ne	1.0×10^{-5}	2.57×10^{-5}
^{23}Na	2.28×10^{-6}	$4.28 - 10^{-5}$
^{24}Mg	3.11×10^{-5}	6.42×10^{-4}
^{25}Mg	4.01×10^{-6}	8.14×10^{-5}
^{26}Mg	4.43×10^{-6}	8.49×10^{-5}
^{27}Al	3.18×10^{-6}	5.35×10^{-5}
^{28}Si	3.46×10^{-5}	6.42×10^{-4}
^{29}Si	1.80×10^{-6}	3.21×10^{-5}
^{30}Si	1.18×10^{-6}	2.14×10^{-5}
^{32}S	1.80×10^{-5}	1.07×10^{-4}
^{34}S	7.61×10^{-7}	4.71×10^{-6}
^{36}Ar	3.39×10^{-6}	2.14×10^{-5}
^{38}Ar	6.23×10^{-7}	4.28×10^{-6}
^{40}Ca	2.28×10^{-6}	4.28×10^{-5}
^{52}Cr	4.15×10^{-7}	2.14×10^{-5}
^{54}Fe	1.94×10^{-6}	6.85×10^{-5}
^{56}Fe	3.11×10^{-5}	1.07×10^{-3}
^{57}Fe	7.61×10^{-7}	2.57×10^{-5}
^{58}Ni	1.25×10^{-6}	2.14×10^{-5}
^{60}Ni	4.84×10^{-7}	8.57×10^{-6}

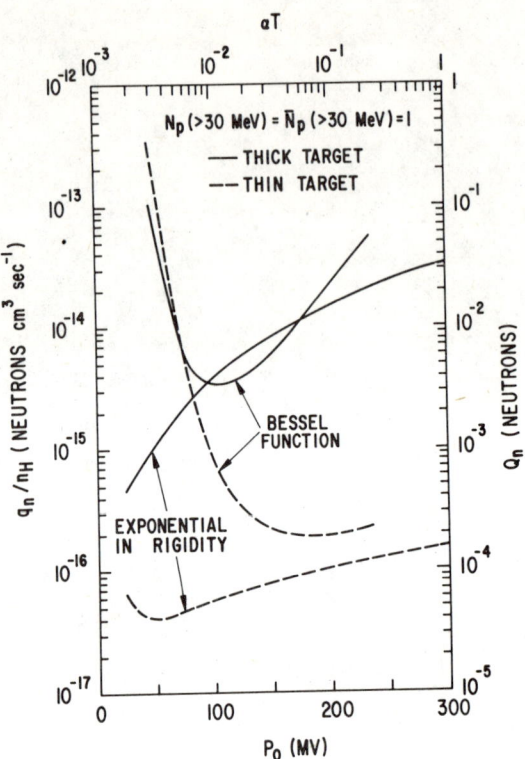

Fig. 1. Total neutron production in the thick- and thin-target interaction models with Bessel function (αT) and exponential (P_0) spectra.

The calculated neutron production rates and yields are shown in Figure 1 for energetic particle densities normalized to 1 proton above 30 MeV. For relatively flat energetic particle spectra, corresponding to large values of P_0 or αT, the bulk of the neutrons are produced in $p\alpha$ reactions. For steep particle spectra, given by the Bessel function at small αT, the large neutron yields result from reactions between α-particles and heavy nuclei. This effect is absent for the rigidity spectra because particles with $Z \geq 2$ have lower energies per nucleon at the same rigidity and hence produce less nuclear reactions. These effects were discussed in more detail in ref. (10) in connection with the comparison of neutron production by particles with spectra that were either power laws in energy or exponentials in rigidity.

The 2.223 MeV gamma ray line emissivity is equal to the neutron production rate times a quantity $\bar{f}$ which is the product of the fraction of neutrons that are captured on hydrogen in the photosphere and the probability of the resultant 2.223 MeV photon escaping from the photosphere. These neutron captures must compete with nonradiative captures, as well as with neutron decay and escape from the sun. Non-radiative capture on ^{3}He is the most important competing reaction[8], even though ^{3}He is only a minor constituent of the photosphere, because the cross section for the reaction ^{3}He $(n,p)^3$H is about four orders of magnitude larger than that for the reaction ^{1}H $(n,\gamma)^2$H. Observations of the intensity of the 2.223 MeV line compared to that of other lines can limit the photospheric ^{3}He/^{4}He ratio. Calculations [8,10,18] assuming ^{3}He/^{1}H of 5×10^{-5}, isotropic neutron production above the photosphere and a flare site away from the limb, give $\bar{f}$ ranging from 0.1 to 0.14, depending on the energy spectrum and interaction model. In the thick target case, however, the neutrons could be produced in the photosphere which could increase $\bar{f}$ to as much as ~0.2.

The 0.511 MeV line from positron annihilation has been observed from several flares. A number of theoretical studies have been made of positron production [5,7,10] and on positron slowing down and annihilation [19,20]. For this paper we give the results of new calculations of positron production based on much greater number of β+ emitters than were considered in the previous calculations. The results are given in Figure 2, where we show the ratio for thick and thin targets, Q_{e+}/Q_n and q_{e+}/q_n respectively.

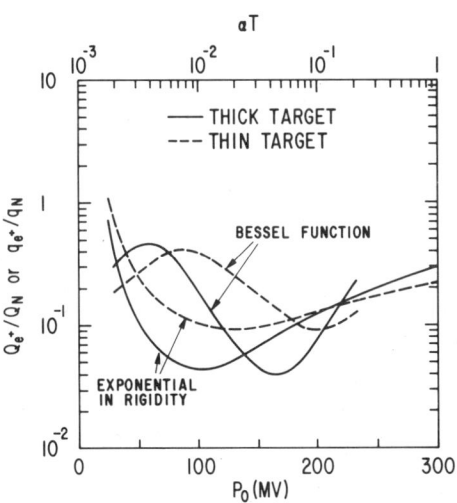

Figure 2. Ratios of the total position yield to the total neutron yield for the thick-and thin-target interaction models and Bessel function (αT) and exponential (P_0) spectra.

Q_{e+} and q_{e+} were calculated using the same equations, abundances and particle spectra as given above for the neutron calculation. The positron yields shown in this figure represent total yields. Because of the finite half-lives of the various β+ emitters in a short observation time of a transient event, fewer positrons than indicated in Figure 2 are available for 0.511 MeV line production.

Flare-accelerated particle interactions also lead to many other gamma ray lines from deexcitation of nuclear levels. The two strongest lines, at 4.44 and 6.13 MeV due to $^{12}C^*$ and $^{16}O^*$ deexcitation respectively, were first observed[1] from the solar flare of August 4, 1972. These and other nuclear lines have been seen in a number of subsequent flares, reviewed by Chupp elsewhere in this volume.

We have treated[12] in detail the production of gamma ray lines in energetic particle reactions. Using this treatment, we have evaluated the prompt gamma ray lines for the abundances, particle spectra and interaction models discussed above. Each of the resultant lines has two components: a narrow ($\Delta E/E \simeq 2.5\%$) line component from deexcitation of the recoiling, ambient-gas nuclei and a much broader underlying component from deexcitation of the fast, accelerated-particle nuclei. The bulk of the gamma ray flux, observed from the August 4, 1972 flare at energies between 4 and 7 MeV, has been shown[11] to result from the superposition of broad and narrow nuclear lines rather than electron bremsstrahlung or other continuum emission processes. This energy band, now referred to in the SMM observations as the "main channel window", can thus provide a direct and sensitive measure of this interaction rate of flare accelerated nuclei.

Because the cross sections of nuclear excitation levels have different energy dependences from those of neutron and positron production, the calculated [5,10] ratios of nuclear deexcitation line emissivities to the neutron production rate depend strongly on the assumed spectra of the accelerated particles. One of the strongest nuclear deexcitation lines is that at 4.44 MeV.

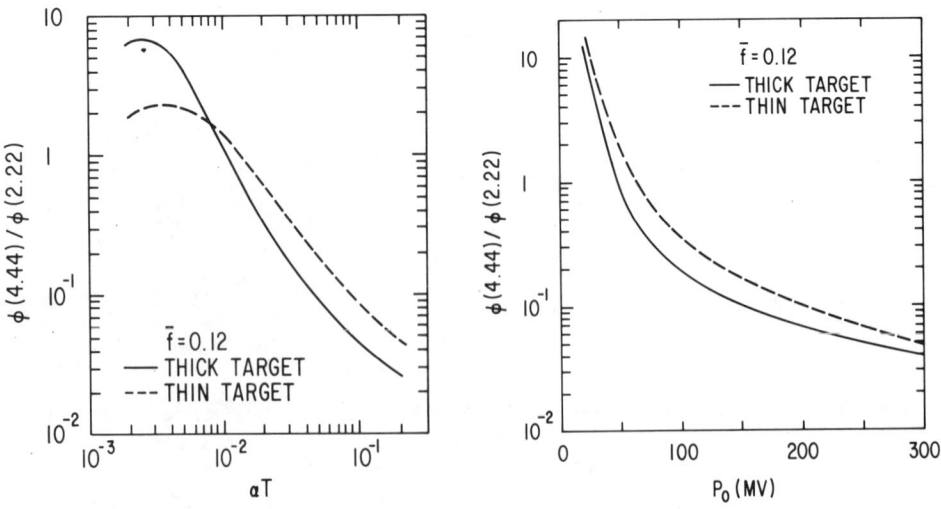

Figures 3 and 4. Ratios of the 4.44 MeV to the 2.22 MeV line intensity in the thick-and thin-target models with $\bar{f} = 0.12$ for Bessel function (αT) and exponential (P_0) energetic particle spectra.

In Figures 3 and 4 we show ratios of the intensity of this line to the 2.223 MeV line intensity for the different interaction models and energetic particle spectra. As can be seen, these ratios generally decrease with increasing energetic particle spectral hardness reflecting an increased neutron production and decreased 4.44 MeV photon production by particles of high energies. Likewise $\phi(4.44)/\phi(2.22)$ is lower for thick targets than for thin targets (except for very steep Bessel function spectra) because the energy losses harden the particle spectra in the thick target. For very steep (small αT) Bessel function spectra, the thick target ratio exceeds that for thin targets, because of the suppression of the heavy particle fluxes by energy losses in the thick target. As mentioned above, the interaction of these particles with helium are the main neutron sources at low energies. This effect is not present for rigidity spectra, because, as explained earlier, the contributions of the heavy particles to the nuclear reaction rates is negligible for such spectra for both thick and thin targets.

Implications of Line Observations

The observations of the intensity, time history and relative strength of gamma ray lines from solar flares allow us to determine important properties of the flare accelerated particles, the interaction and emission region and the particle acceleration process.

Particle Spectrum and Number

In this paper we shall focus on the flare of June 7, 1980 for which there are both gamma ray line observations and interplanetary particle measurements. The combined analysis of these observations and measurements provide important new information. In particular we find that in this flare the bulk of the gamma ray line emission results from thick-target interactions of accelerated particles trapped and slowing down in the solar atmosphere and not from interactions of the particles during their acceleration. Moreover, for this flare the gamma ray line intensities require that the bulk of the flare accelerated particles remained trapped in the solar atmosphere and that only a very small fraction of the accelerated particles escaped into the interplanetary medium.

The June 7 flare was located at N12W74 and hence was well connected magnetically to Earth. Indeed, several observations of energetic particles have been reported[21-23]. Based on these, the number of protons of energies greater than 10 MeV released into the interplanetary medium, Nesc(>10MeV), has been estimated[21] to be ~10^{31}. Furthermore, the spectrum of these protons was well fit with the Bessell function of equation (4) with αT equal to 0.013 (R. McGuire, private communication 1981). This spectral form also fits the α-particle spectrum with essentially the same αT.

In the thin target model, the 2.22 MeV line fluence can be written[13] as

$$\phi(2.22) = \bar{f}(q_n/n_H) n_H T \text{ Nesc } (>30 \text{ MeV})/4\pi d^2 \qquad (7)$$

Where T is the particle escape time from the thin-target interaction region and d is the distance to the sun. If this region is also the acceleration region, the time T in equation (7) is the same as the time T in the parameter αT that defines the spectral hardness of the Bessell function.

Using the proton data ($\alpha T=0.013$, and Nesc($>$10MeV) $\approx 10^{31}$ protons giving Nesc($>$30MeV) $\approx 5\times 10^{29}$) and the gamma ray data ($\phi(2.22) \approx 6.6$ photons cm^{-2} from Chupp in this volume), we obtain from Figure 1 and equation (7) that for a thin target $n_H T \approx 7 \times 10^{14}$ cm^3 sec^{-1}, or a matter traversal for 30MeV protons of 12gmcm^{-2}. The large abundances of spallation products (^{2}H, Li, Be, B) that would result from such a long path length are not observed in solar flares. We conclude, therefore, that the gamma ray lines observed from the June 7, 1980 flare were probably not produced in thin-target interactions. A similar conclusion has been reached[13] for the August 4, 1972 flare.

In the thick-target models, on the other hand, the spallation products, that accompany the production of gamma ray lines, are slowed down in the solar atmosphere and hence are not expected to be seen in the interplanetary medium. In this model the 2.22 MeV fluence is given by

$$\phi(2.22)=\bar{f}Q_n\bar{N}_p(>30\text{MeV})/4\pi d^2 \qquad (8)$$

where Q_n is calculated from equation (2).

If we assume that the same process accelerates both the particles, $\bar{N}_p$, that interact at the sun to produce the gamma ray lines and the particles, Nesc, that are observed in the interplanetary medium, and that both populations therefore have essentially the same energy spectrum, then $\bar{N}p$ should be approximately equal to Nesc (1-F)/F, where F is the fraction of particles escaping into the interplanetary medium. From the Q_n calculated for $\alpha T\sim 0.013$ (Figure 1) and the observed $\phi(2.22)\approx 6.6$ photons cm^{-2}, we then find that $\bar{N}p$ ($>$30MeV) $\approx$ 4.5$\times 10^{31}$ protons or $\bar{N}p$ ($>$10MeV)$\approx 10^{33}$ protons. This exceeds Nesc($>$10MeV) by a factor of $\sim$100, implying that $F\approx 10^{-2}$ or only $\sim$1% of the accelerated protons escaped from the June 7 flare and that the rest remained trapped in the solar atmosphere where they produced the observed gamma ray line emission by nuclear interactions while they slowed down.

Analysis of the prompt gamma ray line emission from the June 7 flare also suggest that the observed emission was produced in thick-target interactions, not thin target. In particular, we consider the ratio of the combined line fluence in the 4 to 7 MeV band to the fluence in the 2.22 MeV line, which in this flare was found to be 1.74±0.27 (from Chupp in this volume).

Using the methods and nuclear data of ref.(12) with the Bessel function spectrum of equation (4) and the abundances given in Table 1, we calculate for $\alpha T=0.013$ that $q(4-7)/n_H=4.4\times 10^{-16} N_p(>30\text{MeV})$ cm^3 sec^{-1} for the thin-target model and $Q(4-7)=1.1\times 10^{-3}\bar{N}_p(>30\text{MeV})$ for the thick-target model. Comparing with the neutron production rates (Figure 1), taking $\bar{f} \approx 0.12$, we calculate $\phi(4-7)/\phi(2.22) \approx 2.6$ for the thick target and $\sim$7.5 for thin target. The observed ratio of 1.74± 0.27 is in much better agreement with the thick-target model than with the thin-target. Moreover, since in the thick-target model the neutrons may be produced in the photosphere where $\bar{f}$ could be as much as $\sim$0.2, the calculated $\phi(4-7)/\phi(2.22)$ could be as low as $\sim$1.6 and thus be in excellent agreement with the observations. Other effects, such as some beaming of the charged particles, could also produce similar agreement in the thick-target case.

Structure of the Interaction Region

Theoretical studies [5,8-10] of the solar flare gamma ray line emission processes suggest that the prompt, deexcitation line emission, the subsequent positron annihilation and the eventual neutron capture emission should each occur at successively deeper mean depths in the solar atmosphere. This results from two effects. First, the range of the ~10MeV protons primarily responsible for the excitation of nuclear levels is only about 1/5th of that of the ~30MeV protons that produce the bulk of the positrons and neutrons. Thus, if these particles are accelerated high in the solar atmosphere and travel down the magnetic field lines into the deeper atmosphere where they lose their energy and are eventually stopped, the bulk of the nuclear line excitation should occur at a higher altitude than the neutron and positron production. Second, since the range of the secondary positrons is less than that of the ~30 MeV protons which produced them, the positrons should annihilate at a depth close to that at which they were produced. But the neutrons on the other hand have a slowing down and capture mean free path that is greater than that of their proton pregenitors, so they can propagate to significantly greater depth before they are captured. On the average, therefore, those neutrons that are captured will do so at a greater average depth than that at which they were produced, since those neutrons which move upward to shallower depths are more likely to escape or decay before they can be captured.

Strong observational evidence for such differences in the mean emission depths of the deexcitation and capture lines appears in the gamma ray line observations of the limb flares of June 21, 1980 and April 27, 1981, described by Chupp in this volume. In these flares the observed ratio of the deexcitation line fluence at 4-7 MeV to that of the 2.22MeV neutron capture line was roughly a factor of 40 times and 10 times that of the average disk flares, respectively. This is clear observational evidence that the 2.22 MeV line is produced in the photosphere with differential attenuation, or limb darkening, of the 2.22 MeV line corresponding to column density differences between the nuclear deexcitation region and the neutron capture region of $(2 \text{ to } 3) \times 10^{25} \text{H cm}^{-2}$. The exact longitudes of these limb flares are not known but these differences probably correspond to radial column depth differences of a few times 10^{24}H cm^{-2}. Such a value for the depth of the mean neutron capture region is also consistent with the average density of $\sim 10^{17} \text{H cm}^{-3}$ in the neutron capture region for the June 7, 1980 flare implied by the $\sim 10^2$ sec decay time of its 2.2 MeV line emission. Detailed studies of the time dependence of the positron annihilation and neutron capture line emission in all of these flares should provide much more information on the structure of the flare emission region.

GAMMA RAY BURSTS

Observations of emission lines and absorption features, seen in the spectra of many gamma ray bursts and transients, are reviewed by Teegarden in this volume.

The absorption features are observed[24] at energies below about 100 keV in the spectra of 20 bursts. If these are due to cyclotron absorption, they require very strong magnetic fields of the order of 10^{12} gauss, such as those expected around neutron stars. The most commonly observed emission line falls in the range from 0.4 to 0.46 MeV, as observed[24] by low resolution NaI detectors in seven gamma ray bursts. In the spectrum of one of these bursts, that of November 19, 1978, a small Ge detector has resolved[25] two lines at ~0.42 MeV and ~0.74 MeV, which the NaI detectors have seen as one broad emission feature from 0.3 to 0.8 MeV. Line emission in the range of 0.4 to 0.46 MeV is most likely due to gravitationally redshifted e^+-e^- annihilation radiation, while the line at 0.74 MeV could be either collisionally excited and gravitationally redshifted 0.847 MeV emission from ^{56}Fe, an abundant constituent of the crusts of neutron stars, or gravitationally redshifted single photon e^+-e^- annihilation[26,27] radiation at 1.022 MeV in a very strong (~10^{13} gauss) magnetic field. In all cases, the implied redshifts of 0.1 to 0.3, are consistent with those expected from neutron star surfaces. Thus, the magnetic fields, the redshifts and the surface composition indicate that the sources of the bursts with observed lines are probably neutron stars.

Positron Annihilation

The principal positron producing mechanism in gamma ray bursts is likely to be pair production in photon-photon collisions. This follows from the enormous photon densities in the burst sources deduced[28,29] from the observed luminosities, likely source distances and sizes. The latter probably are of the dimensions of neutron stars or smaller.

The production of a relatively narrow e^+-e^- annihilation line requires[30] that the pairs annihilate at a temperature which is substantially lower than the burst temperature deduced from the observed continuum spectrum or the temperature required to produce the pairs. A promising cooling mechanism[31,32] is synchrotron radiation in strong magnetic fields ($B > 10^{11}$ gauss). The radiation produced by this cooling could be responsible[31,32] for the continuum emission of the bursts at energies below ~300 keV.

This synchrotron cooling model was specifically applied[31,32] to the March 5, 1979 transient. Although the observation of a ~0.43 MeV emission line from this transient suggests that its source was a neutron star, the origin of this spectacular event remains unresolved. The position of the burst has been determined by triangulation (see review by Cline in this volume) leading to an error box of size less than an arc minute within the supernova remnant N49 in the Large Magellanic Cloud (LMC). Nevertheless, the question remains whether the burst did indeed originate in the LMC or whether its source was much closer. The probability for chance coincidence between the March 5 burst source position and N49 has recently been estimated[33] to be $< 10^{-3}$.

The synchrotron cooling model removes one of the principal difficulties posed by a burst source in the LMC, namely that of the very high brightness temperature (~1MeV) of the observed radiation. For this model, the bulk of the observed continuum emission, rather than being thermal radiation of particles with $kT \lesssim 100$ keV, is optically thin synchrotron emission of the ~MeV electrons and positrons. These particles, after losing their kinetic energy by synchrotron radiation, annihilate and produce the observed ~0.43 MeV line. The minimum magnetic field required for a cooling time shorter than the annihilation time is of the order of a few times 10^{11} gauss. This value turns out, also, to be the minimum B required to confine the e^+-e^- plasma that produces the observed annihilation line. This confinement is crucial, because otherwise the super-Eddingtonian burst luminosity would cause relativistic expansion of the emission region in clear conflict with an observed redshifted line.

An important addition has been recently proposed[34] to the synchrotron model for the March 5 transient. The MeV e^+-e^- pairs, in addition to producing synchrotron radiation, could also Compton scatter their own synchrotron photons and thus produce the observed continuum above ~0.5 MeV. Since both the synchrotron and Compton components of the observed spectrum depend linearly on the same relativistic pair density, the model allows a distance determination. The distance d depends only on B, the mean relativistic particle energy, γmc^2 and the area of the emitting region, A. For $B \sim 2\times 10^{11}$ gauss, $\gamma mc^2 \sim 1$ MeV, and A of the order of a neutron star surface area, d turns out to be approximately the distance to the LMC. If the source were much closer, Compton scattering of the synchrotron photons could not explain the high energy continuum unless the emitting area were very small. For example if d ~ 100 pc, A would have to be only a few millionths of the neutron star surface.

In a separate paper in this volume, by Ramaty, McKinley and Jones, the possibility of gamma ray amplification through stimulated annihilation radiation (grasar) has been proposed. This process could produce a narrow annihilation line without requiring a low temperature annihilation region. Grasar action requires an e^+-e^- pair density which substantially exceeds the thermodynamic equilibrium density at the temperature of the burst source (~10^{30}cm^{-3} at T~3×10^9K). This is equivalent to the population inversion of a regular maser. But it is not clear yet how such an inversion could be produced in a gamma ray burst source. If an inversion can be produced, however, grasar action would lead to a narrow emission line at ~0.43 MeV without the requirement of a large gravitational redshift. One strong observational argument for grasar action would be the detection of narrow annihilation lines of width less than about 0.1 MeV.

Nuclear Line Emission

At least one gamma ray burst, that of November 19, 1978 also shows[25] evidence of possible nuclear line emission. Its spectrum, reviewed by Teegarden in this volume, shows a strong narrow line at about 0.74 MeV, and other features at higher energies. The continuum spectra of gamma ray bursts are suggestive of optically thin bremsstrahlung emission of electrons with temperatures $\gtrsim 10^9$K.

Calculations[35-37] of the expected nuclear deexcitation and radiation capture line emissivities from thermonuclear reactions in high temperature plasms have been made for a temperature range of 10^8 to 10^{12}K.

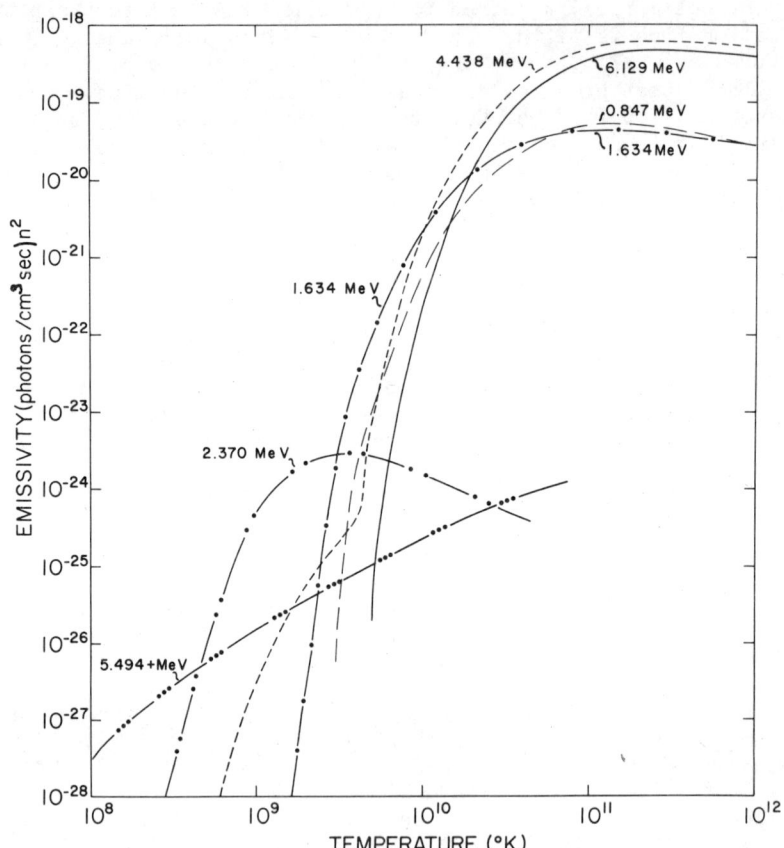

Figure 5. The emissivities of the principal nuclear deexcitation and proton radiative capture lines resulting from thermonuclear reactions in a plasma of local interstellar composition. These lines result from the following reactions: 5.494+ MeV from $^2H(p,\gamma)^3He$; 2.370 MeV from $^{12}C(p,\gamma)^{13}N$; 1.634 MeV from $^{20}Ne^*$; 4.438 MeV from $^{12}C^*$, 6.129 MeV from $^{16}O^*$; and 0.847 MeV from $^{56}Fe^*$.

The calculated emissivities of these lines from a plasma of local interstellar composition are shown in Figure 5. As can be seen, the strongest lines for such abundances at temperatures of 10^8 to $\sim 3 \times 10^9$ K are radiative capture lines from ^{2}H(p,γ)^{3}He and ^{12}C(p,γ)^{13}N. Proton capture on ^{2}H leads to a line at the reaction Q $\sim$ 5.494 MeV plus the kinetic energy available in the center of mass of the interaction which depends directly on the ion temperature. Proton capture on ^{12}C is dominated by a strong, narrow resonance in the capture cross section at about 0.46 MeV/nucleon, leading to a line at $\sim$2.37 MeV whose energy is relatively independent of the ion temperature.

For ion temperatures of about $\sim 3 \times 10^9$ to 3×10^{10} K nuclear deexcitation lines should be the most intense. The principal lines from a plasma of local composition are those at 6.129 MeV from ^{16}O*, 4.438 MeV from ^{12}C*, 1.634 MeV from ^{20}Ne* and 0.847 MeV from ^{56}Fe*. These lines should be relatively narrow, thermally broadened to widths FWHM at $\sim 10^{10}$ K of about 140, 120, 30 and 10 KeV respectively.

At ion temperature $\geq 3 \times 10^{10}$ K neutron production exceeds the emissivity of the deexcitation lines and neutron radiative capture lines could become the most intense. If the capture occurs in the hot plasma ($\geq 3 \times 10^{10}$ K) in which the neutrons are made, however, the capture lines would be greatly shifted and broadened by the large available energy in the center of mass. In particular the 2.223 MeV capture line from ^{1}H(n,γ)^{2}H and the 7.632 and 7.646 MeV lines from ^{56}Fe(n,γ)^{57}Fe could be shifted by a few MeV or more and broadened to a width of more than an MeV.

The $\sim$0.74 MeV line and a possible line at $\sim$1.1 MeV, reported by Teegarden in this volume, in the spectrum of the November 19, 1978 burst could be nuclear deexcitation lines at 0.847 MeV and 1.238 MeV from the first and second levels of ^{56}Fe, redshifted by z$\sim$0.14. Such a redshift is also implied by the broad feature at $\sim$0.42 MeV, if it is redshifted e^+-e^- annihilation radiation. This redshift is within the range of gravitational redshifts expected at the surface of neutrons stars.

But the high relative luminosity of the $\sim$0.74 MeV line, $\sim$9% of the total burst luminosity observed above 200 KeV, puts strong constraints on a nuclear origin of the line. In particular, if this line is redshifted 0.847 MeV emission accompanied by an optically thin, thermal bremsstrahlung continuum, then the line-to-continuum luminosity ratio requires that the emission come from an iron-enriched, two-temperature plasma with Fe/H $\simeq 10^{-2}$, an ion temperature $T_{ion} \geq 2 \times 10^{10}$ K and an electron temperature $T_e \leq 2 \times 10^9$ K. Such a plasma of unit optical depth on the surface on a neutron star at a nominal distance of 100 pc, emitting $\sim 10^{38}$ erg sec^{-1} in bremsstrahlung and $\sim 10^{37}$ erg sec^{-1} in ^{56}Fe deexcitation lines, could produce the observed burst spectrum.

An alternative suggestion[27] that the $\sim$0.74 MeV line is a strongly redshifted (Z$\sim$0.38) line at 1.022 MeV from single photon annihilation of e^+-e^- pairs in very strong magnetic fields ($\sim 10^{13}$ gauss) has not been studied in any detail but such a process could not account for a possible line at $\sim$1.1 MeV or higher energy features.

GAMMA RAY LINE TRANSIENTS

There is apparently another class of gamma ray transients in which essentially all of the observed radiation is in the form of line emission. Such a gamma ray line transient was discovered by Jacobson et al.[38] with a high resolution Ge detector on June 10, 1974 from an unknown source. This event, lasting about twenty minutes, was characterized by strong emission in four relatively narrow energy bands at 0.40-0.42 MeV, 1.74-1.86 MeV, 2.18-2.26 MeV, and 5.94-5.96 MeV with no detectable continuum. A detailed description of this observation is given by Ling et al. elsewhere in this volume.

As pointed out by Jacobson et al.[38] the line identification alone appears to require strongly redshifted emission, since the lines at ~0.41 MeV and ~5.95 MeV cannot be identified with any unshifted lines that would not be accompanied by much stronger companion lines which were not observed. All of these lines however can be identified with lines from the most intense nuclear emission processes expected in nature, positron annihilation and neutron capture on hydrogen and iron, if we assume that the emission comes from a gravitationally redshifted region with a z ranging from ~0.2 to 0.285. In particular the 5.94-5.96 MeV line can be identified with the two strongest lines from neutron capture on the iron crust of a neutron star having a surface gravitational redshift of $z \approx 0.285$. Since ^{56}Fe is expected to be the dominant constituent of neutron star crusts this line alone strongly suggests a neutron star source for the transient emission on June 10, 1974.

The 0.40-0.42 MeV and the 1.74-1.86 MeV lines can similarly be identified as the positron annihilation and neutron capture in a hydrogen atmosphere of density $n_H > 10^{16}$ H cm^{-3} extending from the surface z of 0.285 up to an altitude corresponding to a z of about 0.2. Such an atmosphere is presumably a temporary feature perhaps formed by infall of gas from an accretion disk. This minimum hydrogen density is required in order that a significant fraction of the neutrons are captured before they decay. The width of the ~0.4 MeV line also places a limit on the average temperature of this atmosphere of $<2 \times 10^6$ K. The unshifted ($z \approx 0 \pm .02$) line at 2.22 MeV requires a comparable high density region well above the neutron star surface and can be understood as neutron capture in the atmosphere of a binary companion.

A surface redshift of ~0.285 implies a neutron star mass of about 1.4 to 1.8 $M_\odot$ and a radius of 10 to 13 km, depending on the assumed equation of state. The observed 25 keV full width of the ~5.95 MeV line reduced by the instrumental broadening of 5 keV and the redshifted ^{56}Fe doublet separation of 10 keV gives an emitted width of 10 keV. If the line were produced uniformly over the surface of the neutron star, this width would give a lower limit for the rotation period

$$P > \frac{E_\gamma}{\Delta E_\gamma} \cdot \frac{4\pi r \sin\theta}{c} = \frac{\sin\theta}{4} \text{ sec,}$$

where θ is the angle between the rotation axis and the direction of observation.

If the lines do result from neutron capture near the surface of a neutron star, however, the observed line fluences would then require either rapid mixing or rapid break up of the resultant ^{2}H and ^{57}Fe formed by the capture since these nuclei would quickly accumulate to many optical depths for any reasonable source distance. The column density of ^{2}H and ^{57}Fe, produced at the surface of a neutron star during the June 10, 1974 transient, would amount to $\sim 3 \times 10^{30} D_{100}^2$ ^{2}H cm^{-2} and $\sim 10^{30} D_{100}^2$ ^{57}Fe cm^{-2} where D_{100} is the assumed distance in units of 100 pc.

But even assuming such an identification for the lines, the origin of the neutrons and positrons required to produce them is still problematical. Although one might imagine that it should not be hard to find ways of generating neutrons directly on the surface of a neutron star, the observation of the unshifted neutron capture line on hydrogen would require that the average energy of the neutrons produced at the neutron star surface must be ~150 MeV, equal to the escape energy from the star, so that a significant fraction could escape to a lower z region to be captured. There is no obvious process for generating such energetic neutrons. It also seems unlikely that the neutrons and positrons could have been produced in nuclear reactions in a flare on the binary companion because of energetic considerations. One possibility, however, would appear[36] to be that the neutrons and positrons which were captured and annihilated in the vicinity of the neutron star and its binary companion were all produced by nuclear reactions occurring when gas, flowing episodically from the companion, was gravitationally accelerated to an energy of > 1 MeV and collided with gas in the outer part of the accretion disk around the neutron star. In addition to the observed lines we would also expect other lines and continuum emission, but the expected intensities are not necessarily inconsistent with observations.

Lastly, since the sum of the neutron-capture gamma-ray line intensities for the transient event of June 10, 1974 was ~ 3×10^{-7} erg cm^{-2} sec^{-1}, the neutron capture line luminosity of the source $L_{n,\gamma} \approx 3 \times 10^{35} D_{100}^2$ erg sec^{-1}. Assuming a maximum neutron yield of < 2×10^3 neutrons erg^{-1}, see ref.(36), and an average neutron capture photon energy of ~3 MeV, the total source luminosity L must have been $\geq 10^{38} D_{100}^2$ erg sec^{-1}.

REFERENCES

1. E.L. Chupp et al., Nature, 241, 333 (1973).

2. H.S. Hudson et al., Astrophys. J. Lett., 236, L91 (1980).

3. T. Prince et al., paper presented at Conf. on Cosmic Ray Astrophys. and Low Energy Gamma Ray Astronomy, U. of Minn., Sept. 1980.

4. E.L. Chupp et al., Astrophys. J. (in press 1981).

5. R.E. Lingenfelter and R. Ramaty, in High Energy Nuclear Reactions in Astrophysics, ed. B.S.P. Shen (Benjamin, N.Y., 1967) p.99.

6. R.E. Lingenfelter et al., J. Geophys. Res., 70, 4077 (1965).

7. R. Ramaty and R.E. Lingenfelter, in High Energy Phenomena on the Sun, ed. R. Ramaty and R.G. Stone (NASA SP 342, 1973) p. 301.

8. H.T. Wang and R. Ramaty, Solar Phys. 36, 129 (1974).

9. H.T. Wang, Ph.D. Thesis, Univ. of Maryland (1975).

10. R. Ramaty, B. Kozlovsky and R.E. Lingenfelter, Space Sci. Rev. 18, 341 (1975).

11. R. Ramaty, B. Kozlovsky and A.N. Suri, Astrophys. J. 214, 617 (1977).

12. R. Ramaty, B. Kozlovsky and R.E. Lingenfelter, Astrophys. J. Supp. 40, 487 (1979).

13. R. Ramaty in Particle Acceleration Mechanism in Astrophysics ed. J. Arons et al. (Am. Inst. Phys., N.Y., 1979) p. 135.

14. R.E. McGuire, T.T. vonRosenvinge and F.B. McDonald, 17th Cosmic Ray Conf. Papers, 3, 65 (1981).

15. A.G.W. Cameron in A Festschrift in Honor of Willy Fowler's 70th Birthday (1980)

16. G.M. Mason et al., Astrophys. J., 239, 1070 (1980).

17. L.C. Northcliffe and R.F. Schilling, Nuc. Data Tables, 7, 233 (1970).

18. G. Kanbach et al., 14th Internat. Cosmic Ray Conf. Papers, 5, 1644 (1975).

19. C.J. Cranell et al., Astrophys. J., 210, 582 (1976).

20. R.W. Bussard, R. Ramaty and R.J. Drachman, Astrophys. J., 228, 928 (1979).

21. T.T.vonRosenvinge, R.Ramaty and D.V.Reames, 17th Internat. Cosmic Ray Conf. Papers, 3, 28 (1981).

22. P. Evenson, P. Meyer and S. Yanagita, 17th Internat. Cosmic Ray Conf. Papers, 3, 32 (1981).

23. M.E.Pesses et al., 17th Internat. Cosmic Ray Conf. Papers, 3, 36 (1981).

24. E.P. Mazets et al., Nature 290, 378 (1981).

25. B.J.Teegarden and T.L.Cline, Astrophys.J.Lett., 236, L67 (1980).

26. J.K. Daugherty and R.W.Bussard, Astrophys.J., 238, 296 (1980).

27. J. Katz, preprint (1981).

28. W.K.H.Schmidt, Nature, 271, 525 (1978).

29. G. Cavallo and M.J. Rees, M.N.R.A.S., 183, 359 (1978).

30. R. Ramaty and P. Meszaros, Astrophys. J., 250, (in press 1981).

31. R. Ramaty, R.E.Lingenfelter, and R.W.Bussard, Astrophys. and Space Sci, 75, 193 (1981).

32. R. Ramaty et al., Nature 287, 122 (1980).

33. J.E. Felten, 17th Internat. Cosmic Ray Conf. Papers, (in press 1981).

34. E.P.T.Liang, Nature 292, 319 (1981) and paper in this volume.

35. J.C. Higdon and R.E. Lingenfelter, Astrophys.J., 215, L53 (1977).

36. R.E.Lingenfelter, J.C.Higdon and R.Ramaty in Gamma Ray Spectroscopy in Astrophysics, ed. T.L.Cline and R. Ramaty (NASA,1978) p. 252.

37. W.A. Fowler, G.R. Caughlan and B.A.Zimmerman, Ann. Rev. Astron. and Astrophys., 13, 69 (1975).

38. A.S. Jacobson et al., in Gamma Ray Spectroscopy in Astrophysics, ed. T.L. Cline and R. Ramaty (NASA, 1978) p. 228.

ON THE THEORY OF GAMMA RAY AMPLIFICATION
THROUGH STIMULATED ANNIHILATION RADIATION (GRASAR)[1]

R. Ramaty, J. M. McKinley*, and F. C. Jones
Laboratory for High Energy Astrophysics
NASA/Goddard Space Flight Center, Greenbelt, MD 20771

ABSTRACT

The theory of photon emission, absorption and scattering in a relativistic plasma of positrons, electrons and photons is studied. Expressions for the emissivities and absorption coefficients of pair annihilation, pair production and Compton scattering are given and evaluated numerically. The conditions for negative absorption are investigated. In a system of photons and e^+-e^- pairs, an emission line at ~ 0.43 MeV can be produced by grasar action provided that the pair chemical potential exceeds ~ 1 MeV. At a temperature of $\sim 10^9$K this requires a pair density $\geq 10^{30}$cm^{-3}, a value much larger than the thermodynamic equilibrium pair density at this temperature. This mechanism could account without a gravitational redshift for the observed lines at this energy from gamma ray bursts.

I. INTRODUCTION

The emission and absorption of photons in cosmic sources are governed by many processes. At temperatures of the order 10^9 to 10^{10}K, typical of gamma ray burst sources, two of the most important ones are pair production and annihilation ($e^+ + e^- \rightleftarrows \gamma + \gamma$) and Compton and inverse Compton scattering ($e + \gamma \rightleftarrows e' + \gamma'$). Arguments based on the observed photon intensities of gamma ray bursts and the likely distances and sizes of their sources, lead to the conclusion that the source regions of some of the bursts are optically thick.[1,2]

Photon absorption in $\gamma\gamma$ pair production has been discussed in the literature,[3] but no calculation has included the effects of the stimulation of annihilation or the suppression of pair production due to large photon or particle occupation numbers. When these stimulation and suppression effects are taken into account, the possibility exists for negative absorption.[4] The condition for this is a population inversion, which in the present context is a pair density that exceeds the thermodynamic equilibrium density.

A recent review of gamma ray burst observations has been given by Cline.[5] Of particular interest for the present paper is the existence of an emission line seen from several gamma ray bursts in the energy range from 0.4 to 0.46 MeV.[6-8] These lines are probably due to e^+-e^- annihilation radiation. If so, e^+-e^- pairs should be present in large numbers in the burst sources, and the sources should be sufficiently hot to produce the pairs, but the source regions should not be in equilibrium because no lines can then be

*Also at Oakland University, Rochester, MI 48063

seen. We aim the calculations of the present paper to astrophysical sites where such conditions might exist.

II. EMISSIVITIES AND ABSORPTION COEFFICIENTS IN A RELATIVISTIC PLASMA

We consider systems characterized by temperatures of the order of the electron rest mass energy in which photon-photon collisions can produce much larger pair densities than the ambient electron densities of the astrophysical sites of interest. We therefore consider only cases in which the electrons and positrons have equal densities. As convenient analytical expressions, which allow both equilibrium and non-equilibrium situations, we use Bose-Einstein distributions for the photons and Fermi-Dirac distributions for the pairs.[9]

We assume equal temperatures for the positrons and electrons, $T_+ = T_- = T_\pm$, but allow the photon temperature, T_γ, to differ from $T_\pm$. Since the e^+ and e^- densities are equal, $n_+ = n_- = n_\pm$, these particles must also have equal chemical potentials, $\mu_+ = \mu_- = \mu_\pm$. The photon chemical potential, μ_γ, is zero for a blackbody distribution. We allow non-blackbody photon distributions, but only zero or negative values may be assigned to μ_γ. The pair chemical potential can be positive, zero or negative. If $\mu_\pm = 0$ the pairs are in thermodynamic equilibrium with blackbody photons.

In terms of these temperatures and chemical potentials, the photon and pair densities can be written as[9]

$$n = (4\pi^3 \hbar^3)^{-1} \int d^3p \; \eta, \tag{1}$$

where $\vec{p}$ is the photon or particle momentum and η is the occupation number. These are given by

$$\eta_\gamma = \{\exp[(E_\gamma - \mu_\gamma)/kT_\gamma] - 1\}^{-1} \tag{2}$$

for the photons, and by

$$\eta_\pm = \{\exp[(E_\pm - \mu_\pm)/kT_\pm] + 1\}^{-1} \tag{3}$$

for the particles, where E_γ is the photon energy and $E_\pm$ is the particle total energy (kinetic plus rest mass).

We proceed to define the photon emissivity and absorption coefficient for pair production and annihilation. In particular, we are interested in obtaining a correct expression for stimulated annihilation which has not been taken into account in previous treatments of photon-photon absorption. Stimulated emission has, of course, been taken into account for other processes.[10] But we cannot use the standard expressions for emissivities and absorption coefficients because pair production and annihilation do not fit the usual pattern in which photons are emitted or absorbed singly and the matter has the same form before and after events. We therefore proceed as follows:

The transition rate in vacuum in either direction between photon states in $d^3p_1 d^3p_2$ and pair states in $d^3p_+ d^3p_-$ can be written as

$$w = (4\pi^3 \hbar^3)^{-2} d^3p_1 d^3p_2 d^3p_+ d^3p_- \delta^4(p_1+p_2-p_+-p_-)X. \quad (4)$$

Here $\vec{p}_1$ and $\vec{p}_2$ are photon momenta, $\vec{p}_+$ and $\vec{p}_-$ are momenta of the pair, p_1, p_2, p_+ and p_- are the corresponding 4-momenta, and X is proportional to the squared matrix element of the interaction, summed and averaged over spins and polarization.

To take into account the bath of photons and pairs, we must multiply equation (4) by an appropriate expression in the occupation numbers incorporating the equilibrium conditions $T_\pm = T_\gamma$ and $\mu_\pm = \mu_\gamma$. After doing so and integrating over all four momenta, we obtain

$$\frac{1}{2} \int \frac{n_+ d^3p_+}{4\pi^3 \hbar^3} \int \frac{n_- d^3p_-}{4\pi^3 \hbar^3} \int d\Omega_1 \, (1+n_1)(1+n_2)[Ic\frac{d\sigma}{d\Omega}]_{ann} =$$

$$= \frac{1}{2} \int \frac{n_1 d^3p_1}{4\pi^3 \hbar^3} \int \frac{n_2 d^3p_2}{4\pi^3 \hbar^3} \int d\Omega_+ (1-n_+)(1-n_-)[Ic\frac{d\sigma}{d\Omega}]_{pp}. \quad (5)$$

The left-hand side of equation (5) is the total pair annihilation rate, while the right-hand side is the total pair production rate, both influenced by final state occupation numbers. The invariant product of the flux factor I and differential cross section $d\sigma/d\Omega$ is obtained[11] from X by integration over all final state variables except the angles of one particle:

$$[Ic\frac{d\sigma}{d\Omega}]_{ann} = \int p_1^2 dp_1 \int p_2^2 dp_2 \int d\Omega_2 \delta^4(p_1+p_2-p_+-p_-)X \quad (6)$$

$$[Ic\frac{d\sigma}{d\Omega}]_{pp} = \int p_+^2 dp_+ \int p_-^2 dp_- \int d\Omega_- \delta^4(p_1+p_2-p_+-p_-)X. \quad (7)$$

The factor $1/2$ in equation (5) is introduced so that each pair of photons in either initial or final state is included just once.

To define a photon emissivity and absorption coefficient it is necessary to investigate the balance of reactions involving photons in a chosen increment d^3p_1, rather than integrating over it. It is no longer possible to express the left side in terms of the annihilation cross section, because the necessary integration of equation (6) has not been performed. But we can interchange the order of integration and use equation (7) instead. The required balance is then given by

$$\frac{(1+n_1)d^3p_1}{4\pi^3 \hbar^3} \int \frac{d^3p_2}{4\pi^3 \hbar^3} \int d\Omega_+ n_+ n_- (1+n_2) [Ic\frac{d\sigma}{d\Omega}]_{pp} =$$

$$= \frac{n_1 d^3p_1}{4\pi^3 \hbar^3} \int \frac{d^3p_2}{4\pi^3 \hbar^3} \int d\Omega_+ n_2 (1-n_+)(1-n_-)[Ic\frac{d\sigma}{d\Omega}]_{pp}. \quad (8)$$

Collecting terms in n_1, we obtain

233

$$dE_1 d\Omega_1 j_{\gamma\gamma}(E_1) = \frac{cn_1 E_1^2 dE_1 d\Omega_1}{4\pi^3 (\hbar c)^3} K_{\gamma\gamma}(E_1), \qquad (9)$$

where the rate of spontaneous emission, $j_{\gamma\gamma}(E_1)$, and coefficient of linear absorption, $K_{\gamma\gamma}(E_1)$, are given by

$$j_{\gamma\gamma}(E_1) = \frac{cE_1^2}{4\pi(\hbar c)^3} \int \frac{d^3 p_2}{4\pi^3 \hbar^3} \int d\Omega_+ n_+ n_- (1+n_2) [I \frac{d\sigma}{d\Omega}]_{pp} \qquad (10)$$

and

$$K_{\gamma\gamma}(E_1) = \int \frac{d^3 p_2}{4\pi^3 \hbar^3} \int d\Omega_+ [n_2(1-n_+)(1-n_-) - n_+ n_-(1+n_2)][I\frac{d\sigma}{d\Omega}]_{pp}. \qquad (11)$$

In equation (10) the annihilation emissivity is expressed in terms of the pair production cross section unlike the approach[12] using the annihilation cross section. In the expression for the absorption coefficient (equation 11), the first term in the brackets is due to absorption by the photon bath while the second term is the contribution of induced annihilation.

Equations (10) and (11) are valid also for nonequilibrium situations provided proper nonequilibrium occupation numbers are used. In terms of equations (2) and (3), the most general nonequilibrium distributions are obtained if $T_\gamma \neq T_\pm$ and $\mu_\gamma \neq \mu_\pm$.

With such distributions, the total annihilation and pair production rates are not equal and, moreover, $K_{\gamma\gamma}$ can become negative. This will happen if an appropriate population inversion takes place. By substituting equations (2) and (3) into equation (12) we find that for $T_\pm = T_\gamma$ such an inversion occurs if $2\mu_\pm > \mu_\gamma$. In this case $K_{\gamma\gamma}$ is negative for $E_1 < 2\mu_\pm - \mu_\gamma$.

For the system to exhibit grasar action, however, it is necessary that the total absorption coefficient be negative. For the system of photons and pairs that we consider here, the only important process other than pair production and annihilation is Compton scattering. We ignore the weaker processes of bremsstrahlung and double Compton scattering. We note, however, that synchrotron radiation could potentially be very important, but because we are free to choose an arbitrarily low magnetic field intensity, we ignore synchrotron absorption in the present paper.

For Compton and inverse Compton scatterings ($\gamma_1 + e \rightleftarrows \gamma_2 + e'$) we proceed in essentially the same way as for pair production and annihilation. Using E_1 and E_2 for photon energies and $\vec{p}$ and $\vec{p}'$ for electron and positron momenta, we obtain the Compton emissivity and absorption coefficient in the presence of the bath of photons and pairs

$$j_C(E_1) = \frac{cE_1^2}{4\pi^3(\hbar c)^3} 2\int \frac{d^3 p}{4\pi^3 \hbar^3} \int d\Omega_2 n_2 n'_\pm (1-n_\pm) [I \frac{d\sigma}{d\Omega}]_C, \qquad (12)$$

and

$$K_C(E_1)=2\int\frac{d^3p}{4\pi^3\hbar^3}\int d\Omega_2 [n_\pm(1+n_2)(1-n'_\pm)-n_2 n'_\pm(1-n_\pm)][I\frac{d\sigma}{d\Omega}]_C , \quad (13)$$

where the factors of 2 take into account the contributions of both electrons and positrons. The Compton emissivity, equation (12), represents the scatterings of photons 2 into the element $dE_1 d\Omega_1$. For the absorption coefficient, equation (13), the first term in the brackets is due to scattering of photons out of $dE_1 d\Omega_1$, while the second term represents the stimulated scatterings of photons 2 into $dE_1 d\Omega_1$. $K(E_1)$ can become negative and a necessary condition for this is $T_\gamma > T_\pm$. In our subsequent analysis, however, we shall only consider systems with $T_\gamma = T_\pm$ for which K_C is always positive.

For the numerical evaluations shown below we have used the expressions of Jauch and Rohrlich[11] for the flux factors and differential cross sections. We must also express all quantities in terms of independent variables of integration. The detailed expressions which we used will be published elsewhere.

III. NUMERICAL RESULTS

We have evaluated equations (10) through (13) for various choices of $T_\pm$, T_γ, $\mu_\pm$ and μ_γ. As already indicated, we limit our discussion here to cases with equal pair and photon temperatures, $T_\pm = T_\gamma \equiv T$. We allow, however, arbitrary values for $\mu_\pm$ and μ_γ.

We consider first the case of thermodynamic equilibrium, $\mu_\pm = \mu_\gamma = 0$. The emissivities and absorption coefficients for this

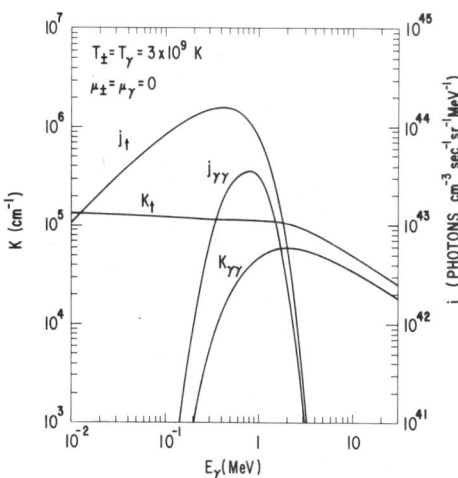

Fig. 1. Emmissivities and absorption coefficients vs. photon energy in a system in thermodynamic equilibrium at 3×10^9K. The Compton emissivity (not shown) is the difference between the total emissivity j_t and the annihilation emissivity $j_{\gamma\gamma}$. The Compton absorption coefficient (not shown) is the difference between the total absorption coefficient K_t and the pair production-absorption coefficient $K_{\gamma\gamma}$. The photon and pair densities in these conditions are 5.5×10^{29}cm^{-3} and 2.4×10^{29}cm^{-3}, respectively.

case and $T = 3\times 10^9$K are shown in Figure 1. As can be seen, the absorption coefficients are positive at all photon energies. The peak of the annihilation emissivity $j_{\gamma\gamma}$ occurs at a higher energy than $mc^2 = 0.511$ MeV, because the annihilation photons must carry away the kinetic energies of the pairs in addition to their rest mass energy.[12-14]

We turn now to the study of cases with inverted populations, $2\mu_\pm > \mu_\gamma$. In this case, $K_{\gamma\gamma}$ is negative for $E_\gamma < 2\mu_\pm - \mu_\gamma$, but since $T_\pm = T_\gamma$, K_C is positive for all E_γ. Grasar action can occur only if $K_t = K_{\gamma\gamma} + K_C < 0$. Since K_C is proportional to $n_\pm$ while the portion of $K_{\gamma\gamma}$ due to stimulated annihilation varies as $n_\pm^2$, a sufficiently large density is needed for $-K_{\gamma\gamma}$ to exceed K_C. This implies a threshold for $\mu_\pm$ which is higher than the threshold required for just $K_{\gamma\gamma}$ to be negative.

To investigate this threshold we have evaluated $K_t = K_{\gamma\gamma} + K_C$ as a function of $\mu_\pm$ for given temperatures and μ_γ. We have carried out calculations in the temperature range $0 < T \lesssim 5 \times 10^9$ K, where the lower limit corresponds to fully degenerate electrons and positrons. We find that the threshold for grasar action is close to $\mu_\pm = 1$ MeV and does not depend strongly on μ_γ and T. This corresponds to a pair density threshold of a few times $10^{30} cm^{-3}$.

We show numerical results in Figure 2 for $T = 3 \times 10^9$ K, $\mu_\gamma = 0$ and $\mu_\pm = 1.1$ MeV. As can be seen, $K_{\gamma\gamma}$ is negative for $E_\gamma < 2\mu_\pm = 2.2$ MeV and positive at higher energies. K_C is positive at all energies. In the energy range from about 0.25 MeV to 0.7 MeV, $-K_{\gamma\gamma}$

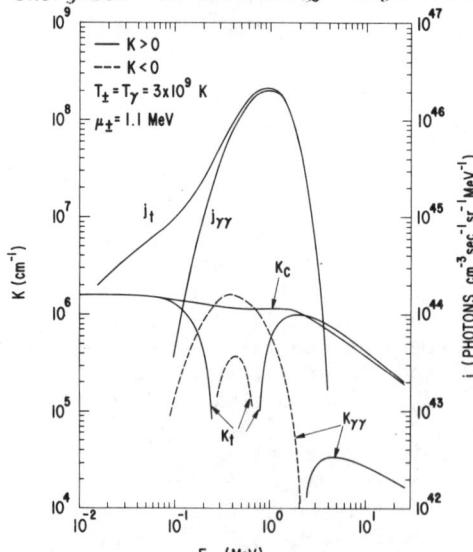

Fig. 2. Emissivities and absorption coefficients vs. photon energy in a system at 3×10^9 K with blackbody photons and an inverted pair population described by pair chemical potential 1.1 MeV. The curves have the same significance as in Fig. 1 except that the Compton absorption coefficient K_C is shown explicitly. Negative values of total and pair production absorption coefficients are represented by dashed curves. The photon and pair densities in these conditions are $5.5 \times 10^{29} cm^{-3}$ and $7.3 \times 10^{30} cm^{-3}$, respectively.

exceeds K_C and hence K_t is negative. If the source is optically thick and K_t is negative over a sufficiently large spatial region, then the radiation intensity has a sharp peak at a photon energy at which $-K_t$ is maximal. The value of this energy, ~ 0.43 MeV from the numerical calculations, is determined primarily from the energy at which $-K_{\gamma\gamma}$ is maximum, shifted somewhat according to the slope of K_C at that energy. From Equation (11) we can also express $K_{\gamma\gamma}$ in the form

$$K_{\gamma\gamma}(E_\gamma) = -\frac{4\pi^3(\hbar c)^3}{cE_\gamma^2} j_{\gamma\gamma}(E_\gamma) + \int \frac{d^3p_2}{4\pi^3\hbar^3}\int d\Omega_+ n_2(1-n_+)(1-n_-)[I\frac{d\sigma}{d\Omega}]_{pp} \quad . \quad (14)$$

Above the grasar action threshold, the first term, due to stimulated emission, is much larger in magnitude than the second term which is due to absorption. As discussed above, $j_{\gamma\gamma}(E_\gamma)$ is broadly peaked at an energy greater than mc^2, reflecting the kinetic energy of the annihilating pairs. The division by E_γ^2 (from the factor of density of states) shifts the peak to an energy somewhat less than mc^2.

From the numerical calculations for other values of T and $\mu_\pm$ ($0 < T < 5\times10^9$K, $0.8 < \mu_\pm < 1.2$ MeV) we find that the peaks of both $-K_{\gamma\gamma}$ and $-K_t$ are in the energy range from about 0.40 to 0.50 MeV.

IV. ASTROPHYSICAL APPLICATIONS

The large photon densities expected[1,2] in gamma ray burst sources should lead to high pair production and Compton opacities. The observation[6-8] of emission lines in the energy range 0.4 to 0.46 MeV, believed to be due to e^+-e^- annihilation radiation, is evidence that e^+-e^- pairs do indeed play an important role in the physics of gamma ray bursts. But it is not immediately obvious how a relatively narrow emission line is produced in a hot and optically thick source region.

Ramaty et al.[15,16] have studied this problem for the transient of March 5, 1979[17-19] from which an emission line was observed[6] at $\sim$ 0.43 MeV.

In their model, the line is formed in the last optical depth of the source region by the annihilation of e^+-e^- pairs that have been cooled by synchrotron radiation prior to their annihilation. The observed[6] upper limit on the width (FWHM ≤ 0.2 MeV), implies[12] a temperature less than 3×10^8K. It also implies an upper limit on the density since even at zero temperature the line is broadened by the motions of the degenerate electrons and positrons. Using equation (10) we have evaluated the FWHM of the annihilation emissivity, $j_{\gamma\gamma}$, as a function of $\mu_\pm$ for $T_\pm = T_\gamma = 0$.

We find that if FWHM < 0.2 MeV, then $n_\pm \lesssim 7\times10^{28}$ cm^{-3}.

The density can be independently calculated[16] from the observed line fluence[6] ($\phi \simeq 10$ photons cm^{-2}). Let $R/(n\pm)^2 \simeq 7.5\times10^{-15}$ cm^{3}sec^{-1} be the annihilation rate coefficient[12] at 3×10^8K, A the area of the emitting region, Δt the time interval in which the observed fluence is produced and d = 55kpc the distance to the March 5 source. If the line is formed in a layer of unit optical depth to Compton scattering, then

$$\phi \simeq 2R\ K_C^{-1}\ A\Delta t\ (4\pi d^2)^{-1}. \quad (15)$$

Since R varies as $n_\pm^2$ and K_C^{-1} as $n_\pm^{-1}$, ϕ is proportional to $n_\pm$. With the above numerical values, K_C from Figure 1, and $n_\pm <$

7×10^{28}cm^{-3}, $A\Delta t$ should exceed 1.5×10^9cm^{2}sec. This condition is well satisfied if the annihilation line is produced over the entire surface of a neutron star, $A \simeq 10^{13}$cm^2, and during the entire impulsive phase[18] of the March 5 event, $\Delta t \simeq 0.15$sec. But the model would face considerable difficulties if future measurements should indicate that the line is narrower than 0.2 MeV, or if $A\Delta t$ should turn out, for other reasons, to be smaller than 1.5×10^9cm^{2}sec.

The advantage of producing the annihilation line by grasar action is that a narrow line can form in a hot optically thick region. To illustrate this, we have evaluated the photon intensity perpendicular to a slab of thickness L in which the emissivity and absorption coefficient do not depend on position:

$$I = (j_t/K_t)\{1 - \exp[-K_t L]\}. \tag{16}$$

Using the j_t's and K_t's of Figure 2, we show in Figure 3 the dependence of I on photon energy E_γ and slab thickness L. As can be seen, grasar action can indeed narrow the line in comparison to the thermal width[12] of ~ 0.8 MeV in an optically thin Maxwell-Boltzmann gas at 3×10^9K.

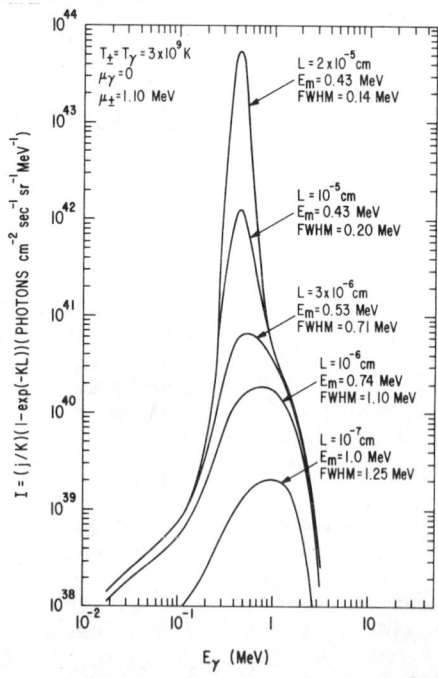

Fig. 3. The development of the annihilation line with increasing thickness of source. The system is the same as that for Fig. 2. The labels on successive maxima indicate the thickness involved, and the peak energy and FWHM of the line.

The peak energy of the annihilation line formed by grasar action, is close to the observed peak energies. Thus, the gravitational redshift of the line due to the compact object which presumably produces the burst should be quite low, $z < 0.1$. This implies that gamma ray burst sources with observed e^+-e^- emission lines could be objects other than neutron stars, or if they are neutron stars, these stars should have small masses.

Assuming that an inverted layer with parameters as in Figure 3 did exist in the March 5 burst source region, the line fluence ϕ is

$$\phi = A\Delta t d^{-2} \int I(E_\gamma) dE_\gamma . \qquad (17)$$

Using the results of Figure 3 with $L = 10^{-5}$ cm and $\phi \simeq 10$ photons cm^{-2}, equation (17) yields $A\Delta t \simeq 1.2 \times 10^6$ cm^{2}sec. By comparing with the minimum $A\Delta t$ deduced above for radiation produced in the last optical depth ($A\Delta t > 1.5 \times 10^9$ cm^2 sec), we see that not only can grasar action produce a narrow line in a hot region, but that the observed line intensity and width are consistent with a much smaller source and/or a much shorter line formation time than required for the optically thin case.

V. CONCLUSIONS

We have carried out a fully relativistic treatment of pair production and annihilation and Compton and inverse Compton scattering in a medium containing photons, positrons, and electrons, with equal e^+ and e^- densities. In the calculation of the emissivities and absorption coefficients we have included the stimulation of transitions caused by the Bose-Einstein nature of the photons and the suppression of transitions due to electron and positron degeneracy. We have shown that for systems in thermodynamic equilibrium the calculations lead to an exact balance between pair production and pair annihilation. For systems not in equilibrium, grasar action is possible. We have evaluated, in particular, the absorption coefficient for equal photon and particle temperatures and positive particle chemical potential ($\mu_\pm > 0$). For this example of population inversion, the total absorption coefficient can become negative due to the much larger probability for stimulated annihilation than for Compton scattering and pair production. Grasar action produces a narrow emission line peaked at an energy of about 0.43 MeV. This energy is lower than the peak of the spontaneous annihilation emissivity, which occurs at energies greater than 0.511 MeV. In a bath of blackbody photons ($\mu_\gamma = 0$) and e^+-e^- pairs at the photon temperature, the threshold for grasar action is at $\mu_\pm \simeq 1$ MeV corresponding to a pair densities of $\sim 10^{30}$ cm^{-3} for $T \simeq 10^9$K. A temperature of $\sim 5 \times 10^9$K is needed to produce this density in equilibrium with blackbody photons.

We have applied our results to gamma ray bursts, in particular to the March 5, 1979 transient from which an emission line at ~ 0.43 MeV was observed.[6] While this line could be produced in a cool skin layer,[15,16] grasar action has the advantage of being capable of producing a narrow line from a hot and optically thick source and from a source region of relatively small emitting area and short duration of line formation. But if grasar action is responsible for the observed 0.40 to 0.46 MeV emission lines of gamma ray bursts, then their sources cannot be a neutron stars of mass larger than about 0.6 M_θ. At the surface of a larger mass neutron star, the gravitational field would shift the line to an energy lower than observed.

There are several difficulties and shortcomings in our treatment. We have not shown how the inversion ($\mu_{\pm}>0$) is produced. It could in principle result from the cooling of the pairs that is faster than their annihilation, or by a rapid external supply of pairs without heating. Cooling by synchrotron emission has already been proposed,[15,16] but for the high densities that we consider here, the required field ($B>10^{12}$ gauss) seems to lead to synchrotron self-absorption that could quench the grasar action. We have ignored other effects of a strong magnetic field as well, by limiting our calculations to isotropic distributions and by using plane wave functions instead of Landau functions. This isotropic treatment also does not allow the study of beaming effects which should be present in a gamma ray maser. Finally, we have not made any attempts to study the spatial and temporal development of a system exhibiting grasar action. We expect this development to be highly nonlinear.

As already indicated, gamma ray burst sources are possible astrophysical sites where grasar action could occur. The most obvious observational test for this would be the observation of a narrow (FWHM << 0.1 MeV) emission line at ~ 0.43 MeV.

REFERENCES

1. G. Cavallo and M. J. Rees, MNRAS 183, 359 (1978).
2. W. K. H. Schmidt, Nature 271, 525 (1978).
3. R. J. Gould and G. P. Schreder, Phys. Rev. 155, 1404 (1967).
4. C. M. Varma, Nature 267, 686 (1977).
5. T. L. Cline, Proc. 10th Texas Symp. on Rel. Astrophys. (1981).
6. E. P. Mazets et al., Nature 282, 587 (1979).
7. B. J. Teegarden and T. L. Cline, Ap. J. 236, L67 (1980).
8. E. P. Mazets et al., Nature 290, 378 (1981).
9. L. D. Landau and E. M. Lifshitz, Statistical Physics (Addison-Wesley, Reading, MA 1958).
10. G. Bekefi, Radiation Processes in Plasmas (Wiley, NY 1966).
11. J. M. Jauch and F. Rohrlich, The Theory of Photons and Electrons (Addison-Wesley, Reading, MA 1955).
12. R. Ramaty and P. Mészaros, Ap. J. (in press) (1981).
13. A. A. Zdziarski, Acta Astron. (in press) (1981).
14. F. A. Aharonian, A. M. Atoyan and R. A. Sunyaev, Yerevan Physics Institute Preprint EFI 432(39)-80 (1980).
15. R. Ramaty et al., Nature 287, 122 (1980).
16. R. Ramaty, R. E. Lingenfelter, and R. W. Bussard, Astrophys. Space Sci. 75, 193 (1981).
17. S. Barat et al., Astron. and Astrophys. 79, L24 (1979).
18. T. L. Cline et al., Ap. J. 237, L1 (1980).
19. W. D. Evans et al., Ap. J. 237, L7 (1980).

EMISSION MECHANISMS OF THE MARCH 5th 1979

GAMMA RAY TRANSIENT

E.P.T. Liang

Lawrence Livermore National Laboratory[†],
University of California, Livermore, CA 94550 and
Institute for Plasma Research[*], Stanford University,
Stanford, CA 94305

ABSTRACT

This paper reviews the spectral and temporal constraints on the emission mechanisms of the initial burst and subsequent pulses of the March 5, 1979 γ-transient. Interpretation of the hard tail above 300 keV as due to inverse comptonization, when combined with the synchrotron-pair annihilation model of Ramaty et al., allows us to derive from first principles the intrinsic luminosity of the source. This is in general agreement with that expected from N49 in LMC. The pulsed emissions have characteristics difficult to explain in terms of conventional models.

INTRODUCTION

The gamma ray transient of March 5, 1979 is probably anomalous among all gamma ray bursts[1]. Yet its many special features, both spectral and temporal, plus the possible identification with a known astronomical object (N49 in LMC) allow us a unique opportunity to understand some of the physics involved. More importantly, various aspects of this event might be universal to all gamma ray bursts (e.g. the appearance of red-shifted pair-annihilation line). The understanding of these features would shed light on the nature of other events. By now many models have been proposed for the origin of this event based on some catastrophe on or inside a neutron star (e.g. thermonuclear explosion[2], sudden internal energy release due to phase transition[3], impact by comet[4], etc.). Instead of speculating further about the origin, here we would like to concentrate on the observables and see how far one can go in constraining the emission models. We will first discuss the burst emission, then the periodic pulsations and finally some ideas on the possibility of

† Supported by DOE contract W7405-ENG-48.
* Supported by NASA NGR-05-020-668.

magnetic field reconnections (or annihilation) as the immediate source of the pairs and gamma rays.

THE BURST EMISSION

If one accepts that the source is in N49 at 55 kpc, then a variety of factors forces one to admit that the optically thin synchrotron and pair annihilation model of Ramaty et al.[5] is the most natural one for the observed spectrum of Mazets et al.[6], at least below $\sim$ 450 keV. There is, however, a hard power law tail extending to $\sim$ 2 MeV not explainable in the Ramaty model. In a recent paper[7], we point out that this hard tail can be naturally interpreted as an inverse Compton continuum produced by the relativistic pairs upscattering the synchrotron X-rays. This interpretation then allows us to derive, from first principles, the intrinsic synchrotron luminosity. The argument is quite simple and worth repeating here.

Suppose we divide the spectrum[1] (Fig. 1) into a soft synchrotron component ($\lesssim$ 300 keV) and a hard Comptonized component (> 300 keV, with the "hump" at $\sim$ 430 keV subtracted off) produced solely by the upscattering of the synchrotron photons by the relativistic pairs. Then we can deduce the pair column density, the mean relativistic Γ ($\equiv \bar{E}/mc^2$) and magnetic field strength B as follows. In the limit when the comptonization (Kompaneet) parameter y<<1 which is obviously the case here, we have $y \simeq L_c/L_s \simeq 0.08$ where L_c and L_s are the total luminosities in the Compton and synchrotron components. The electron scattering depth τ_{es} is also << 1 hence the ratio of total photon numbers in the two components is just the scattering probability:

$$\tau_{es} \simeq \frac{N_c}{N_s} \simeq 7 \times 10^{-3} \tag{1}$$

Since $y \equiv \frac{4}{3} \tau_{es} \Gamma^2$ in the relativistic limit we have $\Gamma \simeq 3$ and the relativistic pair column density is

$$n_e h \equiv (n_+ + n_-)h \equiv \frac{\tau_{es}}{\sigma_{KN}(\Gamma=3)} \simeq 3 \times 10^{22} \text{cm}^{-2} \tag{2}$$

where σ_{KN} is the Klein-Nishina cross-section. From

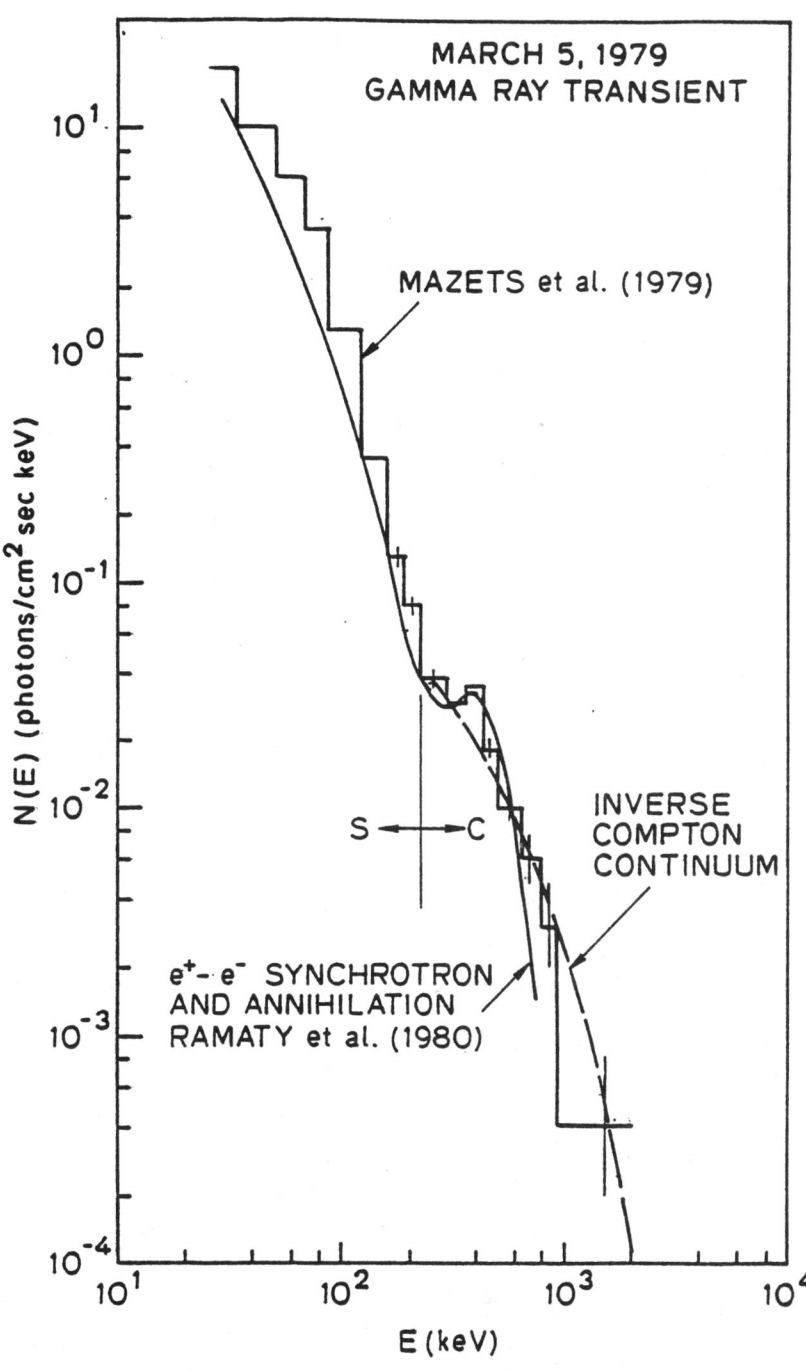

FIGURE 1

the exponential shape of synchrotron component we deduce from the critical frequency

$$h\nu_c \simeq 30 \text{ keV} \propto B\Gamma^2$$

the field strength $B \simeq 2 \times 10^{11}$ gauss. (3)

Hence the (optically thin) synchrotron luminosity becomes

$$L_{syn} \propto B^2\Gamma^2 n_e \times (\text{volume of emission})$$

$$\propto B^2\Gamma^2 n_e \, hf_* \, 4\pi r_*^2$$

$$= 2.5 \times 10^{44} f_* \text{ erg} \cdot s^{-1} \quad (4)$$

where f_* is the fraction of the surface of a 15 km radius ($=r_*$) neutron star emitting. This intrinsic luminosity is consistent with a source distance of 55 kpc. If we also interpret, as other authors do[5,6], the excess above the continuum at ~ 430 keV as a pair annihilation line with a luminosity of 2×10^{41} erg $\cdot$ s^{-1}, more information can be extracted. Since the synchrotron cooling time of the pairs is $\sim 10^{-15}$ sec we expect them to annihilate essentially at low energies so the line appears narrow (FWHM $\lesssim 100$ keV). The annihilation rate (at rest) provides the condition

$$\tilde{n}_+ \tilde{n}_- \tilde{h} \, f_* \simeq 10^{48} \quad (5)$$

where quantities with $\sim$ above refer to nonrelativistic pairs in the annihilation region which need <u>not</u> coincide with the synchrotron and inverse compton region. (This point was not realized in our former paper[7]). In fact, it seems likely that $\tilde{h} \gg h$ since the annihilation time is $\gg$ synchrotron cooling time. A reasonable guess for $\tilde{h}$ would be at least an annihilation mean free path parallel to field lines, unless the confining B field is made up of closed field lines with radius $\lesssim \tilde{h}$ (The gyroradius is $\sim 10^{-6}$ cm). So if

$$\tilde{h} \simeq \frac{1}{\tilde{n}_- \sigma_{2\gamma}} \sim \frac{10^{25}}{\tilde{n}_-} \quad (6)$$

then

$$\tilde{n}_+ \simeq 10^{23} f_*^{-1} \text{ cm}^{-3}. \tag{7}$$

If $\tilde{n}_+ \simeq \tilde{n}_-$ we have

$$\tilde{h} \sim 10^2 f_* \text{ cm} \tag{8}$$

a reasonable scale height. This, however, would mean that the Thomson optical depth of cold pairs is non-negligible:

$$\sigma_{es}(\tilde{n}_+ + \tilde{n}_-)\tilde{h} \simeq 1 \tag{9}$$

The slight deviation of the observed synchrotron spectrum[6] from the theoretical one (Fig. 1) may in fact be due to Thompson scattering by cold pairs. This also means that the width of the annihilation line can never be below the synchrotron peak ($\sim$30 keV).

From particle conservation we expect $\tilde{n}_+ \gtrsim n_+$. Hence

$$h \gtrsim \frac{3.4 \times 10^{22}}{2\tilde{n}_+} = 0.15 f_* \text{cm} \tag{10}$$

The structure of the emission layers would then resemble Fig. 2.

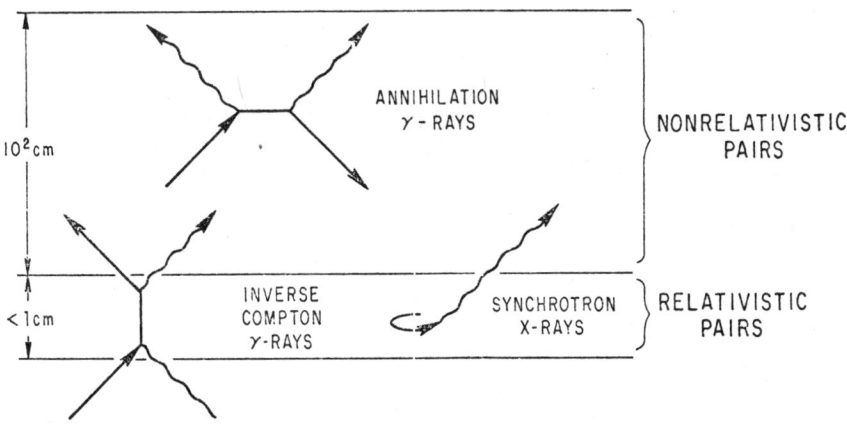

Figure 2. Sketch of the emission layers.

THE PULSED EMISSION

The 8 sec period pulses following the initial burst have several remarkable characteristics which we summarize here.

(a) The spectral shape of the pulses[6] is almost identical to that of the synchrotron component (below ~ 300 keV) of the burst. It would be most natural to interpret it also as synchrotron radiation with same B and Γ. The lower intensity would then imply much fewer emitting pairs in the pulsed phase. This is also consistent with the disappearance of the compton tail and the annihilation line. It is intriguing why their energy remains about the same while their number is greatly reduced, which could not be the case if they were produced thermally.

(b) The pulses are $180°$ out of phase[1,6] while the interpulse is in-phase with the burst. This is the strongest evidence that the burst went off at a magnetic pole of a spinning or processing neutron star, and the pulses are residues of or secondary phenomena induced by the burst. If the burst emission is highly beamed (in which case its duration may be related to the viewing angle rather than the energy supply), bursts could have occurred at both poles and only one pole was pointing at us. Then the pulse and interpulse are remnants of separate events at each pole instead of the results of pole-pole transmission. If that is the case, we could speculate that the unseen burst would even be stronger than the one seen since it left behind stronger pulses.

(c) The initial decay profiles of both pulse and interpulse[1,6] are quite similar and are definitely not exponential. Rather, they resemble power laws followed by periods of constant amplitude. This suggests that a single energy release followed by simple (radiative or adiabatic) cooling models would not work.

(d) The spectrum of the subsequent burst several hours later, though of much lower intensity, retains the same shape as the burst and pulses, suggesting that must be of the same origin. This pretty much rules out any explosion model as the source of the burst energy[2].

MAGNETIC RECONNECTION MODELS

Conceptually it is simplest to try to produce the $e^{\pm}$ pairs and gamma rays via some sort of magnetic field instability. However, if field reconnection or annihilation is the immediate energy source, severe lower limits can be set to the pre-annihilation strength. At strong fields considered here we expect the Alfven speed $v_A \sim c$. Hence the minimum time to annihilate a field of size $\pi f_*^{1/2} r_*$ must be $t_R \sim \pi f_*^{1/2} r_*/c \sim 10^{-5} f_*^{-1/2}$ sec. The maximum rate of magnetic energy release would be

(11)
$$\dot{E}_B \sim \frac{B^2}{8\pi} \cdot \frac{h f_* 4\pi r_*^2}{t_r} = B^2 h f_* \; 5 \times 10^{16}.$$

Since this must supply at least the γ-ray output of 10^{44} erg $\cdot$ s^{-1} we have

$$B \gtrsim 4 \times 10^{13} \text{ guass}.$$

close to the critical field of $B_{crit} \equiv m^2 c^3/e\hbar \simeq 4.4 \times 10^{13}$ guass, when the spontaneous breakdown of the field into $e^{\pm}$ pairs becomes in principle possible. The paradox is then, why we do not see 1-γ annihilation line from the pairs predicted for $B \to B_{crit}$[8,9]. Motion of neutron star disturbance (as in the picture Ramaty et al.[3]) may steepen into a very strong relativistic shock at the neutron star surface, compressing the material and associated magnetic fields to compression ratios unachievable in ordinary Newtonian situations (between 4 and 7 depending on the adiabatic index γ) behind the shock, at which point, if the stress field exceeds B_{crit}, it could decay directly into pairs.

The authors acknowledges valuable discussions wtih Drs. R. Ramaty and S. Antiochos.

REFERENCES

1. T.L. Cline, Comments on Astrophysics 9, 13 (1980).
2. S.E. Woosley and R.K. Wallace, Ap. J., Submitted (1981).
3. R. Ramaty, S. Bonazzola, T.L. Cline, D. Kazanas, P. Mesaros and R.E. Lingenfelter, Nature 287, 122 (1980).
4. S. Colgate and A.G. Petschek, LANL Preprint (1981).
5. R. Ramaty, R.E. Lingenfelter and R.W. Bussard, Ap. and Sp. Sci. 75, 193 (1981).

6. E.P. Mazets, S.V. Golenetskii, V.N. Ilinskii, R.L. Apteker and Yu, A. Guryan, Nature 282, 587 (1979).
7. E.P.T. Liang, Nature 292, 319 (1981).
8. J.K. Daugherty and R.W. Bussard, Ap. J. 238, 296 (1980).
9. J.I. Katz, UCLA preprint (1981).

Sec. I E Gamma Ray Transient Source Models

SURFACE AND MAGNETOSPHERIC PHYSICS OF NEUTRON STARS AND GAMMA-RAY BURSTS

D. Q. Lamb
Harvard-Smithsonian Center for Astrophysics
60 Garden Street, Cambridge, MA 02138

ABSTRACT

This talk reviews some aspects of the surface and magnetospheric physics of neutron stars that may be relevant to gamma-ray bursts if they originate on strongly magnetic neutron stars. The cohesive energy of matter in a strong magnetic field is uncertain, and therefore whether the stellar surface is likely to be liquid or solid is not known. Confinement of the hot plasma by the magnetic field requires that the emission come from near the stellar surface. The burst durations are much longer than the cooling time scale of the hot plasma, whether in a strong magnetic field or not, and require a separate explanation. Cyclotron cooling is more rapid than pair annihilation and can account for the relatively narrow annihilation lines seen in a number of burst spectra. The low energy absorption features seen in many burst spectra are probably the result of time-averaging spectra in which the low energy cutoff varies rapidly. If so, the features cannot be used to infer directly the strength of the magnetic field. The emission region is either extremely optically thick to electron and, especially, cyclotron scattering, or energy and/or matter is continuously replenished during the burst. If the emission region is optically thin, this constrains not only the number density there, but also the aspect ratio and distance of the source. Several important issues are discussed which any theory of gamma-ray bursts must address.

INTRODUCTION

Not long ago there were many different tenable ideas about the origin of gamma-ray bursts (see, e.g., Ruderman[1]). It is now widely believed, however, that the bursts come from strongly magnetic neutron stars. We begin by reviewing the events that have led to this changed perception (for recent reviews of observational developments, see Hurley,[2] Vedrenne,[3] and Mazets and Golenetskii[4]). We then discuss some aspects of the surface and magnetospheric physics of neutron stars that may be relevant to gamma-ray bursts (for more general reviews of the interior and surface physics, see Baym and Pethick[5,6] and Lamb[7]; for a more general review of the magnetospheric physics, see Lamb[8]). We also address the radiation processes expected in strong magnetic fields, emphasizing in particular the question of whether the emitting region is optically thick or thin (for a recent discussion of some radiation processes relevant to gamma-ray bursts, see Katz[9]). We conclude by mentioning several important issues that any attempt to explain gamma-ray bursts must face.

WHY NEUTRON STARS?

Of central importance in focusing attention on neutron stars as the sources of gamma-ray bursts was the famous March 5, 1979 event (for a comprehensive discussion of the observations, see Cline;[10] specific models of the event are discussed by Ramaty, Lingenfelter, and Bussard,[11] and by Hoshi[12]). The burst had a rise time of less than 0.25 msec,[13,14] comparable to the dynamical time scale at the surface of a neutron star. The less intense flux which followed the initial spike exhibited 8 second pulsations[13-16] similar in character to those widely observed in accreting neutron star X-ray sources. The spectrum of the burst showed a feature at ≃ 430 keV [13] that is likely to be an emission line produced by pair annihilation. If so, it is redshifted below 511 keV by just the amount expected for radiation coming from near the surface of a neutron star. Finally, the small positional error box of the event is coincident with the supernovae remnant N49 in the Large Magellanic Cloud.[17] The short rise time and large peak intensity (the burst was briefly 10 times more luminous than the Galaxy) of the March 5, 1979 event are presently unique, indicating that it is distinct from other gamma-ray bursts.[14] Nevertheless, the association of this event with so many features characteristic of neutron stars has had a large impact on the thinking about gamma-ray bursts in general.

Another important observational development has been the detection of spectroscopic features in many gamma-ray burst spectra by Mazets and his colleagues, using the Konus experiments on Venera 11 and 12 [4,18-22] (see also the review by Lingenfelter[23]). Many of the spectra show an emission feature near ≃ 400 keV that is probably again redshifted radiation from pair annihilation. Furthermore, many of the spectra show low energy absorption features which have been widely interpreted as being due to cyclotron absorption in magnetic fields of ≃ 10^{12} G. Such field strengths are similar to those in pulsars and in accreting neutron star X-ray sources.

An important theoretical development has been the recognition that the high spectral temperatures, $kT_{obs} \gtrsim 200$ keV, observed in gamma-ray bursts are possible only if matter is confined by a magnetic field of order 10^{12} G. This fact has been emphasized, in particular, by Colgate and Petschek[24] and Ruderman,[25] and has been used by Ramaty, Lingenfelter, and Bussard[11] in modeling the March 5, 1979 event.

Let us briefly consider the argument, as presented by Colgate and Petschek,[24] in greater detail. The maximum surface temperature that can be achieved by a star in hydrostatic equilibrium occurs when the force of gravity just balances the radiation stress; that is,

$$\frac{a}{4} T^4 = \frac{\mu m_H}{\sigma_T} \frac{GM}{R^2} \equiv \frac{\mu m_H}{\sigma_T} g, \tag{1}$$

where T is the surface temperature, μ is the mean molecular weight per particle, m_H is the atomic mass of hydrogen, σ_T is the Thomson cross section, M and R are the mass and radius of the star, and we have assumed that electron scattering is the dominant opacity. This defines an Eddington temperature (corresponding to the Eddington luminosity, if the entire stellar surface radiates) which is a function only of the surface gravity of the star,

$$T_E \equiv (4 \mu m_H g/\sigma_T a)^{1/4} \simeq 1.8 \mu^{1/4} (M/M_\odot)^{1/4} R_6^{-1/2} \text{ keV}, \tag{2}$$

where $R_6 = R/10^6$ cm (throughout the rest of this paper we will take $M = 1 M_\odot$ and $R_6 = 1$). The Eddington temperature is therefore greatest for neutron stars. Yet for the observed spectral temperatures $T_{obs} \simeq 200$ keV of gamma-ray bursts, the ratio of radiation to gravitational forces is still $\simeq (T_{obs}/T_E) \simeq 10^8$! Gravity cannot restrain the surface layers of the star and they freely expand at the local sound speed, producing an explosion.

However, such explosions are inefficient sources of high temperature radiation (for example, in supernovae only a small fraction of the explosion energy is converted into radiation; see also the discussion of explosions by Cavallo and Rees[26]). The reason is the following. As the matter expands, a rarefaction wave moves inward from the surface; at the same time, radiation is diffusing out. Matter above the rarefaction wave is cold due to adiabatic expansion, and therefore the expanding matter radiates at high temperature only until the rarefaction wave reaches the depth from which photons are diffusing out. For typical neutron star parameters, the duration and total energy of this phase are only $t \simeq 10^{-12}$ s and $E \simeq 10^{33} R_6^2 T_9^4$ ergs. These values are woefully inadequate to explain gamma-ray bursts.

Following the explosion, the photosphere lies at a radius such that equation (2) is satisfied. The plasma thus shrouds the star, and it radiates at a temperature T_E much lower than that observed in gamma-ray bursts. Recent calculations of accretion onto nonmagnetic neutron stars by Newman and Cox[27] and by Howard, Wilson, and Barton[28] show just such behavior.

In contrast, if a strong magnetic field is present and the plasma is radiation-dominated (as will likely be the case), it can be confined perpendicular to the field so long as $\beta_m \equiv P_r/P_m \lesssim 1$; that is, the pressure P_r of the hot plasma is less than the pressure P_m exerted by the magnetic field. Equivalently,

$$\frac{1}{3} a T^4 \lesssim \frac{B^2}{8\pi}, \qquad (3)$$

where B is the magnetic field. The corresponding temperature is

$$T \lesssim 170 \, B_{12}^{1/2} \text{ keV}, \qquad (4)$$

where $B_{12} = B/10^{12}$ G. If the plasma can be confined along the field by closed field lines or by the weight or ram pressure of overlying matter, then for magnetic field strengths of a few times 10^{12} G the plasma temperature can be similar to that observed.

These arguments, which predated the announcement of the discovery of low energy spectral features by Mazets and his colleagues,[4] provide powerful indirect support for the idea that gamma-ray bursts originate from neutron stars with strong magnetic fields.

Gamma-ray burst models that invoke magnetic neutron stars include ones in which the energy source comes from the impact of accreting plasma[29-31] or solid cometary matter[24,32] with the magnetosphere and surface of the star, or from nuclear burning of accreted matter.[33,34] In a third category of models, the energy is derived from a source internal to the neutron star, such as a crustquake[35,36] or a corequake,[37] which may produce a sudden rearrangement or motion of the magnetosphere.[11,37,38] A general review of neutron star models of gamma-ray bursts is given by Woosley.[39]

SURFACE PHYSICS

1. Available energy

A characteristic feature of neutron stars is the large gravitational potential at the surface. Typically,

$$\phi_{grav} = \frac{GM}{R} A m_H \simeq 0.15 \, A \, m_H c^2 \iff 140 \text{ MeV/nucleon}, \qquad (5)$$

where A is the ion atomic number. This provides a bound to the specific energy of the matter that produces the burst, whether it originates from outside the star (as in accretion) or from inside (as in a starquake or in pulsations). If the energy is supplied by nuclear burning, the bound is about a hundred times smaller, since the total energy available from burning hydrogen to iron peak nuclei is $\phi_{nuc} \simeq 0.01\ mc^2 <=> 7$ MeV/nucleon. Given the above bound, the minimum amount of matter required to power a typical gamma-ray burst is

$$\Delta M = m_H (E/\phi_{grav}) \simeq 7.5 \times 10^{17} (E/10^{38}\ ergs)\ g, \qquad (6)$$

where $E \simeq 1 \times 10^{38} (d/100\ pc)^2$ ergs is the burst energy.

2. Small height scale

Another characteristic feature of neutron stars is the large gravitational acceleration at the surface. Typically,

$$g = \frac{GM}{R^2} \simeq 1.3 \times 10^{14}\ cm\ s^{-2}. \qquad (7)$$

As a result, the length scales of phenomena near the surface are small. For example, the pressure scale height at a density of 10^4 g cm^{-3} (a density that may be characteristic of the surfaces of neutron stars with strong magnetic fields) is only a centimeter:

$$h \equiv (dP/dr)^{-1} P \simeq 0.77 (\mu_e/2.15)^{-5/3} \rho_4^{2/3}\ cm, \qquad (8)$$

where we have assumed that the matter is degenerate and composed of iron, and ρ_4 is the density in units of 10^4 g cm^{-3}. This fact is illustrated in Figures 1 and 2, which show the density and temperature profiles near the surface of a 1.54 $M_\odot$ star computed using a moderately soft equation of state by Malone.[40] These profiles are for relatively hot stars (the central temperatures are $T_c = 1 \times 10^9$ and 3×10^8 K, and the corresponding surface temperatures are $T_s = 5.4 \times 10^6$ and 2.3×10^6 K). In cooler stars, the density profile is even steeper.

The energy observed in a gamma-ray burst can, as we have seen, be supplied by a mass $\Delta M \simeq 10^{18}$ g if the matter is accreted or $\simeq 10^{20}$ g if the matter originates in a region of nuclear burning below the surface. Because of the large surface gravity, this mass can reside in a very thin layer at the surface. Again assuming that the matter is degenerate, the pressure at the bottom of the layer is

$$P = -4\pi \int g \rho\ dr = \frac{g}{R^2} \frac{\Delta M}{f}, \qquad (9)$$

where f is the fraction of the surface that the matter covers, and we have used the fact that g and r are nearly constant. Then

$$\rho \simeq 4.0 \times 10^4 (\mu_e/2.15)^{-3/5} (\Delta M_{18}/f)^{3/5}\ g\ cm^{-3}, \qquad (10)$$

where μ_e is the mean molecular weight per electron and $\Delta M_{18} = \Delta M/10^{18}$ g, and

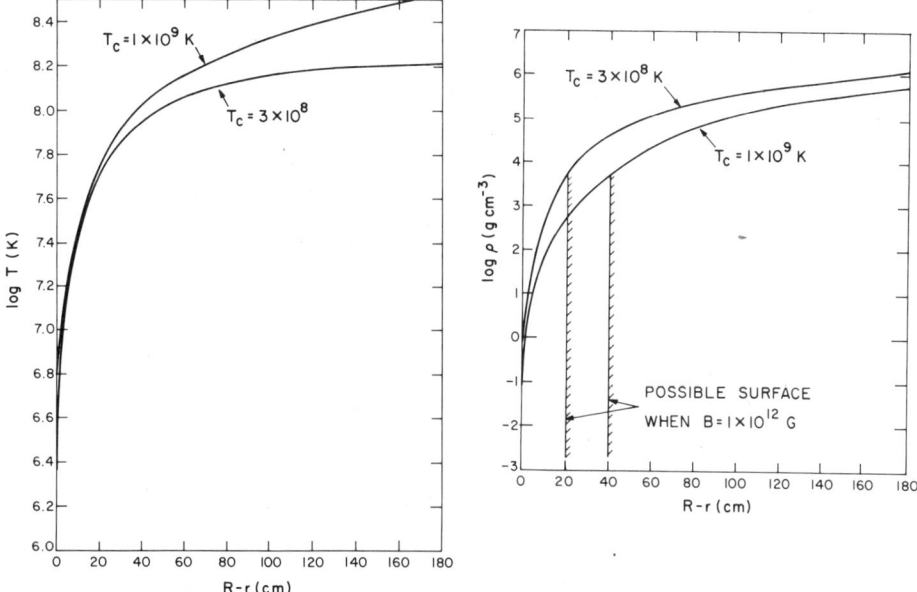

Fig. 1. Temperature profile in the outer layers of a M = 1.54 $M_\odot$ neutron star for central temperatures T_c = 1 x 10^9 and 3 x 10^8 K. The corresponding surface temperatures are T_s = 5.4 x 10^6 and 2.3 x 10^6 K (after Malone[40]).

Fig. 2. Density profile in the outer layers of a M = 1.54 $M_\odot$ neutron star for the same central temperatures as in Fig. 1 (after Malone[40]). The vertical lines with hatching indicate the surface of the star for a magnetic field B = 1 x 10^{12} G, if the surface is a polymerized solid.

$$h \simeq 2.0 \, (\mu_e/2.15)^{-31/15} \, (\Delta M_{18}/f)^{2/5} \text{ cm,} \qquad (11)$$

If the matter is nondegenerate or semidegenerate, as may be the case for accreting matter or matter undergoing nuclear burning, the scale height is larger. Nevertheless, accreting matter penetrates or nuclear burning matter originates at most only 1-10 meters below the surface.

3. Surface structure

The surface of a neutron star with no magnetic field is solid if the surface temperature is sufficiently low. For example, at temperatures below 1808 K iron is solid and has a density of 7.86 g cm^{-3}. This is therefore the density at the surface of a very cold neutron star whose outer layers are iron. At the higher surface temperatures expected for gamma-ray burst sources, the surface is fluid. At a short distance below the surface, however, Coulomb forces are strong enough that the matter forms a solid. This occurs when the ratio of the Coulomb and thermal energies of the ions,

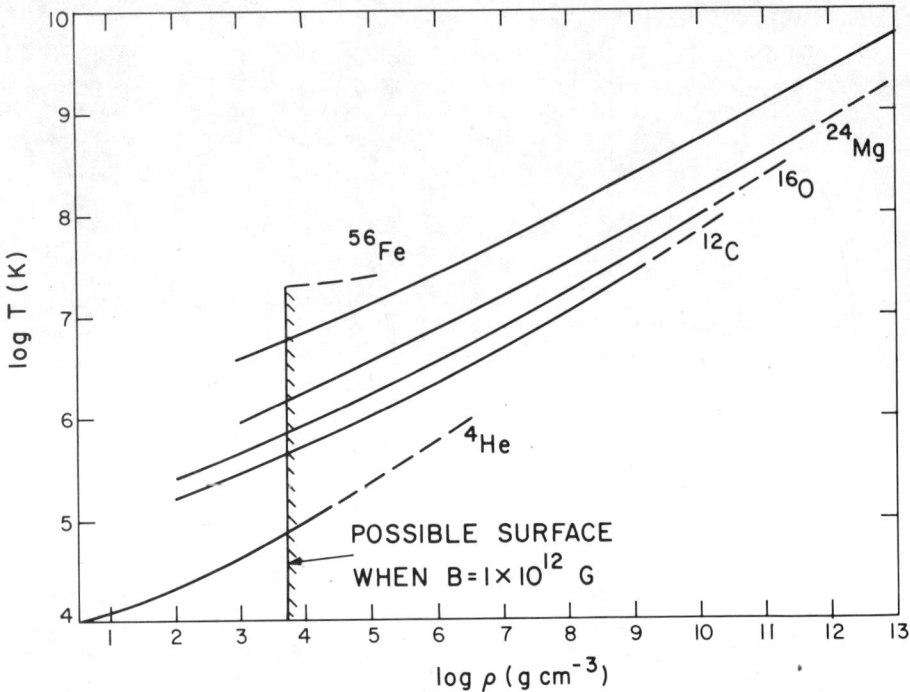

Fig. 3. Crystallization curves for various compositions, ranging from ^{4}He to ^{56}Fe, of the crust of a nonmagnetic neutron star (after Lamb and Van Horn[43]). The vertical line with hatching indicates the surface of the star for an ^{56}Fe composition and a magnetic field $B = 1 \times 10^{12}$ G, if the surface is a polymerized solid. At the surface, the density is $\rho_0 = 4.4 \times 10^3$ g cm^{-3} and the melting temperature is taken to be $T_m = 1.6$ keV $= 1.9 \times 10^7$ K (Flowers et al.[47]). The melting temperatures for compositions other than ^{56}Fe have not been computed, but are a small fraction of the temperature corresponding to the binding energy of condensed matter of that compositon.

$$\Gamma \equiv \frac{(Ze)^2}{akT} \simeq 87 \, (Z/26)^2 \, (A/56)^{-1/3} \, \rho_4^{1/3} \, T_7^{-1} \simeq 160, \qquad (12)$$

where Z is the ion charge, a is the mean spacing between ions, and $T_7 = T/10^7$ K.[41,42] The corresponding melting temperature is

$$T_m = 5.4 \times 10^6 \, (Z/26)^2 \, (A/56)^{-1/3} \, \rho_4^{1/3} \text{ K.} \qquad (13)$$

Crystallization curves for compositions ranging from pure ^{4}He to pure ^{56}Fe are shown in Figure 3 from calculations by Lamb and Van Horn.[43] For a given a temperature in the surface layers of a neutron star, these curves determine the density (and therefore the depth) at which matter becomes crystalline. The resulting crust extends inward to a density of $\simeq 2 \times 10^{14}$ g cm^{-3}.[44,45]

The properties of matter in strong magnetic fields have received considerable attention since the discovery of pulsars 13 years ago. For reviews of the early work, see Ruderman[46] and Baym and Pethick;[5] for a discussion of more recent work, see Flowers et al.[47] and Baym and Pethick.[6] A strong magnetic field constrains electrons to move on quantized Landau orbits, which correspond to spiral motion around cylinders of radius $r_n = (2n+1)^{1/2} \hat{\rho}_Z$, where

$$\hat{\rho} \equiv (\hbar/m_e \omega_c)^{1/2} \simeq 2.6 \times 10^{-10} B_{12}^{-1/2} \text{ cm.} \tag{14}$$

and $\omega_c = eB/m_e c$ is the cyclotron frequency. As a result, electrons in atoms are confined in two directions to radii much nearer the nucleus than in the absence of a field and the binding energy of the atoms is much greater. The effect is characterized by the dimensionless parameter

$$\eta \equiv \frac{a_0}{Z \rho_Z} \simeq 14.7 \, B_{12}^{1/2} \, Z^{-3/2}, \tag{15}$$

which is the ratio of the Bohr radius a_0/Z of the most tightly bound electron when $B = 0$ to the radius $\rho_Z = (2Z+1)^{1/2} \hat{\rho}$ of the innermost Landau orbital.

In the presence of a strong magnetic field, the elongated atoms can potentially bind together covalently to form a solid of polymerized chains in which the electrons are free to move along the length of the chain.[48] A semiclassical calculation of the properties of uniform matter, using the Wigner-Seitz method, gives results that are reasonably accurate and shows the scaling with B and Z.[47] The binding energy per atom at zero pressure is

$$E_0 = -0.517 \, (Z^3 e^2/a_0) \, \eta^{4/5} = -42.6 \, (Z/26)^{9/5} \, B_{12}^{2/5}, \tag{16}$$

and the density at zero pressure is

$$\rho_0 = 11.4 \, A \, m_H \, (Z/a_0)^3 \, \eta^{12/5} = 4.5 \times 10^3 \, (A/56) \, (Z/26)^{-3/5} \, B_{12}^{6/5} \text{ g cm}^{-3}. \tag{17}$$

Figures 2 and 3 show the density at the surface of a neutron star for ^{56}Fe and a magnetic field of 1×10^{12} G, assuming that the matter forms a polymerized solid.

The cohesive energy of such matter is the difference between the binding energy of the polymerized solid and that of isolated atoms in the same field. In spite of considerable effort, its value, and therefore the melting temperature of such matter, is uncertain. In the limit η goes to ∞, the binding energy of the polymerized solid is much greater than that of an isolated atom and is a good estimate of the cohesive energy. However, the difference is small even for H and He (for which $\eta \simeq 33$ and 12 at $B = 1 \times 10^{12}$ G) and detailed calculations are required. Such calculations have not been done for light atoms; however, we can say that the melting temperature T_m must be much less than the binding energy of uniform matter (equation 16). That is, $T_m(H) \ll 100$ eV $= 1 \times 10^6$ K and $T_m(He) \ll 300$ eV $= 3 \times 10^6$ K. The earliest calculations of the cohesive energy for Fe indicated that it is large, of order 10-20 keV;[49] however, an error in these calculations had led to an overestimate.[50] The most reliable calculations to date[47] give cohesive energies ranging from 2.6 keV at $B = 1 \times 10^{12}$ G to 8.0 keV at $B = 5 \times 10^{12}$ G, which are small compared to the binding energy of individual atoms. Figure 3 shows the melting temperature for Fe at the surface of a neutron star with a field of 1×10^{12} G.

Given the fact that the cohesive energy is the difference of two large energies, each calculated variationally, one cannot conclude with certainty that at zero pressure the condensed state formed of polymerized chains is energetically favorable compared with a more ordinary state of elongated atoms with weaker metallic binding. Because of the compression of atoms in strong fields, the latter would still have a density at zero pressure much larger than that of terrestrial solids, but perhaps an order of magnitude or more below that of the polymerized solid. In addition, the melting temperature would be much lower.

From this discussion, we see that it is by no means certain that the surfaces of neutron stars that are sources of gamma-ray bursts are solid, even if the surface layers are composed of Fe. The melting temperatures for H and He are likely to be so small that accreted matter composed of these elements may remain fluid, even for very low accretion rates and correspondingly low surface temperatures.

4. Pinned magnetic field

The conductivity of the crust is large. Even assuming a high concentration ($X_i = 10^{-3}$) of ion impurities with charge differences $\Delta Z = 2$, the decay time scale for any magnetic field threading the crust is[51]

$$t_d = (4\pi\sigma/c^2)(\Delta R/\pi)^2 \approx 1.3 \times 10^7 (\sigma/3 \times 10^{25} \text{ s}^{-1})(\Delta R/10^5 \text{ cm})^2 \text{ yr}, \qquad (18)$$

where ΔR is the thickness of the crust and σ is the conductivity. This has the important consequence that on any time scale of interest for gamma-ray bursts, the field is securely pinned in the crust. As a result, it is able to resist interchange instabilites (see below) involving the field and any hot plasma just above the stellar surface.

MAGNETOSPHERIC PHYSICS

1. Size and shape of the magnetosphere

Little is known for certain about the detailed structure of the magnetic fields of neutron stars. Near the surface of the star, the field may be a simple dipole as illustrated in Figure 4a, or the field may be more complex with concentrations of magnetic flux at several points on the surface as illustrated in Figure 4b. If the field is a simple dipole, then accreting matter will likely be funneled toward one or both of the magnetic poles. However, if the field is more complicated, accreted matter may arrive and accumulate at a number of places on the surface. This would certainly affect the spectra and the time variability of any gamma-ray burst; if accretion occurs, emission might come successively from locations of flux concentration on the stellar surface, or if nuclear burning occurs, a succession of flashes might occur. In either case, the series of outbursts taken together might constitute a burst.[24,34] However, the general simplicity of the pulse profiles of accreting neutron star X-ray sources at higher energies (cf. that of Vela X-1 [52]) suggests that the field near the star may be a simple dipole or, at least, that accreting matter tends to be funneled toward the magnetic poles for both rapidly and slowly rotating stars.

Any concentrations of magnetic field at the stellar surface die away rapidly with distance from the star, so that at sufficiently large distances the field is dipolar.

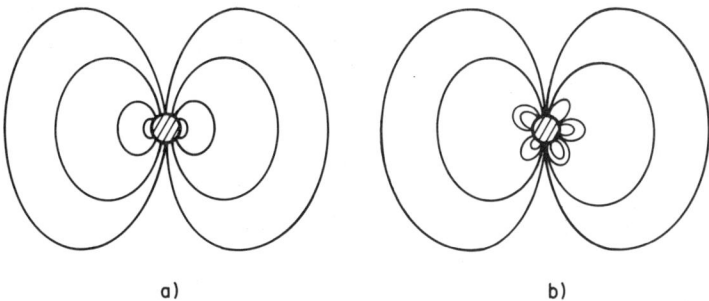

Fig. 4. Possible configurations of the lines of force surrounding a magnetic neutron star. a) Simple dipole geometry. b) Complex field geometry near the neutron star and dipole field at larger distances.

However, even the dipole field falls off as $(R/r)^3$ so that with a surface magnetic field $B = 1 \times 10^{12}$ G, the field at 10^7 cm ($\simeq 10$ R) is 1×10^9 G while at 10^8 cm ($\simeq 100$ R) it is only 1×10^6 G. This places important constraints on the location and volume of hot plasma that can be confined by the field. If the plasma is radiation-dominated, as expected, then from equation (3) the field required to confine the plasma in directions perpendicular to the field is

$$B_c \gtrsim 2.5 \times 10^{11} T_9 \text{ G,} \qquad (19)$$

and the confining radius must satisfy

$$r_c \lesssim 1.6 \times 10^6 \, (B_s/10^{12} \text{ G})^{1/3} \, T_9^{2/3} \text{ cm,} \qquad (20)$$

where B_s is the field strength at the stellar surface. Note that this is more stringent than the requirement that only matter at this temperature be confined, since $\beta_r \equiv P_r/P_g > 1$ where P_r and P_g are the radiation and gas pressures. The latter requirement gives

$$B_c' \gtrsim 1.9 \times 10^{10} \, n_{26}^{1/2} \, T_9^{1/2} \text{ g,} \qquad (21)$$

and

$$r_c' \lesssim 3.7 \times 10^6 \, (B_s/10^{12} \text{ G})^{1/3} \, n_{26}^{-1/6} \, T_9^{-1/6}, \qquad (22)$$

results which have been quoted by a number of authors (cf. ref 9). Here n_{26} is the ion number density in units of 10^{26} cm^{-3}, which is the maximum number density likely in the emission regions of gamma-ray burst sources (see below). For comparison, the scale height at the neutron star surface of matter at such temperatures is only

$$h = kT/m_H \simeq 620 \, T_9 \text{ cm.} \qquad (23)$$

Plasma flowing along the magnetic field can be channeled by the field only so long as the flow is sub-Alfvénic; that is,

$$v_p \lesssim v_A \equiv (B_p^2/4\pi\rho)^{1/2}, \tag{24}$$

where v_p and B_p are the poloidal components of the bulk flow velocity and of the magnetic field. For flow at the sound speed, the confining radius is the same as that given by equation (22). However, if the flow is supersonic, then the confining radius can be much smaller,

$$r_c^* \lesssim 1.1 \times 10^6 \, (B_s/10^{12} \, G)^{2/5} \, (v_p/v_{ff})^{-2/5} \, n_{26}^{-1/5} \text{ cm}, \tag{25}$$

where we have scaled the velocity by its free fall value, $v_{ff} = (2GM/r)^{1/2}$. Beyond this radius the magnetic field is torn away by the streaming plasma.

From this discussion, we conclude that plasma as hot as that observed in gamma-ray bursts can be confined only near the star. This is consistent with the redshift z ≃ 0.2 and narrow width of the 511 keV annihilation line seen in a number of gamma-ray burst spectra.[4,21,22] We also conclude that the magnetospheres of burst sources are likely to be open, at least near the magnetic poles where the dipole field lines extend to large radii in the absence of hot plasma.

2. Parameter regimes of a hot plasma in a strong magnetic field

If the energy of the emitting plasma or the plasma itself are not replenished during a burst, the characteristic number of particles required to power a gamma-ray burst is

$$N = E/(\tfrac{3}{2}kT) \simeq 1.2 \times 10^{45} \, (d/100 \text{ pc})^2 \, T_9^{-1}, \tag{26}$$

where we have taken an energy per particle of $\tfrac{3}{2}kT$. This number and the volume,

$$V = 4\pi f R^2 l = 1.3 \times 10^{19} \, f \, R_6^2 \, l_6, \tag{27}$$

where f is again the fraction of the stellar surface and l_6 is the radial thickness of the burst emission region in units of 10^6 cm, defines a characteristic maximum number density,

$$n = N/V \simeq 9.6 \times 10^{25} \, (d/100 \text{ pc})^2 \, T_9^{-1} \, (f \, l_6)^{-1} \text{ cm}^{-3}, \tag{28}$$

for the emitting plasma. If we again assume no replenishment during a burst, but that the energy is supplied by the annihilation of pairs rather than by thermal energy, we find a somewhat smaller characteristic number density,

$$n_{pair} = (E/m_e c^2)/V \simeq 4.9 \times 10^{24} \, (d/100 \text{ pc})^2 \, (f \, l_6)^{-1} \text{ cm}^{-3}. \tag{29}$$

If replenishment occurs instead and the only requirement is that the plasma supply the observed luminosity, then the number density can be far smaller. For example, if the luminosity of the burst is due to cyclotron emission, the characteristic number density can be as small as $\simeq 10^9$ cm^{-3} (see below).

Figure 5 shows several parameter regimes for such a hot (T ≃ 10^6 - 10^{10} K) low-density (n ≃ 10^{12} - 10^{30} cm^{-3}) plasma in a strong magnetic field. In this example, the composition is taken to be hydrogen and the magnetic field is assumed to be 1 x 10^{12}

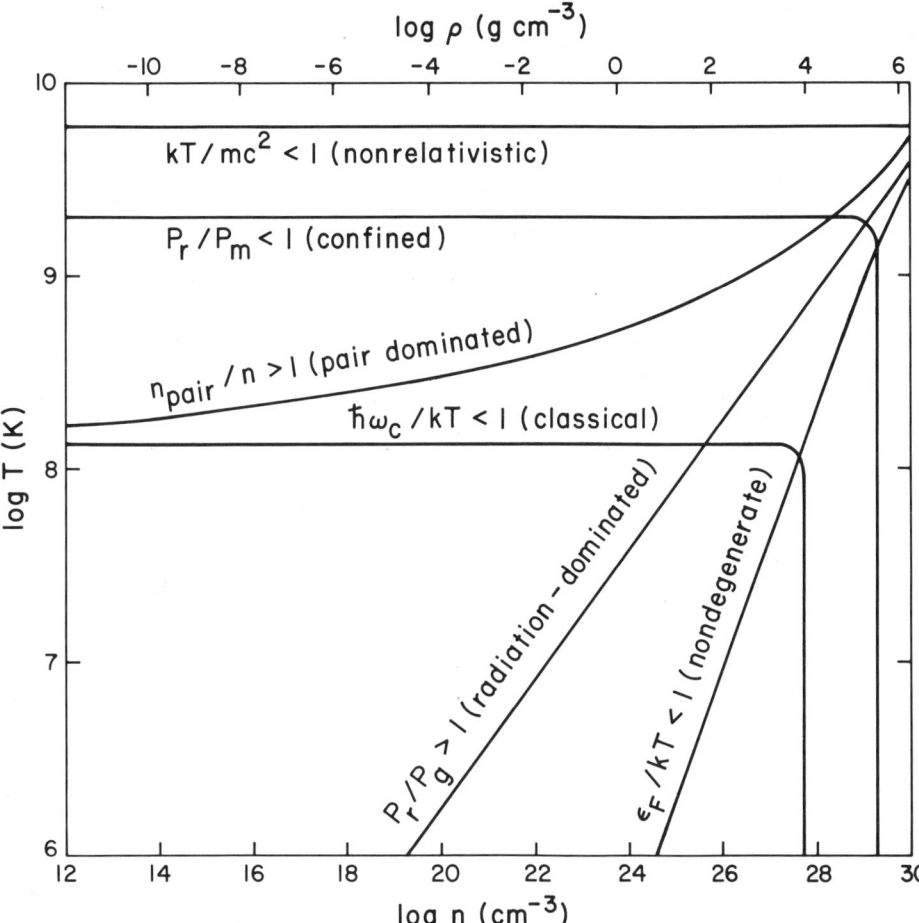

Fig. 5. Parameter regimes of a hot plasma in a strong magnetic field. Above and to the left of the curve labeled $\epsilon_F/kT < 1$, the plasma is nondegenerate. To the upper left of the curve labeled $P_r/P_g > 1$, the plasma is radiation-dominated. Below the curve labeled $kT/M_e c^2 < 1$, the plasma is nonrelativistic. Above the curve labeled $n_{pair}/n > 1$, the plasma is pair-dominated. Above and to the right of the curve labeled $\hbar \omega_c/kT < 1$, a classical treatment of kinetic and radiation processes is valid. The magnetic field is able to confine the plasma for densities and temperatures below and to the left of the curve labeled $P_r/P_m < 1$.

G. The matter is nondegenerate if $\epsilon_F/kT < 1$, where ϵ_F is the Fermi energy of the electrons. In the nonrelativistic regime, this condition is satisfied as long as

$$T \gtrsim 3.3 \times 10^7 \, n_{26}^{2/3} \, K = 2.8 \, n_{26}^{2/3} \, keV. \tag{30}$$

The plasma is radiation-dominated if $\beta_r \equiv P_r/P_g > 1$. In the nondegenerate regime, this condition is satisfied as long as

$$T \gtrsim 1.8 \times 10^8 \, n_{26}^{1/3} = 15 \, n_{26}^{1/3} \, keV. \tag{31}$$

The magnetic field is able to confine the plasma in the directions perpendicular to the field if $P_m > P_r + P_g$. In the nondegenerate regime, this condition is satisfied as long as $\beta_m \equiv P_r/P_m < 1$; that is, as long as equation (4) is satisfied. When the electrons are degenerate, the condition becomes $P_m > P_e = n_e \epsilon_F$; otherwise, the magnetic field is not important in confining the electrons. The electrons in the plasma are nonrelativistic if $kT/m_e c^2 < 1$; that is, as long as

$$T \lesssim 5.9 \times 10^9 \, K = 511 \, keV. \tag{32}$$

We note that the electrons are unlikely to be relativistic since confinement of such a plasma requires a magnetic field strength $B_c \gg 1 \times 10^{13}$ G. If $\hbar \omega_c/kT < 1$ or, equivalently,

$$T \gtrsim 1.3 \times 10^8 \, B_{12} \, K = 12 \, B_{12} \, keV, \tag{33}$$

then the motion of the electrons is not quantized by the strong magnetic field. Quantum effects on kinetic and radiation processes[9,53] can be neglected and, in particular, cyclotron (or gyrosynchrotron) emission can be treated classically.

3. Pair plasmas

One of the most facinating and intriguing aspects of gamma-ray bursts is the implied existence in the burst emission region of plasma in which the electron-positron pair density is significant, if not dominant. The pair annihilation lines observed in many burst spectra[4,21,22] contain $\simeq 0.1$ of the energy in the burst; this represents a lower limit to the fraction of the energy of the plasma that is in pairs, since the line must orginate in a region that is optically thin (see below). For a discussion of radiation processes in pair plasmas, see Ramaty.[54] Here we merely mention that, in equilibrium, the number density of pairs in the plasma exceeds that of the ions ($n_{pair}/n > 1$) at nonrelativistic temperatures if[55]

$$n \lesssim 2.2 \times 10^{30} \, (kT/m_e c^2)^{-3/2} \, e^{-(kT/m_e c^2)} \, cm^{-3}. \tag{34}$$

At relativistic temperatures, the condition becomes

$$n \lesssim 1.5 \times 10^{31} \, T_{10}^3. \tag{35}$$

When $n_{pair}/n > 1826$, the mass density in pairs exceeds that of the ions and the plasma can no longer be confined by gravity. The plasma can, however, be confined by a magnetic field. Note also that at relativistic temperatures, the energy $E_{pair} = 7aT^4/4$

in pairs exceeds the energy $E_\gamma = aT^4$ in photons; that is,

$$E_{pair}/E_\gamma = 7/4. \tag{36}$$

From the foregoing discussion, we conclude that the radiation-emitting plasma in gamma-ray burst sources is likely nondegenerate and nonrelativistic, that the energy density of the plasma is likely dominated by radiation and possibly even by electron-positron pairs, and that a classical treatment of kinetic and radiation processes may be valid.

4. Stability

Hot plasma may enter the magnetosphere by accreting along magnetic flux sheets or tubes[29-31] or by being thrown up in a similar pattern from the stellar surface, either by the impact of accreting matter with the stellar surface[24,29,32] or by a nuclear explosion in the surface layers.[33,34] Hot plasma might also be produced in situ by violent motion of magnetic field lines or by field annihilation.[11,37,38] Whatever the source, such hot plasma is notoriously unstable when confined in a magnetic field. One way in which plasma may rapidly cross field lines is by the interchange (or Rayleigh-Taylor) instability.[56,57] This instability occurs on a time scale,

$$t_{inter} \simeq (gk)^{-1/2} \simeq (\lambda/2\pi g)^{1/2} \simeq 3.5 \times 10^{-5} \lambda_6 r_6 \text{ s}, \tag{37}$$

where $k = 2\pi/\lambda$ is the wavenumber of the fastest growing unstable mode and λ is the corresponding wavelength, here taken to be of order r.

RADIATION PROCESSES

1. Cooling rates and time scales

A hot plasma in a strong magnetic field cools via bremsstrahlung and cyclotron (or gyrosynchrotron) emission. If pairs are present, it may also radiate via pair annihilation. Ramaty, Lingenfelter, and Bussard,[11] Katz,[9] and Bussard and Lamb[53] have recently discussed some of these processes in the context of gamma-ray bursts. Here we compare their importance, assuming that the plasma is nonrelativistic, that pairs do not dominate, and that the emission region is optically thin. (We consider the optically thick case below.) The usual electron-ion bremsstrahlung cooling rate and cooling time scale are (cf. Allen,[58] p. 103)

$$\Lambda_{br} = 1.4 \times 10^{-27} Z^2 n\, n_e\, T_e^{1/2}\, \bar{g} \text{ ergs cm}^{-3} \text{ s}^{-1}, \tag{38}$$

and

$$t_{br} = \frac{3}{2} \frac{n_e kT}{\Lambda_{br}} \simeq 4.4 \times 10^{-11} Z^{-2} n_{26}^{-1} T_9^{1/2} \text{ s}, \tag{39}$$

where $\bar{g}$ is the Gaunt factor, of order unity. At temperatures above 10^9 K, relativistic corrections and the contribution of electron-electron bremsstrahlung are important. These points have been carefully discussed recently by Gould[59]. No fits to observed spectra have been carried out with all of these effects included. We also note that if the plasma is pair-dominated, the bremsstrahlung cooling rate in the nonrelativistic limit is four times larger than that above because the electron and positron radiation fields add coherently.[9]

If $kT \gg \alpha\, m_e c^2 \simeq 5 \times 10^4$ K, Coulomb effects are unimportant, and the power emitted by pair annihilation and the corresponding annihilation time scale are[9]

$$\Lambda_{pair} = 2 m_e c^2 n_e n_{\bar{e}} R \simeq 1.2 \times 10^{32}\, n_{e,26}\, n_{\bar{e},26}\ \text{ergs cm}^{-3}\ \text{s}^{-1}, \quad (40)$$

and

$$t_{pair} = (n_{\bar{e}} R)^{-1} \simeq 1.3 \times 10^{-12}\, n_{\bar{e}}^{-1}\ \text{s}, \quad (41)$$

where $R = \pi r_o^2 c = 7.48 \times 10^{-15}$ cm^3 s^{-1}, $r_o = e^2/m_e c^2 = 2.82 \times 10^{-13}$ cm is the classical electron radius, and $n_{e,26}$ and $n_{\bar{e},26}$ are the electron and positron number densities in units of 10^{26} cm^{-3}.

Finally, the cyclotron cooling rate and cooling time scale are[60]

$$\Lambda_{cyc} = 4 n_e \sigma_T c \frac{B^2}{8\pi} \frac{kT}{m_e c^2} \simeq 5.4 \times 10^{34}\, n_{e,26}\, B_{12}^2\, T_9\ \text{ergs cm}^{-3}\ \text{s}^{-1}, \quad (42)$$

$$t_{cyc} = \frac{3\pi m_e c}{\sigma_T B^2} \simeq 3.9 \times 10^{-16}\, B_{12}^{-2}\ \text{s}, \quad (43)$$

where we have approximated the cyclotron emission by that at the fundamental ω_c. This approximation is valid, even for a semirelativistic plasma, as long as the emission region is optically thin. We note that if the plasma is pair-dominated, the cooling rate is twice as large but the cooling time scale remains unchanged.

For the densities, temperatures, and magnetic field strengths expected in the emission regions of burst sources, the cooling time scales for bremsstrahlung and cyclotron cooling are $\ll 1$ s. If the emission region is extremely optically thick, it is possible to achieve photon diffusion times,

$$t_{diff} = 3l\tau/c \simeq 100\, \tau_6\, l_6\ \text{s}, \quad (44)$$

where τ_6 is the optical depth in units of 10^6. However, the rapid variations on a time scale $\simeq 0.1$ second seen in burst intensities[2,3,61] rule out such an explanation for the duration of the bursts. We therefore conclude that the duration of gamma-ray bursts is not due to the natural cooling time of a hot $T \simeq 3 \times 10^9$ K plasma, whether in a strong magnetic field $B \simeq 10^{12}$ G or not. Rather, it requires a separate explanation.

2. Comparison of radiation processes

Comparison of equations (38) and (42) illustrates the well known fact that cyclotron emission in a strong magnetic field cools a hot plasma much more efficiently than does bremsstrahlung. If the spectra of some gamma-ray bursts, such as the Apollo 16 event,[62] are in fact due to bremsstrahlung, the required physical conditions are

$$B \lesssim 2.9 \times 10^9\, Z\, n_{26}^{1/2}\, T_9^{-1/4}\ \text{G}. \quad (45)$$

If the surface magnetic field is of order 10^{12} g, such a small magnetic field strength and high number density is difficult to achieve. It might occur high in the magnetosphere of a neutron star, where the magnetic field is small, or possibly in the interior of a hot plasma bubble near the stellar surface, where the field strength is reduced by expansion of the field lines.

The relative width of a line due to Doppler broadening is

$$\Delta\omega/\omega \simeq (2kT/m_e c^2)^{1/2}. \tag{46}$$

The widths of the annihilation lines observed in the March 5, 1979 event[13] and in other gamma-ray bursts[4,21,22] then constrain the temperature of the pairs to be (see also the detailed discussion by Ramaty and Mészáros[63])

$$T \lesssim 39 \, (\Delta\omega/200 \text{ keV})^2 \text{ keV}. \tag{47}$$

This temperature is much smaller than that required to produce the continuum spectrum. Ramaty, Lingenfelter, and Bussard[11] have successfully resolved this apparent contradiction by pointing out that if

$$t_{cyc}/t_{pair} = \frac{9}{64} \frac{n_e m_e c^2}{(B^2/8\pi)} \lesssim 1, \tag{48}$$

cyclotron emission can cool the plasma before the pairs annihilate (see also Katz[9]). Equivalently,

$$B^2/8\pi \gtrsim \frac{9}{64} n_e m_e c^2, \tag{49}$$

or

$$B \gtrsim 2.9 \times 10^8 \, n_{e,26} \text{ G}, \tag{50}$$

which is always satisified if the magnetic field is sufficiently strong to confine the plasma. The cooling and annihilation may occur sequentially within the hot plasma itself, or they may occur in spatially distinct regions. For example, the annihilation might occur in a sheath surrounding the cooling plasma.

3. Cyclotron emission spectrum

In gamma-ray burst sources, unlike accreting neutron star X-ray sources, a classical treatment of cyclotron emission may be valid, as discussed earlier. It is convenient to scale the resulting emission spectrum in terms of the expression for the blackbody flux in a single polarization in the Rayleigh-Jeans limit,

$$\frac{1}{2} F_\omega^{RJ} = \frac{kT \omega^2}{8 \pi^2 c^2}, \tag{51}$$

and to introduce the dimensionless size parameter

$$\Lambda \equiv \frac{\omega_p^2 \, l}{\omega_c \, c} = \frac{3}{2} \alpha^{-1} (m_e c^2 / \hbar \omega_c) n_e \sigma_T l, \qquad (52)$$

where ω_p is the plasma frequency, l is the thickness of the emission region, and α is the fine structure constant. The quantity Λ is related to the optical depth of the emission region (here assumed to be small). Then the angle-averaged spectral emissivity and spectral flux are given by[64]

$$j_\omega = \frac{1}{2} B_\omega^{RJ} \frac{\Lambda}{l} \Phi(\omega), \qquad (53)$$

and

$$F_\omega = \frac{1}{2} F_\omega^{RJ} \Lambda \, \Phi(\omega), \qquad (54)$$

where $B_\omega^{RJ} = F_\omega^{RJ}/\pi$ and $\Phi(\omega)$ is a dimensionless spectral flux. Figure 6 shows $\Phi(\omega)$ for electron temperatures T = 20, 50, and 100 keV from calculations by Lamb and Masters[65] (see also Kylafis et al.[66]).

At the high temperatures characteristic of the emission regions of gamma-ray bursts, the Doppler shift due to the velocity of the electrons broadens the individual cyclotron harmonics. This leads to a relatively smooth emissivity, and therefore absorption or scattering opacity, as a function of frequency. The emissivity curves shown in Figure 6 illustrate clearly the smoothing that occurs with increasing temperature. We should note, however, that Figure 6 shows the angle-averaged flux. If the emission region is viewed perpendicular to the magnetic field, the broadening due to the Doppler shift is reduced.[53] Furthermore, the harmonics will be narrower if electron velocities perpendicular to the field are reduced by rapid cyclotron cooling.

4. Cyclotron absorption versus scattering

If the electron Landau levels are largely excited and de-excited by collisions, absorption and emission predominate. On the other hand, if the levels are largely excited and de-excited by radiative processes, scattering predominates and the flux at the cyclotron harmonics is reduced even if the source is isothermal. This point has been emphasized by Bussard and Lamb,[53] and we closely follow their discussion here. The collisional excitation rate per electron is

$$R_{01} \simeq 6.3 \times 10^{-13} \, n \, Z^2 \, B_{12}^{-1} \, T_9^{-1/2} \, \ln(4kT/\hbar \omega_c) \, s^{-1}, \qquad (55)$$

while the Einstein A coefficient for this two-level system is

$$A_{10} = \frac{4}{3} \alpha \, (\hbar \omega_c / m_e c^2) \, \omega_c \simeq 3.9 \times 10^{15} \, B_{12}^2 \, s^{-1}. \qquad (56)$$

Cyclotron scattering predominates over absorption if $A_{10} > R_{01}$; that is, as long as

$$n \lesssim 6.2 \times 10^{27} \, Z^{-2} \, B_{12}^3 \, T_9^{1/2} \, \ln^{-1}(4kT/\hbar \omega_c) \, cm^{-3}. \qquad (57)$$

From this discussion, we conclude that cyclotron scattering, rather than absorption, is

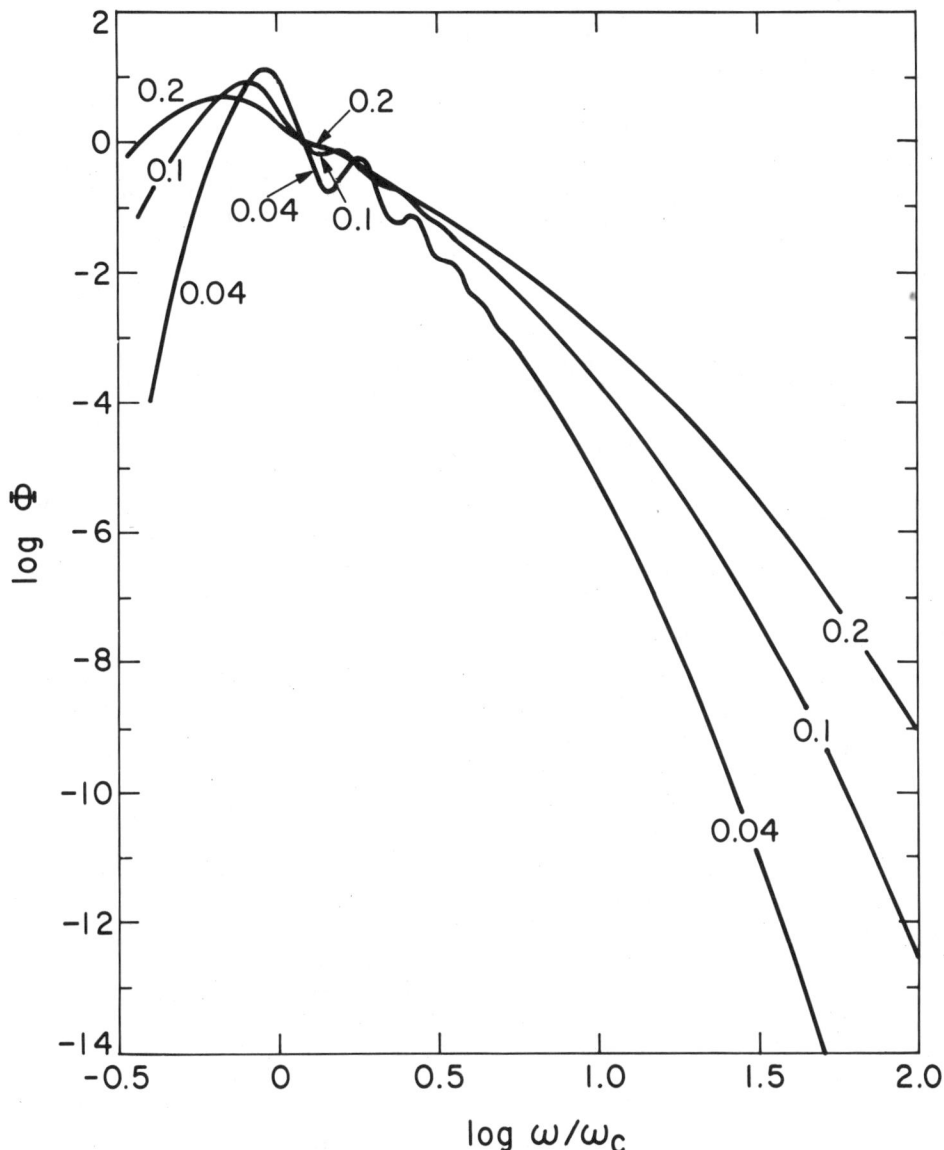

Fig. 6. Dimensionless spectral flux $\Phi(\omega)$ for cyclotron emission from hot plasmas with electron temperatures T = 0.04, 0.1, and 0.2 $m_e c^2$ (20, 50, and 100 keV) (after Kylafis et al.[66]; see also Masters[67]).

likely to predominate in the burst emission region.

THROUGH THICK AND THIN

1. Electron and cyclotron scattering optical depths

So far we have assumed that the emission region is optically thin. However, the electron and cyclotron scattering opacities are sufficiently large than the emission region could be extremely optically thick. Again we assume that the electrons are at most semirelativistic, and the plasma is not pair-dominated. Then the electron scattering optical depth is

$$\tau_e = n_e \sigma_T l \simeq 2.9 \times 10^7 n_{e,26} l_{6'} \tag{58}$$

where l is the thickness of the emission region in units of 10^6 cm. For the emission region to be optically thin, the number density must satisfy,

$$n_e \lesssim 3.5 \times 10^{18} l_6^{-1} \text{ cm}^{-3}. \tag{59}$$

Similarly, the optical depth for cyclotron absorption (or scattering) at the fundamental is[67]

$$\tau_{cyc} \simeq \frac{\pi}{3} \Lambda (\Delta\omega/\omega)^{-1} \simeq \frac{\pi}{3} \Lambda (2kT/m_e c^2)^{-1/2}$$

$$\simeq 4.7 \times 10^{11} B_{12}^{-1} T_9^{-1/2} n_{e,26} l_{6'} \tag{60}$$

where we have used equation (46) to estimate the relative bandwidth of the line. Then the requirement that the emission region be optically thin becomes

$$n_e \simeq 2.1 \times 10^{14} B_{12} T_9^{1/2} l_6^{-1/2}. \tag{61}$$

From equations (58) and (60), the ratio of electron scattering to cyclotron scattering at the cyclotron fundamental is

$$\tau_{cyc}/\tau_{es} = \frac{\pi}{2} \alpha^{-1} (m_e c^2/2kT)^{1/2} \frac{m_e c^2}{\hbar \omega_c}$$

$$\simeq 1.6 \times 10^4 B_{12}^{-1} T_9^{-1/2}. \tag{62}$$

Equation (62) shows that, if a strong magnetic field is present, cyclotron scattering is by far the dominant source of opacity at the cyclotron fundamental. More generally, the ratio of the two opacities is[67]

$$\alpha_\omega^{cyc}/\alpha_T = \frac{3}{2} \alpha^{-1} \frac{m_e c^2}{\hbar \omega_c} \Phi(\omega) \simeq 9.1 \times 10^3 B_{12}^{-1} \Phi(\omega), \tag{63}$$

where the dimensionless size parameter Λ and the dimensionless flux $\Phi(\omega)$ are those defined in equations (52-54). Figure 6 shows that even for electron temperatures $T \simeq 3 \times 10^9$ K, the cyclotron emissivity, and therefore the absorptivity and scattering, decrease rapidly with frequency and, as a result, electron scattering may still dominate at sufficiently high frequencies. Figure 6 can be used, together with equation (63) to determine the frequency at which the two opacities become equal. For example, given an electron temperature $T \simeq 100$ keV, equation (63) shows that the two opacities are equal when $\Phi(\omega) \simeq 10^{-4}$; Figure 6 then shows that this occurs at $\omega/\omega_c \simeq 20$ or, assuming $B = 1 \times 10^{12}$ G, at $\simeq 200$ keV.

Earlier we derived a characteristic maximum number density $n \simeq 10^{26}$ for the emission region of a gamma-ray burst (equation 28). Comparing that result and the constraints represented by equations (59) and (61), we conclude that the emission regions of gamma-ray bursts are either extremely optically thick, or that energy and/or matter is continuously replenished during the burst.

2. Evidence from observation

A number of observations suggest that the emission region of gamma-ray bursts is optically thin. For example, the most detailed burst spectrum obtained to date, that for the Apollo 16 event, is surprisingly well fit by a single-temperature ($T_{obs} = 500$ keV) thin bremsstrahlung spectrum with Gaunt factor over three decades in photon energy.[62] Of course, the fit did not include the contribution of electron-electron bremsstrahlung and relativistic corrections,[59] both of which are important at these energies, or effects of a strong magnetic field.[9,53] Nevertheless, it is remarkable that the fit is so good.

A second, and perhaps even more important, example is the pair annihilation line seen in the March 5, 1979 event[13] and in other gamma-ray bursts.[4,21,22] The recoil from a single Compton scattering of a 511 keV photon is sufficient to scatter it out of the line ($\Delta E/E \simeq \hbar\omega/m_e c^2$), so that observation of a line requires $\tau_{es} < 1$.

Evidence for modest optical depths in the emission regions of some sources is provided by the low energy absorption features that have been observed in a number of burst spectra obtained by Mazets and his colleagues.[4,21,22] These features have been widely interpreted as absorption at the cyclotron fundamental and have been used to infer magnetic field strengths. Instead, the features are likely the result of time-averaging spectra that have a rapidly varying low energy cutoff. For example, the Franco-Soviet Signe experiment on Venera 12, which allows spectra to be measured with finer time resolution than the 4 second resolution of the Konus experiments, shows that the spectrum and the low energy cutoff of the November 19, 1978 event evolve rapidly in time. Initially, the low energy cutoff occurs at $\simeq 500$ keV but decreases rapidly to less than $\simeq 100$ keV within about 4 seconds.[3,68] If our interpretation is correct, the low energy cutoff may reflect continuum absorption or, more likely, scattering, and may then not allow the magnetic field strength to be inferred directly.

3. Constraints on source aspect ratios and distances

If we assume that the burst emission region is, in fact, optically thin, then we may derive constraints on the number density in the emission region, and on the aspect ratio and distance of the source. The emissivity of the emitting plasma must satisfy

$$L = \Lambda_{br(pair,cyc)} V, \qquad (64)$$

where $L \simeq 1 \times 10^{37}$ $(d/100 \text{ pc})^2$ ergs s^{-1} is the observed luminosity of a burst, and V is the volume of the emission region. If we parametrize V by the expression (27) and assume that bremsstrahlung is the dominant emission mechanism, then the number density in the emission region must satisfy

$$n_e \simeq 1.3 \times 10^{20} \, Z^{-1/2} \, (d/100 \text{ pc}) \, (f \, l_6)^{-1} \, T_9^{-1/4} \text{ cm}^{-3}. \tag{65}$$

Similarly, if instead we assume that pair annihilation or cyclotron emission is the dominant emission mechanism, the constraint on the number density becomes

$$n_e \simeq 8.1 \times 10^{18} \, (d/100 \text{ pc}) \, (f \, l_6)^{-1/2} \text{ cm}^{-3}, \tag{66}$$

or

$$n_e \simeq 1.5 \times 10^9 \, (d/100 \text{ pc})^2 \, (f \, l_6)^{-1} \, B_{12}^{-2} \, T_9^{-1} \text{ cm}^{-3}. \tag{67}$$

Equation (67) defines a characteristic number density that is a factor 10^{16} smaller than that given by equation (28)! Its small size reflects the high efficiency of cyclotron emission in a strong magnetic field.

By combining the above bounds on the number density and the expressions for either electron or cyclotron scattering (equations 58 and 60), as appropriate, we can obtain constraints on the aspect ratio and/or distance of the source. Katz[9] gives a clear and comprehensive discussion of these constraints, assuming bremsstrahlung or pair annihilation. Here we give them in a slightly different form and consider also cyclotron emission. Defining the aspect ratio of the burst emission region as

$$\alpha \equiv \frac{l}{fR} = \frac{l_6}{fR_6} = l_6/f, \tag{68}$$

where in the last step we take, as usual, $R = 10^6$, we obtain the constraint,

$$\alpha^{1/2} \, (d/100 \text{ pc}) \lesssim 2.6 \times 10^{-2} \, Z^{1/2} \, T_9^{1/4}. \tag{69}$$

Equation (69) implies that, if the source lies at a distance $\simeq 100$ pc and emits via bremsstrahlung, the burst emission region must have $\alpha \lesssim 6.8 \times 10^{-4}$; that is, an aspect ratio greater than $1:10^3$. (For a source at a distance of 55 kpc, as would be the case for the March 5, 1979 event if it originates in the Large Magellanic Cloud,[10,17] the implied aspect ratio is nearly $1:10^7$!) Similar constraints can be derived from the luminosity in the observed pair annihilation lines,

$$\alpha^{1/2} \, (d/100 \text{ pc}) \lesssim 0.43, \tag{70}$$

or assuming cyclotron emission,

$$f^{-1/2} \, (d/100 \text{ pc}) \lesssim 3.8 \times 10^2 \, B_{12}^{3/2} \, T_9^{3/4}. \tag{71}$$

Equation (71) shows that, if the burst is produced by cyclotron emission, an emission region covering only a fraction $f \simeq 7.0 \times 10^{-6}$ of the stellar surface would suffice. This is again a reflection of the high efficiency of cyclotron emission.

SOME IMPORTANT ISSUES

In concluding, we mention several important issues that any theory of gamma-ray bursts must address.

1. What determines the characteristic duration of a burst? (Is there one?)

Observations tell us that the durations of gamma-ray bursts are as short as 0.1 sec and as long as 100 sec.[2,3,61] In contrast, the radiative cooling time scales for bremsstrahlung and cyclotron emission are much shorter (for example, the time scale for cyclotron cooling can be as short as $\simeq 10^{-16}$ sec; see equation 43). Even were the emission region very optically thick, the observed intensity variations on a time scale 0.1 sec rule out an explanation for the total duration of the burst in terms of a diffusion time scale. Thus, the duration of gamma-ray bursts does not arise from the cooling time scale of the hot plasma, whether in a strong magnetic field or not. Rather, a separate explanation is needed. This issue has been faced in different ways, perhaps not fully satisfactorily, by many of the proposed models of gamma-ray bursts. For example, accretion models[24,29,32] commonly push the explanation back to an earlier stage by attributing the duration of bursts to the interval during which accreting plasma or solid cometary matter rains down on the star. Nuclear burning models[33,34] invoke the time necessary for a deflagration wave to propagate across the lake of nuclear fuel, or the interval during which nearby fuel sites successively ignite. Models that rely on an internal source of energy to power the burst invoke some other time scale, such as the gravitational damping time of neutron star oscillations.[11,37] The situation is not as tidy as it would be, were the burst durations able to be related naturally to a cooling time scale.

2. What determines the characteristic spectral temperature of a burst? (Is there one?)

Observations tell us that the spectral temperatures of gamma-ray bursts range from $T_{obs} \simeq 20$ keV, as in the March 5, 1979 event[13], to several MeV.[2-4,21-23] The characteristic maximum temperature of radiation-dominated plasma confined in a magnetic field of a few times 10^{12} G magnetic field (equation 4) may indeed explain the soft gamma-ray nature of the bursts. This temperature can vary, depending on the strength of the magnetic field and the extent to which the expanding bubble of hot plasma compresses the field, yet it is not clear that this relation can give a temperature range as large as that observed. In addition, a complete model of the bursts must be able to answer the important question: why are the observed spectral temperatures of bursts so much greater than those of the "steady" pulsating X-ray sources, whose emission is also due to the radiation from hot plasma at the surface of strongly magnetic neutron stars? Is the difference due to the impulsive nature of the event, or because the emission is optically thin?

3. Is the radiating region optically thick or thin?

As we have seen, observations suggest that the spectra of gamma-ray bursts are produced by optically thin emission, whether bremsstrahlung, pair annihilation radiation, or cyclotron emission. The discussion in the last section emphasizes that if the emitting region is optically thin, the energy of the emitting plasma and/or the plasma itself must be continuously replenished during a burst. In accretion models,[24,29-32] it could occur through accretion of new matter. In models that rely on nuclear explosions,[33,34] the replenishment might occur, for example, via injection into the magnetosphere of fresh material from successive burning sites.

4. Are all bursts produced by the "same" mechanism?

Occum's razor instructs us to assume that all gamma-ray bursts have a single explanation. However, both their durations and their spectral temperatures encompass a dynamic range of more than 10^3. These ranges are large, even by astrophysical standards, and they raise an important question: are all gamma-ray bursts produced by the "same" mechanism, or are such bursts the manifestations of several transient phenomena occurring in, on, and around neutron stars?

We are grateful to Roger Bussard, Stirling Colgate, Reuven Ramaty, and, especially, Fred Lamb for valuable discussions on various aspects of the physics of gamma-ray bursts.

REFERENCES

1. M. Ruderman, Ann. N. Y. Acad. Sci. 262, 164 (1975).
2. K. Hurley, Adv. Space Explor. 7, ed. R. Cowsik and R. D. Wills (Pergamon Press, Oxford, 1980) p. 123.
3. G. Vedrenne, Phil. Trans. R. Soc. Lond. A 301, 645 (1981).
4. E. P. Mazets and S. V. Golenetskii, Astrophys. Space Sci. (1981).
5. G. Baym and C. J. Pethick, Ann. Rev. Nuc. Sci. 25, 27 (1975).
6. G. Baym and C. J. Pethick, Ann. Rev. Astron. Astrophys. 17, 415 (1979).
7. F. K. Lamb, Ann. N. Y. Acad. Sci. 302, 482 (1977).
8. F. K. Lamb, in Magnetospheric Boundary Layers, Proceedings Sydney Chapman Conference in Alpbach, Austria, Ed. B. Battrick, ESA SP-148 Addendum, 1, 1979.
9. J. I. Katz, Astrophys. J., in press (1981).
10. T. Cline, these proceedings.
11. R. Ramaty, R. E. Lingenfelter, and R. W. Bussard, Astrophys. and Space Sci. 75, 193 (1981).
12. R. Hoshi, Astrophys. J., in press (1981).
13. E. P. Mazets, S. V. Golenetskii, V. N. Il'inskii, R. L. Aptekar', and Yu. A. Guryan, Nature 282, 587 (1979).
14. T. L. Cline, U. D. Desai, G. Pizzichini, B. J. Teegarden, W. D. Evans, R. W. Klebesadel, J. G. Laros, K. Hurley, M. Niel, G. Vedrenne, I. V. Estoolin, A. V. Kouznetsov, V. M. Zenchenko, D. Hovestadt, and G. Gloecker, Astrophys. J. Lett. 237, L1 (1980).
15. C. Barat, G. Chambon, K. Hurley, M. Niel, G. Vedrenne, I. V. Estulin, A. V. Kuznetsov, and V. M Zenchenko, Astron. Astrophys. 79, L24 (1979).
16. J. Terrell, W. D. Evans, R. W. Klebesadel, and J. G. Laros, Nature 285, 383 (1980).
17. W. D. Evans, R. W. Klebesadel, J. G. Laros, T. L. Cline, U. D. Desai, G. Pizzichini, B. J. Teegarden, K. Hurley, M. Niel, G. Vedrenne, I. V. Estoolin, A. V. Kouznetsov, V. M. Zenchenko, and V. G. Kurt, Astrophys. J. Lett. 237, L7 (1980).
18. E. P. Mazets, S. V. Golenetskii, V. N. Il'inskii, V. N. Panov, R. L. Aptekar', Yu. A. Gur'yan, I. A. Sokolov, Z. Ya. Sokolova, and T. V. Kharitonova, "Catalogue of Cosmic Gamma-Ray Bursts from the Konus Experiment Data, Vol. I",

preprint Fiz.-Tekh. Inst. Ioffe Akad. Nauk SSSR No. 618 (1979).
19. E. P. Mazets, S. V. Golenetskii, V. N. Il'inskii, V. N. Panov, R. L. Aptekar', Yu. A. Gur'yan, I. A. Sokolov, Z. Ya. Sokolova, and T. V. Kharitonova, "Catalogue of Cosmic Gamma-Ray Bursts from the Konus Experiment Data, Vol. II", preprint Fiz.-Tekh. Inst. Ioffe Akad. Nauk SSSR No. 637 (1979).
20. E. P. Mazets, S. V. Golenetskii, V. N. Il'inskii, V. N. Panov, R. L. Aptekar', Yu. A. Gur'yan, A. V. Dyatchkov, M. P. Proskura, I. A. Sokolov, Z. Ya. Sokolova, N. G. Khavenson, and T. V. Kharitonova, "Catalogue of Cosmic Gamma-Ray Bursts from the Konus Experiment Data, Vol. III", preprint Fiz.-Tekh. Inst. Ioffe Akad. Nauk SSSR No. 662 (1980).
21. E. P. Mazets, S. V. Golenetskii, R. L. Aptekar', Yu. A. Gur'yan, and V. N. Il'inskii, "Cyclotron and Annihilation Lines in Gamma-Ray Bursts", preprint Fiz.-Tekh. Inst. Ioffe Akad. Nauk SSSR No. 687 (1980).
22. E. P. Mazets, S. V. Golenetskii, V. N. Il'inskii, Yu. A. Gur'yan, R. L. Aptekar', V. N. Panov, I. A. Sokolov, Z. Ya. Sokolova, and T. V. Kharitonova, "Cosmic Gamma-Ray Burst Spectroscopy", preprint Fiz.-Tekh. Inst. Ioffe Akad. Nauk SSSR No. 719, submitted to Astrophys. Space Sci. (1981).
23. R. E. Lingenfelter, these proceedings (1981).
24. S. A. Colgate and A. G. Petschek, Astrophys. J., in press (1981).
25. M. Ruderman, Prog. Part. Nucl. Phys., in press (1981).
26. G. Cavallo and M. J. Rees, Mon. Not. R. Astron. Soc. 183, 359 (1978).
27. M. J. Newman, and A. N. Cox, Astrophys. J. 242, 319 (1980).
28. W. M. Howard, J. R. Wilson, and R. T. Barton, Astrophys. J., in press (1981).
29. D. Q. Lamb, F. K. Lamb, and D. Pines, Nature Phys. Sci. 246, 52 (1973).
30. R. Hoshi, Prog. Theoret. Phys., 56, 542 (1976).
31. F. Takahara and R. Hoshi, Prog. Theoret. Phys. 59, 425 (1978).
32. M. Harwit and E. E. Salpeter, Astrophys. J. Lett. 187, L97 (1973).
33. S. E. Woosley and R. E. Taam, Nature 262, 101 (1976).
34. S. E. Woosley and R. K. Wallace, Astrophys. J., in press (1981).
35. A. C. Fabian, V. Icke, and J. E. Pringle, Astrophys. Space Sci. 42, 77 (1976).
36. G. S. Bisnovatyi-Kogan and V. M. Chechetkin, preprint, Inst. for Space Res. Acad. Nauk SSSR No. 561 (1980).
37. A. I. Tsygan, Astrophys. Space Sci. 44, 21; 49, 159 (1975).
38. F. Pacini and M. Ruderman, Nature 251, 399 (1974).
39. S. E. Woosley, these proceedings.
40. R. C. Malone, Ph.D. Thesis, Cornell University (1974).
41. J. P. Hansen, Phys. Rev. A 8, 3096 (1973).
42. E. L. Pollock and J. P. Hansen, Phys. Rev. A 8, 3110 (1973).
43. D. Q. Lamb and H. M. Van Horn, Astrophys. J. 200, 306 (1975).
44. G. Baym, C. J. Pethick, and P. G. Sutherland, Astrophys. J. 170, 299 (1971).
45. G. Baym, H. A. Bethe, and C. J. Pethick, Nucl. Phys. A175, 225 (1971).
46. M. Ruderman, in Physics of Dense Matter, IAU Symp. No. 53, ed. C. J. Hansen, (Reidel, NY, 1974), p. 117.
47. E. G. Flowers, J.-F. Lee, M. A. Ruderman, and P. G. Sutherland, Astrophys. J. 215, 291 (1977).
48. M. Ruderman, Phys. Rev. Lett. 27, 1306 (1971).
49. Chen, H.-H., M. A. Ruderman, and P. G. Sutherland, Astrophys. J. 191, 472 (1974).
50. W. Hillebrandt and E. Muller, Astrophys. J. 207, 589 (1976).
51. E. G. Flowers and M. A. Ruderman, Astrophys. J. 205, 541 (1976).
52. J. E. McClintock, S. Rappaport, P. C. Joss, H. Bradt, J. Buff, G. W. Clark, D.

Hearn, W. H. G. Lewin, T. Matilsky, W. Mayer, and F. Primini, Astrophys. J. Lett. 206, L99.
53. R. W. Bussard and F. K. Lamb, these proceedings (1981).
54. R. Ramaty, these proceedings (1981).
55. H.-Y. Chiu, Stellar Physics (Blaisdell, Waltham, 1968).
56. J. Arons and S. M. Lea, Astrophys. J., 207, 914 (1976).
57. R. F. Elsner and F. K. Lamb, Astrophys. J., in press (1981).
58. C. W. Allen, Astrophysical Quantities (Athlone Press, London, 3rd ed., 1973) p. 103.
59. R. J. Gould, Ap. J. 238, 1026 (1980).
60. B. A. Trubnikov, dissertation, Moscow, USAEC Technical Information Service AEC-tr-4073 (1960).
61. R. W. Klebesadel, these proceedings (1981).
62. D. Gilman, A. E. Metzger, R. H. Parker, L. G. Evans, and J. I. Trombka, Astrophys. J. 236, 951 (1980).
63. R. Ramaty and P. Mészáros, Astrophys. J., in press (1981).
64. J. L. Hirshfield, D. E. Baldwin, and S. C. Brown, Phys. Fluids 4, 198 (1961).
65. D. Q. Lamb and A. R. Masters, Astrophys. J. Lett. 229, 652 (1979).
66. N. D. Kylafis, D. Q. Lamb, A. R. Masters, and G. J. Weast, Ann. N. Y. Acad. Sci. 336. 520 (1980).
67. A. R. Masters, Ph.D. Thesis, University of Illinois (1978).
68. F. K. Knight, J. L. Matteson, and L. E. Peterson, Astrophys. Space Sci. 75, (1981).

THE THERMONUCLEAR MODEL FOR γ-RAY BURSTS

S. E. Woosley
Lick Observatory, Board of Studies in Astronomy
and Astrophysics, University of California,
Santa Cruz, CA 95064
and
Lawrence Livermore National Laboratory
Livermore, CA 94550

ABSTRACT

The evolution of magnetized neutron stars with field strengths of $\sim 10^{12}$ gauss that are accreting mass onto kilometer-sized polar regions at a rate of $\sim 10^{-13}$ $M_\odot$ yr^{-1} is examined. Based on the results of one-dimensional calculations, one finds that stable hydrogen burning, mediated by the "hot" CNO-cycle, will lead to a critical helium mass in the range 10^{20} to 10^{22} g km^{-2}. Owing to the extreme degeneracy of the electron gas providing pressure support, helium burning occurs as a violent thermonuclear runaway which may propagate either as a convective deflagration (Type I burst) or as a detonation wave (Type II burst). Complete combustion of helium into ^{56}Ni releases from 10^{38} to 10^{40} erg km^{-2} and pushes hot plasma with $\beta > 1$ above the surface of the neutron star. Rapid expansion of the plasma channels a substantial fraction of the explosion energy into magnetic field stress. Spectral properties are expected to be complex with emission from both thermal and non-thermal processes. The hard γ-outburst of several seconds softens as the event proceeds and is followed by a period, typically of several minutes duration, of softer x-ray emission as the subsurface ashes of the thermonuclear explosion cool. In this model, most γ-ray bursts currently being observed are located at a distance of several hundred parsecs and should recur on a timescale of months to centuries with convective deflagrations (Type I bursts) being the more common variety. An explanation for "Jacobson-like" transients is also offered.

INTRODUCTION

Although the theoretical community has historically been far from unanimity in its views concerning the site and specific mechanism(s) involved in the production of γ-ray bursts, observational studies, in particular recent spectroscopic examinations[1], have led to concentrated effort on models whose common underlying theme is a magnetized neutron star. The attractive nature of a neutron star site has long been recognized. The 10^{39} erg needed for a typical γ-ray burst at a distance of, say, 1 kpc, is easily realized by the accretion of only $\sim 10^{19}$ g of material or the thermonuclear burning of $\sim 10^{21}$ g of helium. These small amounts of material can accumulate in a reasonable time and at a reasonable rate and allow the possibility of recurrent outbursts from the same neutron star. Since the total number of neutron stars in the Galaxy is only ~10^9, such repetition is required in order to provide the more than 150 bursts per year already being observed[1] and presumably maintained over the 10^{10} yr lifetime of the Galaxy. The small size of a neutron star is also in good accord with the fine time structure (< 0.1 s) seen in many bursts which limits the dimension of the emitting region to no larger than 3×10^9 cm. As has been emphasized in recent papers[2,3,4], the hardness of the γ-ray burst spectrum requires the confining presence of a strong magnetic field in order to keep the super-Eddington luminosity from blowing the emitting plasma out to great distance where it cools to low temperature. If this occurs, the soft thermal radiation (~ 1 keV) would bear little resemblance to the 200 keV to 500 keV radiation customarily seen in γ-ray bursts (see also Wallace, Woosley, and Weaver[5] and Howard, Wilson, and Barton[6]). In order to magnetically confine 200 keV plasma, one needs $B^2/8\pi >$ aT^4, or a field strength in excess of ~ 2×10^{12} gauss. While this simple argument assumes that all of the hard radiation in γ-ray bursts is produced thermally, which almost certainly is not the case, a field near this strength is still required in models that invoke non-thermal processes. Such strong fields are only expected on neutron stars and it is reassuring to find that spectral line features[1] in γ-ray bursts (attributed to cyclotron absorption and gravitationally red-shifted pair annihilation) seem to confirm this hypothesis. Finally, in the case of the March 5, 1979 event and, with less certainty, the event of October 29, 1977, one has evidence for periodicity in the event (8.0 s and 4.2 s respectively) that is to be associated with the rotation (or perhaps precession) period of an old neutron star.

If γ-ray bursts do in fact originate from magnetized neutron stars, two basic energizing mechanisms can be involved, gravitational collapse and/or thermonuclear explosion. In the case of a gravitational power source, at least three distinct scenarios have been discussed. These include neutron star quakes[7] which energize surface oscillations and produce hot plasma by magnetic interactions[3]; sudden accretion of plasma as a result of a flaring companion star[8] or magnetospheric instabilities[9]; and the accretion of a comet or asteroid[2,6,10,11,12]. Difficulties of varying degree beset each of these models. The underlying cause of a hypothetical neutron star "super-quake" and its efficiency for coupling to large amplitude surface oscillations has yet to be addressed in the Ramaty et al model[7]. Furthermore, if this alleged instability occurs only several times during the life of a neutron star, the requisite number of common γ-ray bursts cannot be provided (although this model might still function adequately for such special events as that of March 5, 1979). One must also wonder why γ-ray bursts are not customarily associated with known pulsar "glitches". Accretion from a flaring companion star does not appear likely to explain the rapid rise time and short duration of such γ-ray bursts as that of June 13, 1979 (ref. 13).

Magnetospheric gating is an attractive hypothesis, not only for γ-ray bursts, but also for a host of other high energy transient phenomena, including Type II x-ray bursts. Unfortunately, the nature of the magnetic instability and a quantitative description of its operation remain to be developed. Similarly, while accretion of a solid object may be a viable γ-ray burst model, the interaction of objects incident with non-zero impact parameter and the frequency of asteroid/neutron star collisions remain great uncertainties (although see ref. 14).

An alternative class of models relies on the sudden release of energy by a thermonuclear explosion[4,15,16,17,18,19] near the magnetic polar cap of a neutron star and is the principal subject of the remainder of this paper. Such models invoke the nuclear combustion, either by detonation wave or convective deflagration, of a highly degenerate layer of helium or carbon in a "nova-like" outburst. The fuel reservoir is a result of slow mass accretion ($\sim 10^{-13}$ $M_\odot$ yr^{-1} km^{-2}), either from a binary companion star or in a supernova remnant or interstellar cloud, and the outburst repeats each time a critical mass of fuel is accumulated. Magnetic confinement of the explosively heated ashes is an essential ingredient. In the next section the results of one-dimensional numerical simulations[5] of thermonuclear explosions on slowly accreting neutron stars are briefly reviewed. Without magnetic confinement, little hard radiation is produced and the observable consequence is a rapid x-ray transient, not a γ-ray burst. The third section then deals with modifications resulting from the introduction of a strong magnetic field ($\sim 4 \times 10^{12}$ gauss). Further observable properties of the γ-ray burst model, as well as instrumental requirements for its proper study, are discussed in the conclusions.

ONE DIMENSIONAL CALCULATIONS

In a recent paper[5], Wallace, Woosley, and Weaver have numerically simulated (non-magnetized) neutron stars accreting compositions of various metallicities at rates in the range 10^{-11} to 10^{-9} $M_\odot$ yr^{-1}. A standard neutron star of radius 14.3 km and mass 1.41 $M_\odot$ was employed in all calculations. For the assumed parametrizations, hydrogen burning in the accreted material is mediated by the β-limited CNO cycle[20] and is therefore stable. Thick helium shells in the mass range 10^{23} to 10^{25} g accumulate, ignite under highly degenerate circumstances, and explode. For specific illustration, I shall consider here only 2 models, called I and II, which exemplify the two qualitatively distinct modes of outburst expected for accretion rates in this range. The numerical results[5] from these two models are summarized in Table 1 which gives the total accretion rate; mass, density, temperature, and thickness of the hydrogen and helium shells; total energy output in electromagnetic radiation and in neutrinos; peak temperature at the base of

TABLE 1

NUMERICAL MODEL CHARACTERISTICS

	Model I	Model II
$\dot{M}$ ($M_\odot$ yr^{-1})	3.5(-10)	2.1(-11)
M_H (g)	8.0(20)	2.4(22)
ρ_H (g cm^{-3})	1.0(5)	1.1(6)
T_H (K)	1.2(8)	7.5(7)
Δ_H (m)	10	23
M_{He} (g)	2.7(23)	1.0(25)
ρ_{He} (g cm^{-3})	9.6(6)	1.2(8)
Δ_{He} (m)	36	73
T_{9peak} (10^9 K)	3.2	6.0
E_γ (erg)	1.3(41)	1.9(42)
E (erg)	3.9(38)	1.5(43)
τ_{rise} (ms)	~ 2	> 1
τ_L (s)	320	6500
τ_{rec} (yr)	0.4	240
$\dot{m}$ (g s^{-1})	5(17)	1(18)
ΔM (g)	2(20)	6(21)
$\Delta M v_{esc}^2/E_{tot}$ (%)	20	5

the burning helium shell; rise time; duration of Eddington luminosity; recurrence time; mass loss rate; total mass lost; and total energy expended in driving mass loss. The low value of metallicity used in Model II is not critical to the results that follow. Other combinations of (more nearly solar) metallicity and (slightly larger) accretion rate can give similar behavior[5]. Owing to the high conductivity of degenerate matter and the thinness of the helium shell, it is assumed that the ignition temperature of the helium shell is set by the steady state burning temperature of hydrogen.

Aside from strictly quantitative differences resulting from its slower accretion rate and cooler hydrogen shell, Model II differs categorically from Model I in the manner of its burning. The runaway propagates as a <u>detonation wave</u> that, with a velocity of roughly 9000 km s^{-1} (Mach 1.5), sweeps around the entire neutron star in about 5 ms (ref. 16), igniting the entire helium layer as it goes. Since the propagation time for this wave to reach the surface from the base of the helium layer is far shorter than this (7.2 µs) and since the surface luminosity rises to near its peak (Eddington) value in only about 1 ms (as a result of the gravitational energy released as the suddenly expanded material falls back to the neutron star surface), this implies a rise time of ~2.5 ms for Model II.

The ignition of the helium layer in Model I, on the other hand, must wait for a convective burning front to move around the star, an occurrence that takes much longer than the sonic transit time. The 1 ms rise time given in Table 1 for Model I is a purely formal result of the one-dimensional calculation and is a gross lower bound. Calculation of a realistic ignition time for the entire star in Model I is a difficult problem since it depends on the rate of convective entrainment of surrounding unburned helium into the burning front. Ignition will proceed as a wave (like a burning fuse or candle) and will be faster than simple conduction. Providing that it does not exceed the 320 s formal (1D) duration of the event, the ignition timescale will once again give the rise time for the transient which could, in principle, be quite long.

In both cases the sudden burning of large amounts of helium to iron-group elements (chiefly ^{56}Ni) leads to an enormously super-Eddington luminosity at the base of the neutron star atmosphere. When we originally carried out these calculations, it was hoped that this large luminosity would translate into a high effective emission temperature, i.e., a γ-ray burst. What we found instead, obvious in retrospect, is that the excess radiation pressure blows the envelope out to a fairly large distance (50 km to 100 km) thereby storing its excess energy in the form of a gravitational "battery". At this large radius the super-heated neutron star emits soft radiation (~1 keV) at a rate that, except during the first millisecond (free-fall timescale), differs very little from the Eddington value. As the event proceeds the stored gravitational potential energy is gradually converted back into heat and radiated away. This energy input, along with continued helium burning, especially in Model II, gives an enduring x-ray transient with properties given in Table 1. During the time that the photosphere is receding back to the "surface" of the neutron star at near constant luminosity, one sees first a hardening of the emission spectrum followed by a softening as the surface of the neutron star cools in the tail of the event. This spectral behavior should be evident in all x-ray transients operating at or above the Eddington luminosity.

An interesting side effect of the enduring and slightly super-Eddington luminosity is the existence of a strong stellar wind of roughly 10^{18} g s^{-1}. Since the radiatively driven material comes off at about the escape velocity of the neutron star, ~.5 c, it consists of very energetic electrons and protons. Indeed, from 5% (Model II) to 20% (Model I) of the total explosion energy goes into driving this mass loss (total mass lost is ΔM in Table 1). Interaction of these energetic particles with either a surrounding (weak) magnetosphere or other plasma in the vicinity (the flow itself is likely to be Rayleigh-Taylor unstable) may lead to hard non-thermal radiation accompanying these soft x-ray transients. This is particularly interesting in light of "Jacobson-like" γ-line transients[21]. Roughly 10^{40} to 10^{42} erg would be produced here in the form of high energy (E > 200 MeV/nucleon) protons and α-particles and the event could easily last as long as the 20 minute observed duration of the June 10, 1974 event. The gravitational red-shift, however, would be no larger than at the photosphere (~ 50 km) or less than about 10%, considerably smaller than at the surface of a cold neutron star. If this explanation is correct, then Jacobson et al saw the high energy portion of a rapid x-ray transient, due primarily to inelastic nuclear scattering and spallation, and missed the softer emission due to the limited response of the detector. A search for γ-ray lines in the spectra of other known x-ray transients would be interesting. Unless the lines originate in a region far above the photosphere where the gravitational field is small or else in a very thin region just above the photosphere it is difficult to understand the narrow line widths. Differential gravitational broadening would be expected to give a width equal to some substantial fraction of the red-shift.

EFFECTS OF A STRONG MAGNETIC FIELD

Now consider the modifications to the above picture resulting from the introduction of a magnetic field of roughly 4 x 10^{12} gauss, consistent with that inferred from the observed cyclotron features. First, accretion will no longer be spherically symmetric but will be focused onto the magnetic polar caps of the neutron star[22,23,24]. The area and geometry of the accreted layer is highly uncertain but I shall adopt a circular region of radius 1 km both for illustration and consistency with the current literature. As was pointed out by Woosley and Taam[15], the principal effects of this focused accretion are a smaller critical mass, explosion energy, and requisite accretion rate for given hydrogen and helium shell parameters, explosion characteristics, and recurrence timescale. The accretion rates for the geometrically modified Models I and II now become 4.3×10^{-13} $M_\odot$ yr^{-1} and 2.5×10^{-14} $M_\odot$ yr^{-1} respectively for accretion on π km^2. These accretion rates give rise to a steady and potentially observable soft x-ray luminosity of 8×10^{32} $\dot{M}_{13}$ erg s^{-1} with $\dot{M}_{13}$ the mass accretion rate in units of 10^{-13} $M_\odot$ yr^{-1}. The emission temperature would be around 5×10^6 K. It is interesting that within the locational error box for the event of November 19, 1978 there is a weak x-ray source[26]. The inferred luminosity, roughly 10^{32} erg s^{-1} at 1 kpc, is consistent with the thermonuclear model although perhaps a little low (depending on the actual distance to the event, the effect of anisotropic emission on the steady x-ray flux, and whether the November 19, 1978 event corresponded to Model I or Model II). Substantially lower limits on x-ray emission, which should be continuing even at the present time, from other γ-ray burst error boxes could greatly constrain, and even rule out, the thermonuclear model for specific events.

Following the anisotropic accretion of matter onto polar caps, it is important to know whether the accreted material remains localized or spreads over the neutron star surface. The heat generated by accretion and hydrogen burning, once it has ignited, melts the neutron star crust in the vicinity of the accretion so that it sags beneath the weight of the accumulated material. Were the crust and accreted material of the same composition then the accreted layer would not project substantially above the surrounding neutron star surface, but because hydrogen and helium have electron mole numbers that differ from the surrounding iron (^{56}Fe) substrate, the surface of the accreted layer floats, in hydrostatic equilibrium, several meters above the surrounding iron. Were it not for the necessity of crossing strong magnetic field lines the accreted material would flow some considerable distance, ultimately covering the entire star. It must be realized, however, that the change in altitude of the neutron star surface does not occur abruptly at any point. Rather there is a lateral pressure gradient and this lateral gradient, if small enough, can be maintained by strong magnetic field lines pinned to the deep interior of the neutron star. Diffusion across the field does occur, but on a very long timescale[4], typically thousands of years for a field strength of 4×10^{12} gauss, and is negligible given the short time spent by the accreted matter in that part of the hydrogen shell above the median neutron star radius. In Model I this distance (which may be computed from Table 1) is 6.7 m and the time spent there by accreted matter is 5 hours. In Model II the corresponding numbers are 16 m and 47 days.

Although it is not quite so obvious, accreted material is in even greater jeopardy of "blowing out" near its base than it is of spreading over the surface (Fig. 1). This is because the hydrogen and helium layers have pressure scale heights that are different from iron and therefore experience a different pressure at a given radius. The absolute value of the pressure difference is greatest about one pressure scale height above the base of the helium layer, far below the surface of the neutron star. It turns out once again that magnetic field lines can inhibit the subsurface spreading in Model I (if one neglects plasma instabilities), but prove inadequate in Model II (ref. 4). Helium in Model II is therefore expected to break out and spread some distance away from the locality of the accretion. This break-out may also occur in Model I if the interchange instability allows material to squeeze between magnetic field lines. Once the material has spread over an additional kilometer or so, however, the temperature of the iron/helium mixture falls below the freezing point and the strong shear modulus of the neutron star crust^{26}, about 10^{24} dyne cm^{-2}, comes into play. This solidification prevents further spreading when the accreted material still covers a small fraction of the neutron star surface^4. Subsurface spreading does, however, introduce an additional degree of freedom into the models in that the region that eventually explodes may be considerably larger than the area of accretion. This gives rise to larger critical masses, greater explosion energies, and longer recurrence timescales than would occur without spreading. Providing that only a small fraction of the neutron star surface is covered, a γ-ray burst is still possible. Indeed, in Model I, the slow speed of the convective ignition wave would allow a γ-ray burst to occur as a precursor to an x-ray transient^4.

Multiplying the critical masses in Table 1 by the ratio of accretion area to neutron star area (1.2×10^{-3} in the cases under discussion) gives the masses involved in the γ-ray outburst. Once these helium masses have accumulated (the recurrence timescale should be little modified from Table 1), an explosion ensues. In Model I the explosion proceeds by convective deflagration, consuming $\sim 3 \times 10^{20}$ g of helium and releasing $\sim 5 \times 10^{38}$ erg. Once again, because of the difficulty in estimating the lateral propagation velocity of the burning front, both the rise time and duration of the outburst cannot be easily estimated for Model I. An upper bound on the ignition timescale is the time required for thermal diffusion to go 1 km at the base of the helium layer, or a few hundred seconds^4. It seems reasonable to associate this time, which could realistically be on the order of 10 s, with the duration of such "long and irregular"^{13} bursts as the April 27, 1972 (Apollo) event. The irregular variations in intensity seen during such bursts would be attributed to "sputtering" of the burning wave (due perhaps to an uneven distribution of fuel on the surface of the neutron star) and to magnetic instabilities in the plasma pushed above the surface of the star by the explosion (see below). The rise time for such events would also be uncertain. Some emission would commence as soon as the convective burning front came

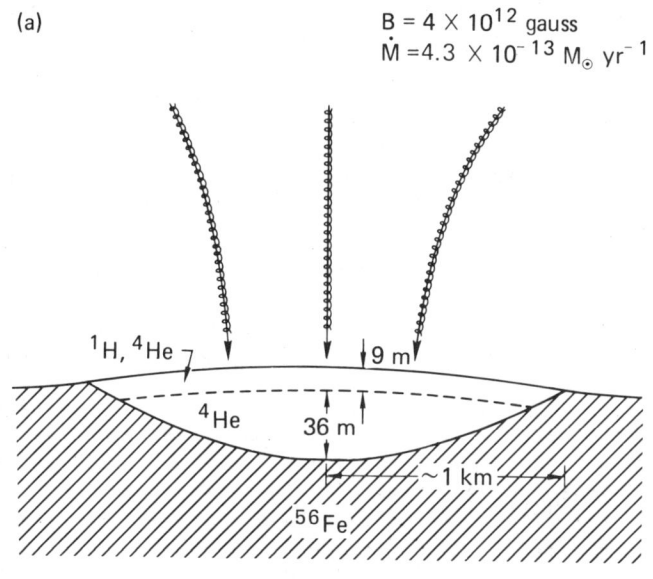

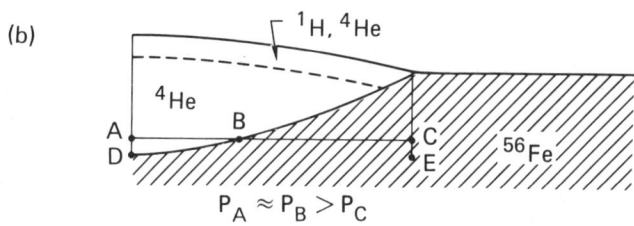

Figure 1. Magnetically focused accretion onto a neutron star gives rise to a blister of hydrogen and helium having dimensions roughly as shown in (a) for Model I. Curvature of the blister surface and interface have been exaggerated for purposes of illustration. In (b) the iron surface of the neutron star has been approximated by a plane and the curvature of the blister again exaggerated. Pressures at equal depths in the helium layer (points A and B) are very nearly equal, a small difference arising due to the varying thickness of the blister. The pressure at an equivalent depth in the iron substrate, however, is substantially less due to its smaller pressure scale height. The pressure gradient from point B to point C must be balanced by magnetic and crustal shear forces. The pressure at points D and E are equal.

close to the surface (~1 ms) but, unless the plasma pushed above
the surface cools very rapidly, emission at a fraction of the maximum luminosity can occur only after a substantial fraction of the
entire helium layer has been ignited. Thus the event would have a
sudden onset, but at a very low luminosity, a gradual rise to
peak, and then a slow decline as the fuel is exhausted and all
thermal energy is extracted from the explosively heated material.
Such behavior is consistent with that observed in γ-ray bursts of
this type.

Model II will ignite by a detonation wave and the rise time,
given by the time required to heat most of the plasma in the polar
cap region and push it above the surface, will be quite short. At
a velocity of roughly 9000 km s^{-1} (ref. 16) the wave will propagate across the 1 km radius of the polar cap in about 0.1 ms.
During this time ~ 10^{22} g of helium is ignited. The burning continues for a considerable time following the passage of the detonation wave, and ultimately releases ~10^{40} erg. Unless stopped by
a physical barrier, such as a region void of thermonuclear fuel,
the detonation wave will ignite the <u>entire</u> layer in one swift passage[16]. Thus one expects a "single spike" event whose duration is
given by the cooling time (or characteristic timescale of some
non-thermal process) at the spectral energy where measurements are
carried out. Examples of such events would be the events of March
5 and June 13, 1979.

Any attempt to examine the spectral properties of these
models leads us to consider the interaction of multi-billion
degree plasma, heated and pushed above the surface by the explosion, with a very strong magnetic field. Thermodynamic conditions
in Models I and II are plotted in Fig. 2 as a function of the
altitude above the surface. This particular figure[4] has been
prepared assuming a field of infinite strength, i.e., β = 0, so
that no lateral expansion can occur. In actual fact, a field of 4
x 10^{12} gauss is not able to effectively confine such hot plasma
and some lateral expansion must occur, especially in Model II. As
the plasma expands, thermal energy in the plasma is converted into
and stored as magnetic field stress. In Model II, for example,
roughly 1/2 of the energy in plasma pushed above the surface,
which in turn is about 20% of the total explosion energy, is
stored in this form. Such highly distorted field configurations
(Fig. 3) are unstable (for example to the interchange instability)
and, without a continuing source of energy input, will return to a
force-free configuration in some uncertain, but presumably quite
brief, time. There may be continuous input, however, from convection and continued nuclear burning below the surface. While never
again so large as when the explosion first occurs, this sustained
input can cause plasma instabilities to continue throughout the
event. One then has a highly complex picture of the plasma configuration and emission properties. In the vertical direction the
plasma rests in near hydrostatic equilibrium, with a small amount
of mass continually driven up the column as a wind (like that

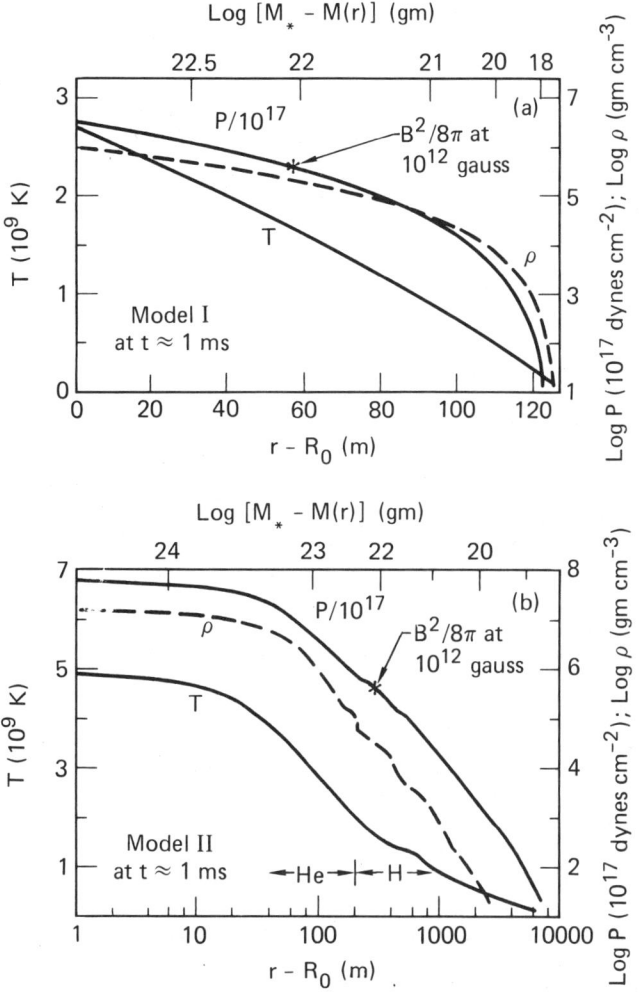

Figure 2. Thermodynamic conditions in plasma pushed above the surface of the neutron star by explosion in Model I (a) and Model II (b). Conditions have been evaluated at a time 1 ms following the sudden increase in surface luminosity due to convective energy transport (Model I) and shock wave break-out (Model II) and any lateral expansion has been neglected. The radius R_0 corresponds to the pre-explosive value of the surrounding iron surface of the star. The top scale gives the mass external to the given altitude. Magnetic "pressure" corresponding to a field strength of 10^{12} gauss is indicated. Plasma expansion is expected to occur in that region where its pressure dominates the magnetic field (a somewhat smaller region if, as is assumed in the text, $B = 4 \times 10^{12}$ gauss).

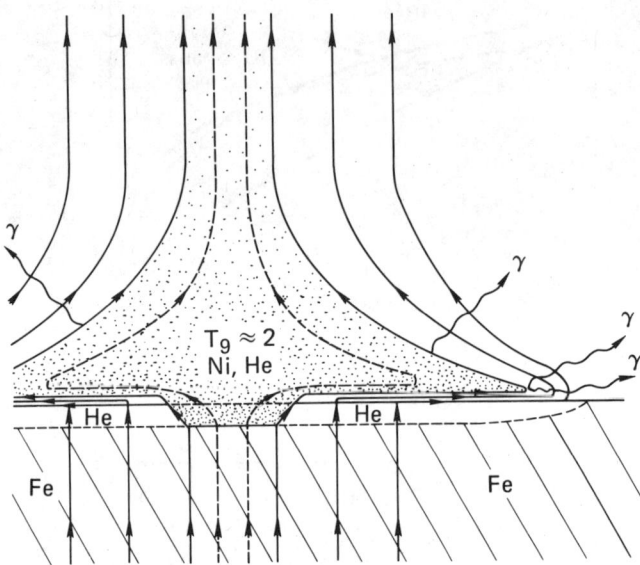

Figure 3. Hot plasma pushed above the surface by the explosion in (in either Model I or Model II) expands causing the magnetic field to become highly distorted. For a field strength of $\sim 10^{12}$ gauss, pressure equilibration ($\beta \sim 1$) occurs when the internal plasma temperature has fallen to about 2 billion degrees. Most of the mass and explosion energy is contained in material still beneath the surface but thermal radiation from the magnetically confined photosphere, as well as non-thermal radiation produced as a result of the magnetic instabilities to which this configuration is subject, is escaping from matter above the surface and producing a γ-ray burst. A small and highly uncertain amount of mass is driven up the magnetic column by the locally super-Eddington luminosity. Enormous electric fields ring the base of this column and may also accelerate electrons to high energy.

obtained in the 1D calculations). In the lateral direction, motion is greatly affected by magnetic interaction, instabilities, and the energy being fed from below. So long as the vertical temperature gradient remains superadiabatic, convection is able to effectively transport heat up the column even when the β of the plasma becomes small[27]. This occurs through a phenomenon known as "overstable convection" in which heat is transported by conduction back and forth between fingers of plasma oscillating in the vertical (field-aligned) direction. Vertical energy transport proceeds on a timescale much more rapid than ordinary conduction.

The spectrum of a γ-ray burst generated by the thermonuclear model will also be complex and will have two components: a <u>thermal</u> component which comes from both the top and magnetically confined sides of the plasma blob, and a <u>non-thermal</u> component which comes from magnetic interactions and/or the radiatively driven wind moving up the plasma column. The thermal component itself is far from simple since there is no unique temperature. Instead, emission from different heights in the plasma column is characterized by different temperatures, with cooler plasma at higher altitudes. A preliminary, and thus far unpublished, calculation by Wilson, Barton, and Woosley shows that the surface of the plasma column cools rapidly and, ignoring non-thermal effects and plasma instabilities, emits in a temperature range roughly 10 to 50 keV with a temperature scale height of roughly 50 m. Although this radiation may account for the bulk of the <u>bolometric</u> emission of a "γ-ray" burst and for such soft-spectrum events as that of March 5, 1979, it is not adequate to explain the hard emission found, especially early on, in the spectra of many common γ-ray bursts.

Non-thermal emission may result from at least three processes. First, because a substantial fraction of the explosion energy is stored in, and continues to be transformed into magnetic field stress, there is the possibility of non-thermal particle acceleration by magnetic recombination. As a specific example, consider plasma breaking out of the foot of the column (Fig. 3) by the interchange instability. This plasma has within it a strong magnetic field and as it squeezes through the surface of the plasma column (photosphere) in finger-like projections, field lines of opposite directionality come into close proximity. If magnetic recombination occurs, then charged particles, chiefly electrons, can be accelerated to high energy and emit synchrotron radiation. Second, consider the fact that the surface of the magnetically confined plasma column is not static For example, following the thermonuclear outburst in Model II, the surface of the star (in the 1D calculation[5]) undergoes rapid, large amplitude oscillations (Fig. 4). One should think of the plasma "bubble" in Fig. 3 as continually "jiggling". As the surface of the superconducting plasma moves back and forth through the strong magnetic field, intense non-thermal radiation will result[3]. This configuration has an added advantage over the model of Ramaty, Lingenfelter, and Bussard[3] (based on a neutron starquake) in that the

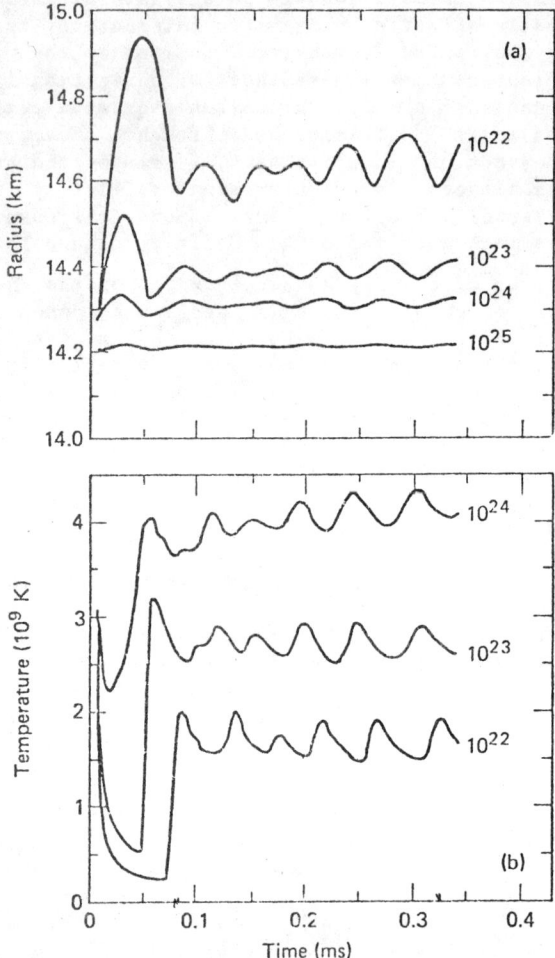

Figure 4. Radius and temperature at the boundary of several representative mass shells during the early evolution of Model II (see Model C, ref. 5). The mass indicated is the mass in grams external to the given zone boundary in the one dimensional calculation (multiplied by 1.2×10^{-3} for equivalent masses in the magnetically focused case). The original radius of the cold neutron star was 14.30 km, so $\sim 10^{24}$ g of material have been pushed above the surface in the 1D calculation. The duration of these oscillations was not determinable with our numerical hydrodynamics code due to the inherent damping in an implicit method.

plasma surface is quite naturally everywhere parallel to the magnetic field lines. This is an assumption that must be introduced artificially in the Ramaty et al model. A third possibility for non-thermal emission involves the "stellar wind" discussed in the previous section. There are particles moving with high velocity along the field lines in an optically thin region above the top of the plasma column. They are accelerated by the super-Eddington radiation pressure streaming up from the neutron star (electrons) and by the strong electric field created by charge separation (nuclei). Direct interaction of these particles (perhaps in a Rayleigh-Taylor unstable region), or acceleration of other non-thermal particles by the strong electric field, can give hard radiation.

The duration of a γ-ray burst produced by a thermonuclear explosion is difficult to estimate since it involves a complicated magnetic interaction, but it is likely to be characterized by different timescales at different spectral energies. In Model I, where the ignition of new fuel is expected to continue throughout the event, one may expect that the spectral hardness would stay relatively constant during (most of) the event. In Model II, the strongest magnetic interactions and hence the hardest radiation, should occur early on during the early "impulsive" phase of the event, with gradual spectral softening expected as the event proceeds and the distorted field sorts itself out into a nearly force-free configuration. In general, the spectral hardness of the event should also be correlated with the strength of the magnetic field. Stronger fields confine the plasma at higher temperatures and are likely to produce more energetic particles by non-thermal processes. An observational study addressing the correlation, if any, between cyclotron absorption energy and spectral hardness would be interesting. So too would a study of the time history of the width of the cyclotron absorption line. Early on, especially in detonating models, considerable broadening is expected due to the highly inhomogeneous field near the surface of the neutron star (Fig. 3). Later, as the field becomes more ordered, the absorption line may become narrower.

Because the plasma column can radiate from its sides, it is not expected to be optically thick at such large radii as obtained in the 1D calculations (~ 50 to 100 km). Roughly speaking, the top of the optically thick plasma column (upper photosphere) should occur at a radius where the time scale for energy transport from below, by conduction and convection, balances the timescale for lateral conduction and escape from the column. Since the flux of energy _up_ the column, along magnetic field lines, is still essentially Eddington limited, the conduction timescale from the center of the column near the upper photosphere should be given by $\varepsilon \Delta M / f L_{Edd} \sim \kappa \rho R^2 / c$, where ε is the average internal energy per gram in the mass ΔM (per unit area) contained in the photosphere (or, roughly, the mass external to the photosphere), $f L_{Edd}$ is the luminosity up the column (i.e., f times the Eddington value), κ

is opacity, ρ the density at the photosphere, R roughly 1 km, and c, the speed of light. From the 1D calculations[5] one finds a self-consistent solution to this equation for Model I at an altitude of several hundred meters. There $\varepsilon \sim 10^{19}$ erg g^{-1}, $\Delta M \sim 5 \times 10^{15}$ g km^{-2}, $f \sim 1$, $\mathcal{K} \sim 0.2$, and $\rho \sim 1$ g cm^{-3}. This corresponds to a conduction time across the column of about 0.1 s. Conditions for Model II are similar but at a somewhat higher altitude, ~1 km.

The time required to transport a substantial fraction of the energy out of the plasma pushed above the surface by the explosion will probably be less than that time required for heat to diffuse ~50 m (the density scale height) at a density of ~2×10^5 g cm^{-3} (density at the base of the column following lateral expansion to $\beta \sim 1$) or about 20 s. This will be accelerated by convection[27], plasma instabilities, and the reduction of the electron scattering opacity along the strong magnetic field. On the other hand, the cooling time will be lengthened by the transport of additional energy from beneath the surface of the star. Further numerical calculations coupling all these effects will be necessary to estimate the actual cooling time.

Since roughly 3/4 of the energy of the explosion is contained in plasma that is never pushed above the surface, there should be a period of softer emission following the hard outburst as these subsurface ashes cool. So long as a locally super-Eddington luminosity is maintained, one expects some continued magnetic interaction. Thus the spectral hardness cannot be accurately calculated. It will, however, be no softer than 2 keV, the Eddington temperature for our neutron star, and probably not much harder than 50 keV, since one expects the principal emission processes in the late stages to be thermal. The luminosity itself will be at least Eddington, or about 2×10^{35} erg s^{-1} for the explosion area we have assumed. The duration will be shorter than the event timescales given in Table 1 for the 1D models since we now have additional and more effective ways of removing energy from the explosively heated plasma. All in all, a timescale of from one hundred to several hundred seconds appears reasonable. As the luminosity falls below the Eddington value there should be a soft thermal tail to the event (~ 50 s in Model I, ~ 1000 s in Model II) during which the temperature gradually declines below 2 keV and the neutron star cools off. Measurements of the characteristics of this tail of the light curve would be especially useful in diagnosing the amount of mass involved in the outburst and in distinguishing thermonuclear and accretion based models.

SUMMARY AND INSTRUMENTAL CONSIDERATIONS

The thermonuclear model is capable of producing a wide variety of high energy transient phenomena including Type I x-ray bursts[28], rapid x-ray transients[5], γ-ray bursts, and, perhaps Jacobson-like γ-line transients. The key characteristics distinguishing among these various possibilities are the accretion rate and magnetic field strength. If the field is weak, then slow accretion ($< 10^{-9}$ $M_\odot$ yr^{-1}) gives large critical masses and x-ray transients shining at the Eddington luminosity with nuclear burning all over the stellar surface. More rapid accretion rates (but still with weak fields) give smaller critical masses, lower energy outbursts that may even be sub-Eddington, and more rapid recurrence timescales. These are the Type I x-ray bursters. With a magnetic field and very low accretion rates it has been argued, both here and elsewhere[4,15], that one expects phenomena resembling γ-ray burst. An interesting, and as of yet unresolved question, centers on the evolution in accreting magnetized systems with accretion rates still smaller than considered here (i.e., $\dot{M} \ll 10^{-13}$ $M_\odot$ yr^{-1}). Such low accretion rates might be typical for single neutron stars in the interstellar medium and may have to be considered if the steady state x-ray fluxes required by our present models are not observed. For such low accretion rates, even with magnetic focusing, hydrogen burning will not be β-limited. However, hydrogen burning can again lose its temperature sensitivity at such high densities that it is mediated by electron capture [the critical rate being p (e$^-$,ν_e)n]. Thus a thick helium shell can still accumulate. Unfortunately, the study of this problem is currently hampered by inadequate knowledge of the pycnonuclear screening correction for the triple-α reaction at densities exceeding 10^8 g cm^{-3}. If, as Ruderman argues[18], the helium burning reaction also loses all temperature sensitivity at such high densities, then it may be that there is no runaway at all and the helium (or carbon) neutronizes directly without ever igniting. If the helium burning reaction retains some temperature sensitivity, however, or if the accretion rate should increase for some reason, say because the neutron star passes through an interstellar cloud, then an energetic detonation-type event may occur[4,19]. One must be concerned, however, that the accretion rate not become so low as to spread over the entire neutron star surface[24]. Further calculations of this interesting range of parameters would be interesting.

Adopting for now our Models I and II as representative, we can examine the demands made on the model by present γ-ray burst statistics. Is it in fact possible to have a sufficient number of magnetic neutron stars accreting at the proper rate? Because of its more rapid recurrence cycle, Model I should be the more common γ-ray burster (even though Model II is visible from a greater distance). The total energy output by Model I is about 5 x 10^{38} erg of which we shall assume about 20% is emitted in the form of hard radiation visible to a γ-ray burst detector. A 10^{38} erg event would give a total flux of 10^{-6} erg cm^{-2} (roughly the threshold of Mazets et al[1]) if located at a distance of approximately 400 pc. Within this distance one expects about 2 x 10^5 neutron stars, which with a recurrence period of 0.4 yr, gives the observed rate of 150 bursts per year if only 0.03% of all neutron stars have strong magnetic fields and are currently experiencing accretion rates in the right range. It does not seem that this small value would pose a severe constraint on the systematics of accreting neutron stars. Events corresponding to Model II would be from 1 to 2 orders of magnitude more energetic than those from Model I and could thus be typically located from 3 to 10 times farther away. This volume of the Galaxy would encompass many more neutron stars and would partly compensate for the long recurrence cycle of detonation events.

Within the context of the thermonuclear model, the March 5, 1979 event would have been a detonation event. Several solutions are possible[4], but a typical case would place the site about 800 pc away with an accretion rate and area the same as in Model II. Due to subsurface spreading, the explosion encompasses an area 10 times larger than the accretion area. The total explosion energy would be 10^{41} erg and the recurrence period 2400 yr. The soft, pulsing tail of the event ($\sim$ 12 keV; ref. 29) would be the cooling surface layer described in the previous section modulated by rotation. The one to three subsequent bursts from the same general location as the March 5 event would be difficult to explain within the context of this model. If the association is real, one would have to invoke "splotchy" accretion with several pockets of fuel concurrently accreting in various localities near the polar caps, or else entirely abandon a thermonuclear origin for this one event. It is also possible, although somewhat artificial, to construct a thermonuclear model that would place the March 5 event in the LMC (55 kpc). The helium layer would, however, have to cover the entire neutron star to a depth of $\sim$ 300 m. Recurrent bursts would again be difficult to explain, but might be attributable to crust quakes following the major outburst (see ref. 4 for further details).

Many uncertainties exist in the theoretical scenario I have outlined and much careful analysis and thought is required before γ-ray bursts can be said to be understood. Numerical studies of magnetically confined, high temperature, optically thick plasma should be carried out to clarify the spectrum expected from such situations. Instabilities in $\beta \sim 1$ plasma in a 4×10^{12} gauss field need examination along with the stability and properties of the wind driven up the plasma column by super-Eddington emission. Realistic two-dimensional calculations of the lateral propagation of a convective deflagration are important, not only for γ-ray burst models, but also for a host of other astrophysical phenomena such as novae and thin helium shell flashes in intermediate mass stars. Pycnonuclear screening corrections for helium at densities exceeding 10^8 g cm^{-3} are required before realistic studies of still lower accretion rates can be carried out. Finally, the nature of convection and "overstable" convection in a strongly magnetized plasma should be explored with numerical models.

On the observational front the principal goals remain much as in the past: 1) accurate source locations, 2) broad band (i.e., x-ray and γ-ray) studies and time histories, and 3) high resolution spectroscopy.

The highest priority at the present time (although possibly not on any space mission planned for the distant future) should be given to obtaining accurate source locations. A beginning has already been made in that direction for 7 events[2,3] using timing considerations. Only with accurate source locations can we respond to such important questions as periodicity, steady state accretion luminosity, nature of binary companion star (if any), etc. The results of such observations will quickly prove or disprove the models theorists have been discussing.

It is my own personal feeling that in the not too distant future, perhaps with GRO, at least one non-ambiguous identification of γ-ray burst site will be made (besides the seemingly unique event of March 5, 1979!) and then that particular type of object will begin to be discovered in or near many previous error boxes. While a gamble of sorts, this suggests that perhaps in the long run, eg. for a proposed transient explorer class satellite that might go up near the end of the decade, the highest priority should be given to high resolution spectroscopy over a broad band of wavelengths and with good time resolution. These studies would yield the highest density of information and should enable us not only to understand qualitatively what a γ-ray burst really is, but also to examine in considerable detail the evolution and physics of an outburst. Since both the thermonuclear and accretion models predict x-ray emission associated with and following the hard outburst (cooling of a hot region on the neutron star surface), it is important that "γ-ray" burst detectors have good sensitivity in the region traditionally assigned to x-rays (1 keV $< E <$ 100 keV). Otherwise a great deal of useful information, including the actual bolometric energy associated with a γ-ray burst, may not be obtained.

This research has been supported in part by the National Science Foundation (AST79-09102 and AST81-08509) and , at LLNL, by the Department of Energy (W-7405-ENG-48). The author is also grateful for interesting and illuminating discussions with Bruce Fryxell, Stirling Colgate, Ron Taam, Rick Wallace, Jim Wilson, and Tom Weaver. He is also indebted to C. Thomas-Duffy for encouragement during and following the meeting.

REFERENCES

1. E. P. Mazets et al, preprint submitted for publication to Ap. and Spac. Sci. (1981). See also Nature, 290, 378 (1980) and A. F. Ioffe Physico-Technical Institute Preprints No. 618, 637, 662, and 687.
2. S. A. Colgate and A. G. Petschek, Los Alamos preprint, submitted to Ap. J., (1981).
3. R. Ramaty, R. E. Lingenfelter, and R. W. Bussard, NASA TM 80674 and Ap. And Spac. Sci. in press (1981).
4. S. E. Woosley and R. K. Wallace, submitted to Ap. J. (1981).
5. R. K. Wallace, S. E. Woosley, and T. A. Weaver, submitted to Ap. J. (1981).
6. W. M. Howard, J. R. Wilson, and R. T. Barton, Ap. J., in press (1981).
7. R. Ramaty, S. Bonazzola, T. L. Cline, D. Kazanas, P. Meszaros, and R. E. Lingenfelter, Nature, 287, 122 (1980).
8. D. Q. Lamb, F. K. Lamb, and D. Pines, Nature Phys. Sci., 246, 52 (1973).
9. P. Kafka and F. Meyer, Max Planck Inst. preprint, submitted to Gravity Research Foundation competition (1981).
10. M. Harwit and E. E. Salpeter, Ap. J. Lettr., 186, L37 (1973).
11. M. J. Newman and A. N. Cox, Ap. J., 242, 319 (1980).
12. D. Van Buren, Bull. Am. Astron. Soc., 12, No. 4, 811 (1980), see also this volume.
13. R. W. Klebesadel, E. E. Fennimore, J. G. Laros, and J. Terrell, this volume (1981).
14. S. A. Colgate, this volume (1981).
15. S. E. Woosley and R. E Taam, Nature, 263, 101 (1976).
16. B. A. Fryxell and S. E. Woosley, submitted to Ap. J. (1981), see also this volume.
17. A. I. Tsygan, Astron. and Ap., 87, 224 (1980).
18. M. Ruderman, Prog. Part. and Nucl. Phys., in press (1981).
19. S. Bonazzola, J. M. Hameury, J. Heyvaerts, and J. Ventura, preprint (1981).
20. R. K. Wallace and S. E. Woosley, Ap. J. Suppl., 45, 389 (1981).
21. A. S. Jacobson, J. C. Ling, W. A. Mahoney, J. B. Willett, in Gamma Ray Spectroscopy in Astrophysics, (eds. T. L. Cline and R. Ramaty), 228, (NASA TM79619, 1978).
22. K. Davidson and J. P. Ostriker, Ap. J., 179, 585 (1973).
23. P. Ghosh and F. K. Lamb, Ap. J., 232, 259 (1979).
24. J. Arons ans S. M. Lea, Ap. J., 207, 914 (1976) and 235, 1016 (1980).
25. K. Hurley, this volume.
26. M. Ruderman, Nature, 223, 597 (1969).
27. H. C. Spruit, Max Planck Inst. preprint MPI-PAE/Astro 228 and Spac. Sci. Rev. in press (1981), see also S. I. Syrovatskii and Y. D. Zhugzhda, Soviet Astron., 11, 945 (1968).
28. W. H. G. Lewin and G. W. Clark, Ann. N. Y. Acad. Sci., 336, 451 (1980).
29. E. E. Fennimore, W. D. Evans, R. W. Klebesadel, J. G. Laros, and J. Terrell, LASL preprint submitted to Nature (1980).

GAMMA-RAY BURSTS FROM YOUNG NEUTRON STARS

K. Brecher
Boston University, Boston, Ma. 02215

ABSTRACT

We argue that young, single neutron stars, not undergoing significant accretion, are plausible candidates for sporadic energetic radiative events such as gamma-ray bursts. We take the March 5, 1979 gamma-ray burst as the prototype source, and examine implications of such an association for future gamma-ray burst studies.

INTRODUCTION

Most of the questions concerning the nature of gamma-ray bursts remain unanswered after nearly a decade of study of the phenomena. Perhaps most strikingly, the distances, and therefore, overall energetics of the events are still debatable and this, combined with the limited information on spectra, time variation, polarization and repetition rate of the events leaves room for incredible speculation as to the nature of the underlying source.

A comparison with another astrophysical phenomena, the pulsars, may be useful in clarifying the nature of the sources. Within a year or two of their discovery, it was thought that the pulsars were rapidly rotating magnetized neutron stars, with rotational energy being the direct energy source, but white dwarfs were also still a logical possibility. However, with the discovery of the Crab pulsar with a period of 33 ms, it became clear that only the neutron star model was viable.

In this paper, I wish to take the point of view that the March 5, 1979 event may play the same role in clarifying the nature of gamma-ray burst sources as the Crab nebula and its pulsar played in establishing the nature of pulsars. Each object is, of course, quite singular, not really the "typical" member of its class. Nonetheless, each may well be the "Rosetta Stone" to understanding the particular objects involved.

One further remark should be made in this analogy. Even with over a decade of theoretical and observational study of pulsars, we do not understand the actual pulsar emission mechanism. Details of the pulses, their spectra and time variations are still only poorly understood. It is not possible even today to go from the pulsar spectra to the nature of the underlying object. It may well be that a similar situation faces us with the gamma-ray burst sources.

THE MARCH 5, 1979 GAMMA-RAY BURST

As has been cogently argued by Felten[1], the association between the March 5 event (hereafter referred to as M5) and the supernova remnant N49 in the Large Magellenic Clouds is significant at the

.04% level. That, is, the odds against a chance superposition are about 2000:1. We take the point of view that it is not due to chance. That being the case, it seems reasonable to associate the event with either some sort of continuing activity in the nebula, or to a remnant neutron star within it. In the absence of a good model for the former, and a plethora for the latter, it again seems reasonable to assume a neutron star source. It should be noted that until the M5 event, no other gamma-ray burst could be identified with any likely candidate. We therefore will take M5 as the prototype of the gamma ray burst source!

What does such an association imply? At the distance of LMC of about 55 kpc, M5 released about 10^{44} ergs with a peak luminosity of about 10^{44} erg/sec. It lasted for some tens of seconds, showed a periodicity of 8.1 seconds and there might be a suggestion[2] of variability on a time scale of 22 ms. What do these numbers suggest? N49 is young (less than 10^4 years old). The two known pulsars in young supernova remnants (Crab and Vela) are both rapidly rotating (33 ms and 79 ms, respectively). It seems likely, therefore, that if N49 contains a rotating neutron star, its rotation period is unlikely to be as long as 8.1 seconds. We therefore take 8.1 sec to be the free precession period of the neutron star, implying a rotation period of 10-100 ms, (perhaps 22 ms?) depending on the composition of the neutron star. It also seems plausible that a young neutron star should still retain a strong magnetic field B as large as a few x 10^{13} gauss.

The energetics of such a source are now easily estimated: the gravitational binding energy is $\sim 10^{53}$ ergs; the rotational energy ($t \simeq 20$ ms) is $\sim 10^{49}$ ergs; the magnetic field energy is $\sim 10^{44}$ ergs. With the possible exception of the last number there is clearly no shortage of energy available for the March 5 event. How might it be liberated?

GLITCHARS

In the early days of gamma ray burst studies, when there were more theories than events, an early suggestion[3,4] was that they might be associated with pulsar "glitches". Such glitches, or sudden decreases in the periods of pulsars can be explained either as a sudden cracking of the crust of the neutron star, or a readjustment in the core of the star. In either case, angular momentum is taken to be conserved, the moment of inertia of the star suddenly decreases, and energy is released in going to the new gravitational equilibrium configuration. Exactly how the energy is released (heating of the neutron star surface, excitation of MHD waves and subsequent radiation from the magnetosphere, even neutrino emission from the interior) is unclear. Whether the energy is released over a period of seconds or days, radiates isotropically or is beamed: all are unanswered questions. Only the change in period is known. For the two Crab events, $\Delta\Omega/\Omega \sim 10^{-8}$. Therefore, $\Delta E \sim I\Omega^2 \Delta\Omega/\Omega \approx 10^{41}$ ergs. At the distance of the Crab nebula, if all of this energy were converted into gamma-rays and radiated isotropically, within a few seconds, it would produce a gamma-ray burst of $\sim 10^{-3}$ erg/cm^2-sec, easily detectable at Earth. Similar figures appear for the Vela pulsar. Why, then, have no gamma-ray bursts been seen accompanying these glitches?

The answer, for 3 of the 5 events seen in the Crab and Vela pulsars, is that no satellite or balloon was observing (or functioning) at the times of the events. In fact, only one event each in the Crab and Vela pulsars in 1975 could have been detected. Again, it is not clear that their absence rules out glitches as being associated with gamma-ray bursts. First, the entire glitch energy may not be radiated as gamma-rays, nor does it have to be isotropic. The fact that gamma-ray bursts do not come from obvious supernova remnants (other than M5) is also not necessarily damning for an association with older pulsars, since several other older pulsars ($\sim 10^6$ years old) have been observed to have glitches (though smaller than those in the Crab and Vela) and it has been estimated that such pulsars might undergo glitches every 10-20 years. Statistics of glitch sizes, event rates, and so forth have not yet been realistically determined, but for the "typical" gamma-ray burst, an association with older pulsars seems possible.

SUPERGLITCHES AND NEUTRON STAR PHASE TRANSITIONS

Such glitches would not explain the M5 event, which requires a much larger release of energy. Is there evidence for such post-neutron star formation "superglitches"? There is one tantalizing piece of evidence based on ancient astronomical records, and one modern record. The well known supernovae of 1006 AD and 1604 (Kepler) may have "recurred" subsequent to the original event, in the years 1016 and 1664, respectively[5]. A more recently observed supernova, in the galaxy M101, also shows tantalizing evidence for the presence of two expanding shells, suggesting two ejective events. Taking these observations to imply some sort of post formation sporadic neutron star activity (vastly larger than the glitches discussed above), several possible explanations come to mind to explain these "recurrent" supernovae. First, it could be that some matter ejected does not escape, is accreted back onto the neutron star, allowing for a flash nuclear event. A second possibility is that the neutron star has formed unstably, and, due to subsequent cooling, magnetic field decay, or spin down tries to readjust to a new equilibrium figure. This may involve only a readjustment in shape or, more speculatively, a transition from one phase to another deep in the interior. A phase transition from hot neutrons to, say pions, quarks, or T.D. Lee's "abnormal nuclear matter" may produce small change in stellar radius ($\Delta R \sim 1$ cm) but enormous energy release ($\Delta E \sim [GM^2/R][\Delta R/R] \sim 10^{47}$ ergs).

YOUNG VERSUS OLD NEUTRON STARS

If neutron stars are the gamma-ray sources, the question arises whether the ultimate "fuel" is intrinsic or extrinsic to the star. In models involving accretion, either of gases, comets or asteroids, the fuel is external. In the nuclear burning case part of the energy released is nuclear, part gravitational. If gamma-ray bursts arise from accreting neutron stars, they are more likely to be old than young, simply because there is a longer period for activity in mass exchange binaries (say 10^6 years) and asteroid pools (10^{10} years?), whereas if large events are intrinsic to the neutron star, they are most likely to occur during a period of, say, only 10^4 years.

We therefore make the following tentative suggestion: that gamma-ray bursts are associated with young neutron stars, and represent post formation glitch, phase transition or other activity intrinsic to the star. (It may well be that the x-ray burst sources are an observed example of neutron star accretion phenomena, making the difference between gamma-ray and x-ray burst sources the difference between internally and externally primed neutron stars.)

PREDICTIONS

If the above picture (not even a "scenario") has any validity there are a number of features which may be testable. First, the M5 event and other energetic gamma-ray bursts should show periodicities on a time scale of 10-100 ms. Second, there of course should be some supernova remnant, gamma-ray burst association found. Third, if the gamma-ray burst sources lie inside supernova remnants, the remnants themselves should show changes in brightness or shape in the years following the burst (as in the changes in wisp and filament brightness in the Crab nebula after pulsar glitches). Future burst monitoring activity should probably clarify the connection between glitch activity in the Crab and Vela pulsars, or other known "glitch active" pulsars such as PSR 1133+16 and PSR 0628+28, and gamma ray bursts.

IMPLICATIONS

We have not considered a detailed mechanism for radiation after a glitch, phase transition or other event mechanism (magnetospheric plasma rearrangement). The vibrating neutron star model[6] is appealing, but neutron star flare-like activity seems equally viable[4]. However, if future gamma-ray bursts show the presence of what appears to be red-shifted positron annihilation line radiation, and if both long (10 sec) and short (10-100 ms) periodicities are found, one can probe the nature of dense matter. For (assuming the periods are due to precession and rotation) by combining the two periodicities one can determine the moment of inertia of the star. This, combined with the redshift determination of the mass to radius ratio can uniquely determine the equation of state of the neutron star.

ACKNOWLEDGEMENT

This research was supported in part by NSF Grant AST 8020756.

REFERENCES

1. J.E. Felten, Conf. Papers, 17th Int. Cosmic Ray Conf. (1981).
2. K. Hurley and T. Cline, private communications, with disclaimers (1981).
3. F. Pacini and M. Ruderman, Nature 251, 399 (1974).
4. A.I. Tsygan, Astron. & Astrophys. 44, 21 (1975).
5. J.M. Wang, Chinese Astron. 4, 407 (1980).
6. R. Ramaty, S. Bonazzola, T.L. Cline, D. Kazanas, P. Meszaros, and R.E. Lingenfelter, Nature, 287, 122 (1980).

A TWO DIMENSIONAL MODEL FOR γ-RAY BURSTS

B. A. Fryxell and S. E. Woosley
Lick Observatory, Board of Studies in Astronomy
and Astrophysics, University of California,
Santa Cruz, CA 95064
and
Lawrence Livermore National Laboratory
Livermore, CA 94550

ABSTRACT

Accretion focused onto a neutron star by a strong magnetic field results in the formation of a lake of helium at the magnetic pole. For the model calculated here, the depth of the lake is 90 m and the radius is taken to be 400 m. After a critical mass of helium has accumulated, a thermonuclear runaway initiates at a point at the base of the helium layer resulting in a detonation wave which propagates across the lake at a constant velocity of 9000 km s^{-1} without dying out, igniting all the helium and eventually liberating 2.4×10^{39} ergs. The interaction of the hot plasma behind the detonation front with the magnetic field produces the hard radiation observed in γ-ray bursts. The rise time of the burst predicted by this model is about 50 μs, the time required for the detonation to propagate across the helium lake.

INTRODUCTION

The presence of gravitationally red shifted pair annihilation lines and cyclotron absorption features in γ-ray burst spectra[1,2] indicates that these events originate on neutron stars with magnetic fields greater than 10^{12} gauss. Assuming a simple dipolar field[3], the slow accretion of a metal deficient mixture of hydrogen and helium onto such a neutron star will be focused onto the magnetic pole. For a suitable accretion rate and metallicity, the hydrogen will burn stably by the β-limited CNO cycle forming a layer of helium[4] which will be confined to a small region (~ 1 km^2) near the magnetic pole by the field and the structural strength of the neutron star crust[5]. After a sufficient amount of helium has accumulated, a thermonuclear runaway initiates at a point at the base of the helium layer. Depending on the conditions at the base of the layer, the runaway could produce either a detonation or a convective deflagration[4]. In this paper, we will be concerned primarily with the detonation model and discuss a two dimensional hydrodynamic calculation of the formation and propagation of the detonation wave in the neutron star atmosphere.

NUMERICAL METHODS

The calculation was performed using a two dimensional explicit Eulerian hydrodynamic code[6,7]. Equation of state data was interpolated from a table generated by the KEPLER computer code[8].

0094-243X/82/770299-09$3.00 Copyright 1982 American Institute of Physics

Nuclear energy generation was calculated using a simplified reaction network containing only helium burning to carbon by the triple alpha reaction and carbon burning by the $^{12}C + ^{12}C$ reaction. The energy produced by carbon burning was calculated assuming that all the carbon burned directly to ^{56}Ni. Screening corrections to both reactions were included[9,10]. The effects of radiation transport are unimportant on the timescales considered here and were neglected. The large magnetic fields associated with neutron stars will be important to the flow near the surface. However, the main emphasis of the calculation was to follow the propagation of the detonation wave at the base of the helium layer where $B^2/8\pi \ll P$, and therefore, no attempt was made to include magnetic fields in the calculation.

The structure of the initial model was taken from a detailed one dimensional calculation[4] of a spherically symmetric detonation on a neutron star shortly before the runaway occurred. The model consists of a 1.4 $M_\odot$ neutron star with a radius $R_* = 14.3$ km accreting a metal deficient mixture of hydrogen and helium ($Z = 4 \times 10^{-5}$) at a rate of 2×10^{-11} $M_\odot$ yr^{-1}. Hydrogen burned stably by the β-limited CNO cycle forming a layer of helium 90 m thick covering the neutron star surface. The temperature and density at the base of the helium layer were 8×10^7 K and 1.15×10^8 gm cm^{-3}.

For the two dimensional calculation, it was assumed that the accreted matter was focused onto a small region at the polar cap by the magnetic field, forming a lake of helium 400 m in radius. The confinement of the fuel to a small region of the surface is very important to the model. If the fuel covers the entire star, it is difficult to make a γ-ray burst (an X-ray burst would be more likely)[5]. Both the accretion rate and the energy in the burst are scaled down by the fraction of the surface area of the neutron star covered with helium. Thus for the model computed here, an accretion rate of only 4×10^{-15} $M_\odot$ yr^{-1} is required. Since the diameter of the lake is only about 1% of the neutron star circumference, the surface was taken to be planar. The finite difference grid contained 50 zones in the vertical (R) direction and 220 zones in the horizontal (X) direction.

RESULTS

The calculation was started by raising the temperature in one zone at the base of the helium layer from 8×10^7 K to 1×10^8 K. Because of the extreme degeneracy in this region, a very violent runaway began which produced a detonation wave. The propagation of the detonation front is illustrated in Figures 1-4. Figure 1 shows the initial expansion of the detonation wave at time 14 μs. The detonation moves horizontally across the base of the helium layer with a constant velocity of 9000 km s^{-1} (Mach 1.8) with an overpressure of about 2.5. The temperature of the gas behind the detonation near the base of the helium layer remains at about 6×10^9 throughout the calculation.

The wave moving up the symmetry axis toward the surface

begins as a weak shock but is rapidly accelerated and strengthened by the steep density gradient in the neutron star atmosphere. The shock is unable to propagate toward the center of the star because of the increasing density gradient and the lack of any helium to fuel the detonation.

In Figure 2, at time 20 μs, the shock has broken through the surface and is pushing matter upward out of the grid at a velocity of 1.5×10^9 cm s^{-1}. The presence of the neutron star gravity is clearly evident from the density contours. Almost immediately after the detonation wave passes, the planar symmetry of the the atmosphere is restored.

About 30 μs after the beginning of the calculation (Figure 3), an interesting feature appeared in the flow. The matter near the surface is accelerated to a high velocity ($\sim 2 \times 10^9$ cm s^{-1}) as the shock travels down the density gradient. As a result, a wave develops which propagates (in the absence of a magnetic field) along the surface of the neutron star with a velocity greater than that of the detonation at the base of the helium layer. At this time the wave is just beginning to break, since the velocity increases with the distance above the surface.

Finally, Figure 4 shows the results at time 41 μs, when the detonation has nearly reached the edge of the helium lake. The breaking of the surface wave is now more apparent. We emphasize again that although the presence of a strong magnetic field will strongly inhibit the propagation of this wave, the propagation of the detonation across the base of the helium layer will not be significantly affected.

The detonation reached the edge of the helium lake at about 50 μs (which corresponds to the rise time for the burst in this model). At this time, the shock strength decreased rapidly as it tried to propagate into the surrounding iron. At the end of the calculation, 5.6×10^{20} gm of helium had been burned into carbon and nickel liberating 9×10^{38} ergs. Assuming that all the helium is eventually burned, the total energy available for the burst is 2.4×10^{39} ergs, or about 5×10^{39} ergs per square kilometer of surface area involved in the explosion.

DISCUSSION

We have described a two dimensional calculation of a point detonation in the atmosphere of a neutron star. Since magnetic fields and radiation transport were not included and only the initial stage of the burst was calculated, no quantitative information about the γ-ray production mechanism could be obtained, although it seems plausible that the interaction of the hot plasma with the magnetic field will produce hard radiation. However, important results concerning the initial stages of the outburst have been obtained. In particular, it was demonstrated that a detonation can ignite the entire helium layer without dying out and will generate 5×10^{39} ergs per square kilometer of surface area involved in the explosion. In addition, this model provides a natural rise time for the event, which is just the time required for the burning front to propagate across the helium lake. For the model calculated here, the rise time was about 50 μs, which is within the current observational limits.

We note that there appear to be two distinct types of γ-ray bursts— those which show a single sharp peak lasting for on the order of 1 s, and those which have a broad irregular peak lasting for tens of seconds. We propose that the detonation model produces the sharp spike event, and that the other type of burst can be explained by a thermonuclear runaway which results in a convective deflagration[4]. In this case, the time required for the burning front to propagate across the helium layer is longer than the timescale for emitting the radiation and thus corresponds to the duration of the burst rather than to the rise time. Although this timescale is very difficult to estimate since the lateral propagation may be mediated by convective entrainment, it should lie somewhere between the conduction time and the sound travel time across the helium layer. The structure observed in such a burst could result from either uneven burning of the deflagration due perhaps to small inhomogeneities in the helium layer or from instabilities in the $\beta \sim 1$ plasma above the neutron star surface which have a longer time to develop in the deflagration model. A third type of burst which is observed consists of a series of peaks which are separated by several seconds. This behavior can be explained if the helium accumulated in several small pockets on the surface of the star instead of in one large lake. After one of these pockets ignites and produces a peak in the burst profile, conduction from the hot ashes can ignite the next pocket producing a second spike. The conduction time in the neutron star crust is comparable to the time observed between peaks. The busts profile for individual spikes withing the same event could be a mixture of the detonation and deflagration types.

The results presented here are directly applicable to the March 5 event. If the burst source was located within the Galaxy, at a distance of about 800 pc, the single peak event can be explained very naturally by the model presented above[5]. If, on the other hand, the March 5, event was located in the LMC, the thermonuclear model could explain the large energy output only if the nuclear fuel covered a major fraction of the neutron star surface. However, since the timescale for the runaway (a few μs) is much shorter than the time to build up a helium layer thick enough to detonate (years), the detonation will always initiate at a single point and then propagate around the star. The results described above represent the early stages of such an event equally well. In either case, it seems clear that if the March 5 event was thermonuclear in origin, it involved a detonation rather than a deflagration.

A thermonuclear model for the March 5 event (if it was in the LMC) would also require a much thicker (~300 m) helium layer[5]. The conditions at the base of such a layer would result in a runaway which might be very different from the one calculated here.

The triple alpha reaction can raise the temperature of the matter to only about 6×10^9 K before nuclear statistical equilibrium is reached. This is too low a temperature to create a large overpressure under such degenerate conditions, and it is possible that a detonation would not result. However, when the burning front propagates upward, it will ignite the material closer to the surface, where the conditions will allow a detonation wave to propagate with greater ease. This wave can then propagate around the entire star with a velocity of about 9000 km s^{-1} as described above and burn the entire helium layer. The rise time in this case would be the time to propagate 1/4 the way around the star, on the order of 1 ms.

REFERENCES

1. B. J. Teegarden and T. L. Cline, Ap. J. Lettr., 236, L67 (1980).
2. E. P. Mazets, S. V. Golenetskii, R. L. Aptekar, Yu. A. Gur yan, and V. N. Ill inskii, A. F. Ioffe Physico-Technical Institute Preprint No. 687 (1980).
3. K. Davidson and J. P. Ostriker, Ap. J., 179, 585 (1973).
4. R. K. Wallace, S. E. Woosley, and T. A. Weaver, submitted to Ap. J. (1981).
5. S. E. Woosley, and R. K. Wallace, submitted to Ap. J. (1981).
6. B. A. Fryxell, Ap. J., 234, 641 (1979).
7. B. A. Fryxell and S. E. Woosley, submitted to Ap. J. (1981).
8. T. A. Weaver, G. B. Zimmerman, and S. E. Woosley, Ap. J., 225, 1021 (1978).
9. H. C. Graboske, H. E. DeWitt, A. S. Grossman, and M. S. Cooper, Ap. J., 181, 457 (1973).
10. N. Itoh, H. Totsuji, S. Ichimaru, and H. E. DeWitt, Ap. J., 234, 1079.

FIGURE CAPTIONS

Figure 1. Density contours (a), temperature contours (b), and the velocity field (c) at time 14 µs showing the initial expansion of the detonation wave. Cylindrical symmetry about the vertical axis (X=0) is assumed. The horizontal line at $R-R_* = 0$ represents the surface of the neutron star. The base of the helium layer, where the detonation initiates, is located at $R-R_* = -90$ m. Regions of different compositions are separated by the dashed lines. Below the line near the bottom of the grid is ^{56}Fe while the area between this region and the surface of the star is ^{4}He. The region behind the detonation front is a mixture of ^{4}He, ^{12}C and ^{56}Ni. This figure (and Figure 2) shows an expanded view of the grid containing only the first 100 zones in the X direction. At this time, the detonation has almost reached the surface. The maximum fluid velocity represented in Figure c is 4×10^8 cm s^{-1}.

Figure 2. Same as Figure 1 at time 20 µs. The shock wave has broken through the surface and is ejecting matter from the grid at a velocity of 1.5×10^9 cm s^{-1}. The detonation is continuing to propagate laterally at a velocity of 9000 km s^{-1}. Note that the strong gravitational field of the neutron star restores the planar symmetry of the atmosphere immediately after the detonation passes.

Figure 3. Same as Figure 1 at time 30 µs. This figure, as well as Figure 4, contain the entire helium layer (the first 200 zones in the X direction). The matter traveling across the surface, which was accelerated to a high velocity by the shock, catches up with (and eventually overtakes) the detonation at the base of the helium layer. If the neutron star were strongly magnetized, as it must be to produce a γ-ray burst, the field would strongly inhibit the propagation of this surface wave, although the detonation at the base of the helium layer would not be significantly affected. The maximum velocity represented in this figure is 2×10^9 cm s^{-1}.

Figure 4. Same as Figure 1 at time 41 µs. The detonation has almost reached the edge of the helium lake and is showing no signs of dying out. Note the breaking of the wave traveling across the surface.

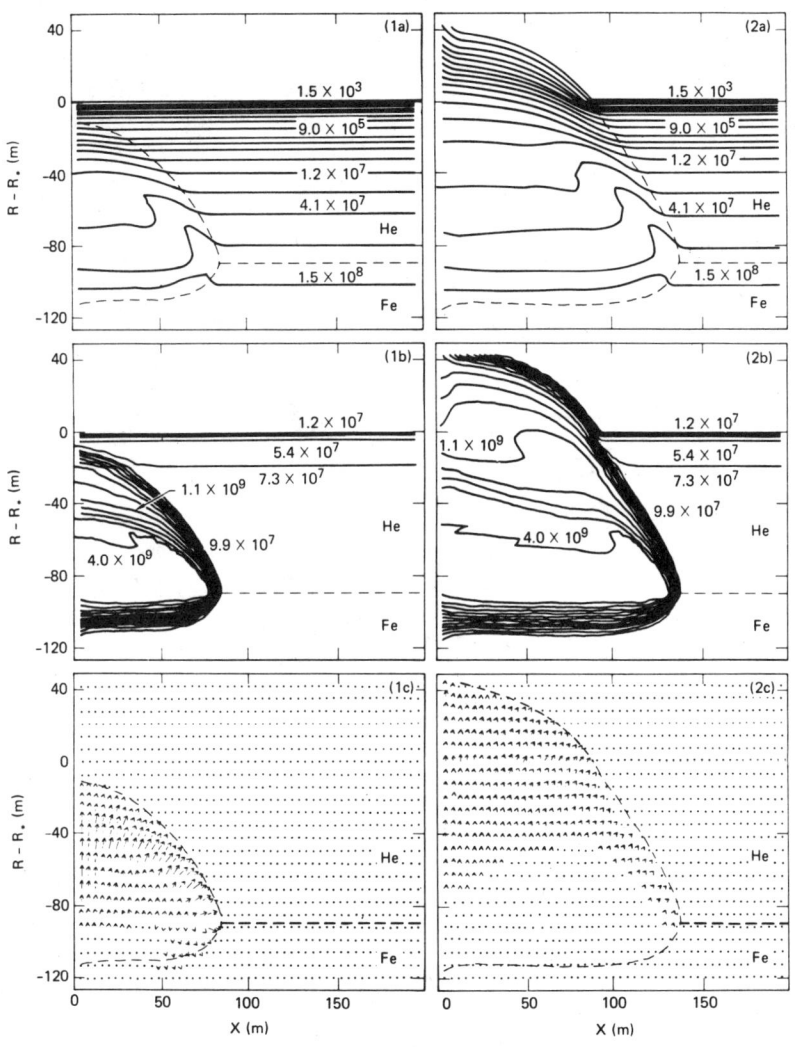

Figure 1   Figure 2

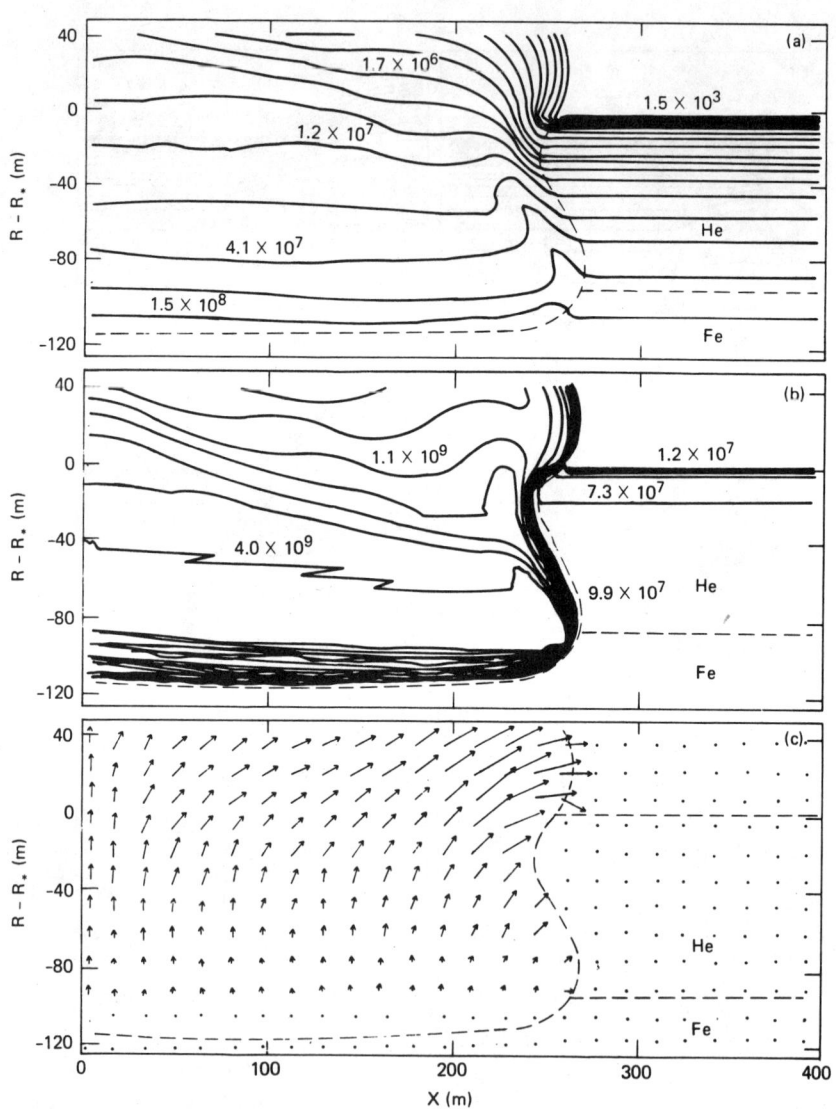

Figure 3

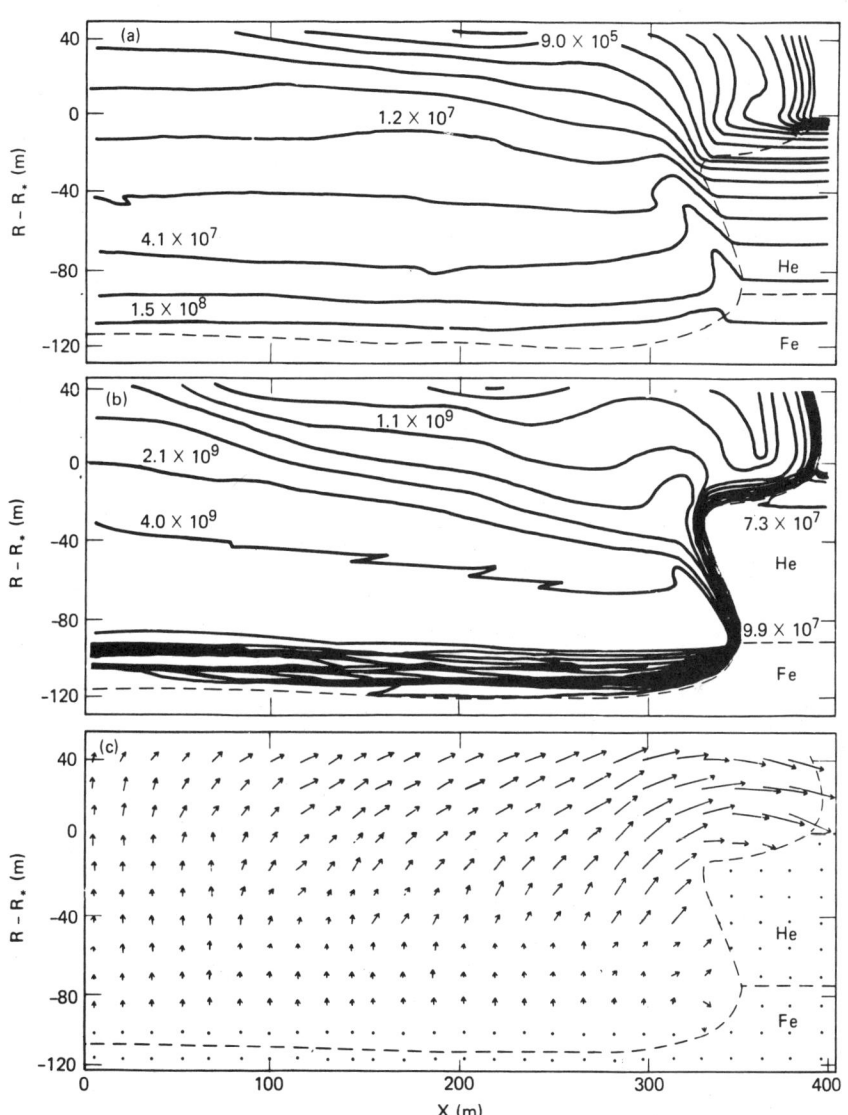

Figure 4

GAMMA BURSTS FROM NEUTRON STARS AND STELLAR FLARES

Stirling A. Colgate
University of California, Los Alamos National Laboratory
P.O. Box 1661, Los Alamos, NM 87545
and
New Mexico Institute of Mining and Technology, Socorro, NM 87801

ABSTRACT

If gamma bursts are locally galactic, then the implied fluxes from a localized region of a neutron star surface are close to the blackbody limit even at the extreme temperatures (of the order of 10^9 degrees) inferred from gamma-burst spectra. One reasonable model is the accretion of an astroid or comet (Harwit and Salpeter 1973) onto a magnetized neutron star. What is frequently described as tidal disruption, instead becomes gravitational compression. Matter landing on a neutron star releases a specific energy density of several times $c^2/10$. This energy density is ample to give rise to the inferred temperatures of 10^8 to 10^9 degrees. However, radiation stress greatly exceeds the gravitational stress even at the neutron star surface and a near instantaneous adiabatic expansion of the hot surface layers, cools them, and terminates the release of any high temperature radiation. The effective temperature of the radiation then becomes roughly the Eddington limit of 2×10^7 degrees. Only by the restraint of the free surface expansion by a strong magnetic field (several times 10^{12} gauss) can the high temperature emission take place. The radiation from such a constrained plasma is not yet understood. The cooling mechanism is analogous to the collapse phase of solar and stellar flares.

Under these circumstances, described in the abstract (Colgate and Petschek 1981) the radiation is emitted essentially normal to the magnetic field lines and the magnetic stress becomes the dominant determination of the geometry. The magnetic field plays a further role in the accretion process. First, matter of a comet or asteroid on a near miss trajectory of a neutron star will collide at larger radius, approximately 30 $r_{neutron\ star}$ resulting in a larger cross section for capture. In the subsequent accretion of this matter, the magnetic field in conjunction with the gravitational field, compresses the body (to $\cong 10^6$ g cm^{-3} and distorts it to a significantly greater extent than the gravitational field alone. The interaction of the matter in the field is initially entirely diamagnetic and the conducting matter parts the lines of force as a diamagnetic fluid. Only subsequently after collisional interaction with a neutron star surface can the matter and field become sufficiently turbulent to interdiffuse and a lower β plasma result. The diffusive cooling of a hot cloud of matter on a neutron star surface constrained by magnetic field is a problem not yet solved.

Many of the features of the neutron star accretion problem are common to the description of stellar magnetic flares. A flare is produced by the convection of magnetic field to the surface of the star (Colgate 1978). As we presently understand, this field must emerge above the surface as a loop and it must then be twisted. When the twist is large enough and the density on the rising loop becomes low enough by gravitational settling, a plasma instability is initiated. This instability will be both hydromagnetic in nature due to the violation of the Kruskal limit, i.e., the number of twists as well as a current carrier starvation instability caused by the reduction in current carriers by gravitational settling. The result is the explosive release of the magnetic energy of the field configuration. Measurements and theory confirm that this energy release is primarily at the top of the loop far above the surface of the star. Electron conduction transports some of this energy downward to the deeper layers of the star where the heating results in the near explosive expansion of the surface layers along the field lines much as described in the gamma burst model.

One asks whether some of these same mechanisms may be playing an important role in gamma bursts. As far as is currently known of the instability of flares, the primary source of instability and energy release is associated with the twist in magnetic field which is created by a current, $J_{||}$, along the field lines. On the other hand, the accretion of a dense body onto a neutron star releases gravitational energy and the interaction with the magnetic field creates diamagnetic currents induced by the perfectly conducting accreting matter. These currents by definition are perpendicular to the field lines and the instabilities inherent to a twisted field from a parallel current from stellar flare formation are excluded. On the other hand, the accretion of a body at larger impact parameter than the radius of the neutron star necessarily implies a relatively large angular momentum and the dissipation of this angular momentum by the Helmholz instability producing eddies in the fluid and field will indeed produce the form of magnetic distortion twist and $J_{||}$ which can lead to the stellar flare type phenomena. This is a problem well beyond our present ability to model, but it is tantalizing in terms of explaining some of the nonthermal high temperature emission in gamma bursts. Finally, in the conclusion of a stellar flare, the heated matter from the surface of a star that has been expelled to higher altitude along the field lines, cools, and must fall back. In the cooling process, a similar geometry to that of the gamma burst problem is encountered, namely, a hot plasma thermally supported against gravitation and confined to motions along the field lines. The frequently observed configuration of matter falling back onto the sun is a striking example of this. On the other hand, at the magnetic field strength of the sun and the density created by the flare matter expansion, the blackbody radiation cooling time constant is short compared to the gravitational free fall time and so the gamma burst phenomena of internal energy trapped by diffusion is not encountered. In some white dwarfs, and possibly M dwarfs, the necessary high field strengths probably exist, and then the problem of the radiation of a magnetically confined gravitationally bound plasma from a stellar flare may exist.

This work was supported by Department of Energy, Los Alamos National Laboratory and the Astronomy Section of the National Science Foundation.

REFERENCES

1. Harwit, M. and Salpeter, E. E., 1973, Ap. J. <u>186</u>, L37.
2. Colgate, S. A. and Petschek, A. G., "Gamma Gursts and Neutron Star Accretion of a Solid Body," 1981, Los Alamos National Laboratory preprint, to be published in Ap. J. September 1981.
3. Colgate, S. A., 1981, Ap. J. <u>221</u>, 1068.

Precursors to γ-Ray Bursts in the Asteroid Impact Scenario

Dave Van Buren
Astronomy Dept., University of California, Berkeley CA 94720

ABSTRACT

The extremely large energy densities in the fireball of a γ-ray burst lead to copious pair production and annihilation independent of the burst mechanism. The spectrum and time behavior of the burst itself thus may not be useful in distinguishing between endogenic and exogenic mechanisms. Asteroid impacts onto neutron stars can give rise to burst precursors as material from the compressed and heated asteroid (now a plasma) is attached to magnetic field lines and free-falls onto the stellar surface while the asteroid orbit decays over a much longer timescale. The properties of these precursors (duration $\simeq 10^4$ seconds, luminosity $\simeq 10^{-5}$ of the main burst, optically thick accretion spectrum) are different enough from any precursors arising from endogenic mechanisms that observations can provide a good test of competing theories. These observations may already be on hand in the x-ray survey satellite scanning data. Future detection of burst precursors can be done efficiently with a coded aperture detector with a wide field of view. An additional test of the hypothesis is that the distribution of burst fluences should mimic the distribution of asteroid masses.

THE IMPACT THEORY

The impact scenario for the γ-ray bursts was originally motivated by the observation that the typical energy released in an event corresponds to a modest fraction of the rest mass energy of a small comet or asteroid if the objects responsible are galactic[1], as seems to be the case[2]. The great burst of 1979 March 5 spurred furthur work on this theory[3,4,5], as well as on others[6].

A quick summary of the theory is in order. By some mechanism, perhaps through scattering by gravitational encounters with planets, an asteroid enters the magnetosphere of a neutron star where electromagnetic drag forces operate[5]. As the asteroid spirals in it reaches a point where it is tidally disrupted and then, a little later, turned into a plasma by ($B^2/8\pi$) forces. These same forces deform the plasma blob into a sheet[6] which then impacts edgewise onto the stellar surface. There temperatures of 10^8 °K are reached[3], the resulting fireball gives us the burst we see.

Regardless of the energy release mechanism, energy densities in the burst source region are of the order of 10-100 MeV per nucleon. At these temperatures, e^+-e^- pair production is ubiquitous[7]. Consequently, spectral information which might have been a clue to the burst mechanism is lost as the pair plasma and attendant γ-rays propagate to form the fireball. Since the fireball looks the same independent of how it was formed, its behavior and spectrum will not be good diagnostics of the burst mechanism.

PRECURSORS

Luckily, the impact scenario provides a diagnostic that endogenic energy sources do not. After the asteroid is turned into a plasma on its way to the surface, material can attach onto field lines and fall down onto the star. Since this material falls on a dynamic time scale, while the asteroid orbit decays over a slower timescale $\tau_{decay}=4\pi\rho_a r_a c/B$, this will give a precursor. Here ρ_a and r_a are the asteroid density and radius; c and B have their usual meanings.

The radius R of the asteroid orbit changes at a rate

$$\frac{d}{dt}\left(\frac{1}{R}\right) = -B^2 r_a^2 / m_a c ,$$

where m_a, the asteroid's mass, is lost according to

$$\frac{d}{dt}(m_a) = 4\pi r_a^2 \rho_a v_{attach} .$$

Here the material attaches onto field lines at an effective velocity v_{attach}. This is generally much smaller than the thermal velocity of the plasma particles because the blob is in pressure balance with the magnetic field. In writing these equations the shape of the blob is taken as spherical.

These equations can be integrated from the point at which the asteroid first becomes a plasma until it impacts on the stellar surface if an equation of state is chosen. Figure I shows the solution in terms of the accretion luminosity $L=.1m_a c^2$ for the case of an adiabatic perfect gas. The two curves give the time history of the precursor luminosity for asteroids with initial radii of 1 and 10 km, a stellar radius of 10 km and surface magnetic field of 10^{12} G. While the peak luminosity of several 10^{34} erg s^{-1} is not very sensitive to the asteroid mass, the precursor duration is roughly 3×10^4 (r_a/1km) seconds. The insensitivity of the peak luminosity to the initial conditions suggests that observed precursor fluxes can be used as distance indicators.

We can estimate the temperature at the accretion point by requiring that $\rho v/B$ be a constant along a flow line. That temperature is $T=(.1c^2\rho_a v_{attach}(B_*/B_p)/\sigma)^{1/4}$ or $\approx 10^7$ °K so the bulk of the radiation is in the x-ray region. The ratio B_*/B_p is the ratio of the magnetic field at the surface to that at the point where the asteroid becomes a plasma, σ is the familiar radiation constant. Because the magnetic field compresses this material faster than the tidal force streches it out, the accretion column at the stellar surface is very optically thick.

With these temperatures and luminosities, burst precursors should be visible to several kpc in scanning data to a flux limit of 10^{-11} erg cm^{-2} s^{-1}. If the burst rate density is 10^{-8} pc^{-3} yr^{-1} (ref. 2) then this corresponds to tens of possible detections per year. Since roughly 2% of the sky was observed at any one time by the HEAO-1 satellite, the chances are not too bad that a precursor was seen.

Burst precursors of this type are not detectable by triggered instruments because of their slow rise times and low flux levels compared to the bursts themselves. Future instruments capable of observing them should have as wide a field of view as possible as well as sensitivity to slowly varying low flux levels. A proposed instrument that satisfies these criteria is the coded aperture camera[8].

A final note about a statistical test of the asteroid impact hypothesis. Since accretion of the asteroid onto the star is responsible for the energy released, the distribution of fluences should mimic the distribution of asteroid masses. The observed distribution of asteroids with masses greater than M is $N(M) \propto M^{-2.84}$ (ref. 9). A power law with this index is consistent with the distribution of burst fluences[10].

I would like to thank the Thacher School for the use of its facilities. This work was supported in part by NSF grant AST 79 23243.

REFERENCES

1. Harwit, M. and Salpeter, E. E. 1973, Ap. J. (Letters), 186, L37.
2. Jennings, M. C. and White, D. S. 1980, Ap. J., 238, 110.
3. Newman, M. J. and Cox, A. N. 1980, Ap. J., 242, 319.
4. Colgate, S. A. and Petschek, A. G. 1981, Ap. J., 249, in press.
5. Van Buren, D. 1981, Ap. J., 249, in press.
6. Woosley, S. E. 1981, these Proceedings.
7. Cavallo, G. and Rees, M. J. 1978, M. N. R. A. S., 183, 359.
8. Gorenstein, P. 1981, these Proceedings.
9. Dohnanyi, J. 1969, J. Geophys. Res., 74, 2531.
10. Jennings, M. C. 1981, these Proceedings.

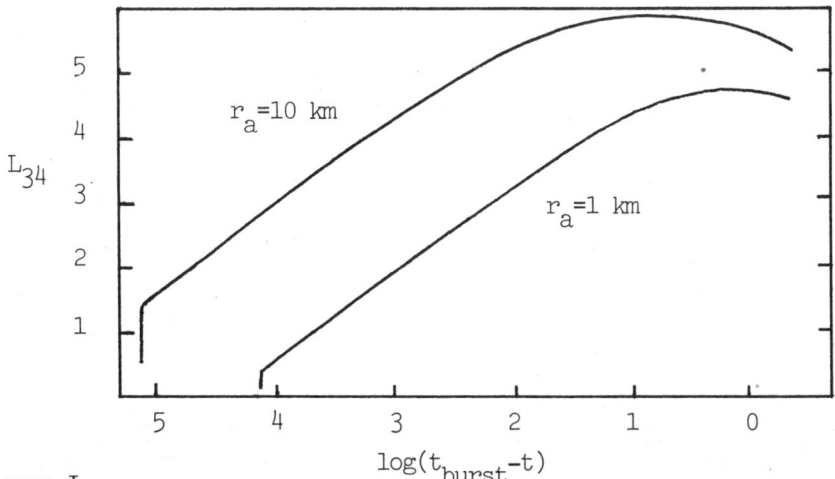

Figure I.

PROTON DECELERATION IN A NEUTRON STAR ATMOSPHERE

J.G. Kirk and D.J. Galloway

Max-Planck-Institut für Astrophysik
Karl-Schwarzschild-Strasse 1
D-8046 Garching bei Muenchen
Federal Republic of Germany

ABSTRACT

The energy loss rate of fast protons in a strongly magnetized electron gas has been calculated using the Fokker-Planck equation to describe the dominant effect of small-angle Coulomb scattering. Stopping lengths or order 1 gm cm^{-2} are obtained for 30 MeV protons and the possibility of maser action at the gyrofrequency is indicated.

INTRODUCTION

The depth to which a freely falling proton, which has a kinetic energy of 30-100 MeV, will penetrate the ionized gas layer immediately above a neutron star surface is of fundamental importance for the construction of models of γ-ray bursters and X-ray pulsars. The very strong magnetic field thought to be present in these objects has a pronounced effect on this penetration depth. Previous calculations[1,2] have arrived at a value considerably in excess of that for an unmagnetized plasma. However, these results were obtained under the assumption that the plasma layer over the surface of the star is cold.

We have performed calculations[3,4] which used the linear response function for a hot magnetized plasma[5] to obtain expressions for the Fokker-Planck coefficients. A knowledge of these coefficients enables the Fokker-Planck equation to be integrated numerically, and one is then able to find the rate of loss of energy of a proton in the plasma. An important aspect of this treatment is that full account is taken of the curvature of the path of the proton. This is important when the Debye screening length of the plasma (i.e. the maximum impact parameter in a Coulomb scattering event) is of the same order as, or greater than, the radius of

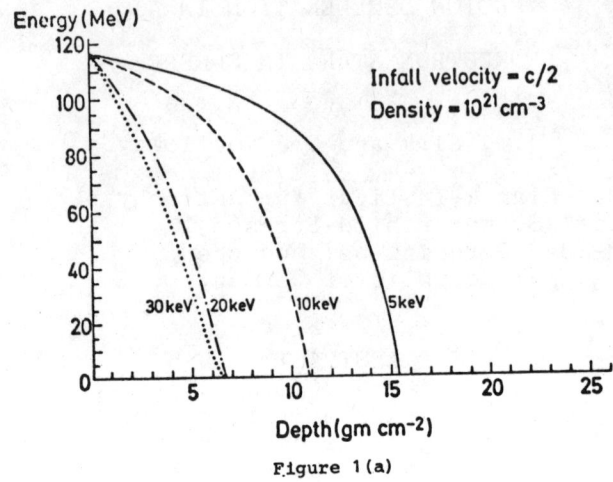

Figure 1(a)

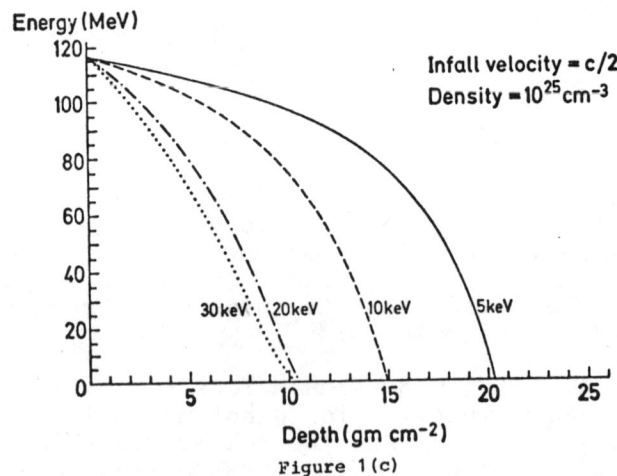

Figure 1(c)

Figure 1 - The stopping length for plasma temperatures 5, 10, 20 and 30 keV and densities (a) 10^{21} cm^{-3}, (b) 10^{23} cm^{-3}, (c) 10^{25} cm^{-3} and (d) 10^{26} cm^{-3}. In each case the initial infall velocity is c/2, and the energy is normalized to the kinetic energy of a single proton.

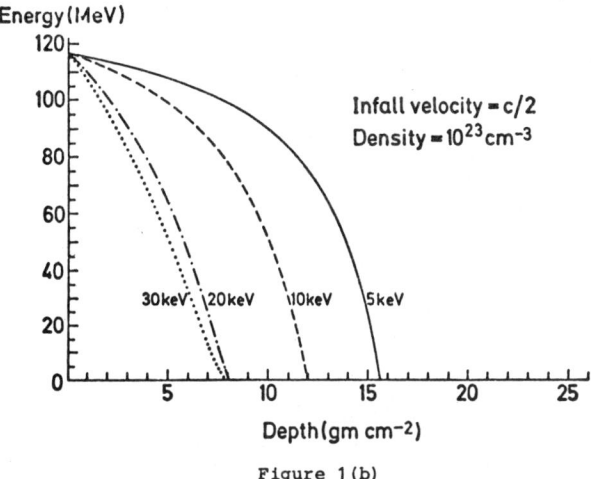

Figure 1(b)

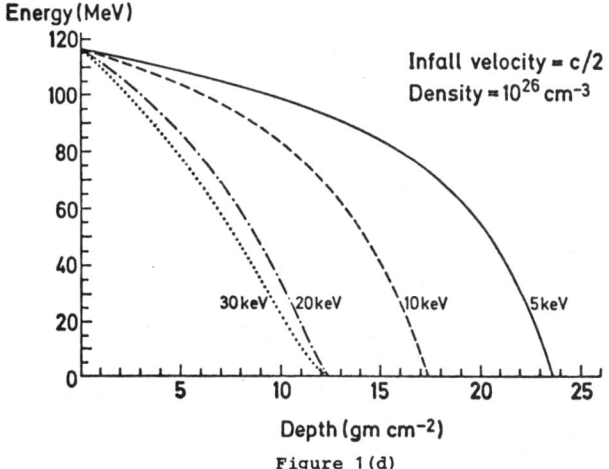

Figure 1(d)

gyration of the proton's orbit in the strong magnetic field. Such conditions apply for typical parameters used in building models of X-ray pulsars and γ-ray bursters.

RESULTS

For an initial velocity of $c/2$, the energy of a proton as a function of depth in the atmosphere is shown in Figure 1. In each case the plasma is taken to be homogeneous. Four values of the plasma temperature are used, and the density is chosen to be 10^{21}, 10^{23}, 10^{25} and 10^{26} particles per cm^3 in the cases a, b, c and d respectively. These stopping lengths are considerably shorter than those for an unmagnetized plasma, particularly if only the later stages of deceleration (from say 50 MeV) are considered. They indicate the possibility that an accretion flow may be stopped within a layer of atmosphere which is optically thin to Compton scattering.

Another interesting aspect of the results is that during the deceleration, protons first <u>gain</u> energy in motion perpendicular to the magnetic field. Thus, in a certain phase of the evolution, the distribution function of the protons contains more particles at higher perpendicular energies than at lower. Such a situation is conducive to stimulated emission of radiation, in this case probably at a frequency equal to the second harmonic of the proton gyro-frequency[6] (60 eV). A significant fraction of the accretion energy would be removed by such a maser, and may be observable if γ-ray bursters are close by objects ($\lesssim 100$ pc), or if the magnetic field of a burster or an X-ray pulsar is low enough to bring the proton's gyro-frequency into the visible spectrum.

REFERENCES

1. M.M. Basko and R.A. Sunyaev, Sov. Phys. JETP <u>41</u>, 52 (1975).
2. G.G. Pavlov and D.G. Yakoviev, Sov. Phys. JETP <u>43</u>, 389 (1976).
3. J.G. Kirk and D.J. Galloway, MNRAS <u>195</u>, 45P (1981).
4. J.G. Kirk and D.J. Galloway, Plasma Physics in press (1981).
5. J.G. Kirk, Plasma Physics <u>22</u>, 639 (1980).
6. R.J. Stoneham and J.G. Kirk, in preparation.

Sec. II X-ray Bursts

SOME TOPICS ON THE X-RAY PULSARS AND THE X-RAY BURSTS OBSERVED BY THE SATELLITE HAKUCHO

Minoru ODA
Institute of Space and Astronautical Science
6-1, Komaba 4, Meguro-ku, Tokyo

ABSTRACT

Several subjects are selected from observational results on X-ray pulsars and X-ray burst sources obtained by the Japanese X-ray astronomy satellite "Hakucho." Results on Vela X-1 and A0535+262 which exhibited transient or time variable pulsar behavior are discussed. We are lead to speculate that the category of Class-II X-ray sources, represented by X-ray sources in globular clusters, very bright sources within $30°$ of the galactic center, GCX sources and soft transient sources, may be characterized by its potentiality for bursting whenever certain, yet unknown, conditions are fulfilled. The results on the X-ray burst may provide keys to evolve the thermonuclear flash model from the simplified original He-flash version to explain much involved dynamic processes in the neutron star atmosphere. Necessity of reservoirs at the neutron star to store matter as the nuclear fuel or the source of gravitational energy appears to be evident in order to explain observational facts of the bursts. Some of the difficulties associated with the model are growing to be definitive even to the extent that the thermonuclear flash model of the Type I burst may have to be rereviewed.

INTRODUCTION

This paper is to report on some selected subjects from the observational results on Galactic X-ray sources obtained by the Japanese X-ray astronomy satellite "Hakucho" (for instrumentation see Kondo et al.(1): general reviews of observational results are given by e.g. Tanaka, Y. (1979(2), Oda, M. (1980(3), Ogawara, Y. 1980 (4), Oda, M. and Tanaka, Y. 1981(5), Hayakawa, S. 1981(6)). I will also describe a couple of collaborative programs under way. What is common along the subjects, perhaps, is their relation, at least as a possibility, to the physics of the neutron star.

Main objectives of "Hakucho" are the wide field of view burst watch, and the observations of pulsars and other variable sources. Directions derived from the wide field of view observation and of pinpointing the source are met by the use of the rotating modulation collimator. Correlation maps in celestial coordinates are produced from the observed modulation of X-ray fluxes. The correlation map exhibits the location of X-ray sources in the field of view. Rotation period of the spacecraft is approximately ten seconds, and the duration of the burst is sufficiently long to generate a correlation map. When an X-ray burst occurs, the map for the quiet sky is subtracted from the map produced for the period of the burst and the map for the burst source is yielded(7). An example of the correlation map which indicates GX5-1 and GX3+1 superimposed on a Palomar Sky Map is shown in Fig. 1. The satellite

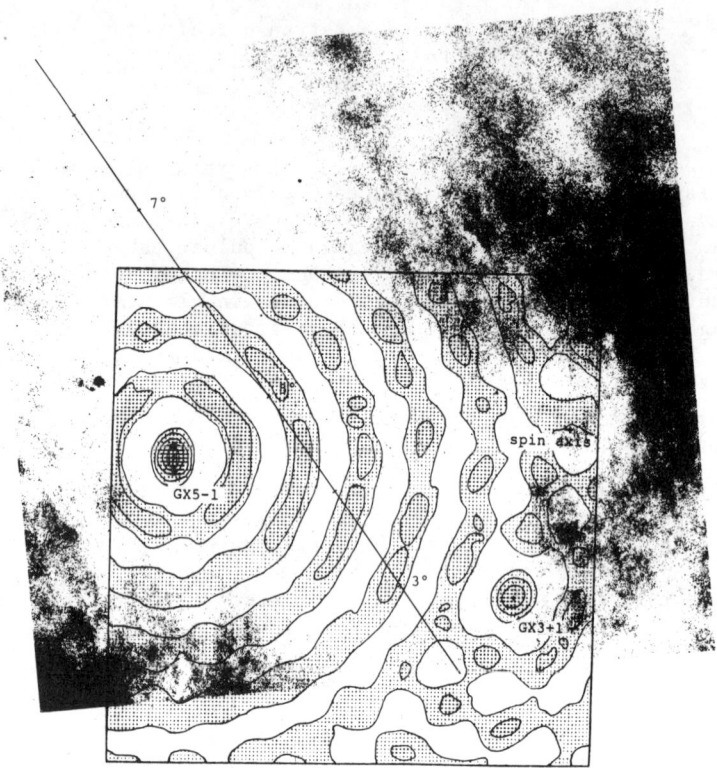

Fig. 1. A correlation map superimposed on a star map is shown in order to indicate the quality of the identification of a source.

has been in operation since March 1979, when SAS-3 and Ariel-5 approached the ends of their lives, and has succeeded in the role of generating the status report of the X-ray sky for various variable sources.

The galactic X-ray sources over the 1∼100 keV energy band may be, at least as a simplified working hypothesis, classified into the following categories: I) supernova remnant (Crab Nebula, Cas A etc.) II) close binary system which consists of a normal star and a collapsed star such as i) a neutron star, (X-ray pulsar, X-ray burster etc.) ii) a black hole (Cyg X-1) and probably iii) a white dwarf. Among various observations I will pick up several topics of interest on the X-ray pulsar and the burster.

THE X-RAY PULSAR

The X-ray pulsar is characterized by its pulsation, eclipse (in most cases) and its hard spectrum: if the spectrum is expressed

by a thermal exponential spectrum, kT is larger than 15 keV. It is
generally believed to be the close binary system of a young massive
star and a strongly magnetized neutron star. It is considered that
the accretion matter funnels into the magnetic poles of the neutron
star and produces two hot spots, whereby the neutron star appears as
a pulsar by its rotation if the spin-axis and the magnetic dipole
are not coaligned. We may call this category Class-I X-ray sources
corresponding to Population I binary systems. I discuss some properties of a couple of X-ray pulsars.

First, Vela X-1. It is known that the pulsation period of a
normal (radio) pulsar gradually increases; i.e. the spin of the
neutron star slows down except for the moment of so-called glitch or
the starquake. In contrast to this, the period of an X-ray pulsar
normally decreases; i.e. the neutron star spins up, probably acquir-
ing the angular momentum from the incoming mass flow. Fig. 2
shows the long term variation of the pulsar period of Vela X-1 (8,
9). The general trend of the spin-up until 1979 is clear, except

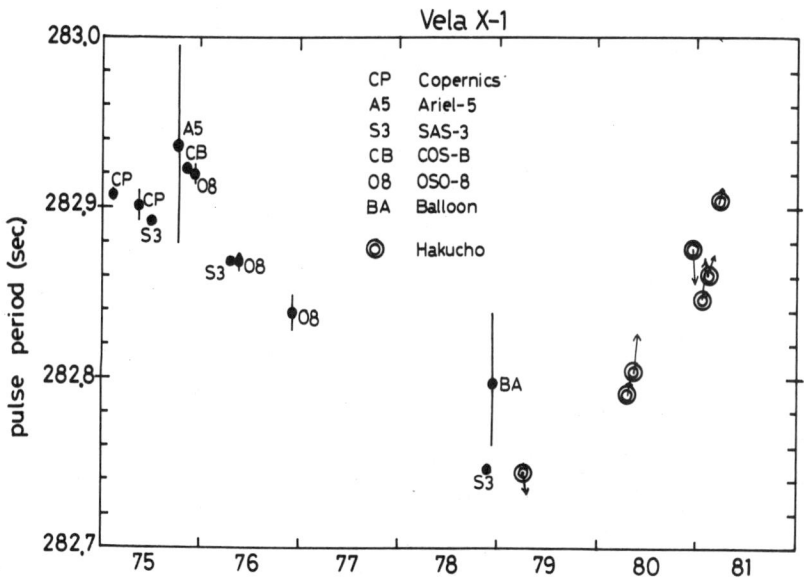

Fig. 2. Long term variation of the pulsar
period of Vela X-1. Five observations
during March 1979 and March 1981 are shown.

for a possible "negative" glitch at the end of 1975. A glitch
of both signs can be qualitatively understood in terms of the re-
formation of the interior structure of neutron star when the stress
by the, say, friction between the inner superfluid core and the
crust of the neutron star reaches a critical limit.

To our surprise, since 1979 the pulsar period has changed the
sign of its variation and has been continuously increasing; i.e. the
neutron star was slowing down right up to our last observation in
April 1981. A secular variation of the pulsar period over a several

year's time scale is suggested for this trend rather than a glitch.
What causes this secular variation is an intriguing question...
If it is of intrinsic origin, some unknown physical processes must
be going on. Makishima (private communication 1981) on the other-
hand suggested that, if this were to be explained in terms of a
Doppler effect caused by a highly eccentic orbital motion of the
binary around the third body, the mass of the latter must be larger
than 100 $M_\odot$ and, since no such bright object is seen, it could be
a giant black hole.

The change of sign of a pulsar period variation was also noticed for GX 301-2, another slow pulsator(10). It is an interesting
question whether these two cases belong to a physically similar
phenomenon or not. To my knowledge this change of sign for GX 301-
2 has not been deeply discussed.

Another fact of interest is the fluctuation of the period as
clearly seen in Fig. 2 which summarizes the results of analysis on
7 data sets between March '79 and March '81. The rapid variations
of day's time scale of both signs, superposed on the slow spindown
trend with an amplitude of $\sim$ 0.02 sec, are noted. The question
whether this fluctuation directly reflects the variation of the mass
accretion rate or is of intrinsic origin has not yet been resolved, but in any case this should have a deep bearing on what is
going on inside the neutron star. Further study of the time variability of X-ray pulsars will be an important probe for the physics
of the neutron star structue (11, 12).

Secondly, we consider the transient pulsar A0535+26. If we plot
epochs of outbursts observed thus far, which last for weeks on a diagram of time scale folded modulo 110 days, the reccurrence with some
jittering is almost convincing. We observed the outburst of A0535+
26 from its onset through the disappearrance in October 1980. The
delay of the pulse arrival times for the "standard" pulsar period,
103.62 sec, for the long span ($\sim$ 25 days) of observation is shown
in Fig. 3. The conventional polynomial fitting of the curve shows
that $\dot{P}/P$ for early October was $2\ 10^{-2}$/yr and then gradually decreased (13). If the intrinsic period is unchanged and the change of
the observed period is attributed to the orbital Doppler effect of
the binary system, severe constraints are derived for the orbital
parameters. The upper limit of the orbital period is $\sim$120 days
with the eccentricity $\sim$0.4, being consistent with the 110-day recurrence period. Is the occurrence of the outburst synchronized
with the orbital phase suggesting that the mass accretion occurs
around the periastron of a highly eccentric orbit? The question has
not yet been answered.

On the other hand, suppose the variation of the period is
intrinsic. In order to explain the approximate constancy of the
period within the accuracy of 0.3 sec for several years we have to
assume the existence of an intrinsic clock which keeps the period
constant and a large despin between successive outbursts, since a
considerable spin-up during each outburst is expected (14). The
analysis and discussion of this object are preliminary and the compilation of the present data with earlier observations is urged.

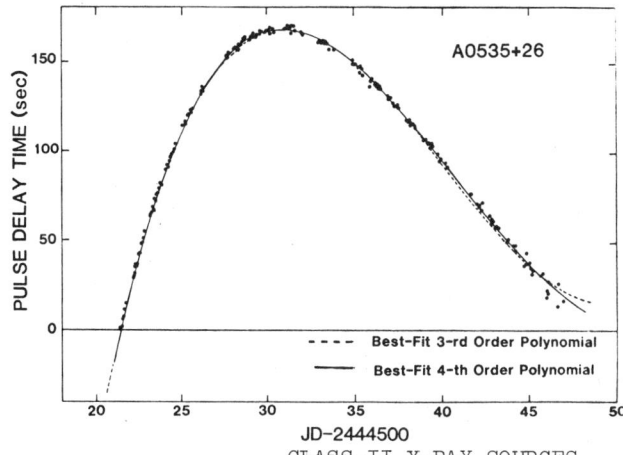

Fig. 3. The delay of the pulse arrival time for the assigned pulsar period, 103.62 sec is exhibited for the whole course of the outburst.

CLASS-II X-RAY SOURCES

We postulate the existence of another class, Class-II, sources in contrast to Class-I sources or X-ray pulsars. They cluster loosely in the vicinity, say within 30°, of the galactic center along the galactic plane. They are often referred to as the "galactic bulge sources" (15) and assumed as members of an old stellar population, or Population II.

X-ray sources in globular clusters, very bright sources called bright bulge sources, and transient (often recurrent) sources with soft spectra, $kT \sim 5$ keV, belong to this class. Sources within a couple of degrees of the galactic center (GCX) may be the same. We are not sure if Sco X-1, Cyg X-2 and 4U0624+09 belong to the same class, while certainly they are not Class-I sources. They commonly exhibit no pulsation, no eclipse and soft spectra. They are assumed to be the combination of a low mass late-type star and a neutron star which by some reason lost most of its magnetic field. Since the field is presumably weak, the matter accretes and is heated over the whole neutron star surface or the equatorial zone of the surface, so that the object does not exhibit the pulsation due to the stellar rotation.

DISTRIBUTION OF BURST SOURCES IN THE GALAXY

An emphasis in the Hakucho observations was placed on the watch of bursts whenever various conditions such as the sun-angle constraint of the spacecraft allowed. (For the X-ray bursts, see e.g. the recent review article by Lewin and Joss (16)).

A preliminary revison of the burster catalogue presented by Lewin and Clark (15), Lewin and Joss (16) and the Hakucho team (3, 5) as of June 1981 is given in Table I. (We use the designation of the X-ray sources following the Uhuru Catalogue: right ascension in hour-min and declination in units of 0.1 degree are indicated.) As careful work has not been completed, the content of this table is tentative. New burst sources were added to the catalogue by the Hakucho observations In addition, some suspected sources in earlier catalogues have been confirmed. We gave these sources the classification burster under the new light that bursters often remain dormant and become burst-active from time to time and that the burst profiles are various; not only the typical burst profile with a fast rise but those with a slow rise in several seconds appear to be common (17, 18). Many of the tabulated bursters are not presently burst-active. They "come and go" or have the experience of bursting at least once in the past. Examples of the varied burst profiles are shown in Figure 4.

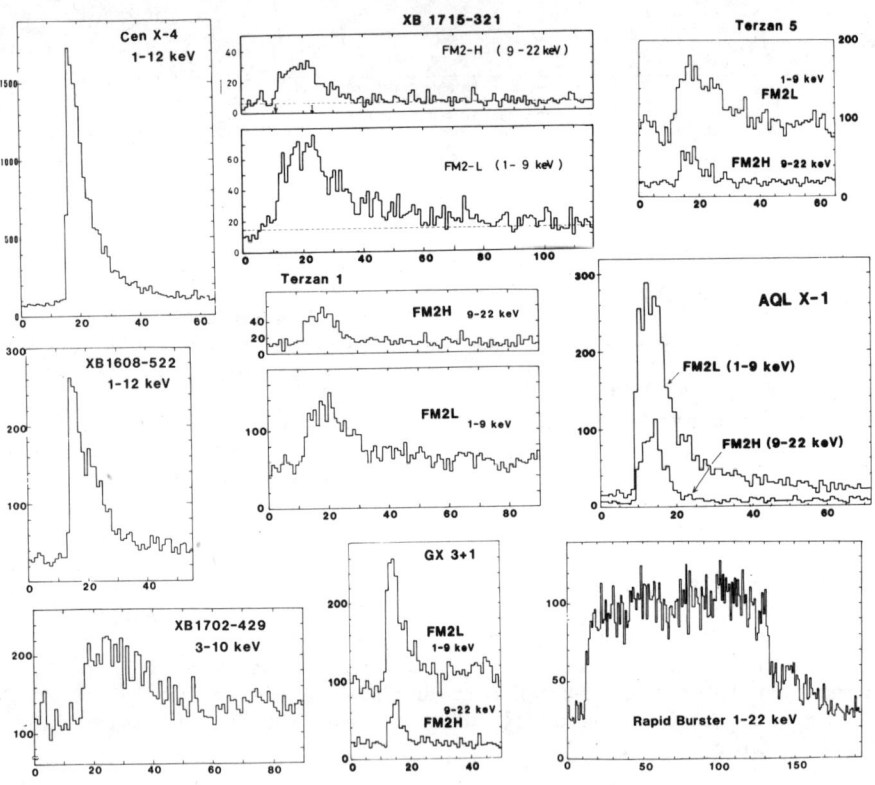

Fig. 4. Examples of burst profiles. Note the variety.

Figure 5 maps the bursters in galactic coordinates. They cluster near the galactic center region. The bursters in the figure are classified in several groups. We find from Table I and Figure 5 the following features of a burster.

1) Considerable portion (approximately one third) of bursters are found in globular clusters.

2) Most of the X-ray sources in globular clusters, if not all, generate bursts (under some, yet unknown, conditions).

3) While several bright bulge sources which are not in globular clusters are known to generate bursts, many of them do not (18,19).

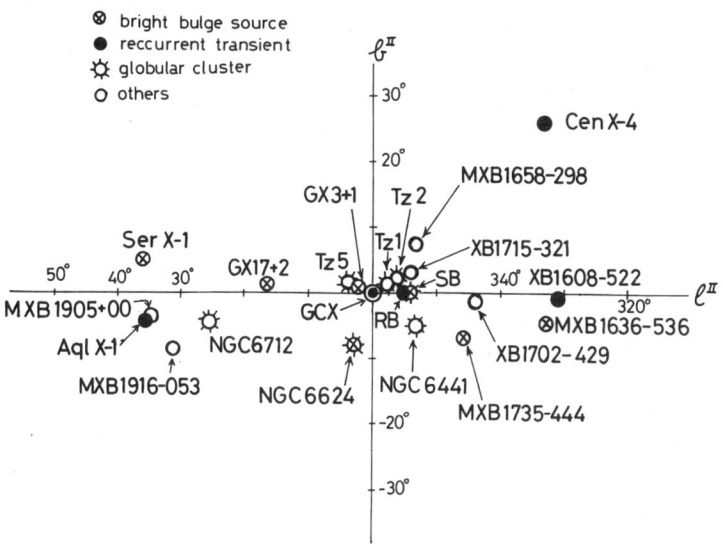

Fig. 5. Sky map of X-ray bursters near the galactic center. The bursters are classified into groups. The GCX sources are indicated by a single mark at the galactic center.

4) Some transient sources (Cen X-4, Aql X-1) which are known to recur for weeks generate typical bursts, probably when the luminosity becomes low (20,21,22) in each outburst. The source 1608-522 may also be of this class (17). (We stress that Aql X-1 is an object for which a well-identified companion star (~K0) is known, hence this is a direct evidence of the burster being associated with a late-type star (23).)

5) Three sources located within ~1° of the galactic center (GCX sources) were observed to burst. The GCX bursters discovered by SAS-3 (24) were quiescent in 1979, but at least one of them was burst-active in 1980 (25). (Whether GCX sources form a physically distinct group from other Class-II sources or there are no differences and they are simply located there is a yet unresolved question).

THERMONUCLEAR FLASH MODEL

A widely postulated interpretation of the burst is, of course, that the neutron star atmosphere is explosively heated by the thermonuclear flash of the accreted matter and cools off by generation of X-rays as the black body radiation. (For historical accounts of the interpretation of the burst, see (16) and references therein.) The radius of the spherical burst source can be calculated from the X-ray intensity and measured temperature on the assumption of the black body radiation, if we know the distance to the source. The radius is known to be always constant through the cooling phase of the bursts from burst to burst for each source inspite of a wide variety of the burst profile and burst peak intensity (26,18). And for GCX, Terzan 1, Terzan 5, GX3-1 and 1728-34 which are all near the GC (hence the distances are known) the radii are all ~10km. Impressive stability of the radius at ~10km for the cooling phase was strongly reconfirmed by the Hakucho observation as exhibited in Fig.6. It strengthens the neutron star hypothesis of the burst and implies

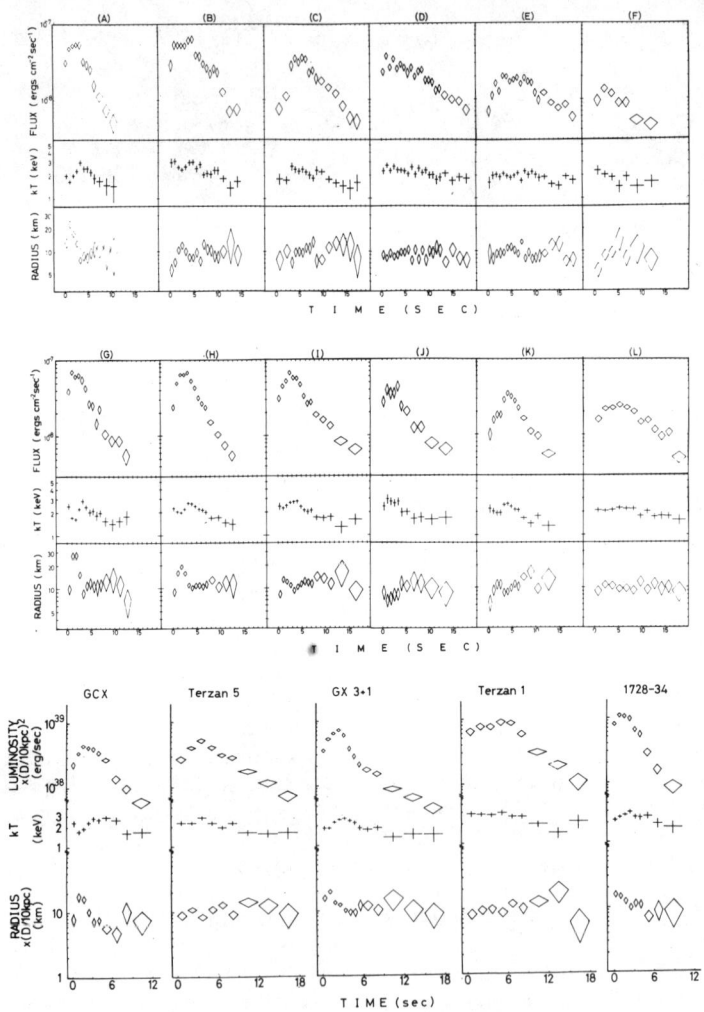

Fig. 6 Evolution of X-ray intensity, temperature and radius during bursts. a) for a number of bursts from 1636-536; and b) from different sources near the galactic center. Stability of the radius at ~10km at the decay (cooling) phase is impressive.

that the nuclear flash occurs always over the entire neutron star surface and not on a localized region of the surface (26, 18).

The parameter α, defined as the ratio:
$$\frac{\text{persistent flux of a burster}}{\text{time averaged burst flux}} = \frac{\text{gravitational energy/nucleon}}{\text{nuclear energy/nucleon}}$$
is of the order of 100 if the thermonuclear flash is due to He (27, 15) and is about 20 for the Hydrogen flash. An origin of the thermonuclear flash model of the X-ray burst was that the numerical calculation on the He flash by Joss (32) beautifully reproduced various properties of some bursts and α was approximately 100 for some bursts.

While for certain sources the time profile of the burst appeared to be stable like their fingerprints, some other sources showed that the profile often varied from burst to burst and from time to time during a long time span. Existence of distinct modes of the burst profile and α, possibly "bimodal" behavior, was once suggested from the finding of clear distinction between two successive episodes for the bursts with a fast rise and a short duration, and those with a slow-rise and a longer duration from 1608-522 (17). However, the next observation after one year revealed possible existence of another mode. The burst properties of 1608-522 are not very clear-cut (28). Nevertheless, it is still true that burst profiles may be classified into a type of a fast rise with a typical exponetical decay (Joss-Type) and that of slow rise with a longer duration.

A wide variety of the burst profiles and that of the α-value over $\sim$20 to beyond 100 even for bursts from a single burster from time to time have been revealed for a number of bursters. It has become clear that, while for some bursts, the origianlly suggested He-flash model fitted well, bursts of different property are not exceptional and the wide variaions of the profile and α-value are common phenomena.

Another variation of the profile, a clear double-peak, has been observed (29, 28). Bursts of this profile are rare but appear for many bursters: 1743-29, NGC 6712, slow burster, Terzan 5, 1608-522, 1636-536, 1906+000 etc. This must be common feature which appears under certain conditions.

I skip these subjects today but I stress that such properties of the burst may reflect the complex nature of the neutron star atmosphere and provide a clue to evolve the theory of the burst under the context of the nuclear flash model from its earlier version in conjunction with dynamic physical processes in the atmosphere (30). Though the present state of the theory is still premature to account for the complex nature of the burst activity, the phenomenon undoubtedly involves very interesting and important problems to be solved.

THE BURSTER AND THE CLASS-II SOURCE

The relationship between the burster and other groups of Class-II X-ray sources may be summarized in Figure 7. It is still premature and several important questions remain unanswered: Do the sets indicated by question marks exist?; e.g. are all recurrent transients at least potentially bursters?; etc.. These features of the bursters, thus summarized, may help in

Table of Bursters (preliminary)

Position & Name	Category	Before March 1979 (Lewin & Clark Review)
0512-401 4U,MXB	NGC 1851*	Out of "bulge", 9.5 kpc
0900-403 4U,MXB,Vela X-1	pulsar	
1455-31 XB,Cen X-4	10yr R.T.	Vela satellite : burst?
1608-522 4U,XB	600d R.T.	UHURU,Vela:burst?,SAS-3:obscure?
1636-536 4U,MXB	B.B.S.	"reliable"
1659-298 MXB	transient?	2~2.6h stable, slow type burst, no burst for high state, α~25
1702-429 4U,XB		MXB1706-43?
1715-322 4U,XB		SAS-3: fast transient?
1724-30 4U,XB	Terzan 2*	OSO-7, OSO-8: fast transient
1728-337 4U,MXB, slow burster	B.B.S.	"reliable", 4~8h, some times double peak
1730-333 MXB,H rapid burster	Liller 1* 180d R.T.	Type I, Type II bursts
1733-30	Terzan 1*	
1735-444 4U,MXB	B.B.S.	"reliable" interval irregular, opt/X
1742-29 MXB,A?,2S?	in GCX	ID with persistent source uncertain
1743-28 MXB	in GCX	3 bursts in succession
1743-29 MXB,A	in GCX nova?	ID with nova uncertain
1744-265 4U,GX3+1	B.B.S.	
1745-25	Terzan 5*	
1746-370 4U,2S	NGC 6441*	probably
1813-140 4U,2S,GX17+2	B.B.S.	
1820-303 MXB,4U	NGC 6624*	
1837+049 MXB,4U,Ser X-1	B.B.S.	"reliable",interval irregular,opt/X
1850-087 MXB,A,2S	NGC 6712*	almost certain
1905+000 MXB,4U		
1908+00 XB,4U,Aql X-1	R.T.	
1916-053 MXB,4U		

* = in Globular cluster
B.B.S. = bright bulge source
R.T. = recurrent transient

HAKUCHO April-September 1979	HAKUCHO 1980 April-September	HAKUCHO 1981 April-Mid July	
no observation	no observation	no observation	
one very long burst (10^3 sec)	a very long burst	a very long burst	#
a very large Type I burst, 1.5kpc?	no observation	no observation	#
active: 22/45d, fast/slow	active: 24/60d	active: 18/36d	#
active: 41/45d, opt/X, fast/slow	active: 35/57d, opt/x	active: 35/37d	
inactive: -/17d	inactive: -/60d	inactive: -/28d	
active: 19/40d structures in profile	3/12d	4/40d	#
slow type bursts: 3/24d	inactive: -/74d	inactive: -/29d	#
inactive: -/24d	inactive: -/74d	-/29d	
17/37d	active: 41/74d	~ 5/29d	
active Aug.8~22, trapezoidal, no type I burst	inactive: -/74d	inactive: -/10d	
-/24d	2/74d	-/31d	#
3/33d	inactive: -/12d	inactive: -/25d	
inactive: at most a few bursts in 25 days	active: 14/72d	~ 2/17d	
-/22d	18/68d	-/8d	#
-/19d	14/52d	-/11d	#
inactive: -/39d	inactive: -/58d	1/26d	#
no observation	no observation	2/12d	#
inactive: -/18d, high state	inactive: -/29d	-/4d	
inactive: -/10d	inactive: -/13d	-/1d	
almost no observation	-/6d	2/8d	#
3/13d, oscillatory structure	3/18d	-/3d	
-/8d	2 bursts in its decay phase of out burst, 2/18d	-/3d	#
1/8d	1/18d	-/3d	

\# = newly found of suspected source confirmed by HAKUCHO

The exposure time of one day observation is 3-4 hours.
Sensitivity to a burst is not always same.

understanding the physics of formation or genetics of the Class-II sources, though arguements still remain speculative. Further observational studies will be necessary and rewarding.

Immediate speculation from impressive correlations among globular clusters, X-ray sources and bursters is that the Class-II binary system in the globular cluster is formed by the capture process (31) and by some reason is in a favorable condition for bursting. The fact that we have not yet seen a globular cluster which contains two or more X-ray sources is still compatible with the hypothesis of the capture process, considering that in one out of approximately ten globular clusters an X-ray source resides and the number of the bursters identified with the globular clusters has increased to 7. (weaker X-ray sources identified with globular clusters discovered by Einstein Observatory are not counted in this statistical arguement.) The capture process may not be applicable to other sources (bright bulge source, soft transient source,) from a statistical arguement and they are likely to be primordial (16).

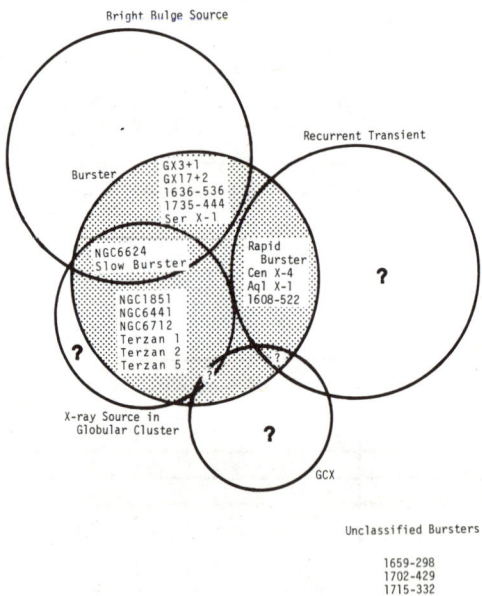

Fig. 7. The relationship among the burster and other Class-II X-ray sources. There are some bright bulge sources which are not the burster. There are no clear cases of globular cluster source, reccurrent transient and GCX which are not the burster.

We stress that, irrespective of probably different genetic origins of the binary system, each kind of Class-II source includes at least a few sources which have been known as bursters. We may speculate and hypothesize that the explicit or potential "burstability" is a common feature of the Class-II sources and that, whenever conditions allow, the source bursts. The conditions for the source to burst are obscure but must be common throughout various kinds of Class-II sources: They may be for example, weak magnetic field, proper matter accretion rate, suitable temperature etc..

PUZZLING FACTS OF X-RAY BURSTS

Over two years, Hakucho observations have provided on one hand a number of results which reenforce the general picture of the X-ray burst under the context of the thermonuclear flash model, and several problems have gradually been formulated for theorists to tackle. On the other hand difficulties for the picture which has already been pointed out have become more definite. I will pick up a few of the problems.

i) SUPERCRITICAL LUMINOSITY?

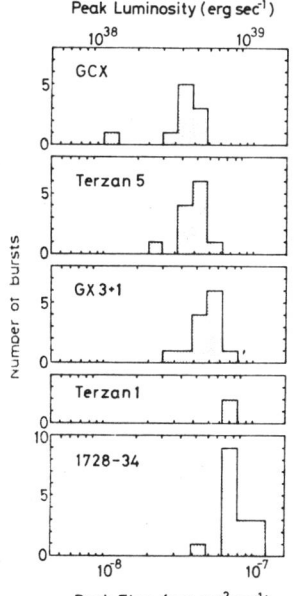

Fig. 8. Distribution of burst peak fluxes for burst sources near the galacatic center. Note that the peak luminosities estimated for 10 kpc distance exceed the Eddington luminosity 2 10^{38} erg/sec for 1.4M by a large factor.

It was once considered that the burst peak luminosity could be utilized as the "standard candle" for all bursts (26). This was supported by the nuclear flash model of bursts (27, 32) which predicts the burst peak luminosity near the Eddington limit; $L_E = 4\pi cGM/\chi$, where M is the stellar mass and χ the opacity for the Thomson scattering. For a generally assumed neutron star mass of 1.4 $M_\odot$, $L_E = 2 \times 10^{38}$ erg s^{-1}. If so, the distance to a burster, D, can be estimated from the observed peak flux ℓ_p, as $D \cong (L_E/4\pi \ell_p)^{1/2}$.

However, it is well established by now that the burst peak luminosity can vary from burst to burst by almost an order of magnitude (2,15, 17, 18). This is clearly against the "standard candle" hypothesis.

Even so, we note that burst sources near the galactic center exhibit similar peak luminosity distributiosn of bursts, as shown in Fig. 8. We may redefine the standard candle by the maximum of the distribution of the bursts for each source. The three dimensional distribution of the bursters constructed on an assumption that the standard candle is equal to the Eddington Luminosity (26, 17) indicates a clear and strong concentration in the direction of the GC and, however, at a distance of 4 ∼ 5 kpc, not at $8 \sim 10$ kpc that is the generally accepted distance to the GC. This implies that the maximum L_p exceeds L_E by a large factor (33, 25). A question is how to explain the luminosity largely in excess of the Eddington limit. This question is related to some fundamental problems such as the physics of the neutron star atmosphere.

ii) BURSTS IN SUCCESSION WITH A SHORT INTERVAL

Among a number of problems, the most difficult one to solve in terms of the nuclear flash model is the existence of bursts in succession with very short intervals. Normally bursts appear with intervals of hours to days or longer. Investigations of the distribution of the intervals indicated clear existence of a pause before each burst, which is understood as the time needed for accumulating the nuclear fuel. There are, however, exceptional cases. A case was found by SAS-3 in the GCX sources where three bursts appeared in 22 minutes (34). We observed two bursts from 1608-522 with an interval of 10 min (18) and recently two bursts from 1636-536 with an interval of 8 minutes were discovered. (H. Pederson observed two optical bursts with a 6 minutes interval from 1636-536 when we were conducting the optical/X-ray simultaneous burst watch: the two bursts occured during a time band with no X-ray coverage, but because of the general association between the X-ray and optical burst we believe that these were also X-ray bursts.) The most remarkable burst pair with the interval of 8 min. was recorded from Terzan 5 (35): both bursts were of similar size and profile and estimated radii of the both were nearly equal and ~ 10 km. With these results we have to admit that, though the bursts with short intervals are rare cases, they are common phenomena for bursters, not for a single exceptional source. With these short intervals we meet a severe difficulty of fuel supply: the amount of accumulated fuel is orders of magnitude short for a thermonuclear flash. The

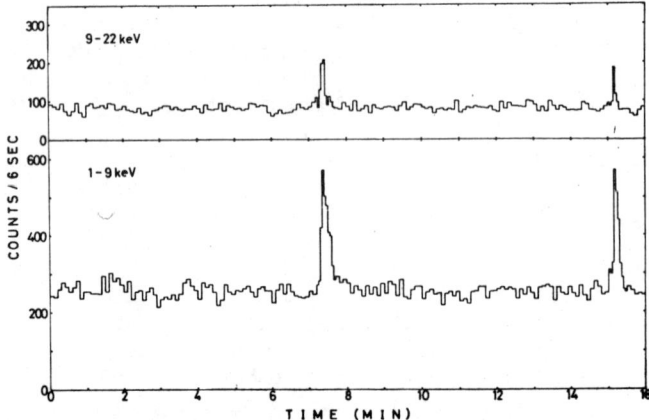

Fig. 9. Successive bursts from Terzan 5 with an interval of 8 minutes.

stored fuel should not be exhausted by the first burst in order to produce the second. It is not possible to attribute successive flashes to two different areas on the neutron star surface, because the estimated radii are the same ($\sim$ 10 km) for both bursts. A reservoir or mechanism must exist by which the fuel for the second burst is reserved, but where? This successive bursts may shake the apparently sound base of the nuclear flash model.

iii) INDEPENDENCE OF BURST ACTIVITY ON THE MASS ACCRETION RATE

It was suggested that when the persistent luminosity of a source, which presumably represents the mass accretion rate $\dot{M}$, exceeds a certain limit, the source can not produce bursts (32, 36). This is consistent with various observational facts thus far [except a recently discovered case for a bright bulge source GX17+2 (IAUC)]. The apparent existence of the upperlimit of the persistent luminosity (mass accretion rate) for a source to burst suggests that the burst activity is controlled by $\dot{M}$.

On the other hand, we observe that once a source becomes burst-active the activity expressed in terms of the time-averaged burst flux and the average burst frequency remain stable for a wide range of the persistent luminosity over a long period of days to months (18, 17, 28). Figure 10 shows an example for 1636-536. In this sense the burst activity is not the function of $\dot{M}$ and hence the α-value fluctuates to a large extent. This

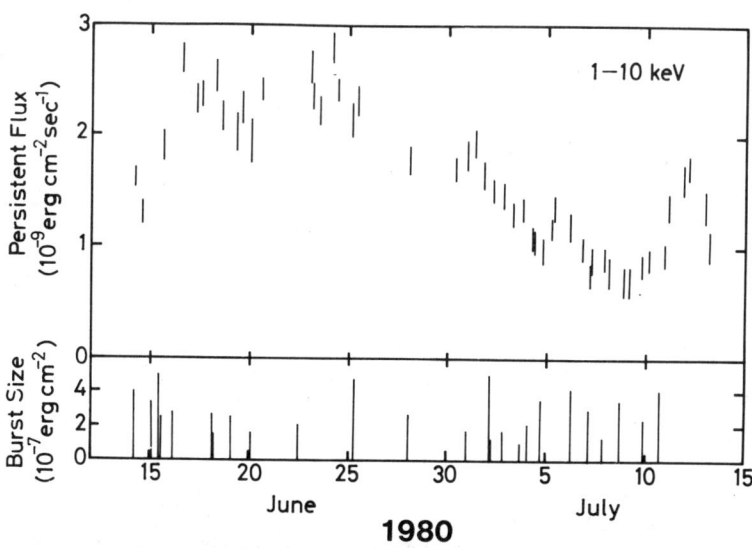

Fig. 10. Persistent fluxes and bursts of 1636-536 observed for June 14 - July 13, 1980. No clear correlations are noted.

independence of the burst activity was noted for bursters observed with Hakucho including 1608-522, 1636-536, GX3+1, 1735-444, and RB It appears to be a common feature. This is a puzzling fact, since the nuclear fuel for flash arises from the mass accretion. An implication is that a reservoir of nuclear fuel exists which buffers change of the fuel supply rate.

RAPID BURSTER (RB)

Apart from the X-ray bursts discussed above, we now turn to another type of burst which was discovered from the source 1730-333 and named Type II burst by the SAS-3 group (34, 33). (They call the former burst Type I.) In contrast to the burst (Type I) which has been discussed so far, features of the Type II bursts may be described as follows:

i) The burst repeats in succession with short intervals, typically 10 seconds to minutes (shorter by orders of magnitude as compared to Type I). From this feature, the source 1730-333 was maned the rapid burster (RB).

ii) Unlike Type I bursts no clear "cooling" is noticed during each burst.

iii) Energy per burst, E_b, is proportional to the waiting time to the succeeding burst, Δt: a kind of relaxation process may be suggested.

The Type II burst is now interpreted as a variation of the persistent flux of the burster (33, 15): The accretion flow, instead of continuous flow, is considered to be interrupted by some instabilities. The property of the relaxation process strongly suggests the presence of a reservoir.

It is known that the rapid burster is a recurrent source: it becomes active once in several months. In August 8 - 23, 1979, the source recurred to be active and showed an unusual behavior. Essentially the whole course of this activity from its onset was observed as shown in Fig. 11. At its onset phase it exhibited a known behavior of Type II. But in its active phase a train of bursts exclusively with profiles of trapezoidal (flat-topped) (37) shape was exhibited as illustrated

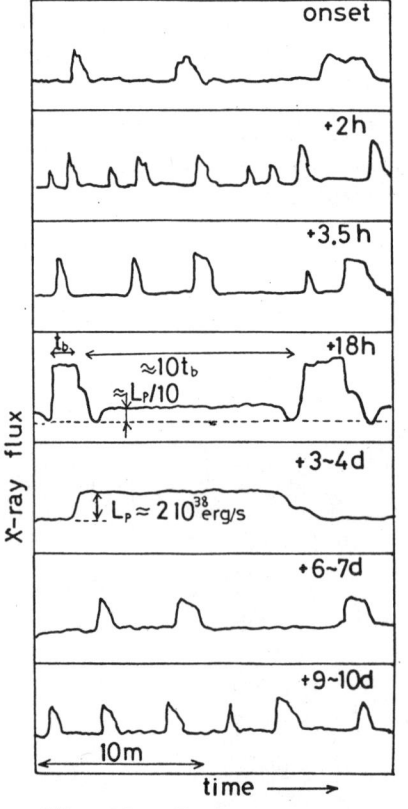

Fig. 11. Evolution of an unusual activity of the rapid burster during Aug 8 - 23, 1979 is illustrated.

in the figure (38, 3). The energy per burst increased by orders of magnitude compared to that of the sharp exponential bursts which appeared in the onset phase and in the previous activities. The approximate proportionality of E_p vs Δt still holds for these trapezoidal bursts. The burst of the longest duration (> 10 minutes) implies the maximum capacity of gravitational energy release to be by a total amount of the matter of 5×10^{20} gr (39,40).

The trapezoidal feature of the burst suggests that the mass flows from the reservoir with a constant drain rate. The procession of the trapezoidal bursts superposed on a low persistent flux may also be understood in terms of a relaxation oscillator which flip-flops with one minute delay switch (see Fig. 11) between high and low states corresponding to higher and lower drain rates relative to the flow-in rate of the reservoir (3, 39). What and where is the reservoir, what determines the drain rate and what is the mechanism of instability or switching are all far from understanding.
Generation of remarkably bright IR bursts and radio bursts from the rpid burster has been reported (see e.g. ref. 3 and references therein). While these may be very important clues in understanding this fascinating object, their association with the rapid burster has not yet been confirmed. An extensive campaign of X-ray/optical/IR/radio simultaneous observation was coordinated by the MIT group during July - September 1980 in an anticipation that the rapid burster would become active if the recurrence period was in the vicinity of six months. However, the source did not recur. The campaign will be repeated this year.

CAN A NEUTRON STAR BE FURNISHED WITH THE RESERVOIR?

We saw that existence of a reservoir of nuclear fuel or (as a source of gravitational energy) is demanded to explain various observational facts about bursters.

It is not easy to estimate the capacity of the reservoir without entering the details of the mechanism. But, for the succeeding bursts with a short interval, we need the stored nuclear fuel at least for one thermonuclear flash, i.e. say 10^{20} gr of H. It should be noted that this reservoir and the reservoir which buffers the variation of $\dot{M}$ and maintains the constant burst activity must be at the neutron star surface. We immediately encounter difficulties in conceptualizing the reservoir at the surface. It can not be that separate areas of the surface store the fuel and burst seperately, because the explosive burning should develop instaneously all over the surface and indeed the two successive bursts of Terzan 5 seem to have occured over the whole surface. It is also hard to conceive a laminar structure of the reservoir in the atmosphere, because it is unlikely that a burst starting at a point does develop throughout the whole fuel. On the other hand, for the rapid burster one speculative idea is that the strong magnetic field still remains at the old neutron star and the Alfven surface of the magnetosphere acts as the reservoir. This idea implicates various involved but interesting questions such as; how the spin axis and the magnetic dipole of the neutron star are coaligned so that the pulsation due to the rotation as in the case of the X-ray pulsar is not seen?, how can a configuration of the magnetic dipole field provide the stability and the capacity of the reservoir?, why the X-ray pulsar cannot burst if the existence of the field for the reservoir does not suppress the flash? etc..

OPTICAL/X-RAY SIMULTANEOUS OBSERVATION OF THE BURST

Now, I describe a program of the optical/X-ray simultaneous observation of bursts which is underway as a successful large-scale collaboration among different nations, institutions and scientific disciplines. The rationale of the project is clear. When the optical burst was first detected simultaneously with the X-ray burst by the MIT group, it was interpreted that the X-ray burst energy is absorbed by some object near the X-ray source and reradiated as the optical emission (41, 42). From the delay time of the optical burst with respect to the X-ray burst one can estimate the distance between the neutron star and the optical source. The smearing involved in the optical burst, if present, corresponds to the spatial spread of the optical source (if the local delay of optical emission such as photon diffusion is neglected). Whether the delay time is constant or scatters from burst to burst is a crucial question, because it deals with the structure of the reprocessing source: If the delay time scatters, it is very likely that the X-ray source and the optical source are orbiting around each other and, if not, the source is likely to be the accretion disk fixed around the neutron star.

We learned in the past 2 years that 1636-536 is a good target for this study. The MIT group took the role of coordination between the X-ray observation by "Hakucho" and the optical observation from the European Southern Observatory and CTIO in Chile (43). Thus far in 1979 and 1980 altogether ten cases of simultaneous records have been obtained. There is a clear one-to-one correspondence between X-ray bursts and optical bursts as long as the simultaneous coverage exists, except a couple of marginal cases. Prelimianry results show that the optical burst delays by a few seconds. Further detailed analyses are in progress. Further coordinated observations are underway.

SUMMARY

I think it is evident that from the study of X-ray pulsars and bursters we will learn a lot about neutron stars a cosmic laboratory under extreme conditions.

On one hand, the summary of updated observational results under the context of the thermonuclear flash model is gradually pinpointing various problems to be resolved. The burst appears to be a common nature of various Class-II sources: The source may become burst-active under certain common conditions inspite of a variety of other apparent features. One of the key questions is what makes the potential burster burst. The reservoir is another important question in conjuction with the physics of the neutron star.

On the other hand difficulties of the picture in explaining some observational facts became more definite to the extent that the picture may have to be reconsidered from its basis. Although the burst is still far from completely understood, we are approaching a new step so that either a more definitive picture will be reached or entirely new developments will come about.

ACKNOWLEDGEMENT

I thank members of the Hakucho team for their discussions and help with this review paper. The member list of the Hakucho Team is as follows:
H.Inoue, K.Koyama, K.Makishima, M.Matsuoka, T.Murakami, M.Oda, Y.Ogawara, T.Ohashi, N.Shibazaki, and Y.Tanaka
   Institute of Space and Astronautical Science.
I.Kondo
   Institute for Cosmic Ray Research, University of Tokyo
S.Hayakawa, H.Kunieda, F.Makino, K.Masai, F.Nagase, and Y.Tawara
   Department of Astrophysics, Faculty of Science,
   Nagoya University.
S.Miyamoto, H.Tsumeni, and K.Yamashita
   Department of Physics, Faculty of Science, Osaka University
M.Yoshimori
   Department of Physics, Faculty of Science, Rikkyo University.
The help of Dr. K.Makishima, Mr. D.Mitsuda, and Ms. F.Iwata in the preparation of the manuscript should be acknowledged.

References
1) I.Kondo et al. Space Sci.Instru. (1981) in press, ISAS RN No.109.
2) Y.Tanaka, Proc.Symp.on the Results and Future Prospects of X-ray Astronomy, August 1979, Tokyo.
3) M.Oda, Proc. HEAD/AAS meeting, January 1980 (in press), Cambridge, Mass. : ISAS Research Note No.119 (1980).
4) Y.Ogawara, Proc.Symp.on Space Astrophysics, July 1980, Tokyo.
5) M.Oda and Y.Tanaka, Proc.Japan-Italy Symp.on Fundamental Physics, January 1981, Tokyo (in press) : ISAS Research Note No.150 (1981).
6) S.Hayakawa, Space Science Reviews, 1981 (in press).
7) S.Miyamoto, T.Murakami, J.Nishimura et al. ISAS RN No.35 (1977).
8) F.Nagase et al. Nature 290,572 (1981).
9) F.Nagase, Proc.15th ESLAB Symposium, June 1981, Amsterdam.
10) R.Kelly, S.Rappaport and R.Petre, Ap.J.238,699 (1980).
11) F.K.Lamb, D.Pines and J.Shaham, Ap.J.224,969 (1978).
12) P.C.Joss and S.A.Rappaport, Proc. Tokyo Symposium (1980).
13) N.Kawai et al. Proc.Symp. on the Result of Hakucho, October 1980, Hakone.
14) F.K.Li, S.Rappaport, G.Clark and G.Jernigan, Ap.J.228,893 (1979).
15) W.H.G.Lewin and G.W.Clark, Annals N.Y. Acad.Sci. 336,451 (1980) : Proc. Tokyo Symposium (1979).
16) W.H.G.Lewin and P.C.Joss, Space Science Reviews 28,3 (1981).
17) T.Murakami, H.Inoue, K.Koyama et al. Ap.J.Lett.240 L143 (1980).
18) T.Murakami, H.Inoue et al. Pub.Astron.Soc.Japan 32,543 (1980).
19) K.Makishima et al. Proc. Hakone Symposium (1980).
20) M.Matsuoka, H.Inoue, K.Koyama et al. Ap.J.Lett.240 L137 (1980).
21) K.Koyama et al. ISAS RN No.135 (1980) : Ap.J.Lett. in press (1981).
22) K.Koyama et al. Proc. Tokyo Symposium (1980).
23) J.Thorstensen, P.Charles and S.Bowyer, Ap.J.Lett.220,L131 (1978).
24) W.H.G.Lewin, J.A.Hoffman, J.Doty et al. M.N.R.A.S.177,83P (1976).
25) H.Inoue et al. ISAS Research Note No.140 (1980):Ap.J.Lett.submitted.
26) J.van Paradijs, Nature 274,650 (1978).
27) P.C.Joss, Nature 270,310 (1977).
28) T.Murakami et al. Proc. Hakone Symposium (1980).
29) J.A.Hoffman, L.Cominsky and W.H.G.Lewin, Ap.J.Lett.240,L27 (1980).
30) M.Fujimoto, T.Hanawa and S.Miyaji, Proc. Tokyo Symp. (1980).
31) G.W.Clark, Ap.J.Lett.199,L143 (1975).
32) P.C.Joss, Ap.J.Lett.225,L123 (1978).
33) J.A.Hoffman, H.L.Marshall and W.H.G.Lewin, Nature 271,630 (1978).
34) W.H.G.Lewin, J.Doty, G.W.Clark et al. Ap.J.Lett.207,L95 (1976).
35) H.Inoue et al. Proc. Hakone Symposium (1980).
36) P.C.Joss and F.K.Li, Ap.J. 238,287 (1980).
37) E.Basinska, W.H.G.Lewin, L.Cominsky et al. Ap.J.241,787 (1980).
38) H.Inoue, K.Koyama, K.Makishima et al. Nature 283,358 (1980).
39) Y.Tawara, Ph.D.Thesis Univ.of Tokyo (1980): ISAS RN No.103 (1980).
40) H.Kunieda and Y.Tawara, Proc. Hakone Symposium (1980).
41) J.E.Grindlay, J.E.McClintock, C.R.Canizares et al. Nature 274, 567 (1978).
42) J.E.McClintock, C.R.Canizares et al. Nature 279,47 (1979).
43) H.Pedersen, M.Oda, Hakucho Team et al. IAU Circ. No.3399 (1979), No.3568 (1980).

ASSOCIATION OF RECURRENT SOFT X-RAY TRANSIENTS WITH X-RAY BURST SOURCES

T. Murakami
Institute of Space and Astronautical Science
4-6-1 Komaba, Meguro-ku, Tokyo, Japan

ABSTRACT

Hakucho observations have revealed that the soft transients, 2S1608-522, 2S1715-321, CenX-4 and AqlX-1, all generate X-ray bursts. This result makes it convincing that these sources involve neutron stars. Furthermore, the optical counterpart of AqlX-1, when X-ray quiescent, is identified to be a $\sim$ K0 dwarf, thus providing evidence that AqlX-1 is a binary of a late dwarf and a neutron star.

Soft transients in the flare state are similar in X-ray and optical properties to so-called bulge sources. This and the observed high association of soft transients and bulge sources with bursters seem to indicate similarities of both to AqlX-1. Flare-up of the soft transients is considered to be caused by episodic mass accretion.

INTRODUCTION

Galactic X-ray sources, except for supernova remnants, are believed to be binary systems including compact objects. They may be classified grossly into two different classes. One is the hard-spectrum sources whose optical counterparts are early type stars, mostly massive giants. Most of them are X-ray pulsars, implying the compact objects being strongly magnetized neutron stars.

The other class is the soft spectrum ($kT \lesssim 7$ keV) sources which comprises so-called bright bulge sources and the sources in globular clusters. Besides the soft spectrum, they share the common nature that no regular pulsations are observed and that the optical counterparts are faint blue objects with the lack of eclipses as well as periodic variations [1,2,3]. More than half the sources of this class are known to generate bursts. Following the discovery of X-ray bursts with ANS [5], it has become convincing that the compact objects in the burst sources are neutron stars and therefore, although not established, we believe that the sources of the second class are binaries of a neutron star and a late dwarf.

Transient X-ray sources were once considered to be a distinct class of object. However, Kaluzienski noted the existence of two classes of X-ray transients, soft transients and hard transients, from the Ariel-5 ASM results [4]. These soft transients look in many respects similar to the persistent soft-spectrum sources.

I shall discuss here the recurrent soft-spectrum transients, mainly based on the Hakucho observations. The important finding is that four soft transients we observed, 2S1608-522, Cen X-4, 2S1715-321 and Aql X-1, all generated X-ray bursts [13-16]. This fact makes the strong case that the compact objects of these soft transients are neutron stars. With these results, properties of the soft transients and their relation to bulge sources will be discussed.

OBSERVATIONAL RESULTS

The results of the Hakucho observations presented here were obtained with the x-ray burst monitor system. The system consists of two sets of rotating modulation collimators; two coarse (CMC-1,2) and one fine (FMC-1) one covering circular fields of view of $17.6°$ and $5.8°$ FWHM, respectively. The CMC's are aligned in the push-pull way so that the sum of the two give an unmodulated signal. Another detector (FMC-2) having a $5.8°$ FWHM field of view with a honeycomb tube collimator provides the source flux with a better signal to noise ratio. Further details of the instrument are given elsewhere[17]. Thanks to a wide field of view and the moderate spatial resolution ($\sim 0.5°$) of the CMC, a panoramic burst survey has been successfully conducted.

a) Norma Transient; 2S1608-522

The brightening of 2S1608-522 was reported in January 1979 from Ariel-5 ASM[18]. Due to the sun-angle constraint, Hakucho could not observe this source until April. When we began to observe this transient on April 8, the intensity was still increasing and reached a maximum level of half the Crab intensity. Soon after we started this observation, an intense x-ray burst with a peak intensity of about 5 times the Crab was recorded. From April to July, 22 bursts were detected. The locations of the observed bursts and the persistent component were in good agreement with each other with an accuracy of 3 arcminutes using the FMC-1 and also consistent with 2S1608-522. In Figure 1 we show the intensity record of the persistent component and the peak flux of the observed bursts. Interesting characteristics of the observed burst peak flux and profiles together with the persistent x-ray flux of this occasion were reported by Murakami et al.[13]

Two more flaring episodes of 2S1608-522 in 1975 and 1977 have been previously reported[19-21]. The x-ray light curve suggests a quasi-periodic recurrence with a period of ~600 days. Detection of large x-ray bursts in the Norma region was first reported by the Vela satellite[22]. Uhuru also recorded three bursts in the same region[23] when the recurrent transient was in a low state. In 1977, SAS-3 monitored the transient in a flare state but did not find the burst activity[24]. Hakucho monitored 2S1608-522 three times from April to July in 1979, 1980 and 1981. Except for 1979, the source was in a low state. However, in each case it was burst active.

b) Cen X-4

The Ariel-5 ASM discovered a reappearance of Cen X-4 in May 1979[27]. The previous x-ray outburst of this source was observed in July 1969[25], when the light curve exhibited a plateau of 24 times the Crab flux, following a giant precursor-like burst (60 times the Crab at the peak) about 50 hours before the initiation of the x-ray outburst[28, 29]. The intensity reached a peak value of about 4 times the Crab on May 17 and

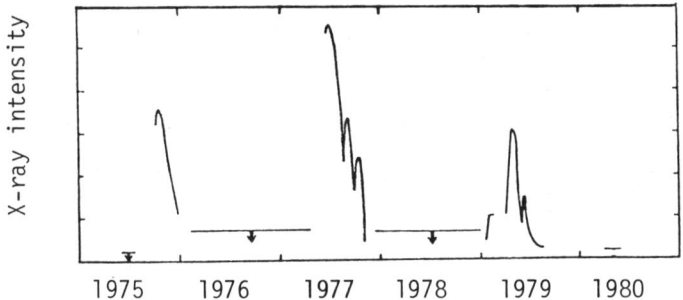

Figure 1a The X-ray light curve of 2S1608-522. The recurrence looks quasi-periodic.

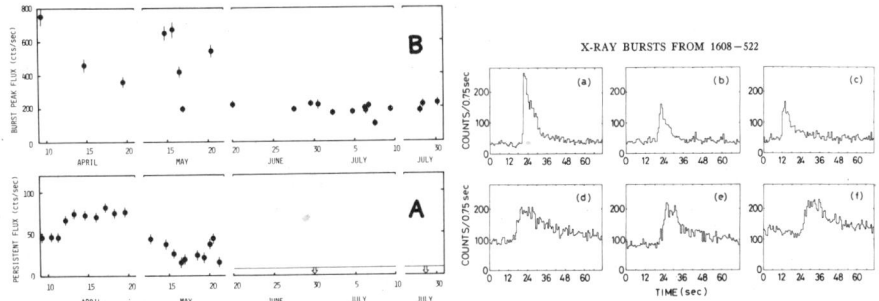

Figure 1b (A) The X-ray intensity of the persistent component and (b) the peak flux of the osbserved bursts as a function of time in 1979. (a)-(f) show examples of the burst profile from 2S1608-522 in 1979.

the optical counterpart, which brightened to 13 mag, was discovered[26]. We watched Cen X-4 from May 28 to June 16, when the x-ray intensity was in its decay phase, and detected a significant x-ray flux until June 9. An intense x-ray burst having a peak x-ray flux of 25 times the Crab was observed on May 31. The x-ray light curve and the x-ray burst are shown in Figure 2[16]. In the decay phase, Kaluzienski et al[27] discovered a periodic modulation in the intensity with a period of 8.2 $\pm$ 0.2 hours. This suggests a binary nature of an x-ray burst source.

c) MX/2S1715-321

A region of the sky near $\alpha=27h$, $\delta=-32°$ was observed in order to monitor the bursters MXB1730-335 and MXB1728-337 during four separate periods in 1979. Three x-ray bursts which were not generated by the known bursters were detected with the CMC system. They were located within an error circle of about 0.3 square degrees in area centered at $\alpha=17h$ 15m, $\delta=-32.1°$ leading to the designation of XB1715-321[15]. The fast transient

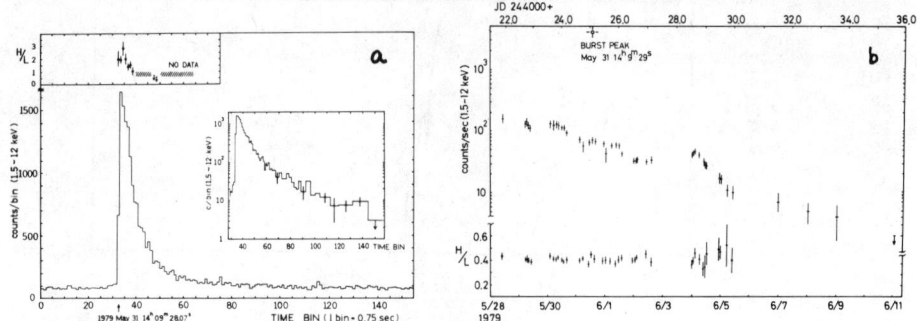

Figure 2 The X-ray light curve of Cen X-4 and the time profile of its X-ray burst on May 31. The upper insert of Fig. 2a shows the hardness ratio of the observed burst. Spectral softening in the decay phase is apparent. The hardness ratio of the persistent component shown in Fig 2b does not change significantly throughout the observation.

MX/2S1715-321 which was discovered by OSO-7 is at the center of this error region. Peculiar burst-like events with much longer duration than the usual x-ray burst were reported from this source twice in the past[31,32].

d) Aql X-1

Aql X-1 is a well known recurrent transient whose optical counterpart has been identified in x-ray quiescence with a dwarf of spectral type between G7 and K3[33]. Following the detection of an optical flare of Aql X-1 by Margon[34], Hakucho observed an x-ray outburst from May 20 to June 9, 1980. In this observation, we detected two x-ray bursts from this source in its declining phase[14]. The x-ray light curve of Aql X-1 and the profiles of the two bursts observed with the CMC in two energy bands are shown in Figure 3 together with the hardness ratio.

Recently, Cominsky et al. reanalyzed a single burst from Aql-MXB detected by SAS-3 on July 20, 1976[35] and found that it was consistent with being from Aql X-1. This burst was also observed during the decay phase of the Aql X-1 outburst in 1976[36,37].

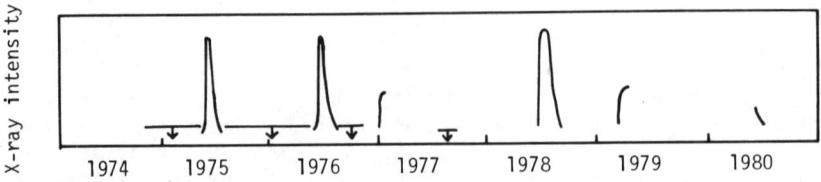

Figure 3a The X-ray flare history of Aql X-1.

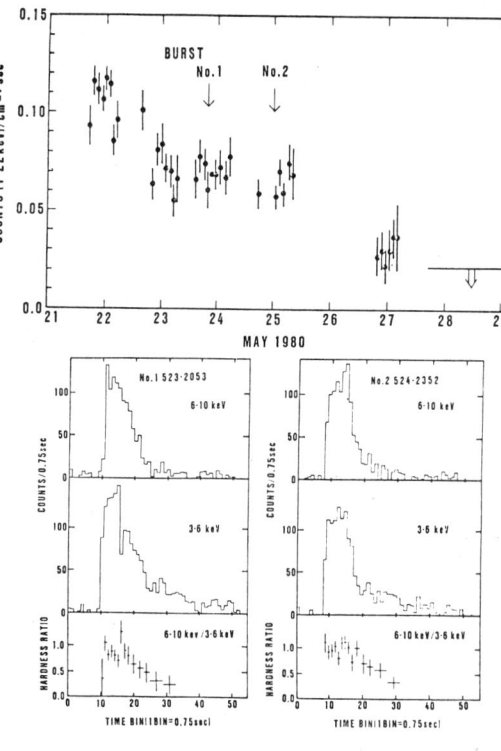

Figure 3b
The X-ray light curve of Aql X-1 in the energy range 1-22 keV observed by Hakucho in 1980. Arrows indicate the onset times of the bursts.

Figure 3c
The time profiles of the X-ray bursts together with the hardness ratio for the two energy bands, 3-6 and 6-10 keV. Spectral softening in the decay phase is clear.

DISCUSSION

We found from Hakucho observations that the soft transients, 2S1608-522, 2S1715-321, CenX-4 and AqlX-1, generated x-ray bursts[13-16]. Among them 2S1608-522, CenX-4 and AqlX-1 exhibit clear recurrent high states. The rapid burster, MXB1730-335, which also shows recurrent high states is reported to produce normal x-ray bursts[11,43]. Other soft transients, MXB1659-29, MXB1743-29 and MXB1742-29 are also bursters[38-42].

In the current understanding, the burst sources include neutron stars. This is supported by the fact that the black body radius derived from the decay phase of bursts is consistent with the size of a neutron star[44]. Therefore, it is convincing that these soft transients are systems involving a neutron star[45].

As already mentioned, it is important that the optical counterpart of AqlX-1 when x-ray quiet was identified with a dwarf of a spectral type ~K0[33]. This clearly indicates that the AqlX-1 system is composed of a neutron star and a late dwarf companion.

Table 1 lists all the bursters established as of June 1981[6,7]. Tables 2 and 3 list soft transients and bright bulge sources. Table 4 lists globular cluster sources, including the recent results from the Einstein Observatory[8-12]. It is remarkable that those soft-spectrum sources all show high association with bursters; 7 bursters out of 15 bright bulge sources, 9 of 15 globular cluster sources and 8 of 25 soft transients.

Table-1   X-RAY BURST SOURCES

Position	Name	Category and Comments
0512-401	4U,MXB	Glob.cluster NGC1851
1455-31	XB,CenX-4	10yr recurrence;Vela satellite burst?
1608-522	4U,XB	~600 days R.T.;UHURU,Vela burst
1636-536	4U,MXB	B.B.S.;opt/X
1659-298	MXB,H	Transient; no burst in high state
1702-429	4U,XB	MXB1706-43?
1715-322	4U,XB	Transient OSO-7,SAS-3;fast transient
1724-30	4U,XB	Glob.cluster Terzan-2
1728-337	4U,MXB,SB	B.B.S. in glob.cluster
1730-333	MXB,H,RB	Liller-1,~180 days R.T.trapezoidal burst
1733-30	XB	Glob.cluster Terzan-1
1735-444	4U,MXB	B.B.S.;opt/X,binary period 4.3 hr 4)
1742-29	MXB,A	in GCX;Ariel-5 transient
1743-28	MXB	in GCX
1743-29	MXB,A	in GCX;Ariel-5 transient
1744-265	4U,XB,GX3+1	B.B.S.
1745-25	XB	Glob.cluster Terzan-5
1746-37	MXB,2S	Glob.cluster NGC6441
1813-14	XB,4U,GX17+2	B.B.S. 3)
1820-303	MXB,4U	Glob.cluster NGC6624
1837+049	MXB,4U,SerX-1	B.B.S.;opt/X
1850-087	MXB,A	Glob.cluster NGC6712
1905+000	MXB,4U	
1908+00	XB,4U	AqlX-1
1916-053	MXB,4U	Binary period 50±0.5 min.

References   Lewin and Clark [2][3],Lewin and Joss[11],Oda [6]

R.T.; Recurrent transient B.B.S; Bright Bulge Source
3) Murakami et al.[7], 4) McClintock and Petro [52]
opt/X; simultanious observation with X-ray and optical

Table-2   Soft Transient sources

Name		Optical	Comments
0499-55		M dwarfs noted ?	
0620-003	Nova Mon'75	B~12,K5-7V #	58 yr recurrence, p~7.8d
1455-31	TrA X-1	B~17 #	
1524-517			
1543-475			
1608-522		I~18 #	~600d recurrence, burster 1)
1630-472			Recurs 615±5 days
1658-298		V=18.3,uv, λ4686	Burster
1705-250	Nova Oph'77	B=16.5,λ4686 #	Fast flare 2-10min., burster 2)
1715-321			
1735-28			
1742-289			Near g.c. ,burster?
1743-322			
1743-288			Near g.c. ,burster?
1744-361			
1745-203	NGC6440		Glob. Cluster
1803-245			
1807-10			
1908+005	AqlX-1	B=16.5,KO,λ4686 #	Burster 3) ,recurs 12-16 months,p~1.3d
1918-14			
CenX-2			
1455-31	CenX-4	V=12.8,λ4686 5)	Burster,recurs 10yr 4), p~8hr 6)
1730-22			
0656-07			
0836-42			
1730-335	Rapid burster Liller-1		Glob.Cluster,burster

References   Bradt et al.[9], Cominsky et al.[8], 1) Murakami et al.[13]
2) Makishima et al.[15], 3) Koyama et al.[14], 4) Matsuoka et al.[16], 5) Canizares et al.[26], 6) Kaluzienski et al.[18]
Brightened coincident with X-ray inclease

Table-3 Bright Bulge Sources

Name		X-ray	optical
1636-536		Burster	uv,V=17.5,λ4686 #
1642-455	GX340+0		
1659-487	GX339-4		V=16.6,λ4686
1702-363	GX349+2		
1705-440			
1728-337		Burster	Globuler cluster Grindley-1 1)
1728-169	GX9+9		uv,V=16.6,λ4686
1735-444		Burster	uv,V=17.5,λ4686,binary period 4.3hr 5) #
1744-265	GX3+1	Burster 2)	
1758-250	GX5-1		
1758-205	GX9+1		
1811-171	GX13+1		
1813-140	GX17+2	Burster 3)	
1820-303	NGC6624	Burster	Globular cluster
1837+049	SerX-1	Burster	uv,B~19,weak λ4686 #

References Bradt et al [9], 1) Grindlay [10], 2) Makishima et al. [53]

5) McClintock and Petro [52]
Simultanious observation with X-ray burst and optical

Table-4 Globular Clusters

Name	X-ray
NGC 104	
NGC 1851	Burster,MXB0512-40
NGC 1904	
NGC 6440	Transient 1)
NGC 6441	Burster,MXB1746-37
NGC 6517	
NGC 6541	
NGC 6624	Burster,MXB1820-30
NGC 6712	Burster,MXB1850-08
NGC 7078	
Liller 1	Burster (Rapid), recurrent transient?
Grindlay 1	Burster (Slow),MXB1728-34
Terzan 1	Burster,XB1733-30 2)
Terzan 2	Burster,XB1724-30
Terzan 5	Burster,XB1745-25 2)

References Grindlay 1980 [10], 1) Cominsky et al. [8]
2) Oda 1980 [6], 3) Middleditch et al. [47].

Soft transients when they are active are similar to bulge sources in x-ray and optical characteristics[26]. This and the observed association with bursters seem to indicate that soft transients and bulge sources are similar astronomical systems. With the evidence of AqlX-1, it is very likely that they are all binaries of a neutron star and a low mass star.

As regards the nova-like outbursts of CenX-4 and AqlX-1, the possibility of carbon flash seems to be excluded[46]. If this were the case, accreted matter should accumulate during a low state to the amount required for the observed energy in an outburst. Such a time-averaged gravitational energy release should be quite observable in x-rays because of the low conversion efficiency of carbon burning as compared to the gravitational energy release. The outbursts are most likely caused by an episodic mass accretion onto a neutron star, either by direct mass flow from the companion or due to instability of the accretion disk.

It may be reasonable to consider that the transient appearance of other sources is also due to the same process. Finally, we should ask the question as to why soft transients exhibit their large intensity excursions such as a factor of 10^5 in the case of NGC6440, in contrast to the persistent sources[10].

REFERENCES

1. J. P. Ostriker, 8th Texas Symp., Ann. N.Y. Acad. Sci., 302, 229 (1977).
2. W. H. G. Lewin and G. W. Clark, in Symp. on the Results and Future Prospects of X-ray Astronomy, Aug. 3-4, Tokyo, ISAS (1979).
3. W. H. G. Lewin and G. W. Clark, Ann. N.Y. Acad. Sci, 336, 451 (1980).
4. L. J. Kaluzienski, Ph.D. Thesis, University of Maryland (1977).
5. J. E. Grindlay and H. Gurskey et al., Ap.J. 205, L127 (1976).
6. M. Oda, in Proceedings of the HEAD/AAS Meeting, Cambridge, Mass., 1980 in press.
7. T. Murakami, Discovery of Burst Activity from GX17+2, in preparation (1981).
8. L. Cominsky, C. Jones, W. Forman and H. Tananbaum, Ap.J. 224, 46 (1978).
9. H. V. Bradt, R. E. Doxsey and J. G. Jernigan, Advance in Space Exploration, 3 (1979).
10. J. E. Grindlay, X-ray Sources in Glob. Clusters preprint (1980).
11. W. H. G. Lewin and P. C. Joss, Space Sci. Rev., 28, 3 (1981).
12. G. Fabbiano and G. Branduardi, Ap.J. 227, 294 (1979).
13. T. Murakami et al., Ap.J. 240, L143 (1980).
14. K. Koyama et al., Ap.J. Letters in press, ISAS RN 136. (1981).
15. K. Makishima et al., Ap.J. 244, L79 (1981).
16. M. Matsuoka et al., Ap.J. 240, L137 (1980).
17. I. Kondo et al., Space Sci. Inst. in press (1981).
18. L. J. Kaluzienski and S. S. Holt, IAUC 3328, 3349 (1979).
19. L. J. Kaluzienski and S. S. Holt, IAUC 3099, 3108, 3129 (1977).
20. L. J. Kaluzienski et al., IAUC 2859 (1975).
21. G. W. Clark and F. Li, IAUC 3090 (1977).
22. R. D. Belian, J. P. Conner and W. P. Evans, Ap. J. 206. L135 (1976).
23. J. E. Grindlay and H. Gursky, Ap.J. 209, L61 (1976).
24. G. Fabbiano et al., Ap.J. 221, L49 (1978).
25. L. J. Kaluzienski and S. S. Holt, IAUC 3362 (1979).
26. C. R. Canizares, J. E. McClintock and J. E. Grindlay, Ap.J. 236, L55 (1979).
27. L. J. Kaluzienski, S. S. Holt and J. H. Swank, Ap.J. 241, 779 (1980).
28. W. D. Evans, R. D. Belian and J. P. Conner, Ap.J. 159, L57 (1970).
29. R. D. Belian, J. P. Conner and W. D. Evans, Ap.J. 171, L87 (1972).
30. T. H. Markert, D. E. Backman and J. E. McClintock, Ap.J. 208, L115 (1976).
31. J. A. Hoffman, W. H. G. Lewin et al., Ap.J. 221, L57 (1978).
32. J. Thorstensen, P. Charles and S. Bowyer, Ap.J. 220. L131 (1978).
33. B. Margon, IAUC 3478 (1980).
34. W. H. G. Lewin, F. K. Li et al., MNRAS, 177, 93 (1976).
35. L. Cominsky, A. Lawrence and W. H. G. Lewin, Private Communication (1980).
36. M. G. Watson, MNRAS, 176, 19 (1976).
37. W. H. G. Lewin, J. A. Hoffman and J. Doty, IAUC 2994 (1976).
38. W. H. G. Lewin, et al., IAUC 3190 (1978).
39. R. Doxsey, J. Grindlay et al., Ap.J. 228, L67 (1979).
40. C. R. Canizares et al., Ap.J. 234, 556 (1979).
41. W. H. G. Lewin et al., MNRAS, 177, 83 and C. J. Eyles et al., Nature, 257, 291 (1976).

42. Y. Tawara, Ph.D. Thesis University of Tokyo (1980).
43. H. Inoue et al., Ap.J. submitted, ISAS, RN140 (1981).
44. P. C. Joss and S. A. Rappaport, Astron. Astrophys, $\underline{71}$, 217 (1979).
45. P. C. Joss, Comm. Ap., $\underline{8}$, 109 (1979).
46. J. Middleditch et al., Ap.J. in press (1980).
47. F. W. Walter, N. E. White and J. Swank, IAUC 3611 (1981).
48. M. G. Watson, MNRAS, $\underline{176}$, 19 (1976).
49. T. Matilsky et al., Ap.J. 210, L127 (1976).
50. K. Makishima et al., Ap.J. Letters in press, ISAS RN 137 (1981).
51. J. E. McClintock and L. D. Petro, IAUC 3615 (1980).
52. K. Makishima et al., Discovery of Burst Activity from GX3+1, in preparation (1981).

ACKNOWLEDGEMENTS

It is a pleasure to thank Professors M. Oda, Y. Tanaka, Y. Ogawara and the Hakucho team for valuable comments on the manuscript.

NUCLEAR FLASH MODELS FOR X-RAY BURST SOURCES

P. C. Joss
Department of Physics
and
Center for Theoretical Physics
Massachusetts Institute of Technology
Cambridge, Massachusetts 02139

ABSTRACT

Since the discovery of cosmic X-ray bursts in 1975, a wealth of observational information concerning this phenomenon has been obtained. X-ray bursts have rise times of $\lesssim 1$ s, decay timescales of $\sim 3-30$ s, and intervals between bursts that are usually in the range of $\sim 10^4-10^5$ s. The burst spectra can often be well fitted by a blackbody spectrum from an emission region of maximum temperature $\sim 3 \times 10^7$ K and roughly constant size. If X-ray burst sources are typically at distances of ~ 10 kpc, as indicated by their concentration in the direction of the galactic center, then they have maximum luminosities of $\sim 10^{38}$ ergs s^{-1}, total emitted energies of $\sim 10^{39}$ ergs per burst, and effective blackbody radii of ~ 10 km. The observational properties of X-ray burst sources have recently been reviewed by Lewin and Joss[1].

Thermonuclear flashes in the surface layers of accreting neutron stars are one of several mechanisms that have been proposed to account for X-ray bursts[2,3]. Computations of the evolution of such surface layers[4] demonstrate that under many circumstances, the helium-burning shell is thermally unstable and should undergo flashes that result in the emission of X-ray bursts from the neutron-star photosphere. The calculated properties of these bursts are remarkably similar to those of bursts from most observed X-ray burst sources.

If the surface magnetic field of the neutron star is sufficiently strong to funnel the accretion flow onto the magnetic polar caps, then the rotation of the neutron star and its associated accretion pattern should result in the emission of periodic X-ray pulsations. However, the magnetic funneling should also alter the structure of the surface layers so as to tend to suppress thermonuclear flashes[5]. Hence, the apparent dichotomy between X-ray burst sources and X-ray pulsars may be readily understood.

A number of theoretical and phenomenological problems remain to be resolved (see Lewin and Joss[1] for a review). Among the theoretical issues currently under active investigation are the complex interactions between the helium-burning and hydrogen-burning shells[6-10], the role of general relativistic effects[11,12,10], and the conditions required to maintain the core of the neutron star in thermal equilibrium[13,10]. Other considerations, such as violations of spherical symmetry and the role of dynamical effects in the outer surface layers, may also turn out to be important.

This work was supported in part by the National Science Foundation under grant AST78-21993.

REFERENCES

1. W.H.G. Lewin and P.C. Joss, Space Sci. Revs. $\underline{28}$, 3 (1981).
2. S.E. Woosley and R.E. Taam, Nature $\underline{263}$, 101 (1976).
3. L. Maraschi and A. Cavaliere, in "Highlights of Astronomy," ed. E.A. Müller (Reidel, Dordrecht, 1977), Vol. 4, Part I, p. 127.
4. P.C. Joss, Astrophys. J. (Letters) $\underline{225}$, L123 (1978).
5. P.C. Joss and F.K. Li, Astrophys. J. $\underline{238}$, 287 (1980).
6. R.E. Taam and R.E. Picklum, Astrophys. J. $\underline{224}$, 210 (1978).
7. R.E. Taam and R.E. Picklum, Astrophys. J. $\underline{233}$, 327 (1979).
8. R.E. Taam, Astrophys. J. $\underline{241}$, 358 (1980).
9. R.E. Taam, Astrophys. J. $\underline{247}$, 257 (1981).
10. S. Ayasli and P.C. Joss, submitted to Astrophys. J. (1981).
11. Y. Goldman, Astron. Astrophys. $\underline{78}$, L15 (1979).
12. J. van Paradijs, Astrophys. J. $\underline{234}$, 609 (1979).
13. D.Q. Lamb and F.K. Lamb, Astrophys. J. $\underline{220}$, 291 (1978).

A NUMERICAL SIMULATION OF THE MAGNETOSPHERIC GATE MODEL FOR THE X-RAY BURSTERS

S. Starrfield
Department of Physics, Arizona State University
Tempe, AZ 85287
and
Theoretical Division
Los Alamos National Laboratory, University of California
Los Alamos, NM 85745

S. Kenyon and J. W. Truran
Department of Astronomy, University of Illinois
Urbana, IL 61801

W. M. Sparks
Laboratory for Astronomy and Solar Physics
N.A.S.A. - Goddard Space Flight Center
Greenbelt, MD 20771

ABSTRACT

We have used a Lagrangian, fully implicit, one-dimensional hydrodynamic computer code to investigate the evolution of a gas cloud impacting the surface of a 20 km, 1 $M_\odot$ neutron star. This gas is initially at rest with respect to the surface of the neutron star, extends to 185 km above the surface, and is optically thick. The infall results in a burst which lasts about 0.1 seconds and reaches a peak luminosity and effective temperature of 2.4×10^5 $L_\odot$ and 9×10^6 K; respectively. The burst was followed by a phase of oscillations with a period 0.2 seconds.

INTRODUCTION

One of the more interesting scenarios for the cause of the X-ray Bursts is the magnetospheric gate theory of Lamb, et al.[1] In this scenario we have a rapidly rotating neutron star with a strong magnetic field in a binary system with a companion that is losing mass toward the neutron star. This material is impeded by the magnetic field in such a fashion that accretion is not continuous but occurs in episodic events. To our knowledge there have been no previous numerical simulations of this phenomena.

We use the same hydrodynamic code described in the last paper[2]. Because it is a Lagrangian, stellar interior, computer code which assumes radiation transport by diffusion; we cannot place a hole in the mesh and so assume that the blob of gas extends from the surface of the neutron star to a radius of 185 km. This blob is optically thick, has a mass of 10^{-11} $M_\odot$, and is initially at rest with respect to the surface of the neutron star. The mass of this blob was chosen so as to simulate a much smaller amount of material falling only onto the polar caps. The gas initially has a temperature of $\sim 10^7$ K and a surface luminosity of 0.1 $L_\odot$. We begin the calculation by allowing the material to fall onto the surface.

RESULTS

Once the infall has begun, an accretion shock forms at the boundary of the core. After about 100μ sec of evolution, the infall velocities have reached free fall and the temperature behind the shock has reached 10^9 K. It takes 3.5 milliseconds for the outer edge of the shock to reach the surface of the blob which has now fallen inward to 80 km. The entire envelope bounces imparting enough energy to the two outermost zones (2×10^{-14} $M_\odot$) for them to reach escape velocity. The light curve for the burst is shown in Figure 1. The peak luminosity and effective temperature produced by the infalling material is 2.4×10^5 $L_\odot$ and 9×10^6 K, respectively. Although, this effective temperature is too low for any of the

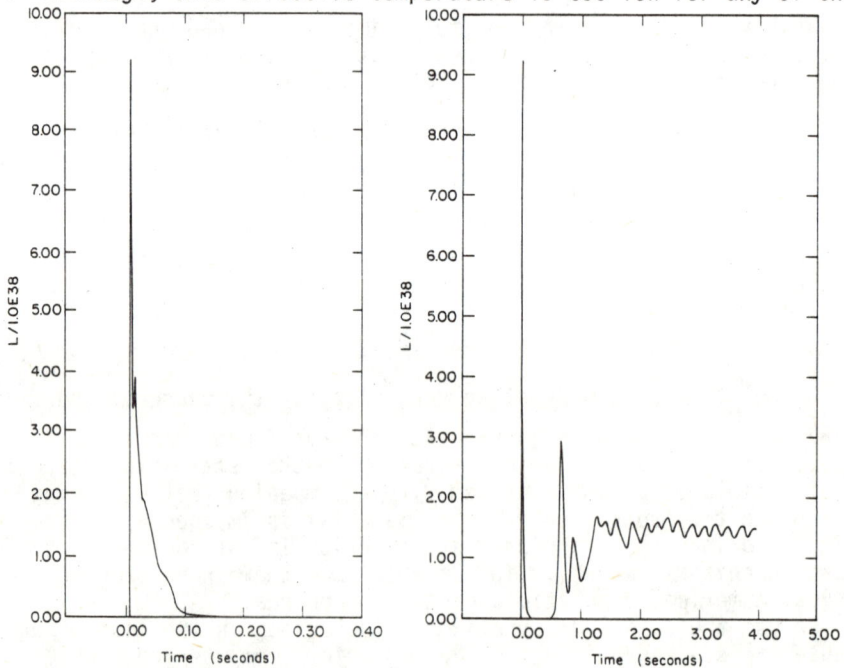

Fig. 1. The luminosity of the infalling material as a function of time.

Fig. 2. The luminosity of the surface as a function of time.

observed bursts, we could raise this temperature by reducing the inner radius of the neutron star and reducing the amount of infalling material.

The temperature in the shell source at the core-envelope interface (CEI) has remained at 10^9 K and enough triple-α reactions are occurring to keep the rate of nuclear energy generation at 10^{18} erg gm^{-1} sec^{-1}. In addition, the deeper layers of the envelope are convective, which is mixing fresh unburned nuclei into the shell source which also is maintaining the large rate of nuclear burning. However, the outer layers are expanding too rapidly for the energy produced in the shell source to reach them and their luminosity drops rapidly. For a short period the β^+-decays in the surface layers are able to halt the decreasing luminosity but the rapid expansion again overcomes this energy source. The entire burst lasts about 0.3 sec.

The decline in luminosity continues until the escaping shells become optically thin. At this time the energy from the interior can now escape and the luminosity begins to increase. Now that we can see into the deeper layers we find that the accreted material remaining on the neutron star is pulsating with a period of $\sim$ 0.2 sec (Figure 2). The average luminosity of this object is $\sim 10^4$ $L_\odot$ and the effective temperature is 0.3 kev. It has a radius of 150km (average).

These pulsations are a direct result of the infall and are not caused by a partial ionization mechanism such as in the Cepheids or ZZ Ceti variables. We also find that these pulsations must be occurring in high overtones since we find a number of nodes in the eigenfunction and the largest amplitudes are in the surface zones.

We closely followed these pulsations for 4 seconds of star time. However, the computer time was becoming prohibitive and we ended this part of the calculation. We removed the escaping zones and followed the resulting evolution for another thousand seconds. No further changes occurred and we ended the calculation. It is estimated that at least one day of further evolution would be necessary to completely burn the hydrogen in the envelope.

We also studied a purely cooling sequence in which we followed the contraction of the extended envelope assuming only neutrino losses - no nuclear energy production. The envelope contracts from about 200 km to 100 km in 10^3 sec with the luminosity climbing to 2 x 10^5 $L_\odot$ and the effective temperature to about 1 kev. It takes another 40 minutes to contract to 20 km. During the cooling phase the surface luminosity reaches a peak luminosity of 2 x 10^5 $L_\odot$ and a peak effective temperature of 1.9 kev.

CONCLUSIONS

We find that our Burst was too cool and too rapid to resemble the normal bursts although the time scale is in agreement with some of the rapid bursts that have been observed for some objects[4]. It is certainly possible that more realistic initial conditions, such as assuming the material was already infalling, could improve the correspondence between our calculations and the observations.

We also found that the infall produced a ringing of the envelope which showed short period pulsations. Such pulsations have been observed in at least one burst event[5].

REFERENCES

1. F.K. Lamb, A.C. Fabian, J.E. Pringle, D.Q Lamb, Ap.J. <u>217</u>, 197 (1977).
2. S. Starrfield, S. Kenyon, J.W. Truran, W.M. Sparks, this volume (1981).
3. S. Starrfield, S. Kenyon, W.M. Sparks, Ap.J. (submitted 1981).
4. W.H.G. Lewin, P.C. Joss, Nature <u>270</u>, 211 (1977).
5. T. Murakami, private communication (1981).

THERMONUCLEAR RUNAWAYS IN THICK HYDROGEN RICH ENVELOPES OF NEUTRON STARS

S. Starrfield
Department of Physics, Arizona State University
Tempe, AZ 85287
and
Theoretical Division
Los Alamos National Laboratory, University of California
Los Alamos, NM 85745

S. Kenyon and J. W. Truran
Department of Astronomy, University of Illinois
Urbana, IL 61801

W. M. Sparks
Laboratory for Astronomy and Solar Physics
N.A.S.A. - Goddard Space Flight Center
Greenbelt, MD 20771

ABSTRACT

We have used a Lagrangian, fully implicit, one-dimensional hydrodynamic computer code to evolve thermonuclear runaways in the accreted hydrogen rich envelopes of 1.0 $M_\odot$ neutron stars with radii of 10 km and 20 km. Our simulations produce outbursts which last from about 750 seconds to about one week. Peak effective temperatures and luminosities were 2.6×10^7 K and 8×10^4 $L_\odot$ for the 10 km study and 5.3×10^6 K and 600 $L_\odot$ for the 20 km study. Hydrodynamic expansion on the 10 km neutron star produced a precursor lasting about 10^{-4} seconds.

INTRODUCTION

We have studied thermonuclear runaways in the accreted hydrogen rich envelopes of 10 km and 20 km neutron stars using a fully implicit, Lagrangian, hydrodynamic computer code which incorporates a nuclear reaction network[1]. We have assumed that the Bursters and Transient X-ray sources occur as a result of mass transfer from a secondary onto a neutron star in an analogous fashion to the nova phenomena[2]. Reviews of published work on this subject can be found in the literature[3,4].

The published work[3] has produced simulations of the Burster phenomena which are in reasonable agreement with the observations but have not yet reproduced the full range of observed behavior. By this we mean that observed Bursts show a very wide range of time scales lasting from seconds to minutes and the published work has, so far, only addressed the characteristics of the shortest bursts. If one wants to prolong the time scale, then there are three possibilities all of which involve increasing the amount of fuel available for nuclear burning. First, one can assume that the luminosity of the neutron star is low and the mass accretion rate is low and build up a thick hydrogen envelope. Second, one can assume a rapid inflow of material and include the accretion energy in the radiative losses; and, third, one can assume that at some accretion rate the hydrogen will burn stably and a thick layer of helium can be built up on the neutron star.

Because of the success of the studies of Joss[3] and Taam[4] reviewed in this volume by Joss, we have concentrated on a regime of the $(\dot{M},L)$ plane not yet studied by these investigators: that of low internal neutron star luminosity and low mass accretion rates. In our case we are trying to both model the long time scale phenomena observed from some Bursters and to understand the cause of the very long time scale outbursts of the Transients.

Our computer code has previously been used in studies of the nova outburst,[5,6,1] thermonuclear runaways in the accreted hydrogen rich envelopes of white dwarfs, and is ideally suited to this

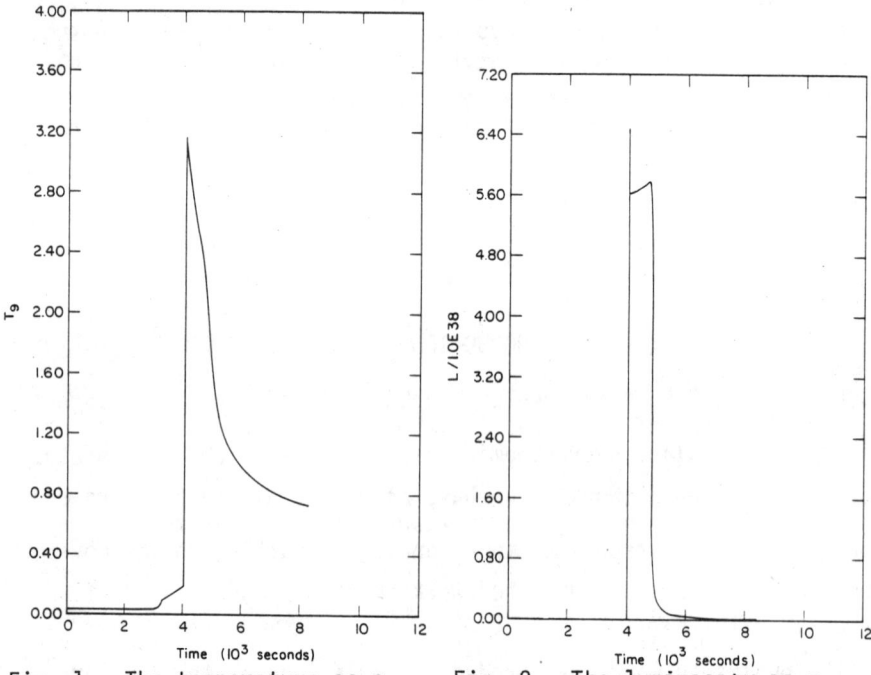

Fig. 1. The temperature as a function of time.

Fig. 2. The luminosity as a function of time.

study. The physics that we include is described elsewhere[7]. We include the p-p chain, the CNO reactions, the triple-α reaction, the $^{12}C(\alpha,\gamma)^{16}O$ reaction and, finally, assume that the $^{14}O(\alpha,p)^{17}F$ and $^{15}O(\alpha,\gamma)^{19}Ne$ reactions are acting in the fashion described by Wallace and Woosley[8]. We use these last two reactions to provide a measure of the rate of depletion of the CNO nuclei during the evolutionary sequences and then use this rate of depletion to calculate an energy generation assuming that a few further proton captures will occur.

RESULTS

The radius of model 1 was 10 km and it had an envelope mass of 1.5×10^{-11} $M_\odot$. This is a thick envelope for a neutron star of this mass but can be obtained for low accretion rates onto neutron stars with low internal temperatures[4]. Although some nuclear burning could have occurred in the deeper layers during the accretion process, we have neglected this and assumed a sharp composition interface. We shall refer to this boundary as the core-envelope

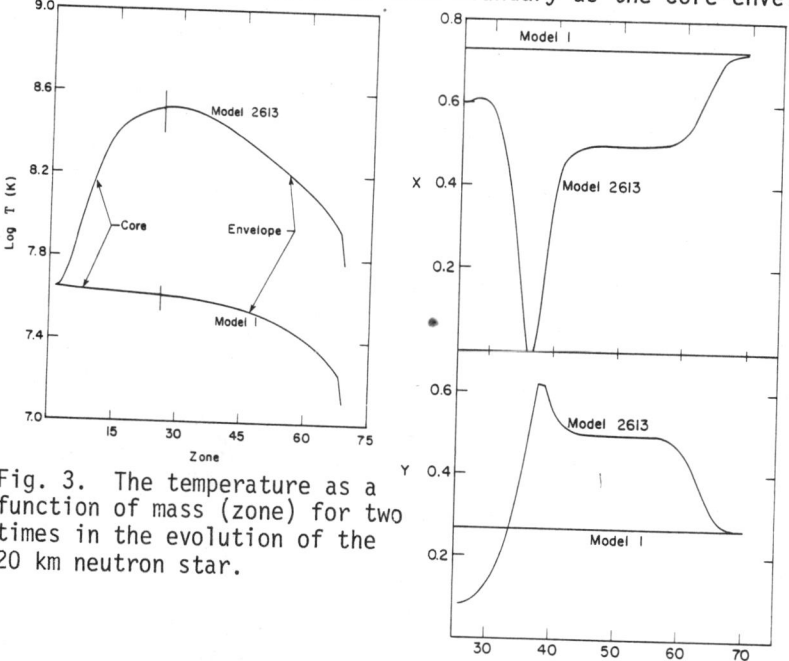

Fig. 3. The temperature as a function of mass (zone) for two times in the evolution of the 20 km neutron star.

Fig. 4. The hydrogen (x) and helium (y) abundances (mass fraction) as a function of mass at two times in the evolution of the 20 km neutron star.

interface (CEI). We chose a luminosity for the neutron star of 0.1 $L_\odot$ which results in an effective temperature of 8×10^5 K. The temperature and density at the CEI are 3.3×10^7 K and 3.1×10^6 gm cm^{-3} respectively.

We begin the evolution by turning on the nuclear reactions and it takes this sequence about 760 seconds of evolution to reach a temperature of 2.45×10^9 K and an energy generation of 10^{21} erg $gm^{-1} s^{-1}$. The surface luminosity has reached 8×10^4 $L_\odot$ and the effective temperature is 2.6×10^7 K (kT $\sim$ 2.2 kev). The entire envelope has become convective and is mixing fresh unburned nuclei into the shell source from the surface.

Because of our initial conditions, which may be physically realizable, the initial region of peak temperature is not at the CEI but about 10^{-12} $M_\odot$ closer to the surface. The reason for this effect is that the electron degenerate conductivity is large enough to transport the energy produced in the deepest layers of the accreted envelope into the core. This effectively keeps these

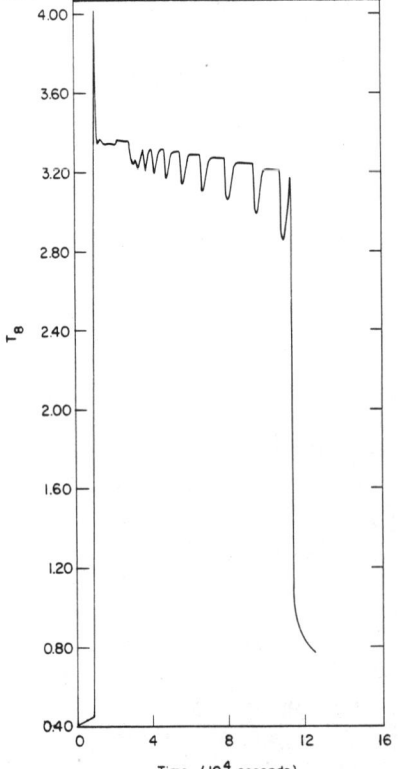

Fig. 5. The temperature at the Core-Envelope-Interface as a function of time.

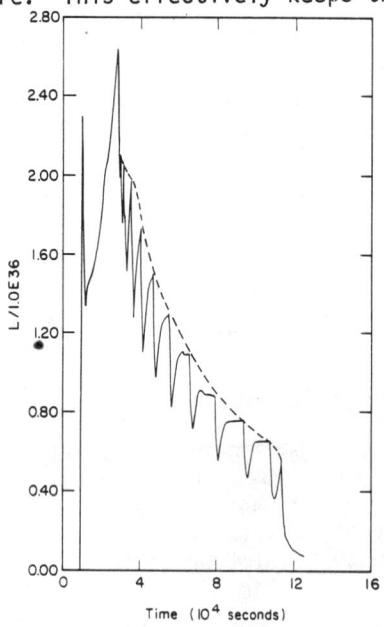

Fig. 6. The luminosity of the neutron star as a function of time.

regions cooler than the less degenerate regions closer to the surface. However, the energy produced by the shell source is now heating the hydrogen rich layers just below it and their temperature slowly climbs until they reach about 3×10^8 K. At this point they flash to a temperature of 2.8×10^9 K and an energy generation rate of 10^{25} erg gm^{-1}s^{-1}. This flash produces an overpressure of a few percent which results in a shock wave that reaches the surface some 5×10^{-5} sec after the flash occurred. The surface luminosity climbs to 2×10^5 $L_\odot$ and the effective temperature to 3×10^7 K. The shock causes the surface zones to expand at velocities of 2×10^3 km sec^{-1}. However, this expansion lasts for only about 10^{-6} sec and a total excursion in radius of 1.5 meters.

The shell source slowly moves inward and these layers flash to high temperatures although no more shocks occur. The temperature as a function of time for the deepest hydrogen rich shell is given in Figure 1. The hydrogen burning layers stay hot until all of the helium and most of the hydrogen is consumed. It takes about 400 seconds for this to occur and during this entire phase of nuclear burning the envelope has slowly expanded to about 12 km and remained at virtually constant luminosity. The light curve for this simulation is given in Figure 2.

Once the fuel has been burnt, the radius of the envelope slowly shrinks which maintains the high luminosity and causes a slow increase in the effective temperature of 3×10^6 K. The sequence then cools rapidly and reaches equilibrium in about an hour.

We extended our study to neutron stars with a radius of 20 km; bracketing the published work. In this case the reduced gravity should produce outbursts with lower peak luminosities and effective temperatures but the time scale for the outburst should be increased over that of the 10 km evolution. Such was the case.

For this evolution we chose an envelope mass of 2×10^{-11} $M_\odot$ which gave a temperature and density at the CEI of 4×10^7 K and 4×10^5 gm cm^{-3}, respectively. Because of the lower density, the evolution proceeds more slowly than for the 10 km case and it takes nearly 10^3 sec for the peak temperature in the envelope to reach 10^8 K. As before, this does not occur at the CEI, but a few zones closer to the surface. In addition, the lower gravity allows the envelope to expand to nearly 40 km during the evolution which prevents the temperature at the CEI from exceeding 4×10^8 K. At this temperature the $^{14}O(\alpha,p)^{17}F$ and $^{15}O(\alpha,\gamma)^{19}Ne$ reactions are removing catalytic nuclei from the CNO reaction sequence and are producing very little additional energy. This results in a most interesting phenomena. Once the shell source has reached inward to the CEI, the temperatures are high enough for a reasonable number of α-reactions on ^{14}O and ^{15}O to occur. This is not true farther out in the envelope where the temperatures are lower. This results in the CNO reactions being able to cycle faster at lower temperatures causing an inversion in the hydrogen abundance. This can be seen in Figures 3 and 4 which show the temperatures, and hydrogen abundances at two different times in the evolution. Therefore, some 6 hours after the evolution began, the hydrogen abundance drops to zero about one-third of the way from the CEI to the surface. Because helium is nearly depleted at the CEI the α-reactions on ^{14}O and ^{15}O also become less important and the principle source of energy comes from the β^+-decays followed by proton captures. Because the CNO catalytic nuclei have their highest abundances at the edge of the region where hydrogen has become completely depleted, this is just the region where they are cycling the most rapidly and the burning front slowly moves inward and outward. However, the hydrogen abundance is low and the runaway time scale for each zone is more rapid than the time scale for the inward diffusion of heat. This allows each zone to burn out before the next inner zone can flash, producing a very ragged curve for both the temperature versus time (Figure 5) and the luminosity versus time (Figure 6). If we had been able to use more zones, these curves would have been smoother.

The peak rate of energy generation reached in this simulation was only 10^{15} erg gm^{-1}s^{-1}. The peak luminosity was 700 $L_\odot$ and the peak effective temperature barely exceeded 0.5 kev. It takes this sequence about one day to burn out all of the hydrogen in the shell source and for the luminosity to begin dropping. It then takes about 45 days to return to minimum. All of the hydrogen and helium are burnt to higher mass nuclei except for a thin shell of H of about 10^{-13} $M_\odot$ which is still burning but on a much longer time scale.

Because the $^{14}O(\alpha,p)^{17}F$ and $^{15}O(\alpha,\gamma)^{19}Ne$ rates are theoretical, we also investigated the effects of major changes in these rates by evolving one additional sequence with the identical initial conditions as in the previous sequence but with these rates set to zero. It takes this simulation 2×10^3 sec for the peak temperature in the envelope to reach 5×10^8 K. The rate of energy generation has reached 10^{15} erg gm^{-1}sec^{-1}. At this temperature a significant number of triple-α and $^{12}C(\alpha,\gamma)^{16}O$ reactions are occurring and the triple-α reaction is feeding new catalytic nuclei into the CNO cycle. This increases the value of Z and counteracts the effects of the increasing number of nuclei being trapped as ^{14}O and ^{15}O. The temperature continues to increase finally reaching 10^9K. The peak luminosity is 2.5×10^4 $L_\odot$ and peak effective temperature is 1.2 kev.

About 2×10^3 sec later, hydrogen and helium burn out and the temperature starts to drop. It takes nearly 3 days for the effective temperature to fall below 0.1 kev.

CONCLUSIONS

We have found in this study that we can produce long time scale outbursts on neutron stars if we assume low accretion rates and "cool" neutron stars. The time scales for these outbursts range from 10^3 sec for the 10 km neutron star to about one day or longer for the 20 km neutron star. The peak temperatures and luminosities were inversely proportional to the radius of the neutron stars and our calculations (plus those noted earlier) suggest that the actual radii of most neutron stars must be closer to 10 km than 20 km. On the other hand, the fast, soft, X-ray transients can be produced on larger radii neutron stars if such a wide range in neutron star radii is possible.

We also produced flat topped outbursts similar to some of those observed. Such a theoretical outburst results when the accreted envelope has had enough time to reach thermal equilibrium before the outburst begins.

Finally, we have been able to achieve outbursts in hydrogen rich material because the $^{14}O(\alpha,p)^{17}F$ and $^{15}O(\alpha,\gamma)^{19}Ne$ reactions act to remove catalytic nuclei from the CNO reaction cycle and, at high temperatures, restores the temperature dependence of the CNO reactions. In addition, at these same temperatures, the triple-α reaction is feeding new nuclei into the CN cycle which also keeps the rate of energy generation elevated over what one would predict if these reactions were not occurring.

REFERENCES

1. S. Starrfield, J.W. Truran, W.M. Sparks, Ap.J. <u>226</u>, 186 (1978).
2. J.S. Gallagher, S. Starrfield, Ann. Rev. Ast. and Ap. <u>16</u>, 171 (1978).
3. P.C. Joss, Ann. N.Y. Acad. Sci. <u>336</u>, 479 (1980).
4. R.E. Taam, Ap.J. <u>241</u>, 358 (1980).
5. G.S. Kutter, W.M. Sparks, Ap.J. <u>175</u>, 407 (1972).
6. S. Starrfield, W.M. Sparks, J.W. Truran, Ap.J. Supp. <u>28</u>, 247 (1974).
7. S. Starrfield, S. Kenyon, W.M. Sparks, submitted to Ap.J.(1981).
8. R.K. Wallace, S. Woosley, Ap.J. Suppl. in press (1981).

Sec. III Solar Transients

SOLAR ENERGETIC PHOTON TRANSIENTS
(50 keV -- 100 MeV)

E. L. Chupp*
University of New Hampshire, Durham, NH 03824

ABSTRACT

Understanding the origin of transient emissions of energetic photons associated with solar flares and other cosmic sources is a key problem in high energy astrophysics. These impulsive photon emissions span the energy range from a few keV to $\sim$ 100 MeV and in some cases the temporal and spectral characteristics of solar and cosmic transients are similar. In this review, we compare the salient observed properties of the solar and non-solar transients and discuss the principal reasons for continued study of the solar flare transient emissions. The observational data obtained primarily on solar flares from the SMM Gamma Ray Spectrometer, but also from some other recent satellites is then reviewed. A major direct observational conclusion from the latest solar flare measurements is that both ions and electrons are accelerated (or released) and interact with matter simultaneously or within seconds of one another. This one fact alone dramatizes the need for a concrete theoretical model of a solar flare, satisfying observations, which incorporates the mechanism(s) which accelerate both the ions and electrons to subrelativistic and relativistic energies. We finish with a brief discussion of future instrumentation requirements for this field.

---

*Invited Paper - The material for this review is based largely on the work of the SMM Gamma Ray Spectrometer Team which consists of: Principal Investigator E. L. Chupp, and Co-Investigators D. J. Forrest, J. M. Ryan, University of New Hampshire; C. Reppin, E. Rieger, G. Kanbach, K. Pinkau, Max-Planck Institute for Physics and Astrophysics, Institute for Extraterrestrial Physics, Garching, Federal Republic of Germany; G. Share, R. L. Kinzer, J.D. Kurfess, W.N. Johnson and M. S. Strickman, Naval Research Laboratory.

INTRODUCTION

This talk is mainly concerned with observations of the impulsive emissions of energetic photons associated with solar flares. Traditionally photons with energies between roughly 50 keV and 200 keV have been referred to as hard X-rays and those above 200 keV as gamma rays with no limit at the upper end. We have chosen the title "Solar Energetic Photon Transients" because reference to hard X-rays or gamma rays in the literature usually implies photons produced by electrons or ions respectively, whereas in fact both species can contribute to the electromagnetic spectrum in the range of interest. Thus rather than focusing on arbitrary distinctions determined by experimental convention we are ultimately interested in determining the process(es) which energize both electrons and ions and the signatures they give us in the form of energetic photons.

Since this workshop is concerned with photon transients in the broadest astrophysical sense, we compare, in Table I, the important known physical properties of a few solar and non-solar energetic photon transients. The data are in general self-explanatory. Clearly both types of transients have many features in common. Even though not shown, there are solar flares whose time duration is much shorter than or longer than for the two flares listed in the table. While it is known that the solar flare photon transients in question are well correlated with emissions at other wavelengths such as for soft X-rays, optical and radio, such is not the case for the "cosmic photon transients". Observational bias is undoubtedly an important factor here since one does not know where to look in the sky to find a cosmic burst. We feel it is important to make the comparisons shown in Table I for two reasons: first, because some features of the two types of transients are similar, there could be common physical processes involved in both and second, it seems clear that many of the instrumentation requirements for studying the two types of transients are the same.

Observations of apparent non-solar transients with the SMM gamma ray spectrometer are described in another paper in this volume.

Table I

COMPARISON OF SOLAR AND "COSMIC" ENERGETIC PHOTON TRANSIENTS

	MARCH 29, 1980	JUNE 7, 1980	APRIL 27, 1972 (APOLLO 16)	MARCH 5, 1979	APRIL 19, 1980
E_γ (MAX)	~ 0.6 MEV	~ 8.0 MEV	~ 2.0 MEV	~ 1.0 MEV	~ 7 MEV
CONTINUOUS SPECTRAL SHAPE ($CM^{-2}S^{-1}MEV^{-1}$)	$1.0 \times 10^{-1} E^{-4.2}$ $E > 275$ KEV	$1.4 E^{-2.5}$	$\frac{3.0}{E} G(E) e^{-(E/.5)}$	$\frac{2.9 \times 10^3}{E} e^{-(E/.03)}$	$.4 E^{-2.2}$
TIME DURATION	~ 41 SEC	50 SEC FOR FIRST PHASE	40 SEC	200 MS INITIAL PEAK FOLLOWED BY LESSER PEAKS FOR 180 SEC	~ 2 SEC
TIME STRUCTURE	8 PEAKS ~ 1 SEC WIDE SEPARATION TIME 1 – 2 SEC e-FOLDING RISE-TIME OF 2.9 ± 0.2 S	SERIES OF QUASI-PERIODIC PULSES FOUR IMPULSIVE PHASES EACH SEPARATED BY ~ 1 MIN.	STRONGEST PEAK 1000 $\frac{PHOTONS}{SEC}$ COUNT FELL TO ZERO FOR 148 MS THEN RETURNED TO ALMOST PREVIOUS LEVEL 3 MAIN PEAKS 5 LESSER PEAKS	INITIAL BURST RISE TIME < .2 MS FWHM 120 MS FOLLOWED BY $8\pm.5$ SEC PERIODICITY LASTING UNTIL END OF BURST	~ 10 SEC BEFORE BURST, EMISSION IN 10–15 KEV RANGE REMAINING HIGH 3 MIN AFTER BURST, BURST, 2 MAIN PULSES EACH WITH DURATION 360 MS.
LINE FEATURES	NONE APPARENT	EMISSION LINES $2.223\pm.012$ MEV FLUX$(7.1\pm1.2)\times 10^{-2} CM^{-2}S^{-1}$ PROMPT LINES 4.4 MEV FLUX$(2.7\pm1.3)\times 10^{-2} CM^{-2}S^{-1}$ 6.13 MEV FLUX$(2.7\pm.8)\times 10^{-2} CM^{-2}S^{-1}$	NONE APPARENT	EMISSION FEATURE ~ 420 KEV FLUX $3 \times 10^{-6} \frac{ERGS}{CM^2-S}$ ASSUMING LINE RADIATION EMITTED ONLY IN INITIAL PULSE FLUX $5 \times 10^{-5} \frac{ERGS}{CM^2-S}$	NONE APPARENT
ENERGY FLUX AT PEAK > 100 KEV					
FLUX ERGS/CM^2-S	1.15×10^{-6}	1.26×10^{-5}	4.85×10^{-5}	2×10^{-3}	2.88×10^{-6}
FLUENCE ERGS/CM^2	3.8×10^{-5}	5.09×10^{-4}	8.75×10^{-4}	8×10^{-3}	4.21×10^{-5}
SOURCE	SOLAR AR-2357	SOLAR AR-2495	?	LMC N49	SOLAR OR COSMIC?
REFERENCES	25, 26	14, 15, 27	31, 32	28, 29, 30	33

SCIENTIFIC OBJECTIVES FOR SOLAR FLARE STUDIES

Since the discovery of "solar cosmic rays" by sea level neutron monitors in 1942[1,2] following a large solar flare, it has been expected that fluxes of both gamma rays and neutrons produced at the Sun by accelerated protons could also be detectable at the Earth and hence provide direct data on the conditions in the flare region. Detailed estimates on the gamma ray and neutron fluxes expected from flares were given by Lingenfelter and Ramaty[3] who also reviewed earlier theoretical and experimental work. Subsequently Chupp[4] gave a comprehensive review of the (unsuccessful) experimental efforts to search for both neutrons and gamma rays from the Sun during both quiet and active periods. It was not until the period of extreme solar activity in August 1972 that definitive evidence was obtained for the impulsive emission of gamma ray lines and continuum during two large flares; on the rising phase of the 3B (Hα) flare from (0623:49 to 0633:02 UT) on 4 August 1972 and during the decay phase of a second 3B (Hα) flare from (1538:20-1547:33 UT) on 7 August 1972[5,6]. Delayed gamma ray lines were observed in both flares at 2.223 MeV from the reaction $^1H(n,\gamma)^2H$ and at 0.511 MeV from positron-electron annihilation. The nuclear reactions of the accelerated ions producing the neutrons and positrons also produce excited states of ambient nuclei ^{12}C and ^{16}O which give prompt gamma ray lines. Their intensity time variations during a flare give directly the time history of the nuclear interactions at the Sun. In the case of the two flares mentioned above weak transient gamma ray features were observed at 4.43 and 6.13 MeV corresponding to deexcitation of excited states of ^{12}C and ^{16}O, respectively. These discoveries inspired detailed theoretical calculations on the yield of gamma ray lines and continuum for generalized energy spectra of flare accelerated ions (cf. Ramaty et al.)[7] assuming standard abundances (Cameron)[8]. It is explicity assumed in these calculations that the solar flare mechanism accelerates all ions in proportion to their solar abundance for a given energy/nucleon. In this case, the prompt gamma ray line spectrum, assuming no Doppler shifts must consist of narrow lines at their proper energy, produced by accelerated protons interacting with the ambient solar atmosphere and Doppler broadened lines resulting from accelerated ions interacting with the ambient hydrogen atmosphere. Using this model, Ramaty, Kozlovsky and Suri[9] have concluded that all the energetic photons in the 4-7 MeV range observed in the 4 August 1972 flare were due to nuclear reactions as described above. Therefore, a study of time variations of the prompt gamma ray lines should be a powerful diagnostic tool for determining the characteristics of the accelerated ions at the Sun. Crannell et al.[10] have also emphasized the importance of studying the intensities and line shapes of the ^{12}C dexcitation lines at 4.44 and 15.11 MeV in order to determine the spectra and directivity of the accelerated particle motion.

The possibility of making observations such as mentioned above was one of the primary motives for under taking measurements of solar flare gamma ray spectra with the SMM gamma ray spectrometer. In Table II, we summarize the measurement objectives for this experiment. Considering our previous remarks the first three items in the table are self explanatory. The fourth item referring to the composition of energetic ions and target species can be, in principle, determined by a study of the relative intensities of the gamma ray lines in a suitably "prolific flare" and we will discuss such an example later. The fifth item is of a serendipitous nature since gamma ray flare observations are still new enough that we may have some surprises. First we must realize that we may be concerned with two classes of "solar cosmic rays"; Class I, those that produce observable gamma ray lines at the Earth and Class II, those which produce observable energetic particles, "solar cosmic rays", at the Earth. It is, of course, fundamentally important to determine if and how these classes of "solar cosmic rays" are related. We expect this to be determined by correlation of gamma ray flare and interplanetary particle observations.

Table II
Objectives for SMM Gamma Ray Spectrometer Flare Events

I. Determine Energy Spectra of Energetic Nuclear Species.

II. Determine Time History of Energetic Nuclear Species.

III. Determine Angular Distribution of Energetic Nuclear Species.

IV. Determine Composition of Energetic Ions and Target Species.

V. Establish Other Gamma Ray Event Characteristics and Correlations.

OBSERVATIONAL SITUATION

In Table III we summarize the most significant observations of solar flare associated gamma ray line events starting with the 4 August 1972 event mentioned above. The 7 August 1972 is not listed because it was observed well after the impulsive phase of the flare was over. In all cases, the duration of the impulsive phase given in column two is meant to give a measure of the total time span during which impulsive bursts are occurring. Column 3 gives an approximate power law fit to the continuum spectrum over the energy range indicated which extends in all cases to ~ 1 MeV. In columns 4, 5, and 6 the fluences are given for gamma ray lines at the indicated energies. In the 4-7 MeV energy band, several events clearly show the 4.4 and 6.1 MeV lines mentioned earlier and evidence for other unresolved lines. Forrest et al.[11] have presented evidence that all the excess counting rate above the extrapolated continuum in this energy band is due to gamma ray lines produced by nuclear interactions of accelerated ions which are produced within seconds of, or simultaneously with the electrons producing the bremsstrahlung continuum.

Table III

SOME FLARES WITH DETECTABLE LINE FLUXES
PRELIMINARY ESTIMATES

FLARE	IMPULSIVE PHASE (SEC)	CONTINUUM FIT FOR IMPULSIVE PHASE $\gamma (MeV\,cm^2\,sec)^{-1}$	0.5 MEV γ/cm^2	2.2 MEV γ/cm^2	4-7 MEV γ/cm^2
4 AUGUST 1972 (OSO-7)	≥ 600	$0.4\,E^{-3.42}$ (0.36–0.70) MEV	35 ± 11	≥ 150	114 ± 12
11 JULY 1978 (HEAO-1)	90 (+150)	$10\,E^{-3}$ (1–5) MEV	--	200 ± 60	220 ± 90
7 JUNE 1980 (SMM)	50	$1.26\,E^{-2.6}$ (0.29–1.0) MEV	$< 2\,(3\sigma)$	6.6 ± 1	11.5 ± 0.5
21 JUNE 1980 (SMM)	66	$11.0\,E^{-2.3}$ (0.29–1.0) MEV	5.8 ± 0.2	3.1 ± 0.2	121 ± 2
1 JULY 1980 (SMM)	60	$0.195\,E^{-3.0}$ (0.29–1.0) MEV	0.9 ± 0.4	3.3 ± 0.5	3.1 ± 0.4
6 NOVEMBER 1980 (SMM)	64 (+160)	$1.24\,E^{-2.5}$ (0.29–1.0) MEV	$< 2\,(3\sigma)$	10.3 ± 1.3	14.8 ± 0.8
10 APRIL 1981 (SMM)	345	$0.139\,E^{-2.6}$ (0.29–1.0) MEV	$< 6.6\,(3\sigma)$	13.5 ± 1	18.6 ± 1.6
27 APRIL 1981 (SMM)	~ 720 SEC	$\sim E^{-2.3}$		11.7 ± 2	118 ± 2

Prior to the SMM launch on 14 February 1980, there were four observed gamma ray line flares, the properties of two of which are shown in Table III. We have not listed the first reported high resolution observation (Prince et al.)[12] of the 2.223 MeV gamma ray line with the high resolution germanium spectrometer on HEAO-3 during the 0306 UT flare on 9 November 1979 since only an unpublished 2.223 MeV fluence value is currently available. We also have not listed the 0445 UT flare of 22 November 1977 suggested to be a gamma ray line flare since the instrumentation did not have adequate energy resolution to directly demonstrate the presence of lines[13]. The remaining six flares, all observed by the gamma ray spectrometer[14] on SMM, produced the largest gamma ray line fluxes of the more than 50 SMM flares yielding energetic photons > 300 keV observed through June 1981. We will now describe some of the important observations for three of these flares.

In Figure 1 we show the net gamma ray spectrum for the long duration flare beginning at 0805 UT on 27 April 1981. This is a preliminary spectrum and a complete unfolding of the spectrum using the full instrument response function has not yet been completed. However, the strong lines at 4.4 and 6.1 MeV are clearly identified and have average intensities of 1.9×10^{-2} and 1.3×10^{-2} photons $cm^{-2}s^{-1}$ respectively. The 2.223 MeV is

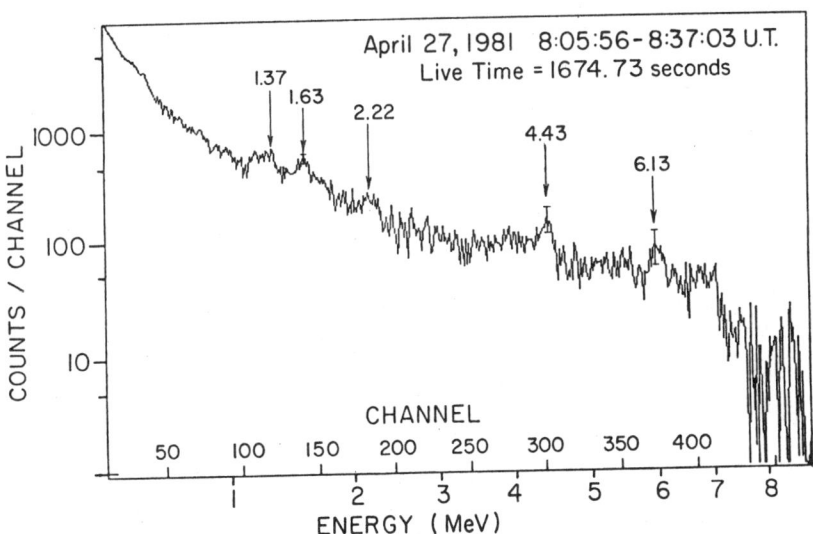

Figure 1. Net energy loss spectrum for the flare of 27 April 1981. Preliminary data.

only weakly evident because the flare occurred on the west limb and this line, of photospheric origin, is strongly attenuated. Near the energies marked 1.37 and 1.63 MeV there are actually several lines which could be present, such as ^{56}Fe(1.238 MeV), ^{32}S(1.249 MeV), ^{52}Cr(1.334 MeV), ^{24}Mg(1.369 MeV), ^{55}Fe(1.370 MeV), ^{14}N(1.632 MeV), ^{20}Ne(1.634 MeV), and ^{23}Na(1.636 MeV). The time history of this event, as recorded in several data channels of the instrument is shown in Figure 2. In the upper two panels the hard X-rays show several impulsive phases extending over ∼ 600 s. The extended behavior from (600-1600 s) of the counting rate in the nuclear line region (extending from 4.1-6.7 MeV) is noteworthy since the hard X-ray flux has decayed. This appears to be a case of delayed acceleration of ions alone or else a release of ions from some storage region since only ions could have produced the excess counting rate in the 4.1 to 6.7 MeV band without X-ray photons. It can be seen from Figure 2 that background at this energy has a strong rigidity effect so care must be taken in analysis.

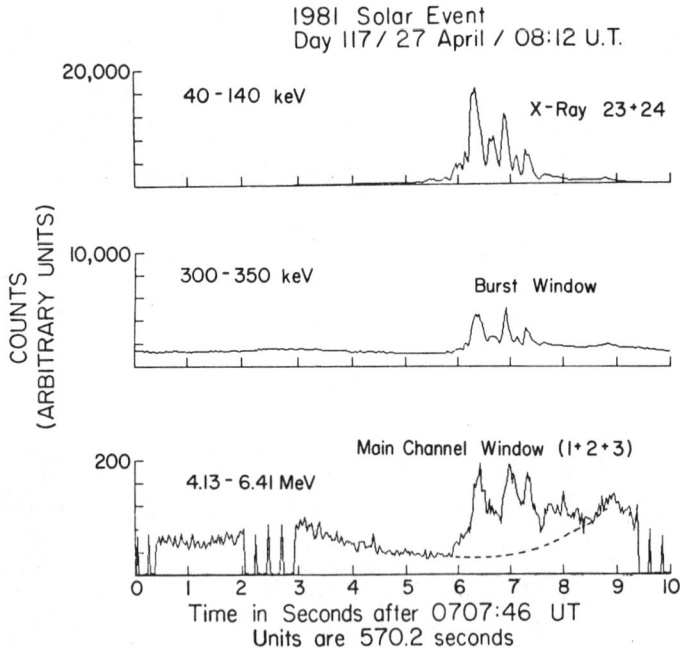

Figure 2. Time history for the flare of 27 April 1981. The background represented by the dashed line is from an orbit on the previous day which closely duplicates the geomagnetic parameters and Earth aspect during the flare orbit.

In Figure 3, we show the time history for the 0312 UT flare on 7 June 1980. Results on the 2.223 MeV line in this flare have been previously published (Chupp et al.)[15] and Forrest et al.[11] have discussed the evidence for ion acceleration in the event. One important aspect of this event is that the time history in the main phase of the flare is very similar for both hard X-rays ($\sim$300 keV) and the highest energy gamma rays observed ($\sim$ 6.4 MeV). There are, in fact, seven impulsive spikes occurring within $\lesssim$ 60 s each separated by 8 s and each spike has a width of only a few seconds. Forrest et al.[11] have concluded that this event demonstrates that relativistic electrons and > 30 MeV ions must have been accelerated together. In the case of the gamma ray line flare on 11 July 1978, Hudson et al.[16] have shown that the MeV gamma ray emission lagged the hard X-ray and microwave emission by less than 20 s; thereby, concluding that the "second stage" acceleration of high-energy solar particles must commence promptly after the impulsive phase. The essential new result from the SMM gamma ray

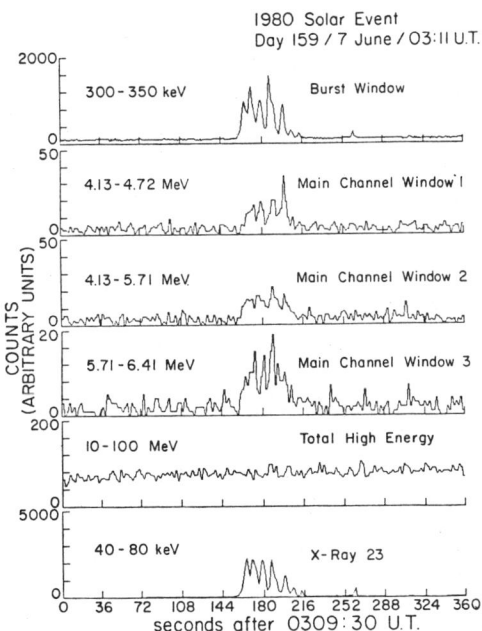

Figure 3. Time history for the flare of 7 June 1981.

observations is that there is no evidence at all for a second stage acceleration in this, 7 June 1980, flare and the data requires that electrons and ions be repetitively accelerated to produce the spikes shown in Figure 3. The apparent quasi-periodicity in this event fits the picture of "elementary flare bursts" first noted by Frost[17] and Van Beek et al.[18]. These bursts are nearly symmetric in time and each may be characteristic of the shape of the accelerator pulse injected into dense solar material, generating the X-rays and gamma rays.

In Figure 4, this near periodic behavior may be seen more clearly where the detailed time histories of the hard X-ray and gamma ray line channels are compared; the former being due to near relativisitic electrons $\beta \sim 0.9$ and the latter due to subrelativistic protons $\beta \sim 0.1$. At least for the fourth and fifth pulses it is clear that the peak of the gamma ray emission lags the hard X-ray emission by a few seconds. A simple explanation of this fact results if the acceleration region for both electrons and ions is at the top of a loop and the energetic photons are due to interactions after the particles have traveled down to a more dense atmosphere, the delay being simply due to the lower velocity of the protons. A trap model like that discussed by Bai and Ramaty[19] in which the release time is energy dependent could also explain the time delays. It is also interesting to note that the microwave emissions follow a time history very close to that of the hard X-rays (Rust)[20] consistent with the picture that the microwaves are generated by gyrosynchrotron emission of $\sim$ 100 keV electrons in $\sim$ 100 gauss fields at the top of a loop. While this picture may explain

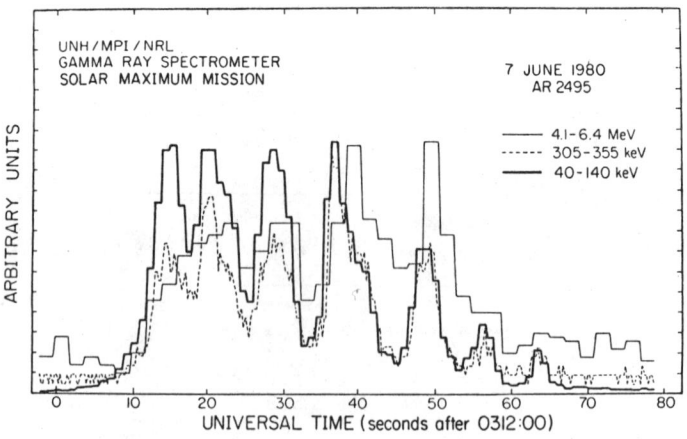

Figure 4. Overlay of the 40-140 keV, 305-355 keV, and 4.1-6.4 MeV time histories for the flare of 7 June 1980.

the time delays, it does not obviously predict the quasi-periodicity. This feature is suggestive of a damped oscillation whose period may, for example, be controlled by the Alfven speed V_A and a loop dimension L giving a period $T_A \sim L/V_A \sim L/B\,(4\pi\rho)^{1/2}$. For a loop dimension $L \sim 10^{10}$ cm, $\bar{B} \sim 300$ gauss and $T_A = 8$ sec, we find $\rho \sim 5 \times 10^{-13}$ grams cm^{-3} or $n \sim 10^{11}$ particles cm^{-3}. According to this picture, the oscillating disturbance would trigger repeated accelerations of energetic particles.

Another explanation suggested by Kiplinger et al.[21] is that the repeated pulse trains are due to successive firings of seven adjacent loops. This idea is based on the interacting loop model of Emslie[22] in which expansion of a given loop triggers an instability in an adjacent loop. The expansion of a loop is caused by a momentary increase in the gas pressure which then exceeds the magnetic energy density. Thus the apparent periodicity requires a proper matching of loop separations and expansion speeds of individual loops (see Hoyng et al.)[23].

A further illustration of a solar flare gamma ray line spectrum, and hence ion signature, is shown for the impulsive phase of this flare in Figure 5. The upper curve in the figure is the total counting rate spectrum during the impulsive phase of the flare multiplied by $E^{2.5}$ so as to compensate for the steep continuum below 1 MeV. Also shown below in the figure is the background counting rate spectrum just before the impulsive phase.

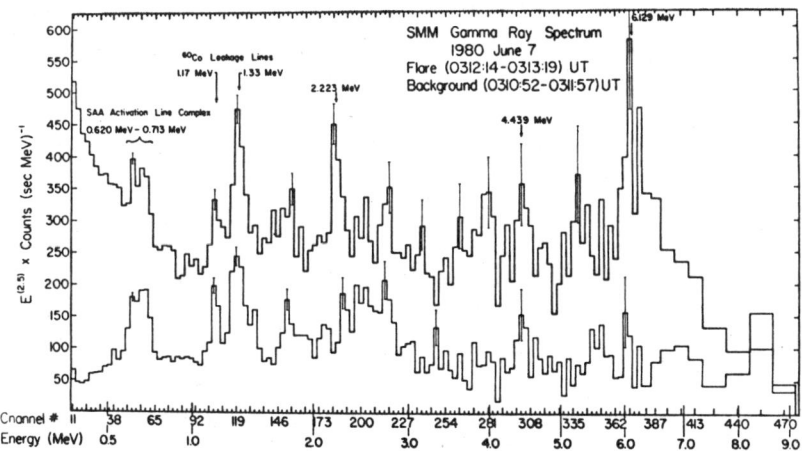

Figure 5. Impulsive phase energy loss spectrum for the flare of 7 June 1980 and a preflare background spectrum. Both spectra are multiplied by $E^{2.5}$ for display purposes (see text).

In the flare spectrum, we see strong lines at 2.223 MeV, 6.129 MeV and a weaker line at 4.439 MeV. In addition, the flare spectrum falls off steeply at about 6.4 MeV characteristic of a photon spectrum dominated by nuclear lines without any detectable evidence for electron bremsstrahlung or π° gamma radiation.

The X2/3B solar flare on 10 April 1981, which reached a maximum in Hα at $\sim$1655 UT, produced considerable geophysical effects with the geomagnetic field disturbed for over six hours, a situation which apparently has not been repeated since August 1972. The SMM Gamma Ray Spectrometer (GRS) observed the early phase of this event just before the satellite sunset which occurred before the optical maximum. The time history for several data channels is shown in Figure 6. The impulsive photon emission (> 20 keV) is observed to have two maxima separated by 250 s with energetic photons observable to $\sim$ 7 MeV. It is not known if there was further impulsive emission after eclipse; however, the total impulsive duration observed was $\sim$5 min, considerably longer than the majority of events observed by our instrument in 1980. Of particular interest in this event is the clear evidence for two separate bursts of neutron production as required by the 2.223 MeV time history as shown in the middle

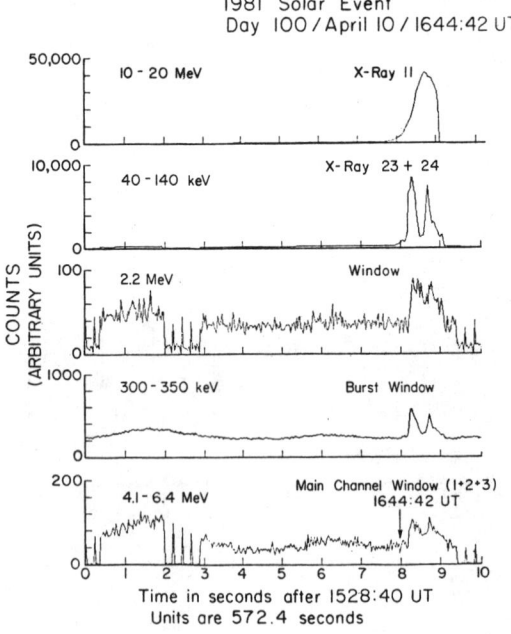

Figure 6. Time history for the flare of 10 April 1981.

panel of Figure 6. This is expected if ions are accelerated in each burst and consequently neutrons produced; however, this is the first case where the multiple neutron production has been observed. The 2.223 MeV time history has been modeled for this event by assuming that the neutron production time history does indeed follow the counting rate in the 4-7 MeV energy band as shown in Figure 7. Then using the method similar to that described by Chupp et al.[15] for analyzing the 2.223 MeV time history for the 7 June 1980 event described above, we have found the neutron loss time that best fits the observed 2.223 MeV line time history. The result is shown by the curve indicated by the solid segments in the middle panel of Figure 7. For both

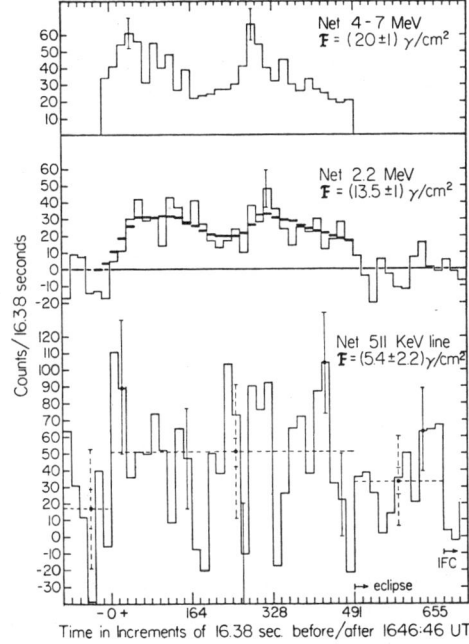

Figure 7. **Middle panel:** 2.2 MeV line time history for the flare of 10 April 1981 with model fit. **Upper panel:** 4-7 MeV time history used as the neutron production time history in the 2.2 MeV line model. **Lower panel:** .511 MeV line time history. _Preliminary data._

neutron injections, a characteristic neutron loss time of ~ 50 s is indicated. This time appears to be nearly a constant for all observed events. Kanbach et al.[24] find approximately the same value in Monte Carlo results on solar 2.223 MeV production — if the ^{3}He/H abundance in the photosphere is $\sim 5 \times 10^{-5}$.

One of the basic results from the SMM gamma ray spectral observations is that whenever there is a measurable excess above the continuum in the 4-7 MeV band, ions had to be accelerated in the impulsive phase(s). The 2.223 MeV line is the only one observable in the weaker gamma ray line events. Only in stronger gamma ray flares can the lines at 4.43 and 6.13 MeV be observed and usually there is only weak evidence for the presence of other gamma ray lines. However, the gamma ray spectrum from the flare on 27 April 1981 gives also clear evidence for other prompt nuclear lines (see Figure 1). It is interesting to compare the spectrum from this strong gamma ray line event with the summation of the spectra for six weaker flares. This is shown in Figure 8 in which only the net impulsive phase spectra are used after preflare background is

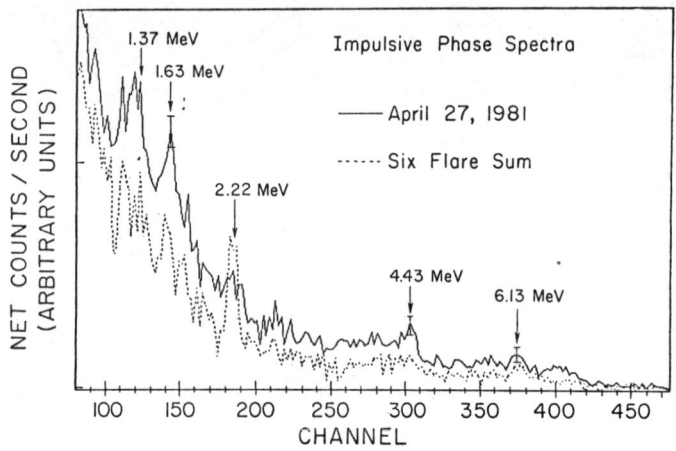

Figure 8. The energy loss spectrum measured during the impulsive phase of the large limb flare on 27 April 1981 compared to a composite of impulsive phase spectra from six weaker flares. <u>Preliminary data</u>.

subtracted. The stronger identifyable line features are indicated. Error bars due to counting statistics alone are indicated at several channel positions. By a careful comparison of the single flare spectrum with the composite flare spectrum it can be seen that there is a one to one correspondence of the stronger features in both cases. It is tempting to conclude that all flares in which ions are accelerated produce the same, or very similar, gamma ray line spectra.

It is also interesting to compare the measured fluences for the 2.223 MeV line, the 4-7 MeV band and for the continuum at 1 MeV for several SMM flares. The result is displayed in Figure 9 where it can be seen there is an approximately linear correlation of the 2.223 MeV fluence to the 4-7 MeV fluence as shown by the symbol x, except for the two limb flares which are identified by the flare dates. It can be seen that there is no correlation of the electron bremsstrahlung fluence at 1 MeV with the prompt (4-7 MeV) gamma ray fluence as shown by the open circles. This is an indication that the relativistic electron flux does not scale with proton flux at the Sun.

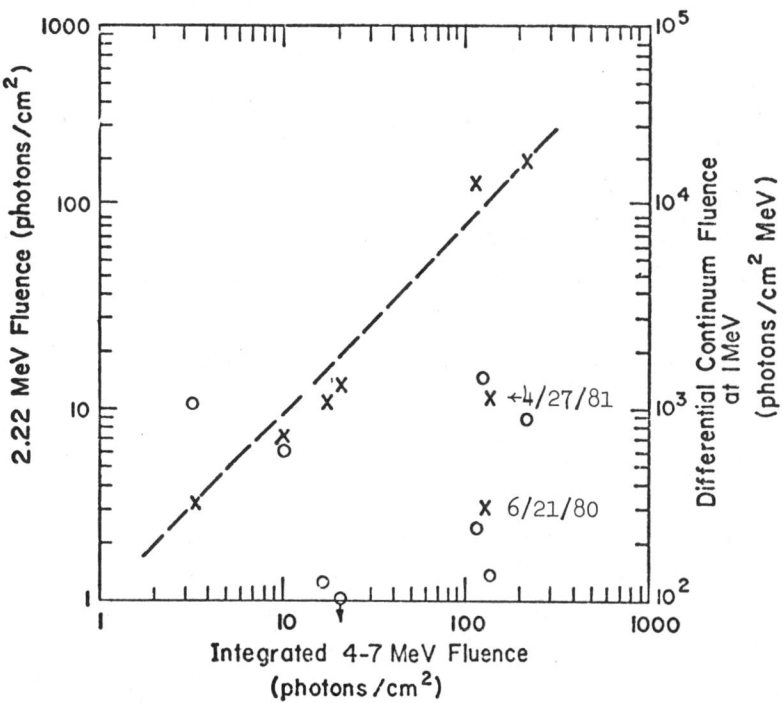

Figure 9. Scatter plot showing the correlation of 2.2 MeV line fluence (X's) and differential continuum fluence at 1 MeV (O's) to 4-7 MeV fluence. The limb flares where the 2.223 MeV line is strongly attenuated are indicated by dates. Preliminary data.

DISCUSSION

The experimental observations of solar gamma rays has given us direct knowledge on the behavior of the energetic particles at the Sun. In all of the gamma ray line events observed on SMM, we must conclude that the impulsive emission of energetic photons is due to <u>both</u> ion and electron interactions with matter and that both species must have been accelerated simultaneously or within seconds of one another. Further, the general time structure for the gamma ray events follows that for the microwave bursts, where data has been compared, in a similar manner to that previously established for hard X-rays. Bursts of all three components have been observed to repeat with a quasiperiodicity of < 10 s. Many gamma ray flares have been observed to have durations less than a minute while a few have durations greater than 10 mins. The 2.223 MeV neutron capture line has now been observed in several events occurring at different heliographic longitudes. In all cases the decay time of this line's intensity is ~ 50 s which, according to Monte Carlo calculations, is consistent with a <u>finite</u> photospheric ^{3}He/H abundance of 5×10^{-5}. In one very <u>intense</u> west limb flare, impulsive phase photons with energies above 50 MeV were observed.

The observational facts just discussed lead directly to some important conclusions about solar flares contrary to statements now existing in the literature. The "classic" two stage acceleration model consists of an impulsive primary acceleration of electrons to energies of ~ 100 keV followed by a gradual second stage acceleration of electrons and ions to much higher energies. We find that this model is not, <u>in general, valid</u>. Also, contrary to previous conclusions ions and relativistic electrons <u>are observed</u> in small impulsive flares. Finally, the observations of "solar cosmic rays" and gamma ray line flares now require a re-definition of a "proton flare". We suggest the two following classifications to separate the events:

Class I – in which the acceleration of protons in a flare is based on gamma ray line observations.

Class II – in which protons are directly observed in space and are causally associated with a solar flare(s).

Of course, it is important in the future to determine the physical connection between the two types of events.

FUTURE INSTRUMENTATION FOR ENERGETIC
SOLAR PHOTON TRANSIENT OBSERVATIONS

There will be detailed discussions of instrumentation in other talks later in this workshop; however, a few remarks are appropriate here concerning the solar observations. Even though we now have a large body of data on solar gamma ray line events, there have only been a few cases with strong lines and only in one case in which a partial study of the prompt line time history is possible. There has not yet been an event in which we could study line profiles and, hence, Doppler broadening. Thus there is a continuing need for higher sensitivity measurements obtainable with scintillation spectrometers. An order of magnitude improvement over the SMM spectrometer's capabilities is highly desirable. Also it is clear that high energy resolution measurements are needed in order to unravel all the complex line features that are evident in the flare spectrum above ~ 0.8 MeV. For example, use of a germanium spectrometer (such as was planned for SMM) would make possible isotopic composition measurements. A study of the widths of the lines such as the 511 keV annihilation line would permit a determination of the flare region temperature. Finally, if gamma imaging, to arc second accuracy, were to be developed the actual source of the gamma rays could be determined and compared with the spatial structure of a flare at other wavelengths. However, such capability is not yet within the state of the art.

The next generation spectrometers that are necessary to extend our knowledge on energetic photon transients should utilize both scintillation and solid state detector elements. An instrument of adequate size will also give the greatest chance for discovery and should remain in orbit over a fair portion of the next solar cycle.

ACKNOWLEDGEMENTS

The author wishes to thank Mary Chupp for her help in editing this paper and Celeste Dietterle for its preparation. He is also grateful for the data analysis and other assistance from John Heslin, John Lanigan, and Steve Matz. (Work was supported by NASA contracts NAS 5-23761 (UNH) and S.70926A (NRL) and in the Federal Republic of Germany by BMFT contract 010K 017-ZA/WS/WRK 0275:4.)

REFERENCES

1. A. Ehmert, Z. Naturforsch 3a, 264 (1948).
2. I. Lange and S. E. Forbush, Terrest. Magetism Atmos. Elec. 47, 185, (1942).
3. R. Lingenfelter and R. Ramaty, in B. S. P. Shen (ed.), High-Energy Nuclear Reactions in Astrophysics, W. A. Benjamin, Inc., New York, p. 99, (1967).
4. E. L. Chupp, Space Sci. Rev. 12, 486 (1971).
5. E. L. Chupp, D. J. Forrest, P. R. Higbie, A. N. Suri, C. Tsai, and P. P. Dunphy, Nature 241, 333 (1973).
6. E. L. Chupp, D. J. Forrest, and A. N. Suri, Solar Gamma-, X-, and EUV Radiation, IAU Symposium No. 68, in S. R. Kane (ed.) (D. Reidel Publishing Company, Dordrecht, Holland, 1975) p. 341.
7. R. Ramaty, B. Kozlovsky, R. E. Lingenfelter, Space Sci. Rev. 18, 341 (1975).
8. A. G. W. Cameron, Space Sci. Rev. 15, 121 (1973).
9. R. Ramaty, B. Kozlovsky, and A. N. Suri, Astrophys. J. 214, 617 (1977).
10. C. J. Crannell, H. Crannell, and R. Ramaty, Astrophys. J. 229, 762 (1979).
11. D. J. Forrest, E. L. Chupp, J. M. Ryan, C. Reppin, E. Rieger, G. Kanbach, K. Pinkau, G. Share, and R. Kinzer. To be presented in the late volumnes of the 17th International Conference on Cosmic Rays, Paris, France, 13-15 July 1981.
12. T. Prince, J. C. Ling, W. A. Mahoney, G. R. Riegler, and A. S. Jacobson, Paper presented at the Conference on Cosmic Ray Astrophysics and Low-Energy Gamma-Ray Astronomy, Minneapolis, Minnesota, 3-6 September 1980.
13. G. Chambon, K. Hurley, M. Niel, R. Talon, G. Vedrenne, I. V. Estuline, and O. B. Likine, Solar Phys. 69, 147 (1981).
14. D. J. Forrest, E. L. Chupp, J. M. Ryan, M. L. Cherry, and I. U. Gleske, C. Reppin, K. Pinkau, E. Rieger, G. Kanbach, R. L. Kinzer, G. Share, W. N. Johnson, and J. D. Kurfess, Solar Phys. 65, 15 (1980).
15. E. L. Chupp, D. J. Forrest, J. M. Ryan, C. Reppin, G. Kanbach, E. Rieger, K. Pinkau, G. Share, R. Kinzer, and M. Strickman, Astrophys. J. 244 Astrophys. J. 244, L171 (1981).
16. H. S. Hudson, T. Bai, D. E. Gruber, J. L. Matteson, P. L. Nolan, and L. E. Peterson, Astrophys. J. (Letters) 236, L91 (1980).
17. K. J. Frost, Astrophys. J. (Letters) 158, L159 (1969).
18. H. F. Van Beek, L. D. de Feiter, and C. de Jager, in D. E. Page (ed.), Correlated Interplanetary and Magnetospheric Observations (Reidel), p. 533.

19. T. Bai and R. Ramaty, Astrophys. J. 227, 1072 (1979).
20. D. Rust (private communication) 1980.
21. A. L. Kiplinger, B. R. Dennis, K. J. Frost, and L. E. Orwig, to be submitted to Astrophys. J. (1981).
22. A. G. Emslie, An Interacting Loop Model of Solar Flare Bursts, submitted to Astrophys. Letters (1981).
23. P. Hoyng, Z. Svestka, and A. Kiplinger, Proceedings Third European Solar Activity Meeting, C. Jordon (ed.), Oxford, England, 13-6 April 1981.
24. G. Kanbach, K. Pinkau, C. Reppin, E. Rieger, E. L. Chupp, D. J. Forrest, J. M. Ryan, G. H. Share, and R. L. Kinzer. To be presented in the late volumnes of the 17th International Conference on Cosmic Rays, Paris, France, 13-25 July 1981.
25. J. M. Ryan, D. J. Forrest, E. L. Chupp, M. L. Cherry, C. Reppin, E. Rieger, K. Pinkau, G. Kanbach, G. H. Share, R. L. Kinzer, M. S. Strickman, W. N. Johnson, and J. D. Kurfess, Astrophys J. (Letters) 244, L175 (1981).
26. B. R. Dennis, K. J. Frost, and L. E. Orwig, Astrophys. J. (Letters) 244, L167 (1981).
27. L. E. Orwig, K. J. Frost, and B. R. Dennis, Astrophys. J. (Letters) 244, L163 (1981).
28. T. L. Cline, The Unique Cosmic Event of 1979 March 5 (GSFC Report 80630, Goddard Space Flight Center, Greenbelt, MD., 1979).
29. E. P. Mazets, S. V. Golenetskii, V. K. Il'inskii, R. L. Aptekar', and Yu. A. Gur'yan, Nature 283, 587 (1979).
30. J. Terrell, W. D. Evans, R. W. Klebesadel, and J. G. Laros, Nature 285, 383 (1980).
31. D. Gilman, A. E. Metzger, R. H. Parker, L. G. Evans, and J. I. Trombka, The Distance and Spectrum of the Apollo Gamma-Ray Burst (Jet Propulsion Laboratory, Pasadena, CA, 1979) preprint.
32. A. E. Metzger, R. H. Parker, D. Gilman, L. E. Peterson, J. I. Trombka, Astrophys. J. (Letters) 194, L19 (1974).
33. G. H. Share, M. S. Strickman, R. L. Kinzer, E. L. Chupp, D. J. Forrest, J. M. Ryan, E. Rieger, C. Reppin, and G. Kanbach, To be presented in the late volumnes of the 17th International Conference on Cosmic Rays, Paris, France, 13-25 July 1981.

HIGH-ENERGY OBSERVATIONS OF STELLAR FLARES: COMPARISON WITH THE SUN

Hugh S. Hudson
Center for Astrophysics and Space Sciences, UCSD

ABSTRACT

This paper reviews recent observations of flaring activity on flare stars of the UV Ceti type, concentrating on X-ray and radio data in comparison with possible solar analogs. Although detailed differences exist, we conclude that similar mechanisms may work for both cases. Extending the analogy, we estimate the hard X-ray and γ-ray fluxes expected from typical stellar flares. In the solar case, these radiations give information about accelerated particles in the flare region. Hard X-ray observations of stellar flares may be possible, eventually, but the predicted γ-ray fluxes are prohibitively weak.

I. INTRODUCTION

Our knowledge of the physics of solar flares has been transformed in the last three decades by results obtained in "new" wavelengths: in particular, radio and X-ray. As instrumentation has become more sensitive, analogous advances are being made in the study of stellar flares. This review describes the stellar observations with special reference to the newer information on solar physics. Particular progress in the present (1980) maximum of solar flare activity has come from the Solar Maximum Mission satellite (SMM) and continues to come the radio observations of the Very Large Array (VLA). We present estimates of hard X-ray and gamma-ray fluxes expectable from stellar flares based upon extrapolation from the solar values, updating the approach of previous authors[1,2,3] The basic optical observations of stellar flares have been reviewed by Kunkel[4], with emphasis upon the astronomy, and by Gershberg[5] with emphasis on the physics. Mullan[6] has recently published a brief review of the entire subject. The bulk of the observations of flares on stars other than the sun come from the UV Ceti stars, of spectral classification dMe, and the present review will emphasize the high-energy aspects of these phenomena. But it must be noted that the discovery of coronae in an unexpectedly large range of stellar spectral types[7] suggests quite strongly that flare-type variability may also be a rather widespread phenomenon. At the time of writing, however, there are no observations of X-ray bursts known definitely to come from solitary stars other than the classical flare stars of the UV Ceti type.

In this review we summarize and comment upon the X-ray and radio observations of stellar flares in section II. This discussion includes a renormalization of the best observations to a common basis. The stellar flares closely resemble solar flares in these rough comparisons across the spectrum. We use the solar analogy to predict hard X-ray and γ-ray fluxes from stellar flares (section III). Only sketchy reference is made to solar phenomenology, and for further details recent reviews[31,32] should be consulted.

II. OBSERVATIONAL MATERIAL

II.1. X-ray Observations of Stellar Flares

There have been several extensive searches for X-ray emission from stellar flares, at first based upon scanning data and chance coincidence[8] and later with observing campaigns built around pointed X-ray instruments such as the ANS, SAS-3, Apollo-Soyuz, and HEAO-1 spacecraft[9,10,11,12]. The most recent and most sensitive searches have been with the Einstein soft X-ray telescope. Thanks to these efforts there now exists a fairly large body of data on simultaneous flaring at radio, visible and X-ray wavelengths. Of the data presently available, the "prize" event is probably the Einstein observation of a flare on YZ CMi reported by Kahler et al.[13], since it describes simultaneous observations in all three wavelength bands. Figure 1 summarizes these data.

Selected data are shown in Table I for four stars. The observed ratios of X-ray to optical luminosity have been reduced to a standard quantity, the bolometric X-ray luminosity divided by the luminosity in the B photometric band (essentially a spectral luminosity). The X-ray adjustment of course requires knowledge of the temperature. The inference of a bolometric X-ray luminosity requires certain model assumptions (e.g. isothermality) and depends upon theoretical cooling curves of the type introduced by Cox and Tucker[14], but this step away from the pure observational material is not a dangerous one if the analogy with solar flares holds, since the X-ray total luminosity is a weak function of temperature. The data selected all involve

Table I
Ratios of peak fluxes (selected data), X-rays

Star	Date	L_X/L_{opt} (quoted)	L_X/L_B (standardized)	source of data
YZ CMi	1979.10.25 (impulsive)	0.08	0.2	(13)
YZ CMi	1979.10.25 (gradual)	1-10	2.5-25	(13)
YZ CMi	1979.10.27	–	< 1.6	(13)
YZ CMi	–	–	< 0.3	(10)
UV Cet	1975.1.8	≤ 0.03	≤ 1.1[1]	(9)
Prox Cen	–	< 0.08	< 2.4[2]	(11)
Solar reference			2	see text

[1]Conversion factor $L_X/L_{.2-.28\ keV}$ = 30.
[2]Assumed flare color U - B = -1.

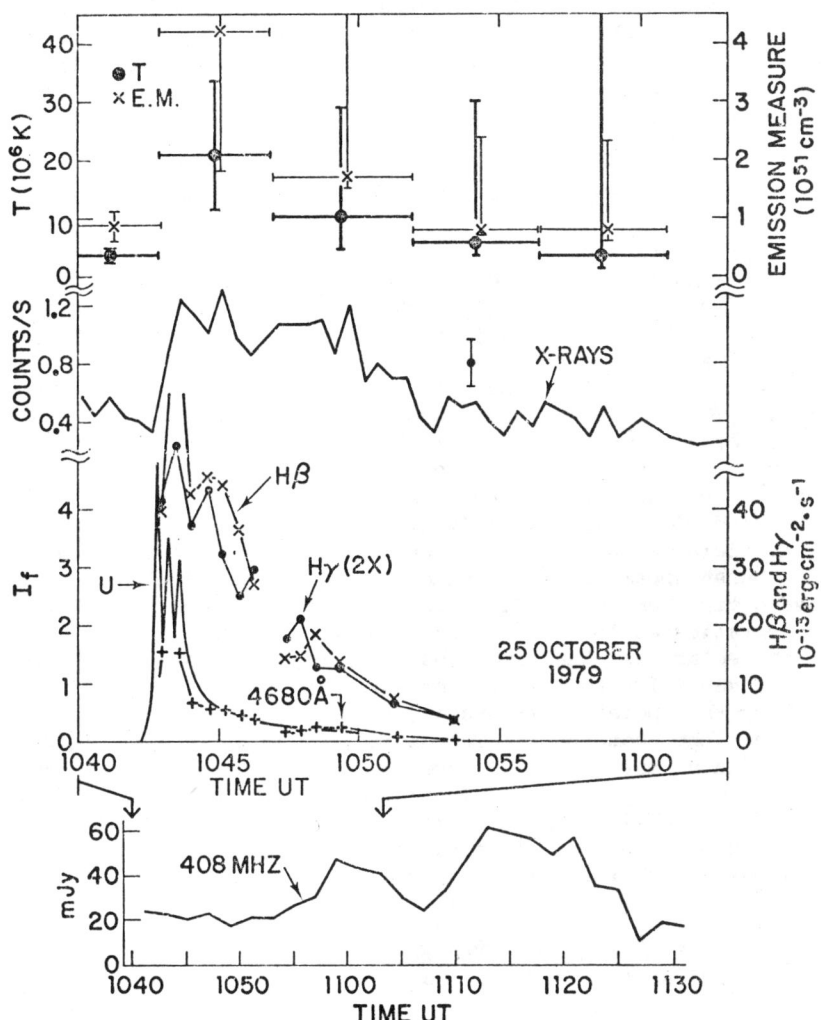

Figure 1. X-ray, optical and radio fluxes from the YZ CMi flare of 1979 Oct. 25[13]. Fits of temperature and emission measure, and 0.2-4 keV counting rate are from the Einstein observations; U-band (Cloudcroft), Hα, Hγ, and 4680 A (McGraw-Hill), and 408 MHz (Jodrell Bank) are shown below.

simultaneous measurements of X-ray and optical emission; hence we do not show for example the observations of Kahn et al.[12], for which indirect arguments were used to estimate L_x/L_{opt} for AT Mic and AD Leo. The soft X-ray emission is assumed to be strictly thermal; there are at present no observations of hard X-rays (> 10 keV) or γ-rays from stellar flares.

II.2 Radio Observations of Stellar Flares

Radio observation of stellar flares proved to be quite difficult with early single-dish techniques because of weak fluxes and terrestrial interference. More recently, however, interferometric observations[14-15] and observations with a single dish[17] have provided a good measure of immunity to these problems and have confirmed some of the main characteristics of bursts observed with the earlier techniques. The bulk of the successful radio observations are at frequencies of a few hundred MHz or less. These observations show long-duration events whose starting times lag considerably behind the times of the optical flares, suggesting a natural identification with the same kind of behavior in solar bursts at comparable frequencies. The YZ CMi flare shown in Figure 1 exhibits this behavior, with the 408 MHz burst apparently lagging by as much as 15 minutes after the impulsive optical emission. Table II gives the crude radio/optical luminosity ratios analogous to those of Table I.

In the solar case the time lags for the longer-wavelength bursts are due essentially to the remoteness of the plasma-frequency level (selected by the operating frequency) from the site of the energy release near the photosphere. The exciter of solar type II, Type IV, or late continuum may be an MHD wave disturbance with speeds[18] on the order of a few thousand km/sec. The interpretation of stellar flares on this model will eventually result in some indirect information regarding the structure of the stellar corona, and in the absence of direct images such information may be invaluable.

Table II
Ratios of peak fluxes (selected data), radio[1]

Star	Date	Radio Band	L_r/L_B	source of data
AT Mic	1980.10.25	5 GHz	< 1.6×10^{-6}	(19)
Prox Cen	1977.05.16-18	5 GHz	< 4.4×10^{-6}	(11)
YZ CMi	1979.10.25	408 GHz	1.2×10^{-4}	(13)
Solar reference				
"coronal"		408 MHz	1×10^{-6}	see text
"µwave impulsive"		5 GHz	1.2×10^{-5}	see text
"µwave gradual"		5 GHz	4×10^{-6}	see text

[1]Radio fluxes calculated from assuming $\Delta\nu = \nu$

The observations at 5 GHz reported recently by Slee et al.[19] are the "most convincing" observations obtained in microwaves, with a dual-beam chopping system to aid in rejecting spurious variations. The observations (Figure 2) show near-simultaneity between the optical and microwave light curves, as would be expected from a solar analogy with a "post-burst increase" emission due to optically thin bremsstrahlung from the soft X-ray source. In this observation the optical observations consist of photographic measurements at 5-minute intervals and hence do not reveal much detail. The stellar radio source has a much higher brightness temperature and the comparability of flux ratios (Table I) may therefore be coincidental.

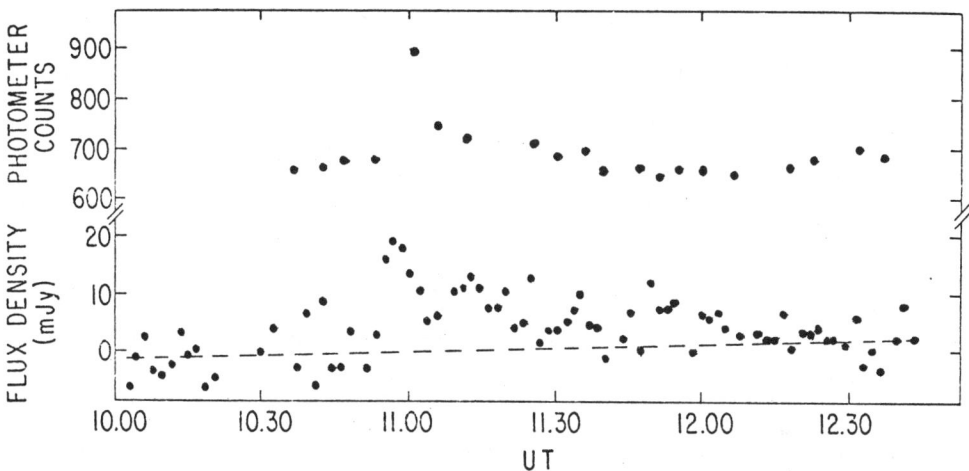

Figure 2. Optical and 5-GHz light curves for a flare of 1980 October 25 on AT Mic[19]. The points at the top show photographic exposures of 60 s at ~5 min intervals from Mt. Stromlo; lower data are 5 GHz (60 s averages, 1.7 min spacing) from Parkes.

II.3 Solar Reference Luminosity Ratios: X-rays

As a practical basis for intercomparisons, we attempt here to estimate the visual luminosity of a solar flare in stellar terminology. The observations are of course in a heterogeneous assembly of different units, often related to the parameters of the observational technique. We make an attempt to standardize these observations here in terms of the Johnson B photometric band (λ_o = 4200 A; $\Delta\lambda$ = 960 A) and the integral X-ray luminosity for for the observed or estimated temperature (typically $1 - 2 \times 10^7$ K). The different photometric colors appear to be fairly tightly correlated[20]. The B band avoids the Balmer jump and therefore might be less model-dependent.

We now need estimates of the B magnitudes for different solar flares. Since there apparently exist no broad-band photometric observations of solar flares, the value of L_B for the solar case must be an estimate derived indirectly from a combination of spectroscopic and imaging data, and theoretical considerations. Note that the no-filter observations of Lacy et al.[20], expressed as L_C, are related to L_B empirically by $L_C/L_B = 2.5$.

The light curves of solar flares may conveniently be divided into <u>impulsive</u> and <u>gradual</u> phases, which are especially clearly distinguished at X-ray energies[21] The available observations of visible continuum ("white-light flares") probably occur during the impulsive phase, and our knowledge of the gradual-phase optical luminosity is restricted basically to scaling from observed emission-line strengths. For the well-observed YZ CMi flare of 1979 October 25, the data warrant[13] separate consideration of separate analogous phases.

For impulsive emission, the observations of the white-light flare of 1972 August 7 provide the best information. Rust and Hegwer[22] estimate a total continuum luminosity of 4.4×10^{27} erg/sec over the range 3900-6900 A. Scaling this to the B bandpass, we find $L_B = 1.2 \times 10^{27}$ erg/sec. A typical importance 3B flare has a peak Hα luminosity of 1.1×10^{27} erg/sec[23], whereas Zirin and Tanaka[24] find $L_{H\alpha} = 1.2 \times 10^{27}$ erg/sec at the peak of this particular event. Thus for the impulsive-phase optical emission of solar flares, we can simply assume $L_B \simeq L_{H\alpha}$.

The gradual-phase luminosity ratio can probably be estimated more accurately from theory than from observations, given the lack of photometric solar observations. We can for example simply take the Case B recombination spectrum of hydrogen[25] and integrate the higher Balmer lines over the B passband, to find crudely that $L_B/L_{H\alpha} \simeq 0.66$. Within an uncertainty perhaps of a factor of two, we can therefore also conclude for the gradual phase that $L_B \simeq L_{H\alpha}$. Mochnacki and Zirin[26] find for a YZ CMi flare that the ratio $L_{H\alpha}/L_{H\beta}$ is about unity, compared with the Case B value of 2.87 at $T = 10^4$ K. This deviation probably reflects the order of magnitude of the uncertainties involved in estimates of this type - a factor of two or three.

To scale these estimated optical luminosities to solar soft X-ray emission, we assume that $L_X = 0.6 L_{H\alpha}$ for an 8 - 12 A passband[27] The results of these estimations for solar flares can be stated as

$$L_X/L_B \simeq 2 \qquad (1)$$

where the X-ray flux has been increased by a factor ~3 to give an approximate match to the bolometric X-ray luminosity of an isothermal solar-composition plasma at a temperature of $10 - 20 \times 10^6$ K[28]; note however that a single carefully-analyzed solar flare[29] gave the result $L_X/L_{H\alpha} \simeq 10$. Uncertainties of this order are presumably still present in the solar data.

II.4 Solar Reference Luminosity Ratios: Radio

Solar radio emissions are so complex and variable that simple scaling does not make too much sense. There are many types of solar radio bursts, corresponding to several emission mechanisms and a wide variety of environments[30]. Nevertheless for the sake of uniformity we will calculate the ratio L_r/L_B for three model events: 10^4 w/m^{2}Hz at 408 MHz and 5 GHz, representing respectively the "coronal" and "microwave impulsive" cases, and 3×10^{-20} w/m^{2}Hz at 5 GHz, representing the "microwave gradual" case, all referred to a flare of importance 3B. These values are rather arbitrary and may be have an uncertainty of an order of magnitude owing to the strong solar variability, especially at longer wavelengths.

II.5 Gross Energetics of Solar/Stellar Comparisons

The data in Tables I and II summarize the recent significant observational material that constrains the crude values of the ratios L_x/L_B and L_r/L_B. We define these ratios of peak X-ray and radio fluxes to the peak optical flux independent of the detailed variations within the light curves of the flares. This obviously suppresses most of the physics, but the simultaneous data are not detailed enough to warrant more sophisticated analysis at present. An inspection of Table I shows that departures from a strict solar analogy appear to be present. In the well-observed YZ CMi flare of 1979 Oct. 25, the impulsive optical emission was about an order of magnitude in excess of the solar value, relative to the soft X-ray flux. This tendency of stellar flares to produce a bright, short-lived continuum has previously been noted[4]. In the gradual phase, the opposite tendency may prevail: the stellar soft X-ray source may be relatively bright compared with the solar case. Once again, however, this conclusion is really based upon the same single well-observed event.

With reservations about the generality of the result, we can conclude (as have numerous authors[3,5,31]), that solar-type mechanisms are appropriate to the stellar case, but that certain parameters are probably substantially different. In the case of solar flares[32,33] the soft X-ray source is confined in magnetic loop structures at high pressure. A large part of the optical emission is secondary to the presence of the high-temperature material, and results from its conductive cooling. A deficiency in the gradual-phase optical emission of stellar flares may result from radiative cooling of the X-ray source[3]. This would be consistent with relatively large or high-density loops.

The interpretation of the impulsive emission is more difficult on the solar model, since no workable theory for the "white-light flare" has been demonstrated. However the preferred "scenario" for white-light continuum probably involves impulsive heating by non-thermal particles penetrating into the atmosphere of the star. Therefore a bright visible continuum could be taken to confirm the importance of energetic particles in stellar as well as solar flares[34]. On the other hand, the bright continuum may simply represent an early phase of conductive cooling of the stellar-flare loops, prior to their gradual radiative cooling. At present the solar mechanisms for optical radiation resulting from conductive cooling are not worked out well enough to extrapolate to the stellar case.

II.6 Detailed Light Curves

The original ANS observation of YZ CMi[9] probably gives the best information on time variations of the flare X-ray spectrum, since a significant detection was made both with the soft (0.2 - 0.28 keV) and medium-energy (1 - 6 keV) detectors. Other time-resolved data[13,33] show the same soft-hard-soft spectral evolution, which is characteristic also of solar flares. Beyond this resemblance it would be difficult to comment upon the details, since solar flares exhibit a great variety of time profiles.

III. PREDICTIONS OF HARD X-RAY AND GAMMA-RAY FLUXES

Scaling from solar flares appears to fit the stellar-flare soft X-ray/optical behavior, so we are encouraged to estimate hard X-ray and γ-ray fluxes in this way also. We take the soft X-ray flux as a reference (a) because it is better measured than the optical flux, and (b) because it is a high-energy phenomenon and may therefore be more closely related to some fundamental mechanism of a flare.

The peak hard X-ray and soft X-ray fluxes have a definite correlation but with a scatter exceeding an order of magnitude[35,36] We estimate a central point in this correlation to be characterized by a 1-8 A flux of 0.1 erg/cm^{2}sec, corresponding to a total X-ray flux F_X = 1.2 erg/cm^{2}sec, with a hard X-ray flux of 100 ph/cm^{2}sec keV at 20 keV. The relationship is given by

$$F_{20\ keV}/F_X = 100 \text{ ph/keV per erg.} \qquad (2)$$

Thus a stellar-flare soft X-ray flux of 2×10^{-9} ergs/cm^{2}sec, approximately the value for the UV Cet flare observed by Heise et al.[9], would according to solar scaling have produced a peak hard X-ray flux (at 20 keV) of 1.6×10^{-7} ph/cm^{2}sec keV, roughly four orders of magnitude below modern detection thresholds at this photon energy.

For γ-radiation, there is considerably less information. Furthermore there is clear evidence that the acceleration of high-energy protons does not scale proportionally with the total flare energy[37]. Nevertheless if we assume that scaling works for the purpose of estimation, we can take the brightest solar γ-ray flux as a starting point. The HEAO-1 observations of the flare of 1978 July 11 gave a peak 2.2 MeV line flux of ~1.0 ph/cm^{2}sec, and a comparable γ-ray continuum in the few-MeV range. This flare had a peak soft X-ray flux estimated at 3 ergs/cm^2 sec, 1-8 A, resulting in the estimate:

$$L_{2.2\ MeV}/L_X = 0.03 \text{ ph/erg} \qquad (3)$$

For the UV Cet flare cited above, we therefore find a γ-ray flux of only 6×10^{-11} ph/cm^{2}sec in the 2.2 MeV line.

These estimates, together with additional data for two representative flares, are summarized in Table III. The second flare taken as a model is the Einstein detection of a YZ CMi flare[13], and therefore represents the faint end of the spectrum of flare total energy. Although brighter flares than the UV Cet flare in Table III may occur, we can conclude that hard X-ray fluxes from flares on ordinary dMe flare stars will be very difficult to detect, while γ-ray fluxes will be impossible.

Table III
Predicted X-ray and γ-ray fluxes

Quantity	Major stellar flare (UV Cet 1975 Jan. 5)	Minor stellar flare (YZ CMi 1979 Oct. 25)	The sun (3B flare)
L_X^1	1.8×10^{30}	8×10^{28}	8×10^{27}
F_X^2	2×10^{-9}	1.7×10^{-11}	3
F_{20}^3	1.6×10^{-7}	1.4×10^{-9}	100
$F_{2.2}^4$	6×10^{-11}	4×10^{-13}	0.1

[1] ergs/sec
[2] ergs/cm^{2}sec
[3] ph/cm^{2}sec keV (at 20 keV)
[4] ph/cm^{2}sec (in the 2.2 MeV lines)

IV. CONCLUSIONS

The recent observations of stellar flares have tended to confirm the conclusions drawn by many commentators that solar-type mechanisms can explain many of the phenomena observed in UV Ceti stars. The stellar-flare properties least resembling those found on the sun include the higher occurrence rates and (possibly related) the larger total energy of the greatest stellar flares. This large total energy corresponds to a large emission measure in soft X-ray thermal emission, which may exceed 10^{53} cm^{-3} as compared with some 10^{51} for the most energetic solar flares.

The estimates in Table III show fairly clearly that γ-ray observations of stellar flares are not feasible if solar-type mechanisms are at work. Thus the direct information they provide on the interactions of high-energy protons in flares will not be available. On the other hand the hard X-ray fluxes are not beyond the range of potential future technological developments. We anticipate learning from hard X-ray observations whether or not non-thermal electrons play an analogously important role in stellar flares; quite conceivably the strong early optical continuum in stellar flares is related to these electrons.

The activity represented by flares on UV Ceti stars, both the maintenance of steady coronal emission and also the generation of flares, shows that these solar processes occur commonly on late-type stars. Almost certainly such activity results from stellar magnetism, and the natural speculation is that many solitary stars of other types also exhibit such phenomena. The more violent high-energy phenomena occurring on compact stars may or may not have much relationship to the models derived from analogy with the sun, but the intermediate cases of X-ray emission from binary stars of ordinary types – RS CVn systems, for example – may involve solar analogs in strong measure.

Interestingly, some of the greatest difficulty in intercomparing solar and stellar observations comes from gaps in our knowledge of solar behavior. Among these can be listed our almost complete lack of photometric solar flare observations, due of course in part to the relative brightness of the solar surface. In addition the mechanical "luminosity" of a solar flare, that part of the total energy that flows out into the solar wind in the form of waves and material ejecta, is typically not well measured. Finally it cannot be overemphasized that, while descriptive models of all solar phenomena certainly exist, in many cases the theory is far from the level of being actually predictive. In many such cases, it is quite difficult to have much faith in an extrapolation of the solar theory to a stellar range of parameters, and it is probably wisest to rely upon intercomparisons of the actually observed parameters to guide the theory, rather than the other way around. In this sense the stellar observations may have a profoundly beneficial influence upon our understanding of the way the sun works.

Acknowledgements. I would like to thank several persons for helpful discussions, notably S. W. Kahler and D.J. Mullan, and D.Worrall for a reading of the manuscript. NASA supported this work under grant NSG-7161.

REFERENCES

1. Kahler, S., and Shulman, S., Nature Phys. Sci. 237, 101 (1972).
2. Crannell, C.J., McClintock, J.E., and Moffett, T.J., Nature 252, 659 (1974).
3. Mullan, D.J., Astrophys. J., 207, 289 (1976).
4. Kunkel, W.E., in IAU Symposium No. 67, "Variable Stars and Stellar Evolution," ed. V. Sherwood and L. Plaut (Reidel, Dordrecht, 1975), p. 15.
5. Gershberg, R.E., in IAU Symposium No. 67, "Variable Stars and Stellar Evolution," ed. V. Sherwood and L. Plaut (Reidel, Dordrecht, 1975), p. 47.
6. Mullan, D.J., Solar Physics 54, 183 (1977).
7. Rosner, R., and Vaiana, G.S., in R. Giacconi and G. Setti (eds.), X-ray Astronomy (Reidel, Dordrecht, 1980), p. 129.
8. Tsikoudi, V., and Hudson, H.S., Astron. Astrophys. 44, 273 (1975).

9. Heise, A.C., Brinkman, J., Schrijver, J., Mewe, R., Gronenschild, E., and den Boggende, A., Astrophys. J. (Lett.) 202, L73 (1975).
10. Karpen, J.T. et al., Astrophys. J. 216, 479 (1977).
11. Haisch, B.M., Linsky, J.L., Slee, O.B., Hearn, D.R., Walker, A.R., Rydgren, A.E., and Nicolson, G.D., Astrophys. J. (Lett.) 225, L35 (1978).
12. Kahn, S.M., Linsky, J.L., Mason, K.O., Haisch, B.M., Bowyer, C.S., White, N.E., and Pravdo, S.H., Astrophys. J. (Lett.) 234, L107 (1979).
13. Kahler, S. et al., Astrophys. J. (to be published, 1981).
14. Cox, A.N., and Tucker, W., Astrophys. J. 157, 1157 (1969).
15. Davis, R.J., Lovell, B., Palmer, H.P., and Spencer, R.E., Nature 273, 644 (1978).
16. Nelson, G.J., Robinson, R.D., Slee, O.B., Fielding, G., and Walker, W.S.G., Mon.Not.R.astr.Soc. 187, 405 (1979).
17. Slee, O.B., and Page, A.A., Proc. IAU Colloquium 46, 150 (1979).
18. Wild, J.P., Smerd, S.F., and Weiss, A.A., Ann. Revs. Astron. Astrophys. 1, 291 (1963).
19. Slee, O.B., Tuohy, I.R., Nelson, G.J., and Rennie, C.J., Nature 292, 220 (1981).
20. Lacy, C.H., Moffett, T.J., and Evans, D.S., Astrophys. J. (Suppl.) 30, 85 (1976).
21. Kane, S.R., Astrophys. J. (Lett.) 157, L139 (1969).
22. Rust, D., and Hegwer, F., Solar Phys. 40, 141 (1975).
23. Thomas, R.J., personal communication (1981).
24. Zirin, H., and Tanaka, K., Solar Phys. 32, 173 (1973).
25. Osterbrock, D.E., Astrophysics of Gaseous Nebulae (Freeman, San Francisco, 1974), p. 65.
26. Mochnacki, S.W., and Zirin, H., Astrophys. J. (Lett.) 239, L27 (1980).
27. Thomas, R.J., and Teske, R.G., Solar Phys. 16, 431 (1971).
28. Raymond, J.C., Cox, D.P., and Smith, B.W., Astrophys. J. 204, 290 (1976).
29. Canfield, R.C., Cheng, C-.C., Dere, K.P., Dulk, G.A., McLean, D.J., Robinson, R.D., Jr., Schmahl, E.J., and Schoolman, S.A., in "Solar Flares," ed. P.A. Sturrock (Colorado, Boulder, 1980), p. 452.
30. Kundu, M.R., Solar Radio Astronomy (Interscience, New York, 1965).
31. Kodaira, K., Astron. Astrophys. 61, 625 (1977).
32. Svestka, Z., Solar Flares (Reidel, Dordrecht, 1976).
33. Sturrock, P. (ed.), Solar Flares (Colorado, Boulder, 1980).
34. Lin, R.P., and Hudson, H.S., Solar Phys. 50, 153 (1976).
35. Datlowe, D.W., personal communication (1975).
36. Dennis, B.R., personal communication (1980).
37. Hudson, H.S., Solar Phys. 57, 237 (1978).

SOLAR HARD X-RAY IMAGES OBSERVED BY ASTRO-A

K. Ohki
Tokyo Astronomical Observatory, Mitaka, Tokyo 181, Japan

S. Tsuneta, N. Nitta, and T. Takakura
The University of Tokyo, Bunkyo-ku, Tokyo 113, Japan

K. Makishima, T. Murakami, Y. Ogawara, and M. Oda
Institute of Space and Aeronautical Science, Megro, Tokyo, Japan

ABSTRACT

The Solar X-ray Telescope (SXT) on board the Astro-A has observed many events since its launch on February 21, 1981. Several of the largest events, with counting rates >10^4 c/s, have been analyzed to reveal very compact sources for the large hard X-ray bursts. Although a few limb events show some extended features up to about one arcmin, most events have linear dimensions less than the FWHM of the SXT triangular response, which is about 30 arcseconds. This compactness of the largest events may conflict with traditional models of hard X-ray sources, including thin and thick target models. In this paper, two typical large events are presented. A disk event on April 2, 1981, shows a single source with a very small diameter, while a April 27, 1981, limb event shows a double source structure with unbalanced intensities.

INTRODUCTION

Although many solar X-ray imaging experiments have been carried out in space, e.g. by the OSO series satellites and by Skylab, almost all experiments had their energy limited to the soft X-ray range. The only exceptions were a balloon experiment carried out by our group in 1969[1] and the HXIS experiment[2] on the SMM. The former had only a one-dimensional scan while the latter one had its energy range restricted to less than 30 KeV. The HXIS efficiency in hard X-ray detection in the 16 - 30 KeV range is inferior to that of the SXT. Therefore, concerning the pure hard X-ray energy range, only a little information has been obtained by these experiments.

The SXT experiment on board the Astro-A has an energy range 17 - 60 KeV. It is very efficient in this pure hard X-ray range due to the imaging technique of the rotational modulation collimator. Moreover, since the SXT can always observe the whole sun with fine resolution, it never misses large flare events as long as they occur during the satellite observing time.

Since the launch on Feb. 21, 1981, SXT has observed more than 20 large events of the X-class with sufficient hard X-rays to draw two dimensional images which may supply fruitful information about flare hard X-rays, when all analysis is completed. A preliminary analysis has been done for two of them. One is a typical disk event, and the other is a limb event. As we shall see, even this preliminary analysis of the events places some restrictions on the traditional models of hard X-ray sources.

0094-243X/82/770395-05$3.00 Copyright 1982 American Institute of Physics

INSTRUMENTATION

The details of the instrumentation of SXT will be presented elsewhere in the near future. In this paper we briefly outline the instrument in Table I.

Table I.
SXT Main Characteristics

Field of view:	total Sun 2.16' (redundancy-free field of view)
Angular resolution:	20 arcsec (using higher Fourier components in case of large flares) 35 arcsec (in case of medium flares)
Temporal resolution:	3 seconds (flare onset - 768 seconds) 6 seconds (769-1024 seconds) 60 msec (for one-dimensional scan)
Energy range:	17 - 40 KeV (initial set) 5 - 10 KeV (only 769 - 1024 seconds after the flare onset)

1981 APRIL 2 EVENT

The Hα importance 1B flare of 1981 April 2 began at $10^h 57^m$, peaked at $11^h 07^m$, and ended at $11^h 19^m$ UT. The Hα location was S43, W68. The flare was classified as X2.2 on the GOES scale. The hard X-ray monitor on Astro-recorded 31383 counts/sec at $11^h 06^m$ UT and an unusually steep spectrum in the hard X-ray range. This hard X-ray burst was one of the strongest event observed during the first several months after launch.

The most characteristic feature of this strong event observed by SXT i its spatial compactness. As indicated in Figure 1, even a single scan by SXT of this source has sufficient counts to show a very clear one-dimension image. Various one-dimensional scans from different position angles also show fairly good triangular shapes as seen in Figure 1, indicating a very small point-like source.

We have done a preliminary analysis of this source by using the Maximu Entropy Method, which was recently introduced into X-ray astronomy. The result shows that this point-like source extends at most 15 arcsec. Most probably, the FWHM diameter of this source is less than 10 arcsec.

1981 APRIL 27 EVENT

The limb flare of 1981 April 27 occurred when Region 3049 was just crossing the solar west limb. The lowest energy (17 - 60 KeV) channel of the Hard X-ray Monitor on the Astro-A increased up to about 60000 counts/se the largest count rate observed by the Astro-A until then. This hard X-ray event was a very slow-rising event. The flare triggering system on the

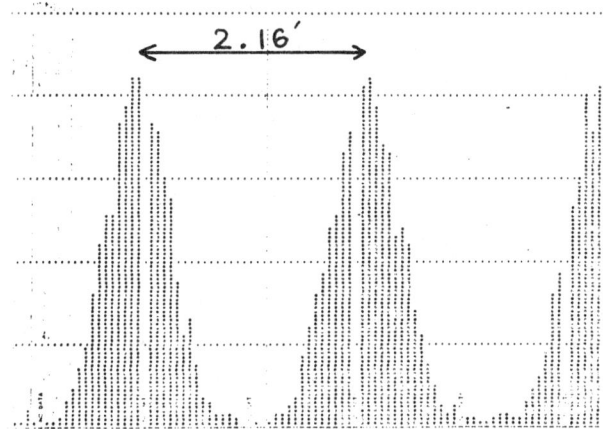

Fig. 1 One-dimensional scan data for 1981 April 2 event. One bin corresponds to 3.9025 msec of time.

satellite shifted the on-board data recorder into the flare mode at $07^h\ 50^m$ UT; the X-ray count of the lowest channel then increased monotonically from 1024 counts/sec to 60000 counts/sec without any obvious X-ray spikes.

The spin period of the satellite was about 16.2 seconds. Therefore, we can obtain all position angles of one-dimensional scans by SXT in 8.1 seconds. In Figure 2, five one-dimensional scan data from different position angles are presented for the time interval between 07:59:12-20, when the HXM count was about one-third of its peak value. In these scan data, a double source is apparent. In the top four scans shown in the Figure, a weaker source indicated by arrows appears preceding the main stronger source. The separation between these two sources changes with position angle. On the left side of this Figure, the relationships among the solar limb, two point sources, and the response patterns subtended by the modulation collimator of the SXT are indicated schematically. Combining the data from the SXT modulation patterns and the SXA (Solar X-ray Aspectmeter) modulation patterns, we can obtain the location of the X-ray source with respect to the center of the sun and the spin axis of the satellite. In the case of the 1981 April 27 event, the hard X-ray sources was situated almost on the line between the solar center and the spin axis. In this case, the separation between two sources on the scan data should be maximum at $90°$ of position angle and minimum at $0°$ and $180°$ if both sources were situated just on the limb. However, the observational results differ from this situation. Combining the five scan data, we can conclude that the maximum separation comes at less than $90°$ position angle. Therefore, the line connecting the two sources is a little tilted from the tangent to the limb at the X-ray source position. In other words, if the stronger source is situated just on the limb, then the weaker source should be situated a little bit above the limb as indicated schematically in the Figure. A preliminary calculation shows that the weaker source is situated about 0.2 arcmin above the solar limb. However, this conclusion is only preliminary,

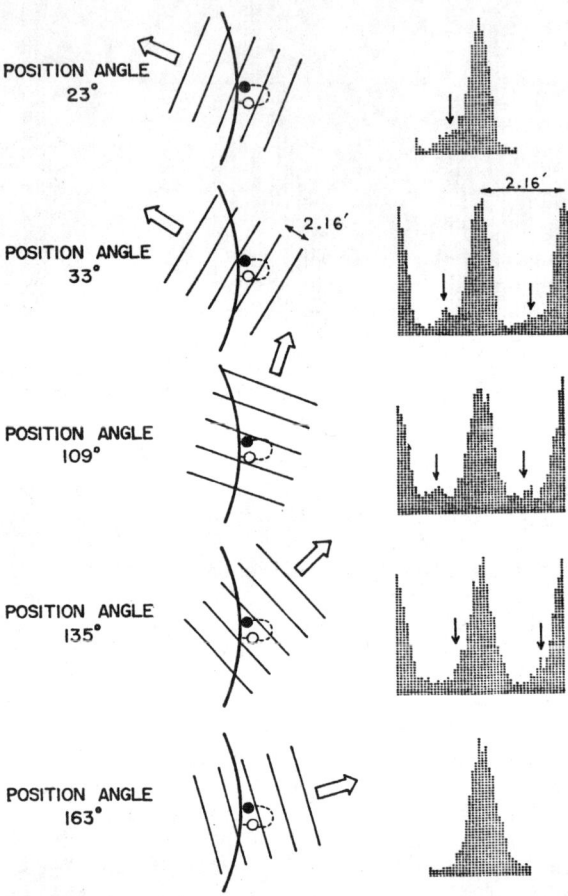

Fig. 2. Five scan data for 1981 April 27 event from different position angles. On the left side, schematic illustrations are drawn to indicate the positions of two sources. The solid circle is the stronger source, and the open circle is the weaker source indicated by an arrow in the scan data on the right side. The arc indicates the solar west limb. The open arrow indicates the direction of each scan by the rotational motion of the grid pattern around the spin axis.

because the absolute positions of both weak and strong sources have not been calculated yet. This calculation will be done shortly when all the computer software of the data analysis are completed. The only thing safely concluded at the present time is that the altitudes of the weak and strong sources in the solar atmosphere are certainly different by 0.2 arcmin, namely about 9000 km.

DISCUSSION

Two kinds of hard X-ray flares have been presented here; one with a single point-like feature on the disk, and another with a double source structure. These show the variety of the hard X-ray images observed by the SXT. Future analysis of many other strong events already observed by the SXT will soon reveal the nature of this variety. However, already at the present time, we can say it seems difficult to give a simple interpretation to hard X-ray sources by adopting a specified source model.

A point-like single source like April 2 probably has a diameter less than 10 arcsec (≈ 7000 km). If a thermal model is assumed, the emission measure is about 6.2×10^{46} cm^{-3} from HXM data. Therefore, the electron density N_e is $\gtrsim 2 \times 10^{10}$ cm^{-3}. On the other hand, when the thin target model is adapted to this event, the density for field particles should be much greater than 2×10^{10} cm^{-3} because the density of accelerated particles must be lower than that of field particles. Collision times for the accelerated particles in such a high density plasma are much less than one second. Thus the assumption for the thin target model becomes invalid. A thermal model seems suitable for the hard X-ray source of April 2.

In the double source on April 27, the X-ray counts of the weaker source and the stronger source seem to change simultaneously, suggesting that two sources occupy the same magnetic tube. In this case, the thick target model is preferable. However, this conclusion is still far from certain. More exact analysis of this event and many other events will bring a more definite conclusion. Comparison between the location of the hard X-ray sources on the disk and that of Hα flare or other kinds of radiation will be important.

We wish to acknowledge the efforts of many people who are engaged in various aspects for the Astro-A. Especially, we thank Prof. Y. Tanaka for his general management of the Astro-A mission, and also thank Prof. K. Tanaka for his organizing the observation team for the Astro-A.

REFERENCES

1. Takakura, T., Ohki, K., Shibuya, N., Fujii, M., Matsuoka, M., Miyamoto, S., Nishimura, J., Oda, M., Ogawara, Y., and Ota, S., 1971, Solar Phys., **16**, 454.

2. Hoyng, P., Machado, M. E., Duijveman, A., Boelee, A., de Jager, C., Fryer, R., Galama, M., Hoekstra, R., Imhof, J., Lafleur, H., Maseland, H. V. A. M., Mels, W. A., Schadee, A., Schrijver, J., Simnett, G. M., Svestka, Z., van Beek, H. F., van Tend, W., van der Laan, J. J. M., van Rens, P., Werkhoven, F., Willmore, A.P., Wilson, J. W. G., and Zandee, W., 1981, Ap. J., **244**, L153.

EVIDENCE FOR DELAYED SECOND PHASE ACCELERATION IN SOLAR FLARES

J. B. Willett, J. C. Ling, W. A. Mahoney,
G. R. Riegler, and A. S. Jacobson
Jet Propulsion Laboratory
California Institute of Technology
Pasadena, California 91109

ABSTRACT

Twenty-two (22) solar flares with emission at or above 0.5 MeV have been observed by the HEAO-3 gamma-ray spectrometer during the period 1 October 1979 through 1 July 1980. In two of these flares, the peak emission in a band around 0.5 MeV follows the peak of the impulsive hard x-ray emission at energies greater than 0.1 MeV by as much as 30 seconds. In both of these flares the 0.5 MeV flux risetime was 30-40 seconds followed by a decay time of less than 10 seconds. The onset of the 0.5 MeV flux was not measurably different (<10 seconds) from the onset of the integrated emission greater than 0.1 MeV. The remaining 20 flares gave no indication of a delay in either peak emission or onset time between the high energy and low energy components (<5 seconds), nor were there any decay times shorter than risetimes. Thirteen (13) of the flares had a measurable flux above 3.8 MeV and have been characterized by a two power law fit with an artificial break at 0.5 MeV. The remaining flares were characterized by a single power law up to 0.5 MeV.

INTRODUCTION

Prior to the launches of HEAO-3 and the Solar Maximum Mission there were very few solar flare data in the gamma-ray portion of the electromagnetic spectrum. Chupp, et al.[1,2] and Hudson[3] have reported on gamma-ray flares in which time profiles of the gamma-ray portion of the spectrum were similar to those of the hard x-ray part of the spectrum, although some time delay has been noted between the high and low energy portions of the spectrum. In the 4 August 1972 flare[1,2], the flux in the energy band from 0.35 - 8 MeV follows the flux in the 29 - 41 keV band by more than a minute. Simultaneous production of the nuclei responsible for the 2.2 MeV line emission and the high energy gamma-rays is consistent with the observed data.[4] In the 11 July 1978 flare[3], the emission greater than 1 MeV was delayed from the emission greater than 40 keV by 20 seconds. Theoretical models[5] give some support to the idea of two different acceleration processes in at least some flares, but experimental evidence for this is only circumstantial. In this paper we present data on some flares measured in both the hard x-ray and gamma-ray spectral regions, two of which have a time separation of up to 30 seconds between the hard x-ray and gamma-ray

0094-243X/82/770401-08$3.00 Copyright 1982 American Institute of Physics

peaks of the flares, and which exhibit significant differences in time profiles between the x-ray and gamma-ray components.

INSTRUMENT

The measurements reported here were made with the HEAO-3 High Resolution Gamma-Ray Spectrometer during the period 1 October 1979 through 1 July 1980. The instrument sensor is composed of four high purity Ge crystals surrounded by a 6.6 cm thick CsI anticoincidence shield as shown in Figure 1. The shield is split into four equal segments on the sides and bottom and capped by a collimator with a hole drilled over each Ge crystal to define a viewing aperture. The nature of the experiment (extrasolar gamma-ray line survey) nominally constrained the field-of-view of the instrument to be perpendicular to the instrument-sun line so that the high resolution Ge detectors were never directly exposed to the sun. However, the collimator and two of the four shield side pieces projected an area of over 900 cm^2 to the sun. Several levels of discriminators sampled at 1 to 10 second intervals provided a time profile for all flares and a first order spectral shape for some of the more energetic events.

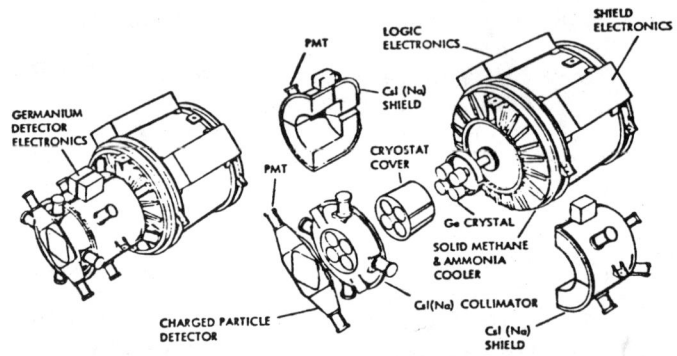

Figure 1. Exploded view of the HEAO-3 Gamma Ray Spectrometer.

MEASUREMENTS

During the period 1 October 1979 through 1 July 1980 at least 37 flare-like events were observed by the HEAO-3 gamma-ray spectrometer with measurable flux at energies greater than 80 keV. Figure 2 shows a sample of flares which occurred during the last two months of the survey. For each flare, time profiles are shown for discriminator countrates greater than 80 keV in the top panel, between 420 and 580 keV in the middle panel, and greater than 3.8

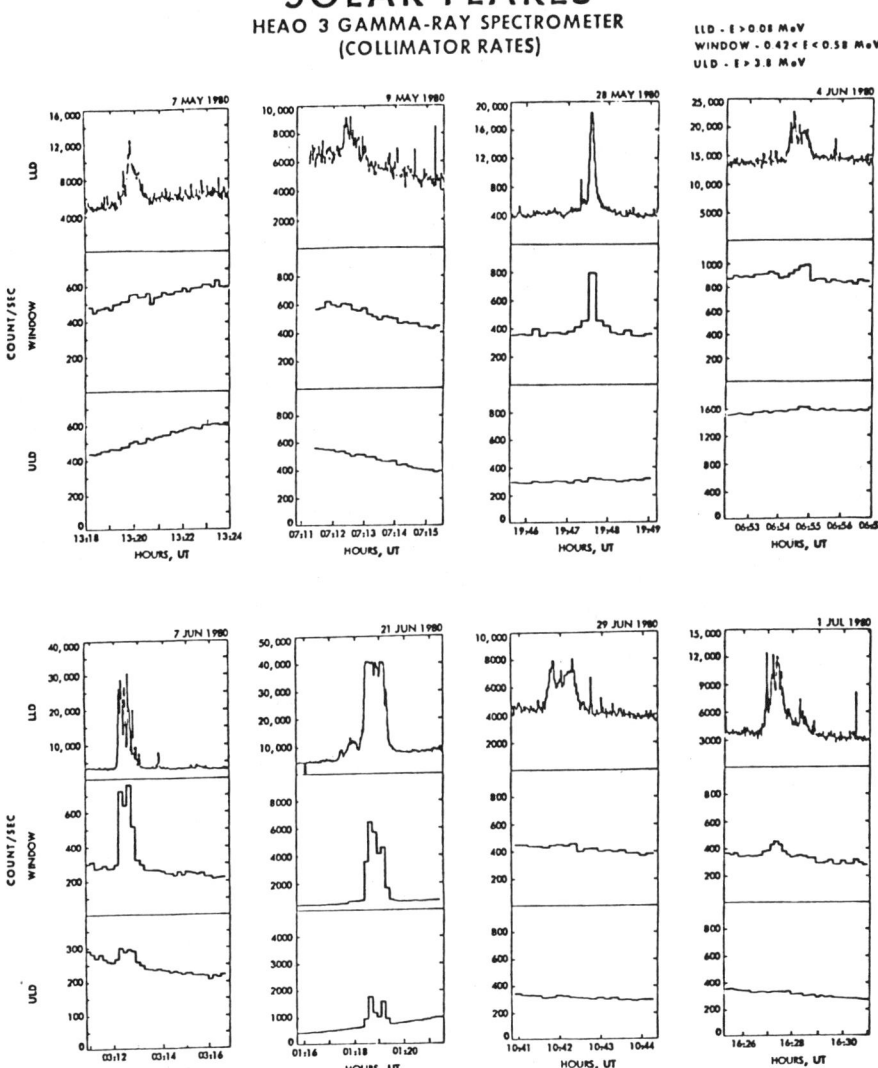

Figure 2

MeV in the bottom panel. The flare of 21 June 1980 was so intense that it saturated the lower level discriminator counter. Another intense flare reported by Prince, et al.[6] occurred on 9 November 1979 and was accompanied by strong, very narrow line emission at 2.22 MeV which penetrated the shield and was measured by the high resolution Ge detectors. The intrinsic line width was less than 5 keV, and the intensity averaged over 80 seconds was about the same as the flux reported for the 2.2 MeV line in the 4 August 1972 flare. Figure 3 shows the comparative time profiles of both the continuum and line fluxes. The delay of some 20 seconds for the build up of the line flux is thought to be due to the thermalization time of the neutrons prior to their capture of hydrogen, and therefore does not signify a delay in the accelerating process.

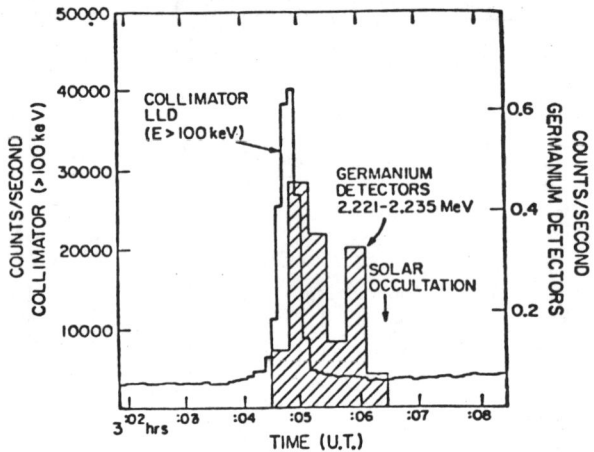

Figure 3. Comparative time profiles of the 2.22 MeV gamma-ray line emission and the continuum emission from the 9 November 1979 flare.

SPECTRAL SHAPE

All of the flares observed to have flux at energies greater than 420 keV were fit with power law spectra. Figure 4 shows the frequency of occurrence of spectral indices for both low and high energy portions of the power law spectra. For some of the flares, determination of shape was taken both at the beginning and the end of the flare, and the second measurements can be seen in Figure 4 as shaded boxes. There is no noticeable change in distribution of indices between early and late development of the flares, but we can immediately see that the high energy indices are more tightly clustered than the low energy indices. For most of these flares, the x-rays and gamma-rays appear to be produced simultaneously.

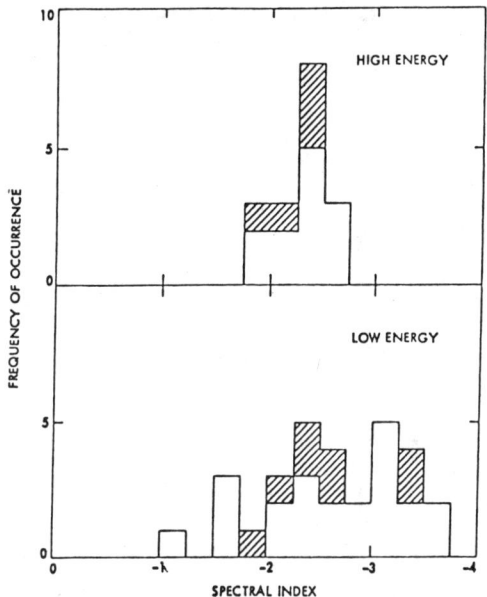

Figure 4. Frequency of occurrence of spectral indices. The spectral indices for the power law fit to the 0.08 - 0.5 MeV energy band are shown in the low energy panel. The indices for the fits to the data at energies greater than 0.5 MeV are shown in the high energy panel. Spectral fits at two different times were obtained for some flares, and the indices from the latter of the fits are shown as shaded boxes.

DELAYED SECOND PHASE ACCELERATION

Two of the flares, those on 4 June 1980 and 29 June 1980, are different in that they show strong evidence for time separation of up to 30 seconds between the peak hard x-ray emission and the peak gamma-ray emission. They also show significant gradual spectral hardening. The 4 June 1980 flare is shown in Figure 5, indicating quasi-periodic impulsive bursts on top of a slowly rising and relatively quick decaying high energy component. The low energy portion of the spectrum softens appreciably during the course of the flare, while the high energy spectral index remains unchanged. This behavior may be indicating that the x-rays and gamma-rays are being produced by two distinct acceleration processes. The first and most common process is thought to accelerate electrons to hundreds of keV creating impulsive x-ray bursts which primarily trigger the

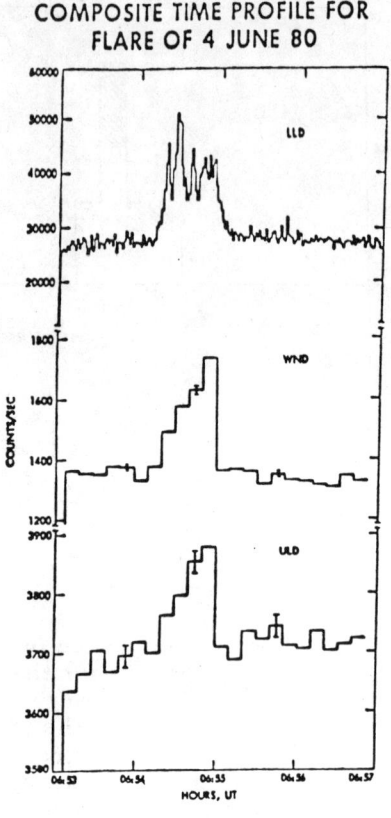

Figure 5

lowest level discriminator of the HEAO-3 instrument. The second proposed process begins at the same time as the first, but builds more slowly in intensity to peak approximately 30 seconds after the peak of the x-ray emission. It is likely that the gamma-rays are produced by ions accelerated to tens of MeV.

The flare of 29 June 1980 behaves in a similar fashion to that of 4 June 1980 and is shown in Figure 6. This type of behavior has been observed in about 10 percent of the HEAO-3 flares and seems to indicate two distinct phases of acceleration which reach peak intensity at different times. There is as yet no explanation for the sudden termination of these flares immediately after the high energy component reaches maximum intensity.

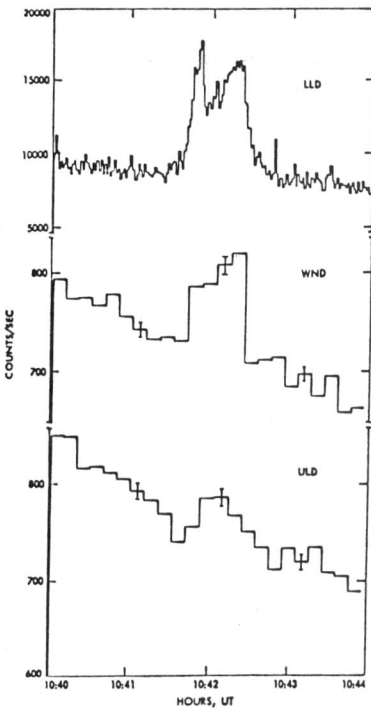

Figure 6

CONCLUDING REMARKS

In some solar flares there is a distinctly different time development of the gamma-ray emissions when compared with that of hard x-rays, although both emissions begin at essentially the same time. It is likely that the flux in the 0.42 - 0.58 MeV energy band is of electron origin, while the flux greater than 3.8 MeV is nuclear in nature. It appears from the data presented in this paper that the electrons producing the greater than 80 keV x-rays during the impulsive phase of the flare may be accelerated by a different mechanism than the high energy electrons and ions which produce the x-rays and gamma-rays observed in the above-stated energy bands.

ACKNOWLEDGEMENT

We would like to express our appreciation to Veronica O'Brien for the preparation and editing of this manuscript for publication.

REFERENCES

1. Chupp, E. L., Forrest, D. J., Higbie, P. R., Suri, A. N., Tsai, C., and Dunphy, P. P., 1973, Nature, 241, p. 333.

2. Chupp, E. L., Forrest, D. J., and Suri, A. N., 1975, "Solar Gamma-, x-, and EUV Radiation", ed. S. Kane, IAU Symp. 68, p. 341.

3. Hudson, H. S., Workshop on Flare Research and Solar Maximum Mission, Solar Physics Division of AAS, Ann Arbor, Mich., Nov. 1978.

4. Bai, T., and Ramaty, R., 1976, Solar Phys. 49, p. 343.

5. "Solar Flares", 1980, ed. P. Sturrock, Chaps. 4 and 9.

6. Prince, T. A., Ling, J. C., Mahoney, W. A., Riegler, G. R., and Jacobson, A. S., 1980, Conference on Cosmic Ray Astrophysics and Low-Energy Gamma-Ray Astronomy, joint AAS/APS meeting at the University of Minnesota.

SECOND-PHASE ACCELERATION VERSUS
SECOND-STEP ACCELERATION IN SOLAR FLARES

T. Bai
University of California, San Diego
La Jolla, California 92093

ABSTRACT

Observations and theoretical interpretations leading to two phases of acceleration and two steps of acceleration are reviewed. A distinction is made between the second-phase acceleration and the second-step acceleration: The former takes place several minutes later than the impulsive phase and probably in the high corona, but the latter is very much impulsive-like (but delayed by several seconds) and takes place in the same flare loop as the impulsive acceleration. The second-step mechanism is delayed because it is a secondary effect due to the energy deposition of the first-step electrons. It is proposed that in flares producing impulsive gamma-rays (e.g., those on 9 Nov. 1979 and 7 June 1980), energetic protons and heavy nuclei are accelerated by the second-step mechanism.

I. INTRODUCTION

Energetic electrons accelerated during solar flares emit bremsstrahlung X-ray continua, which sometimes continue to gamma-ray energies. Energetic protons and heavy nuclei produce narrow gamma-ray lines and a 4-7 MeV continuum made of Doppler-broadened gamma-ray lines. Therefore, by studying these radiations, we can learn about the mechanisms accelerating electrons and ions during solar flares. Especially by studying time profiles of these radiations, we can learn about the relative timing of acceleration of different species of particles, the density of the interaction region, and so forth.

Over the past years, many interesting observations have been made of the solar hard-X-ray and gamma-ray emissions, and some theoretical work has been done in this area. The detectors aboard the Solar Maximum Mission (SMM) have recently yielded time profiles of hard X-ray and gamma-ray emissions with considerably better time resolutions. The SMM results will undoubtedly enhance our understanding of the acceleration processes operating in solar flares. In order to view the SMM results from the proper perspective, it seems necessary to review the earlier observations and theoretical work in this area.

The plan of the paper is as follows. In Section II, I review the historical backgrounds. In this section, I make a distinction between the second-phase (or second-stage) acceleration and the "second-step acceleration", as introduced by Bai and Ramaty[1]. In Section III, I discuss the "impulsive acceleration" of protons. Especially I discuss in detail the hard X-ray and gamma-ray time profiles of the 7 June 1980 flare, and propose that the 2-s delay of the gamma-ray time profile is an indication that during this flare, protons and heavy ions were accelerated by the second-step mechanism.

II. HISTORICAL BACKGROUNDS

(a) <u>Two Phases of Acceleration</u>

Energetic protons were observed in interplanetary space after major solar flares[2,3,4]. From a good association of the proton events with type IV radio bursts, which occur after the flash phase of the H_α flare, the idea that energetic solar protons are accelerated during the second phase was proposed[5]. De Jager[6] proposed that (sub-) relativistic electrons are also accelerated during the second phase. These authors proposed that the second-phase acceleration occurs about 10-20 minutes later than the impulsive phase acceleration, and probably takes place in the corona higher than the impulsive flare site.

In the 30 March 1969 flare, Frost and Dennis[7] found two separate intervals of hard X-ray emission exhibiting distinguishable characteristics. The first phase (impulsive phase) reached its maximum strength 1.5 minutes after the onset and lasted for $\sim$3 minutes altogether. The second phase emission started simultaneously with type II radio bursts immediately following the impulsive phase, reached its maximum strength about 2 minutes later, and lasted for about 40 minutes, decaying rather smoothly. The impulsive phase X-ray spectrum was steeper at energies greater than 100 keV. On the other hand, the second-phase X-ray spectrum was flat (E^{-2} power-law spectrum) all the way to the highest energy channel. From these observations, Frost and Dennis suggested that the hard X-rays observed during the later interval are due to the electrons accelerated by the second-phase (or second-stage) acceleration.

In studying the hard X-ray emissions from the 4 August 1972 flare, Bai and Ramaty[8] noticed that the X-ray continuum above 350 keV (observed by Suri et.al. [9]) reached the peak strength a few minutes later than the continuum above 30 keV (observed by van Beek et.al.[10]). From this delayed build-up of high energy continuum and the flattening of the X-ray spectrum above 0.6 MeV, Bai and Ramaty concluded that the (sub-) relativistic electrons were accelerated by the second-phase acceleration. They also found that the 2.2 MeV time profile was better explained when the neutron-production time profile was assumed to be similar to the time profile of the high energy (>350 keV) continuum rather than that of the low energy continuum. From this they concluded that protons as well as mildly relativistic electrons were accelerated by the second-phase mechanism. However, they noticed that although the build-up of the second-phase acceleration is slower, it is not clearly separated in time from the impulsive-phase acceleration.

(b) <u>Second-Step Acceleration</u>

By comparing the timing of the impulsive hard X-ray spikes of the 4 and 7 August 1972 flares in further detail, Bai and Ramaty[1] found that the impulsive spikes of higher-energy channels of the TD 1 data[11] are delayed with respect to those of the lowest-energy channel. The delay time increased progressively to ~ 5 s as the energy increased from ~ 30 keV to ~ 150 keV and they jumped to ~ 15 s for energies above 150 keV. Bai and Ramaty were able to fit the time profiles of X-rays below 150 by assuming the delay to be solely due to the increase of energy loss time of electrons with increasing energy. The ambient density which could account for the observed delay was $3 \times 10^{10} cm^{-3}$ for the 4 August flare and $2.5 \times 10^{10} cm^{-3}$ for the 7 August flare.

The large delay of the time profiles of channel 6 and 7 (>150 keV), however, could not be explained in this manner. Though they were delayed, the hard X-ray time profiles of channels 6 and 7 closely resembled that of channel 5. From these characteristics, Bai and Ramaty proposed a second-step acceleration model. According to this model, the second-step mechanism accelerates further some of the high-energy-tail electrons accelerated by the first-step mechanism. The interpretation that the first-step electrons are mostly trapped in the flare loop of the density $\sim 3 \times 10^{10} cm^{-3}$, where turbulence is presumably well developed, is compatible with the idea that the second-step acceleration is likely to be due to shocks and turbulence.

Notice here the differences between the second-step acceleration of Bai and Ramaty[1] and the second-phase (or second-stage) acceleration (see refs. 5,6,7). The latter is separated in time from the impulsive phase, and is believed to occur in the corona higher than the impulsive flare site. On the other hand, the second-step acceleration is very close to the first-step acceleration both in time (lags of only several seconds) and in space (the same electron population is involved).

The 11 July 1978 flare observed by HEAO-1[12] was in many respects similar to the 4 August and 7 August 1972 flares. It produced the 2.2 MeV line and the MeV -range continuum; it lasted about 5 minutes and its time profiles of hard X-ray and gamma-ray continua showed multiple impulsive spikes. An interesting point I would like to emphasize here is that the time profile of the continuum above 1.1 MeV lagged behind the time profile of the continuum above 40 keV by ~ 20 s, similarly to the August 1972 flares. This suggests that the second-step acceleration was also in operation in this flare.

(c) **Purely Coronal Hard X-rays**

Hudson[13] reported an OSO 7 observation of "purely coronal" hard X-rays from the 14 December 1971 flare. The parent flare which caused this radiation was thought to be located $\sim 25°$ over the limb[14] with corresponding occultation height $\sim 7 \times 10^9$ cm. The characteristics of the X-ray emission are as follows: (1) It lasted for a long time (~ 40 minutes). (2) The flux increased and then decreased gradually, with the rise time shorter than the decay time, and the time profile had no rapidly varying structure. (3) The X-ray spectrum was flat (power-law index ~ 2).

Hudson, Lin and Stewart[15] also reported an OSO 7 observation of hard X-rays associated with a flare that occurred $\sim 20°$ over the limb on 22 July 1972. The 30 March 1969 flare observed by OSO 5 was also an over-the-limb event (by 15°-20°). The characteristics of the hard X-radiation from the 22 July 1972 flare and the second-phase of the 30 March 1969 flare were exactly the same as those of the 14 December 1971 flare mentioned above.

The 4 August 1972 flare was a disk flare, but during the decay phase (after 0635UT), the X-ray intensity decreased smoothly and gradually without the impulsive characteristic exhibited during the earlier period, and the spectrum became considerably harder than during the earlier period (cf. ref. 11). These are exactly the same characteristics as shown by the hard X-radiation observed from the limb-occulted flares - a fact suggesting that during the decay phase the coronal hard X-radiation was probably the dominant component.

III. IMPULSIVE GAMMA-RAY FLARES

(a) The 9 November 1979 Flare

Definite evidence for impulsive acceleration of protons was first obtained from the 9 November 1979 flare observed by HEAO-3 (ref. 16). The time profile of the hard X-ray continuum above 80 keV showed an impulsive spike burst with a total duration of $\sim$ 40 s. The time profile of the continuum above 3.8 MeV showed a similar general time dependence, although its time resolution was a rather coarse one, 10.24 s. If the gamma-ray continuum above 3.8 MeV was dominated by Doppler-broadened gamma-ray lines[17], this indicates impulsive acceleration of protons and heavy ions. The time profile of the 2.2 MeV line is also consistent with the assumption that the neutron-production time profile was the same as the time profile of the gamma-ray continuum.

(b) The 7 June 1980 Flare: Observations

Further evidence for impulsive acceleration of protons and heavy ions was obtained from SMM observations. Especially interesting are the time profiles of the hard X-ray and gamma-ray emissions of the 7 June 1980 flare [18,19,20]. This flare produced seven quasi-periodic impulsive bursts, with the time intervals between peaks varying from 7 to 11 s. No apparent delay was found even from the time profile of the highest energy channel (228-386 keV) of the Hard X-Ray Burst Spectrometer (HXRBS) on SMM[20]. The time profile of the 4.1-6.4 MeV continuum shows that at least for two of the impulsive bursts, the peaks of gamma-ray emission lag behind the peaks of the hard X-ray emission by about 2 s. (see Fig. 4 of ref. 18).

(c) The 7 June 1980 Flare: Interpretation

By studying the gamma-ray spectrum of the 4 August 1972 flare[21], Ramaty, Kozlovsky and Suri[17] concluded that the gamma-ray continuum between 4 and 8 MeV could be entirely of nuclear origin. Abundant CNO nuclei, when excited, give rise to de-excitation gamma-ray lines

in this energy interval. When the ambient CNO nuclei are excited
by collisions with the accelerated protons and α-particles, they
give rise to narrow lines. When the accelerated CNO nuclei are excited by collisions with the ambient protons and helium nuclei, they
give rise to Doppler-broadened lines (with FWHM ∿ 1 MeV), which are
blended together to form a gamma-ray continuum. The ratio between
the 4-7 MeV fluence and the 2.2 MeV line fluence was found to be
about 1 for all the disk flares from which SMM detected the 2.2 MeV
line[18]. This is also consistent with the interpretation that the
4-7 MeV gamma-ray continuum is mostly of nuclear origin.

In view of the above-mentioned facts, we can infer that the
impulsive spikes of the 4-7 MeV time profile indicate impulsive
acceleration of protons and heavy ions. However, an alternative
interpretation is also plausible: although the bulk of the 4-7 MeV
photons are of nuclear origin, the rapidly varying component may
be bremsstrahlung by relativistic electrons. A detailed analysis of
the gamma-ray spectrum will be able to discriminate between these
two interpretations. But the former interpretation, which is more
interesting and more likely, will be adopted in the following.

What can we learn from the time profiles of the 7 June 1980
flare? First, the rapid increase of the 4-7 MeV flux obviously
indicates rapid acceleration of protons and heavy nuclei. Second,
the thick-target interaction model is the correct model for this
flare. In other words, the energetic particles move down toward
the photosphere and interact mainly below the transition region.
If the energetic electrons of this flare had been trapped in a
flare loop of the density $10^{10} - 10^{11}$ cm^{-3}, the hard X-ray time
profiles of the high energy channels of HXRBS would have been delayed
because of the longer energy loss time of higher energy electrons.[1]
The energy loss time of energetic protons[23] is

$$T = -E/(dE/dt) = \frac{1.2 \times 10^{11}}{n} E_{MeV}^{1.5} \text{ s.} \quad (1)$$

The energy loss time in a medium of 10^{11} cm^{-3} is 38 s for 10 MeV
protons, and 1200 s for 100 MeV protons. Both the rapid decay time
(∿5s) of the 4-7 MeV flux and the 2.2 MeV decay time of ∿50s are
therefore contradictory to the trap model. Because the pitch angle
scattering of energetic protons and heavy nuclei due to Coulomb
interaction is negligible, the thick-target interaction means these
particles were accelerated to small pitch angles. These two points
(rapid acceleration and thick-target interaction) are important
constraints for identifying acceptable acceleration mechanisms.

The 2 s delay of the gamma-ray time profile with respect to
the hard X-ray time profiles may be due to the longer travel time
of the protons from the acceleration region to the interaction
region[18]. However, this puts the acceleration region at about 10^{10}
cm above the photosphere, which is much higher than the estimate by
Rust et.al.[22] Moreover, such a high acceleration region will also
result in an observable delay of the time profiles of low-energy
hard X-rays, which was not the case. For these reasons, I suggest
that the delay of the gamma-ray time profile is an indication of the

delayed acceleration of protons and heavy nuclei. Since we have already seen similar delays in the time profiles of high-energy hard X-rays from the 4 and 7 August 1972 flares and the 11 July 1978 flare, I propose that the delay of the gamma-ray time profile of the 7 June 1980 flare is evidence that the energetic protons and heavy nuclei were also accelerated by the second-step acceleration.

IV. SUMMARY AND DISCUSSION

(a) Second-Phase Coronal Acceleration

As we have seen in Section II, there exists ample observational evidence for the second-phase coronal acceleration of electrons. The hard X-ray emissions observed from the limb-occulted flares occurring on 30 March 1969, 14 December 1971 and 22 July 1972 were produced by the second-phase electrons[7,13,15]. The hard X-ray emission after 0635UT on 4 August 1972 might have been dominated by the bremsstrahlung of the second-phase electrons. The characteristics of the X-ray emissions from the second-phase electrons are: (1) long duration of the order of tens of minutes, (2) gradual rise and decay devoid of rapid fluctuations, and (3) flat spectrum. The second-phase coronal acceleration appears to take place only during major flares, possibly because its mechanism requires the coronal medium to be unusually turbulent.

A coronal acceleration mechanism is known to accelerate protons and heavy ions[24,25,26]. It is also known that protons are accelerated in the interplanetary medium[27,28].

(b) Second-Step Acceleration

Energetic electrons above a few hundred keV are also accelerated by the second-step acceleration[1]. The main characteristic of the X-ray time profiles of the second-step electrons is that they are very much like the X-ray time profiles of the first-step electrons, but they lag behind the latter by about several seconds. The second-step acceleration is delayed because its mechanism is the shockwaves and turbulence generated by the energy deposition of the first-step electrons. The time profiles are similar because the high-energy tail of the first-step electrons serve as the injection electrons for the second-step acceleration. Evidence for the second-step acceleration is found from the 4 and 7 August 1972 flares[1] and the 11 July 1978 flare[12].

The second-step acceleration of the energetic nuclei is evidenced from the hard X-ray and gamma-ray time profiles of the 7 June 1980 flare observed by Chupp (ref. 18). The gamma-ray (4-7 MeV) time profile was impulsive, having distinguishable peaks and valleys similarly to the hard X-ray time profiles, but it was delayed by $\sim$2s with respect to the latter. This is exactly the signature of the second-step acceleration. The small delay time is possibly due to a small size of the flare loop.

(c) Acceleration Mechanisms

From the beginning, the mechanism for the second-phase acceleration was thought to be the second-order Fermi acceleration[5,6,29] (however, see ref. 30). But it is not a suitable mechanism for the second-step acceleration, because it cannot satisfy the two constraints discussed in Section IIIc.

If energetic particles are confined in a flux tube between the two shock fronts approaching each other, these particles gain energy every time they collide with the shock fronts. This is the so-called first-order Fermi acceleration. If shock waves propagate upward from the footpoints of a flare loop, the first-order Fermi acceleration can operate[31]. But as Wentzel pointed out, at every collision mainly the parallel component of the particle velocity increases, and therefore after several collisions its pitch angle becomes small enough so that it can overcome the shock fronts. This puts a limit to the efficiency of the acceleration. If there is a mechanism for pitch-angle scattering, this acceleration mechanism can be very efficient.

Notice that in this scheme, proton acceleration is rapid and protons attaining small pitch angles eventually impinge on the chromosphere to produce gamma-rays in a thick-target interaction. It therefore satisfies the two constraints discussed in Section III.

Because the second-step acceleration operates in closed flare loops, the second-step protons are less likely to escape into the interplanetary medium, consistent with the interplanetary proton observations (cf. refs. 16, 32). Because the second-phase acceleration takes place in the high corona, on the other hand, the second-phase protons are more likely to escape into the interplanetary medium[33,34]. Chupp[21] classified "proton flares" into the following two classes:

Class I - in which the knowledge of proton acceleration in a flare is based on gamma-ray line observations.

Class II - in which protons are directly observed in space and are causally associated with a solar flare(s).

This phenomenological classification fits into the theory of the present paper. Class II protons are mainly produced by the second-phase acceleration; Class I protons are mainly produced by the second-step acceleration. These two classes of proton events are, however, not mutually exclusive; in major flares (with long duration and intense radiations) such as the 4 August 1972 flares, both classes of protons are accelerated. In a medium flares, such as the flares on 9 November 1979 and 7 June 1980, only the second-step acceleration operates.

In the 7 June 1980 flare, the time profile of the highest HXRBS energy channel (300-350 keV) was not delayed, suggesting that the second-step acceleration did not influence electrons with energies up to $\sim$400 keV. In the August 1972 flares, the second-step acceleration did accelerate further electrons with energy above 200 keV. This difference is explained as follows: During the August 1972 flares, the energetic electrons were trapped in flare loops of the density 3×10^{10} cm^{-1}, giving ample time for the second-step acceleration, but in the 7 June 1980 flare, the energetic electrons were beamed down (Section IIIb), not giving much time for the second-step acceleration.

The delay of the gamma-ray continuum (4-7 MeV) has been reported only for one flare (7 June 1980). Observations of such delays from other flares will enhance the confidence on the second-step acceleration of energetic protons, and will enhance our understanding of it.

ACKNOWLEDGEMENT

This research was supported by NASA Grant NSG-7161. The author wishes to thank Dr. Hugh Hudson for useful discussions with him.

REFERENCES

1. T. Bai and R. Ramaty, Astrophys.J. 227, 1072 (1979).
2. I. Lange and S.E. Forbush, Terrestr. Magnetism Atmosph. Elec. 47, 185 (1942).
3. P. Meyer, E.N. Parker, and J.A. Simpson, Phys. Rev. 104, 768 (1956).
4. C.E. Fichtel and F.B. McDonald, Ann. Rev. Astron. Astrophys. 5, 351 (1967).
5. J.P. Wild, S.F. Smerd, and A.A. Weiss, Ann. Rev. Astron. Astrophys. 1, 291 (1963).
6. C. de Jager, in COSPAR Symp. on Solar Flares and Space Sci. Res., ed. C. Jager and Z. Svestka (Amsterdam, North-Holland), p. 1 (1969)
7. K.J. Frost and B.R. Dennis, Astrophys. J. 165, 655 (1971).
8. T. Bai and R. Ramaty, Solar Phys. 49, 343 (1976).
9. A.N. Suri, E.L. Chupp, D.J. Forrest, and C. Reppin, Solar Phys. 43, 415 (1975).
10. H.F. van Beek, P. Hoyng, and G.A. Stevens, World Data Center Rep. UAG-28, Pt. II, 319 (1973).
11. P. Hoyng, J.C. Brown, and H.F. van Beek, Solar Phys. 48, 197 (1976).
12. H.S. Hudson, T. Bai, D.E. Gruber, J.L. Matteson, P.L. Nolan, and L.E. Peterson, Astrophys. J. (Letters) 236, L91 (1980).
13. H.S. Hudson, Astrophys. J. 224, 235 (1978).
14. T. Kosugi, Solar Phys. 48, 339 (1976).
15. H.S. Hudson, R.P. Lin, and R.T. Stewart, Solar Phys., to be

published (1981).
16. T. Prince, J.C. Ling, W.A. Mahoney, G.R. Riegler, and A.S. Jacobson, Astrophys. J., submitted (1981).
17. R. Ramaty, B. Kozlovsky, and A.N. Suri, Astrophys. J. 214, 617 (1977).
18. E.L. Chupp, this volume (1981).
19. E.L. Chupp et.al., Astrophys. J. (Letters) 244, L167 (1981).
20. L.E. Orwig, K.J. Frost, and B.R. Dennis, Astrophys. J. (Letters) 244, L163 (1981).
21. E.L. Chupp, D.J. Forrest, P.R. Higbie, A.N. Suri, C. Tsai, and P.P. Dunphy, Nature 241, 333(1973).
22. D.M. Rust, A. Benz, G.J. Hurford, G. Nelson, M. Pick, and V. Ruzdjak, Astrophys. J. (Letters) 244, L179 (1981).
23. V.L. Ginzburg and S.I. Syrovatskii, The Origin of Cosmic Rays, (New York, Macmillan), (1964).
24. M.A.I. van Hollebeke, L.S. Ma Sung, and F.B. McDonald, Solar Phys. 41, 189 (1975).
25. J.W. Kohl, C.O. Bostrom, and D.J. Williams, in World Data Center Rep. UAG-28, Pt. II, ed. H.E. Coffey, 330 (1973).
26. D.L. Bertsch, S. Biswas, and D.V. Reams, Solar Phys. 39, 479 (1974).
27. G. Gloeckler, Particle Acc. Mechanisms in Astrophys., (New York, AIP), 43 (1979).
28. L.A. Fisk, ibid., 63 (1979).
29. Z. Svestka, Solar Flares, Dordrech, Reidel (1976).
30. P.A. Sturrock, in IAU Symp. 57, ed. G. Newkirk, 437 (1974).
31. D.G. Wentzel, Astrophys. J. 137, 135 (1963).
32. T.T. von Rosenvinge, R. Ramaty, and D.V. Reams, Conference Papers, 17th Internatl. Cosmic Ray Conf., Paris (1981).
33. G. Newkirk, in Symp. on High Energy Phenomena on the Sun (NASA SP-342), ed. R. Ramaty, R.G. Stone, 453 (1974).
34. R. Ramaty and R.E. Lingenfelter, in IAU Symp. 68, ed. S.R. Kane, 363 (1975).

SOLAR FLARE ENERGETICS

R. P. Lin

Space Sciences Laboratory, University of California at Berkeley, Berkeley, CA 94720

ABSTRACT

We review current knowledge of the energetics of solar flares. Recent observations by the Solar Maximum Mission and by balloon-borne instrumentation indicate that the flare hard X-ray emission arises from nonthermal bremsstrahlung — the collisions of fast electrons into a cold ambient medium ($Ee \gg kT$). Under this interpretation, most of the energy released for many flares is initially contained in the energetic electrons. These electrons can produce most of the observed flare phenomena via interactions with the solar atmosphere. In large flares a shock wave may result from explosive heating of the solar atmosphere by these electrons. This shock wave can accelerate nuclei to relativistic energies. We argue that recent SMM observations of fast gamma-ray bursts are consistent with this picture of shock acceleration of nuclei.

Solar flares provide a nearby example of the transient release of energy from an astrophysical object. The radiation emitted by flares commonly extends to energies of order 10^2 keV, and occasionally nuclear gamma-ray emission is observed. Here we review what is known about how energy is released in solar flares, and in particular, what the role of particle acceleration is in solar flares. Several new results in the past year from the Solar Maximum Mission and balloon observations have significantly changed our perception of solar flares and the associated particle acceleration processes. These new results come from the spatial imagery of the flare hard X-ray source, the high sensitivity solar gamma ray observations, and high energy resolution measurements of the hard X-ray spectrum.

I. SOLAR FLARE HARD X-RAYS AND ENERGETIC ELECTRONS

The time profile of a medium size flare in hard (≥ 20 keV) X-rays is shown in Figure 1. The hard X-ray event consists of a series of impulsive bursts, each of a few to a few tens of seconds duration, followed by a long slow decay (visible in the 22 - 33 keV channel) which lasts for an additional ~15 minutes. Temporal variations down to ~100 ms have been observed in flare bursts, while typical total durations range from ~10 to ~10^3 sec or more. The hard X-rays come from optically thin bremsstrahlung emission of energetic electrons in the solar atmosphere. Most of the quantitative information about these electrons at the Sun is derived from hard X-ray observations. Usually the ≥ 20 keV X-ray burst observed during the impulsive phase is assumed to be due to a single component of electrons. The observations are generally consistent with a falling power law spectrum, with some evidence for a change to a more rapidly decreasing spectrum above an energy of 60 – 100 keV [2] although some flare bursts have been reported to be consistent with an isothermal source spectrum with temperatures of $10^8 - 10^{9}$°K[3], [4] The broad energy channel data available from scintillation detectors, however, appear inadequate to distinguish between power law and isothermal spectra (see review by *Kane et al.*).[5]

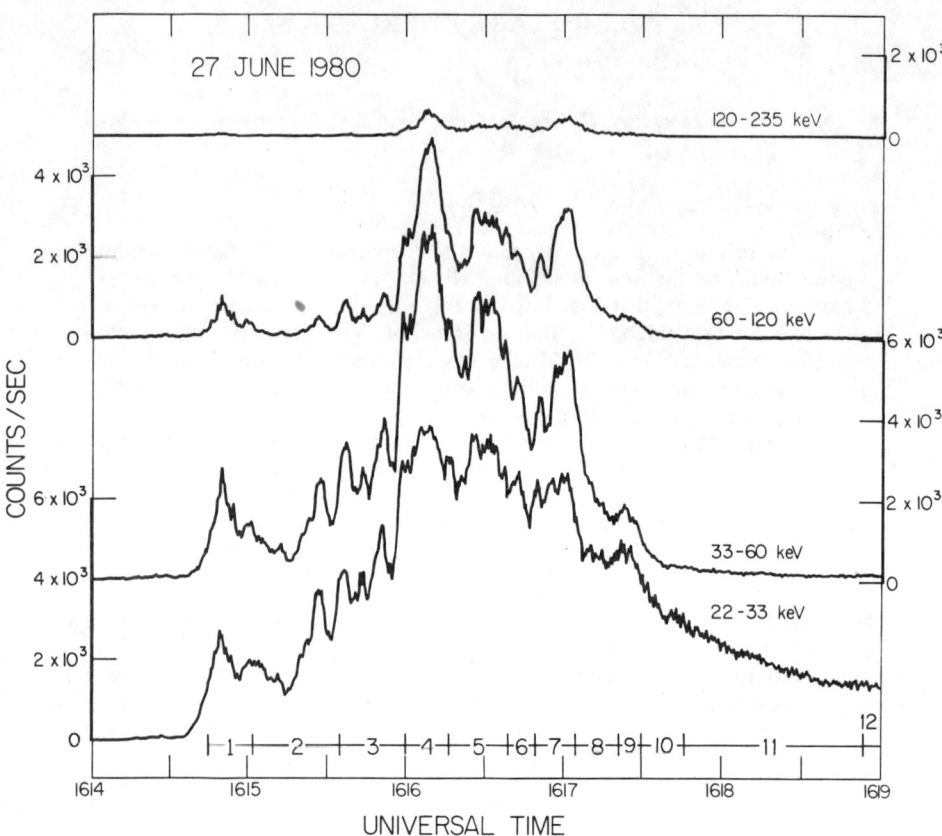

Figure 1. A flare hard X-ray burst observed by a balloon-borne scintillation detector. The numbered intervals at the bottom indicated the times when spectra were accumulated for a germanium detector array (from *Lin et al.*)[1] (See Figure 2).

If the observed X-ray emission is produced by bremsstrahlung from fast electrons colliding with a cold plasma (i.e., fast electron energy $\gg$ kT), then the Coulomb collision energy losses would be $10^4 - 10^5$ times the X-ray losses. Under this non-thermal assumption, a large fraction of the total flare energy in many flares must be initially contained in the $\gtrsim 20$ keV electrons[6,7] the exact amount depends on how low in energy the fast electron spectrum extends. The electron acceleration process would then be intimately related to the flare energy release process. Under alternative, thermal interpretations, the hard X-ray emission is produced by a very hot $T \approx 10^8 - 10^9 \, °K$ plasma[8,9,10] Then electron-electron Coulomb collisions will result in exchanges of energy only among the hot electrons without any net collisional energy loss. Provided the hot plasma could be confined, the energy requirements for the energetic electrons might be drastically reduced (see discussion in *Ramaty et al.*).[11] In the case of a homogenous source (for example, the adiabatic heating model of *Mätzler et al.*[12] or the high density, thermonuclear burning model of *Colgate et al.*[13], a Maxwellian electron distribution would be expected.

Recently some new observations bearing on this important question of thermal vs. non-thermal have been reported. First, very high energy resolution ($\lesssim 1$ keV FWHM) measurements of the spectrum of a solar flare X-ray burst were obtained from a balloon-borne array of cooled germanium detectors[1]. These are shown in Figure 2 for the numbered time intervals in Figure 1. From the start of the event to the maximum, the spectra

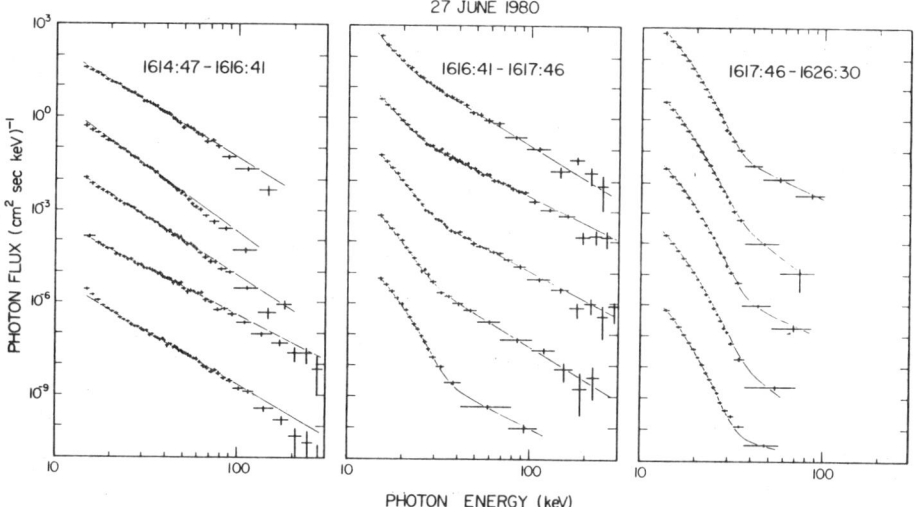

Figure 2. Energy spectra from the germanium array through the event. The vertical scale applies to the uppermost spectrum, with each succeeding spectrum offset downward by two decades. The steep component which dominates below ~ 30 keV (second and third panels) fits closely to the emission from an isothermal distribution of electrons with temperature $\sim 30-35 \times 10^{6 \circ}$K and emission measure of a few times 10^{48} cm^{-3}.

are accurately power law (index ~ 3.5) from 13 keV up to $\sim 60-100$ keV, with a steepening at higher energies. Single temperature isothermal fits to this component can be excluded by chi-square tests at less than 10% likelihood. These measurements show that single temperature isothermal plasmas cannot account for the impulsive hard X-ray emission, at least in this flare, and that the energetic electron spectrum must extend down to at least 13 keV.

Beginning near the peak of the event at ~ 1616, a new very steep component is observed at low energies, superimposed on the power law. At its steepest this new component has a power law index $\gamma \sim 11$, where $(dJ/dE) \propto E^{-\gamma}$. Such a steep spectrum is unresolvable by scintillation detectors. This steep component varies much more slowly than the power law component. Its spectrum fits closely to an isothermal with temperatures $\sim 30-35 \times 10^{6 \circ}$K. We will return to discuss this hot isothermal component later.

As pointed out by *Brown*[14] and others, essentially any X-ray spectrum can be reproduced by a suitable distribution of thermal sources, so that even accurate spectral measurements such as these cannot rule out the thermal interpretation. Recently, however, spatial imaging of solar hard X-rays in the 16 - 30 keV range has been achieved for a few flares by the HXIS (hard X-ray imaging spectrometer) experiment on SMM. *Hoyng et al.*[15] found that during hard X-ray impulsive spikes of the flare of 1980 May 21 (Figure 3) the 16 - 30

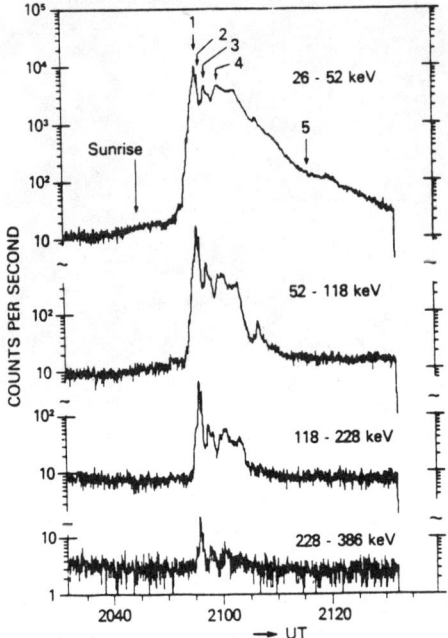

Figure 3. Hard X-ray time profiles as observed by the HXRBS instrument on SMM on 1980 May 21. [From Hoyng et al.][15]

keV hard X-rays came from simultaneously brightening footpoints (Figure 4); footpoints a and b in the first spike and a and c in the second. Upon comparison with the SMM UV spectrometer they find that these footpoints coincide with flare brightenings in Mg II and Hα (Figure 5), lines which are formed in a partially ionized medium. Thus the ambient temperature in the hard X-ray emitting region must be much lower than the fast electron energy, as required for non-thermal bremsstrahlung emission. *Hoyng et al.*[15] also point out that thermal models for the hard X-ray emission where the very hot plasma is generated at the top of a flare loop and then energy is thermally conducted down to the footpoints [16,17] can also be ruled out. Such conduction fronts would take ~10 s to travel down, while no sources were observed located between the footpoints prior to the footpoint frightening themselves. Although it is still possible to construct thermal models to fit these observations, it seems much more likely that the flare hard X-ray emission is due to non-thermal bremsstrahlung and therefore that the accelerated electrons contain a significant fraction of the energy released in many flares. Measurements of hard X-ray polarization and directivity, which are planned for the future, may provide conclusive evidence for non-thermal energetic electron beams.

II. THE FLARE ENERGY OUTPUT

Provided that these recent results are generally applicable, then the basic energy release mechanism for many flares must convert the available energy into fast electrons. These electrons can produce much of the observed flare hard X-ray, EUV, optical and radio emissions through various interactions with the solar atmosphere (e.g., *Kane,*).[18] Large solar flares may release as much as 10^{32} to 10^{33} ergs in a period of order 10^3 s. These flares emit electromagnetic radiation over the entire spectrum from γ-rays to low frequency radio

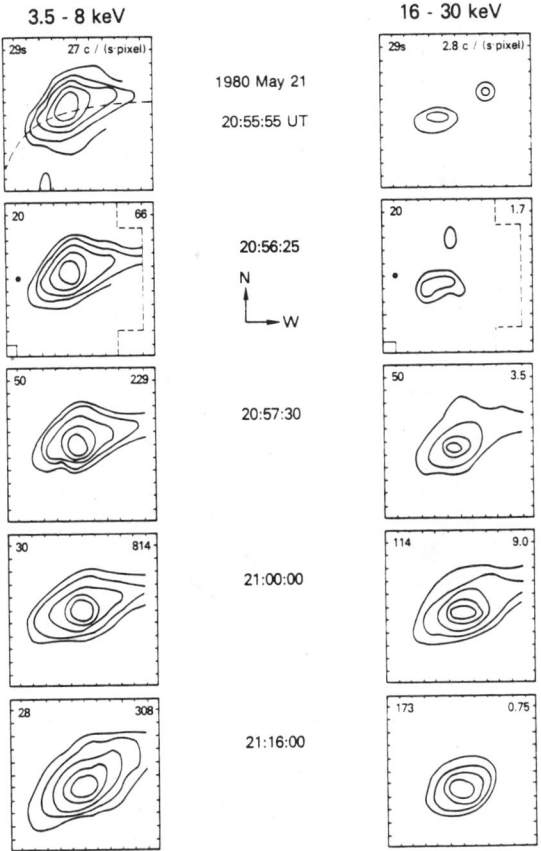

Figure 4. X-ray contour plots from the HXIS instrument on SMM at the times indicated by arrows in Figure 3. The contours are at fractions of 3/4, 1/2, 1/4, 1/8, and 1/16 of the maximum brightness. Both contour plots at $20^h56^m25^s$ show the center (•) and the boundary (---) of the HXIS fine field of view. The marks on the axes are 8" apart. [From *Hoyng et al.*] [15]

bursts, but these emissions account for only a minor fraction of the total flare energy release. Ions and electrons are also accelerated up to relativistic energies by these flares. Most of the energy in large flares, however, eventually ends up contained in matter ejected at high velocity into the interplanetary medium to form a shock wave (see Figure 6).

Table I gives a summary of the energy released in a large solar flare, based on the study of the 4 August 1972 flares. [24] The 4 August 1972 flare is the largest energetic particle-producing flare, of the last solar cycle and comparable to the large energetic particle flares observed in the last two solar cycles. These large particle flares tend to occur at times of the ascending and descending phase of the solar cycle, rather than at solar maximum. The 4 August 1972 flare appears to differ in several respects from the large flares observed by SMM at the maximum in the solar cycle, particularly as regards particle acceleration. It is used as an example here because of the relatively complete set of observations which are

FLARE KERNELS

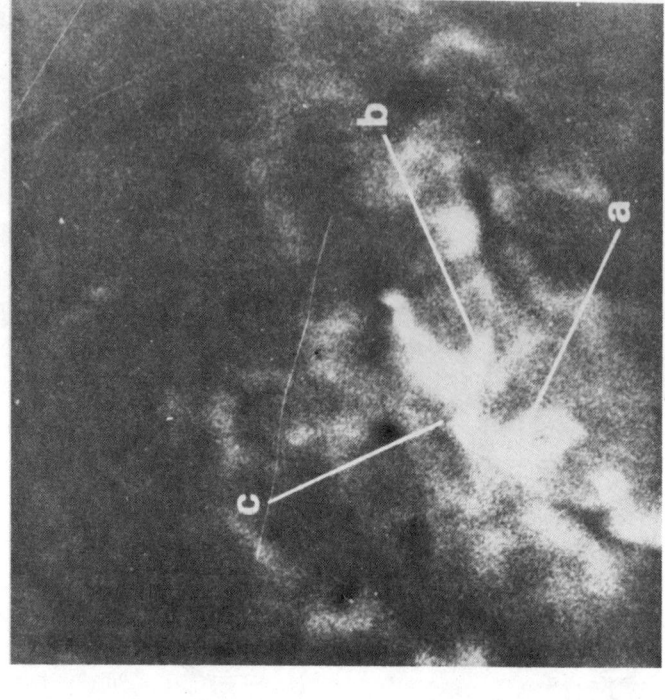

Mg II 2051-2101 UT   **Hα 2054:50 UT**

Figure 5. The flare kernels of 1980 May 21. *Left*: UVSP Mg II spectroheliogram γ2795 measuring about 4' × 4'. Pointers mark the centers of the three brightest 8" × 8" HXIS pixels involved in the 16 × 30 keV emission around $20^h55^m55^s$ and $20^h56^m25^s$. *Right*: flare patrol Hα frame at $20^h54^m50^s$. Points a, b, and c indicate the centers of the three most important HXIS pixels involved in the 16–30 keV emission around 20^h56^m. [From *Hoyng et al.*,]¹⁵

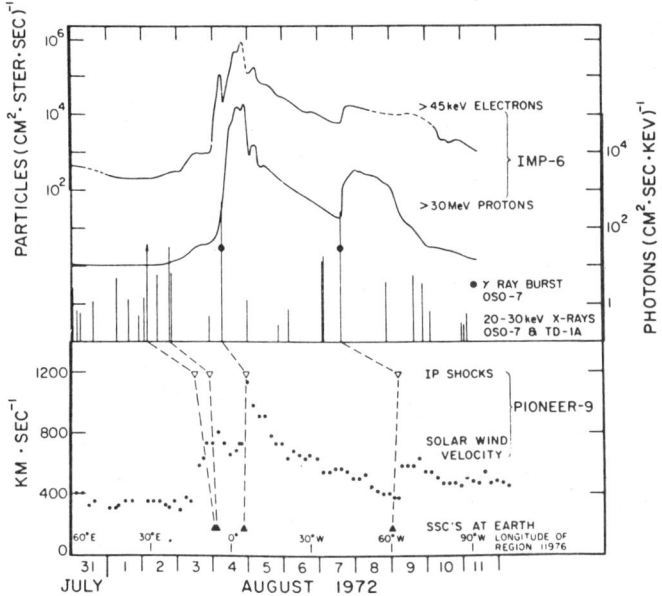

Figure 6. A summary of solar non-thermal phenomena for the 31 July - 11 August 1972 period, taken from the World Data Center "A" report UAG-28 and Solar Geophysical Data monthly reports. Except for the Pioneer 9 plasma observations at 0.78 A.U., all the data are from Earth-orbiting spacecraft. Note the correspondence between energetic particle injections, shock waves, γ-ray bursts, and intense hard X-ray bursts. [From *Lin and Hudson,*][7]

available.

The energy in the >20 keV electron population at the Sun is computed from the observed hard X-ray emission under the assumption of non-thermal, thick target bremsstrahlung. The >20 keV electron and >10 MeV proton populations escaping to the interplanetary medium are estimated from the observed fluxes near the orbit of the Earth.

Nuclear gamma ray lines of solar origin were observed for the first time from the 4 August flare. The flux and spectral parameters of the >10 MeV protons at the Sun were derived from the intensities of the 2.223 MeV neutron capture line and 4.438 MeV line resulting from the prompt de-excitation of C^{12}.[19] No information is available about the proton flux at the Sun at energies below $\sim$7 MeV since ions of these energies are not significant in producing gamma rays unless the spectrum is extremely steep. There are observations of solar flare protons down to $\sim$50 keV in the interplanetary medium near 1 A.U. The observed spectrum tends to flatten at low energies [20] but protons at energies much below $\sim$10 MeV may be subject to substantial energy changes in the propagation from the Sun to 1 A.U. Thus the energy released in protons below $\sim$10 MeV is still unknown; it may well be a significant fraction of the total flare energy release.

Radiative Energy Release

Zirin and Tanaka [21] estimated the total emission in Hα to be $\sim 2 \times 10^{30}$ and 2.5×10^{30} ergs for the 4 and 7 August 1972 major flares, respectively. Ellison [22] estimated that emission in other lines of the Balmer series at about 15 times the emission in Hα, and in all other lines at about 10×Hα, yielding a total of approximately 5×10^{31} ergs in visible line emission.

Table I. Energy Output of a Large Solar Flare
(based on 4 August 1972 flare)

Flare phenomenon	Energy (ergs)
I. Energetic particles	
a. >20 keV electrons at Sun	1.1×10^{32}
escaping to IP medium	1×10^{29}
b. >10 meV protons at Sun	1×10^{29}
escaping to IP medium	8×10^{30}
II. Radiative quasi-equilibrium region	
a. White light continuum	5×10^{29}
b. Hα	2×10^{30}
c. Total all optical lines (estimate from Hα)	5×10^{31}
d. UV (estimate)	$\sim 1 \times 10^{31}$
e. EUV	1×10^{31}
Total optical, UV, EUV radiation	$\sim 7 \times 10^{31}$
III. Flare explosion region	
a. Soft X-ray flare plasma	4×10^{31}
b. Hot isothermal plasma	1×10^{31}
c. Interplanetary shock wave and mass ejection	1×10^{32}
Total flare explosion	1.5×10^{32}
Total flare energy release	$\sim 2.5 \times 10^{32}$

Donnelly et al. [23] and *Donnelly and Hall* [24] found that the flash phase emission was dominated by chromospheric and transition region lines. For the 7 August 1972 flare, Donnelly (private communication) estimated a total of $\sim 10^{31}$ ergs emitted in the $10-10^3$ Å wavelength range during the flare impulsive phase. No measurements are available in the Å wavelength range 1000 - 3000 Å, but we expect that its energy flux is of the same order of magnitude as that of the EUV.

Thermal Flare Plasma

For the August flares the emission measure, $n_e n_i V$, and temperature T, are available from soft X-ray measurements [25, 26] The total energy in the thermal flare plasma can be obtained if the plasma density if known since

$$\epsilon_{plasma} \approx \frac{\frac{3}{2}(n_e n_i V) k T}{n_i}$$

Assuming a typical density of $\sim 3 \times 10^{10}$ cm^{-3} for the X-ray flare plasma we obtain a total energy of 4×10^{31} ergs. Radiation losses are small compared to the total plasma energy content.

Beside the thermal flare plasma at $\sim 10-20 \times 10^{6}$°K recent balloon measurements (Figure 2) have discovered a new hot isothermal component with maximum temperature $\sim 35 \times 10^{6}$°K, and emission measure $\sim 2.5 \times 10^{48}$ cm^{-3}. It appears that this hot isothermal component is often present in flares. Under the *ad hoc* assumption that the plasma pressure is equal for the new hot isothermal component and normal flare plasma, we estimate that the energy in this new component is approximately one third of that contained in the thermal flare plasma.

The Interplanetary Shock Wave

The amount of energy involved in the interplanetary shock wave can be estimated from the plasma density ρ, velocity v, temperature T, and magnetic field B observed near 1 A.U.,

$$\epsilon_{shock} = \Omega R^2 \int [(\tfrac{1}{2}\rho v^3 - \rho_b v_b^3) + \epsilon_g(\rho v - \rho_b v_b)$$
$$+ 3/2 k(\rho v T - \rho_b v_b Tb) + \frac{3}{8\pi}(B^2 v - B_b^2 v_b)] dt$$

where the terms represent the flow, gravitational, thermal, and magnetic energy, respectively, (the subscript b denotes the pre-shock background levels). $\epsilon_g \approx 2$ keV is the energy particle associated with the gravitational potential and ΩR^2 is the area subtended by the shock. Computations by *Dryer et al.* [27] using Pioneer 9 plasma observations and assuming $\Omega = 2\pi$ steradians at 0.78 A.U., show that the energy in the shock wave is $\sim 1 \times 10^{32}$ ergs.

The total energy released in a very large flare is then $\sim 2.5 \times 10^{32}$ ergs, of which $\sim$ half is initially contained in >20 keV electrons. Since recent observations show that the accelerated electron spectrum extends down to at least ~ 13 keV, most of the energy released in the flare could be contained initially in the accelerated electrons. The energy input to the solar atmosphere from the collisional losses of these electrons has been computed under the simplifying assumption that the energetic particles are precipitated from high in the corona, vertically downward into the solar atmosphere in a column of constant cross-sectional area. [28,7] More complex particles injection models, i. e., with non-vertically magnetic configurations, with various injection heights or with various initial pitch angle distributions, could be incorporated into the computation, but the vertical case illustrates the general features. Because only the highest energy electrons will reach the deepest portions of the solar atmosphere and because the electron spectrum falls rapidly with energy, the energy deposition rate decreases rapidly as we go deeper into the solar atmosphere. Furthermore, the ambient particle density increases so that low in the atmosphere the input energy can be radiated away (Figure 7). This radiative equilibrium region is responsible for the optical line and continuum emission and the UV and EUV emission during the impulsive phase. Since the emissivity of the solar atmospheric gas as a function of temperature has a peak, above some critical level the energy input rate will be too rapid for radiative equilibrium and the ambient gas will rapidly increase in temperature and expand upward, giving rise to the thermal flare plasma. At even higher levels of the atmosphere the energy gained by the ambient plasma may exceed that needed to escape the restraining forces of the Sun's gravitational and magnetic fields. The material will then be ejected from the Sun. If the velocity of the ejected material is sufficiently high, a shock wave will form in front. This shock can give rise to type II radio emission, and may be observed later as an interplanetary shock wave. We note here that besides a reduction in the energy outputs (Table I) by 1 - 3 orders of magnitude, small flares are fundamentally different from these large flares in that shock waves and ejected material are not observed for small flares.

III. THE ACCELERATION OF NUCLEI

Interplanetary shock waves are observed at 1 A.U. to be efficient accelerators of particles, particularly ions, to energies up to 1 - 10 MeV. These same shock waves traversing the stronger and more ordered magnetic fields near the Sun, at speeds up to 3000 km/s, could accelerate ions and electrons to relativistic energies in a second non-thermal phase in large flares. The γ-ray line observations for 4 August 1972 indicate that the acceleration of

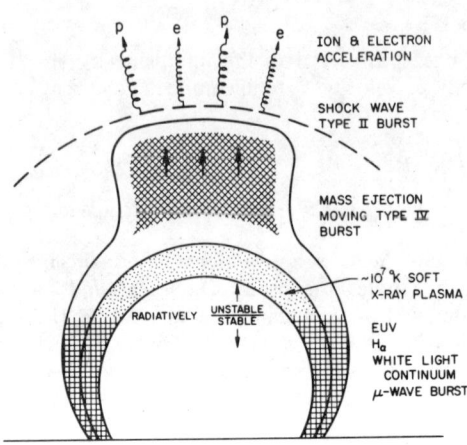

Figure 7. This figure shows a schematic representation of a large flare. In the cross-hatched region, the solar atmosphere is able to radiate away the input energy. Above that is the flare explosion region, divided into flare plasma and ejected material, depending on whether enough energy is dumped in fast electrons to overcome the restraining gravitational and magnetic forces. [From Lin][29]

protons and other nuclei is a separate process from the flash-phase electron acceleration. Most of the protons escape to (or are accelerated in) the interplanetary medium near 1 A.U. and the bulk of the proton acceleration appears to occur a few minutes later than the electron flash phase. (Figure 8)

SMM observations of gamma-rays from flares occurring near maximum of the solar cycle, however, show quite different temporal behavior (see review by Chupp in this book). The nuclear gamma-ray emission appears closely similar to the flash phase impulsive hard X-ray emission, with a delay of only a few seconds as opposed to a few minutes in the case of 4 August 1972. The fluxes of energetic, >10 MeV protons released to the interplanetary medium after these gamma-ray fluxes are observed to be lower by several orders of magnitude relative to the 4 August 1972 flare, even though the nuclear gamma-ray fluences are only ~ one order of magnitude lower. Most of the accelerated >10 MeV ions thus appear to lose their energy at the Sun rather than escape to the interplanetary medium as in 4 August 1972.

It is important to determine whether these 'fast' gamma-ray flares near solar maximum are fundamentally different from the 4 August 1972 type of flare; in particular whether in these flares the energetic protons are accelerated by a different mechanism than in 4 August, possibly by the same mechanism that accelerates the electrons in the flash phase. We note here that these SMM gamma-ray flares in all other respects are similar to the 4 August flare: in particular they are accompanied by strong impulsive hard X-ray emission and type II and IV radio emission. Type II emission, as we noted earlier, is the radio signature of shock waves passing through the corona. We believe therefore that these SMM flares are fundamentally similar to the 4 August flare. The short but significant delay of a few seconds from the impulsive hard X-ray peaks to the gamma-ray peak may mean that

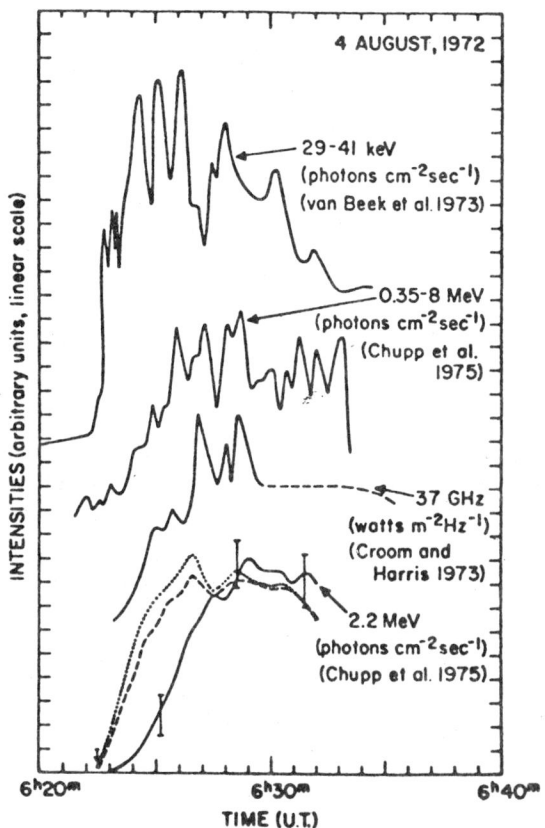

Figure 8. The observed energetic radiation for the 1972 August 4 flare from *Bai and Ramaty*[30] showing the delay of the proton acceleration from the impulsive electron acceleration. The three upper lines are the measured time profiles of X-rays (29 to 41 keV), gamma rays (0.35 to 8 MeV), and microwaves (37 GHz). The error bars in the lower part of the figure represent the measured intensities of the 2.2 MeV line. The solid, dashed, and dotted lines are calculated by assuming that the instantaneous number of energetic nuclei in the flare region has the same time dependence as that of the observed 0.35 to 0.8 MeV continuum. The dashed and dotted lines are obtained by assuming that the time dependence of the nuclei is the same as that of the 29 to 41 keV X-rays. For the solid and dotted lines we use a photospheric ^{3}He/H abundance ^{3}He/H $= 5 \times 10^{-5}$, for the dashed line ^{3}He/H $= 0$.

the shock acceleration of protons occurs within several thousand kilometers — the distance travelled by the shock in a few seconds — of the electron energy deposition region, as opposed to several hundred thousand kilometers for 4 August. If the acceleration occurs so low in the solar atmosphere, the efficiency of escape of the accelerated ions may be greatly reduced from 4 August.

The variation of the characteristics of the acceleration of energetic nuclei may be related to the evolution of active regions over the solar cycle. Active regions become larger and more complex as the solar cycle progresses from maximum to descending phase. The acceleration of ions by the shock waves at the Sun may require a favorable magnetic field configuration. Particle acceleration at the Earth's bow shock or by interplanetary shocks is known to depend on the orientation of the shock to the upstream magnetic field. Assuming that active region magnetic structures are where particle acceleration by shocks occurs, then we would predict both the gamma-ray delays to become larger and the proton escape efficiency to the planetary medium to increase as the solar cycle progresses from maximum to descending phase. Sensitive measurements of the flare gamma-ray bursts and interplanetary protons over the solar cycle are needed to test this hypothesis.

Acknowledgments

This research was supported in part by NASA grant NSG 7527 and contract NAS 5-25980 and NSF grant ATM-7924559.

References

1 Lin, R. P., R. A. Schwartz, R. M. Pelling, and K. C. Hurley, *Astrophys. J. Lett.*, in press (1981).
2 Kane, S. R., and K. A. Anderson, *Astrophys. J.* **162**, 1002, (1970).
3 Crannell, C. J., K. J. Frost, C. Mätzler, K. Ohki, and J. L. Saba, *Astrophys. J.* **223**, 620, (1978).
4 Elcan, M. J., *Astrophys. J. Lett.* **226**, L99, (1978).
5 Kane, S. R., C. J. Crannell, D. Datlowe, U. Feldman, A. Gabriel, H. S. Hudson, M. R. Kundu, C. Mätzler, D. Neidig, V. Petrosian, and N. R. Sheeley, Jr., *Solar Flares,* edited by Peter A. Sturrock, (Boulder: Colorado Associated University Press, 1980) p. 187.
6 Lin, R. P., and H. S. Hudson, *Solar Phys.* **17**, 412, (1971).
7 Lin, R. P., and H. S. Hudson, *Solar Phys.* **50**, 153, (1976).
8 Ramaty, R., S. A. Colgate, G. A. Dulk, P. Hoyng, J. W. Knight, R. P. Lin, D. B. Melrose, F. Orrall, C. Paizis, P. R. Shapiro, D. F. Smith, M. Van Hollebeke, *Solar Flares,* edited by Peter A. Sturrock, (Boulder: Colorado Associated University Press, 1980) p. 117.
9 Chubb, T. A., *Proc. Leningrad Symp. Solar Terrestrial Physic/1970,* part I (edited by E. R. Dyer) (Dordrecht, Holland: D. Reidel Publ.) p. 99.
10 Brown, J. C., *Solar Gamma-, X-, and EUV-Radiation,* (edited by S. R. Kane) (Dordrecht, Holland: D. Reidel Publ., 1975) p. 245.
11 Kahler, S., *Solar Gamma-, X-, and EUV-Radiation,* (edited by S. R. Kane) (Dordrecht, Holland: D. Reidel Publ., 1975) p. 211.
12 Mätzler, C., T. Bai, Crannell, C. J., and K. J. Frost, *Astrophys. J.* **223**, 1058, (1978).
13 Colgate, S. A., Audouze, J., and Fowler, W. A., *Astrophys. J.* **213**, 849, (1977).
14 Brown, J. C., *Coronal Disturbances,* (edited by G. Newkirk) *IAU Symp.* **57**, 395, (1974).
15 Hoyng, P., A. Duijveman, M. E. Machado, D. M. Rust, Z. Svestka, A. Boelee, C. de Jager, K. J. Frost, H. Lafleur, G. M. Simnett, H. F. van Beek, and B. E. Woodgate, *Astrophys. J.* **246**, L155, (1981).
16 Brown, J. C., D. B. Melrose, and D. S. Spicer, *Astrophys. J.* **228**, 592, (1979).
17 Smith, D. F., and C. G. Lilliequist, *Astrophys. J.* **232**, 582, (1979).
18 Kane, S. R., *Coronal Disturbances,* (edited by G. Newkirk) *IAU Symp.* **57**, 105, (1974).
19 Ramaty, R. B. Kozlovsky, and R. E. Lingenfelter, *Space Sci. Rev.* **18**, 341, (1975).
20 Van Hollebeke, M. A. I., *A Close-Up of the Sun,* (edited by R. W. Davies) *JPL publication* **78-70**, 105, (1978).
21 Zirin, H., and K. Tanaka, *Solar Phys.* **32**, 173, (1973).

22 Ellison, M. A., *Planet. Space Sci.* **11**, 597, (1963).
23 Donnelly, R. F., A. T. Wood, Jr., and R. W. Noyes, *Solar Phys.* **29**, 107, (1973).
24 Donnelly, R. F., and L. A. Hall, *Solar Phys.* **31**, 411, (1973).
25 Collected Data Reports on August 1972 Solar Terrestrial Events, World Data Center A, UAG-78, part II (1973), p. 298.
26 Neupert, W. M., R. J. Thomas, and R. D. Chapman, *Solar Phys.* **34**, 349, (1974).
27 Dryer, M., Z. K. Smith, R. S. Steinolfson, J. D. Mihalov, J. H. Wolfe, and J. K. Chao, *J. Geophys. Res.* **81**, 4651, (1976).
28 Brown, J. C., *Solar Phys.* **31**, 143, (1973).
29 Lin, R. P., *Study of Travelling Interplanetary Phenomena/1977*, (edited by M. A. Shea) (Dordrecht, Holland: D. Reidel Publ.) p. 23.
30 Bai, T., and R. Ramaty, *Solar Phys.* **49**, 343, (1976).

THE IMPULSIVE FLUX TRANSFER SOLAR FLARE MODEL

P. J. Baum and A. Bratenahl
Institute of Geophysics and Planetary Physics
University of California, Riverside, Ca. 92521

ABSTRACT

Magnetic Reconnection and electric double layers are combined interactively in the impulsive flux transfer flare model. Changes in the active region flux function drive a separator-aligned sheet current which saturates at $\sim 10^{12}$ amperes producing a magnetic field of ~ 23 gauss. The nascent double layer is initially stabilized by an influx of particles from upstream. We show that this cannot continue indefinitely by proving that the critical runaway threshold is a decreasing function of inflow velocity and separator electric field. Crossing this threshold, the double layer grows catastrophically due to positive feedback effects and develops a peak potential of $\sim 10^{11}$ volts across a layer of thickness $\lambda \sim 900$ km. The time development of flare parameters is calculated using a circuit analog which treats the double layer as an effective resistance. The effective resistance is calculated from the ultra-relativistic Child-Langmuir law. Conditions in the double layer become highly relativistic so that electrons and ions become equally massive and carry equal currents; the model produces $\sim 10^{33}$ energetic particles of which half are ions. The model requires a single stage of acceleration but several stages of decceleration and thermalization. At any instant particles exit the double layer with a nearly monoenergetic spectrum (energy eV(t)) and the spectrum is steepened outside the layer by generation of plasma turbulence. The accelerated particles are quasithermalized by the Buneman instability at the ends of the layer. The electrons produce thermal x-rays and the protons produce neutrons by spallation reactions on heavy nuclei. The neutrons proceed to the photosphere capturing protons to produce nuclear gamma ray emission. The model leads naturally to explanations of other flare features such as separation of the two flare ribbons and to predictions concerning the directivity of particle emissions.

INTRODUCTION

The Current Interruption[1] (CI) solar flare model involves the explosive development of a double layer in a field aligned current circuit of large inductance. The impulsive Flux Transfer[2,3] (IFT) model, on the other hand, involves nonsteady reconnection between two independent field systems, e.g., the field of one bipolar sunspot group intruding into, or "invading", the field of another. One might suppose the two models are quite unrelated, but the facts are otherwise. Baum, Bratenahl, and Kamin[4] (hereinafter called BBK) showed that the transient behavior of both models could be understood in terms of the same LR analogue circuit in which R(t), the effective resistive impedance, acting as a nonlinear resistor, increases rapidly in time. As originally proposed, IFT depended on the rapid development of anomalous resistivity along the separator current path which is a circuit of large inductance L, just as in the case of CI. (Separator is defined in next section). Flare onset in both models occurs when an increasing current crosses the threshold of some form of overcurrent instability (insufficient current carriers). The instability then continues to be further aggravated by the effects of the resulting transient thus initiated (positive feedback). The CI model is applicable to field aligned current along any bundle of field lines. The EMF which drives the current is not specified. In the case of IFTE, on the other hand, a very special bundle of field lines is singled out of the global field's topological structure; the bundle having, as already noted, the separator as magnetic axis. The EMF is also specific: it is generated by changes in strength and/or geometrical arrangement of the sources of the field, e.g., sunspots.

BBK suggested that the rich variety of flare morphologies might be explained by considering (hybrid) combinations of both models, citing as examples, the flare models of Spicer[5] and Colgate[6]. Since then, Spicer[7] synthesized a hybrid model, combining a modified version of CI with a modified version of his original tearing mode model. This hybrid, however, does not conform to the suggestion of BBK. The purpose of this paper is to present a revision of the original IFT that does so conform. In essence, it simply substitutes the CI double layer concept as the nonlinear "resistor" in place of the original concept, anomalous resistivity. The major part of this paper is devoted to an expression of the double layer in relativistic terms which provides a more adequate accounting for nonthermal particle acceleration and the high energy effects thereof as well as the positive feedback effect. The next section attempts to clarify our particular point of view through a discussion of the basic assumptions underlying both Spicer's hybrid and our own.

BASIC ASSUMPTIONS

Both models recognize the strong observational evidence that the preferred flare sites of magnetic energy release are magnetic arch structures in active regions. Since arch structures are everywhere dense in such regions, an important part of the problem is to understand what causes a particular arch to become flare active: what EMF acts on it in preference to its neighbors, thus to drive the requisite current?

Spicer's first basic assumption: an ordinary arch structure is involved. Field aligned current on such an arch produces helically twisted field lines. Since his first assumption leaves no other reasonable choice, his second assumption follows: the current must be driven by EMF's acting directly on the arch's footpoints. (The EMF is assumed to be generated by mechanically rotating those footpoints.) Finally, his third assumption: in CI, replace the double layer with anomalous resistivity. The resulting structure, and its dynamics, resembling a Tokamak, has great appeal to theorists because a large amount of relevant stability analysis is available from fusion energy research. Offsetting this theoretical advantage, however, are some disadvantages. As with other models, the flare starts when the current crosses instability threshold. The resultant heating in this case, however, raises that threshold (negative feedback). Therefore, to maintain instability (i.e., the flare), the EMF is required to continuously increase the current, while working at all times against the effects of large circuit inductance. This flare model which is thus "fueled" from below (there is no preflare energy storage available in the atmosphere), burns like a torch at the whim of rotary motions in the photosphere of a kind very rarely observed. Moreover, it is by no means clear that anomalous resistivity can admit substantial fluxes of nonthermal, let alone relativistic, runaway ion and electrons as is claimed.

In contrast to Spicer, our first basic assumption is that a very special kind of magnetic arch structure is involved: as already noted, its magnetic axis must be a singular field line known as the separator. The separator, connecting between hyperbolic neutral points in the photosphere, forms the line of intersection of four branches of the separatrix surface (Fig. 1). The example shown represents the combined field of two bipolar sunspot groups. Actually, separator lines appear in almost any system of greater source complexity than that of a simple bipolar system. Current along, and in the neighborhood of a separator, represents an excess of magnetic flux in one pair of cells defined by the separatrix and a deficiency of flux in the other pair, relative, that is, to a purely potential field. It therefore represents magnetic energy stored in the atmosphere, and it is this energy that fuels the flare when the resistive impedance of the separator current circuit rapidly increases. The separator current satisfies the analogue equation

$$\dot{I}L + IR = V \qquad (1)$$

where V is an EMF that is a function of changes in the strengths and/or geometrical arrangements of the photospheric sources of the magnetic flux. Under these conditions, the response of I to V represents an indirect coupling via hydromagnetic wave propagation from the sources. It is a familiar fact of a circuit obeying (1) that if R should increase, the IR voltage drop increases on account of the first term. But if, in its unstable mode, R should be an increasing function of the IR drop the result is catastrophic, like the failure of a dam. We will see below that a strong double layer has just that property. It is an obvious case of positive feedback.

In our CI-IFT model, the separator arch commonly occurs in all complex active regions or even simple ones that are in an ambient field. Likewise, are the factors that are responsible for the EMF. The EMF works against the $\dot{I}L$ term mostly during the preflare period, and it is this work that stores the flare energy ahead of time. During the flare, it is the $\dot{I}L$ term that does the driving. On the

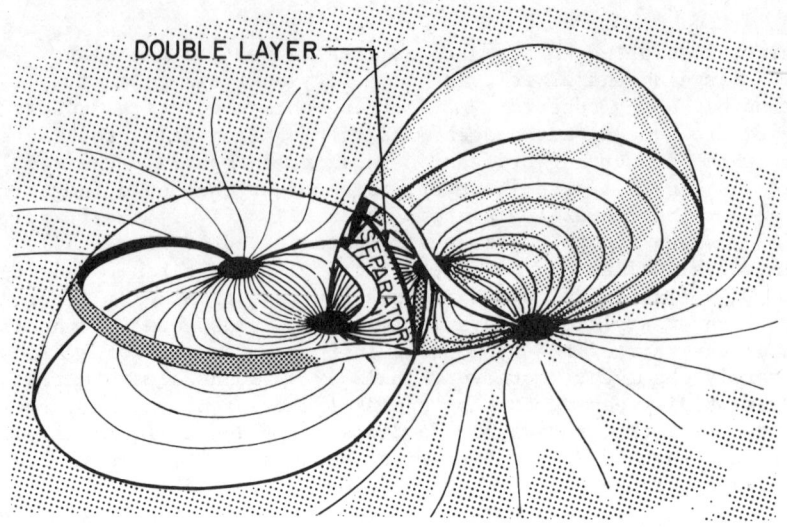

Fig. 1. An electric double layer forms near the top of the arch-shaped separator line. The separatrix surfaces defining the flux cell boundaries are shown for two bipolar sunspot pairs.

other hand, much less theoretical work has been done with magnetic arches of separator type. The subject area, of course, is reconnection, and the bulk studies thus far apply to 2-dimensional problems. We do know, however, that a separator arch current system develops a hyperbolic sheet pinch with no net helical twisting of field lines. This form of pinch sandwiched between slow mode shocks accepts plasma from the pair of excess magnetic flux cells upstream, compresses it, and then energetically ejects it into the pair of deficiency flux cells downstream. By Faraday's law, the flux transfer rate from the upstream to the downstream cells is given by the voltage drop along the separator even if most of that drop occurs over a very small portion of the circuit, e.g., a double layer.

We refer the reader to BBK for the circuit equation analysis and the demonstration that the typical active region EMF drives the separator current up to saturation at $\sim 10^{12}$ amperes in ~ 1.2 days.

THE RELATIVISTIC CHILD-LANGMUIR LAW

In order to derive the proper effective resistive impedance, we need to know the behavior of the current-voltage relation of a double layer under high voltage or relativistic conditions. A double layer's current is space charge limited, obeying the Child-Langmuir Law. We start with Poisson's equation for a single species

$$\frac{d^2V}{dy^2} = \frac{ne}{\varepsilon_o} = \frac{j}{\varepsilon_o v} \qquad (2)$$

where V is the potential across two points separated by a distance y, j is the current density, and v is the terminal velocity of the particle as it exits the higher potential point. Equation (2) is integrable giving

$$\frac{dV}{dy} = \sqrt{\frac{2j}{\varepsilon_o}} \cdot \sqrt{\int \frac{dV}{v}} \qquad (3)$$

Bergmann[8] (his eq. 9.9) finds that the particle exits with a velocity

$$v = c\sqrt{1 - \frac{1}{(1 + \frac{eV}{mc^2})}} \qquad (4)$$

where m is the rest mass of the particle. Integrating with (4) in (3) we find

$$\frac{dV}{dy} = \sqrt{\frac{2jmc}{\varepsilon_o e}} \left[\left(\frac{eV}{mc^2}\right)^2 + \frac{2eV}{mc^2}\right]^{1/4} \qquad (5)$$

In (5) the integration over y (from 0 to λ, the double layer thickness) is trivial and if we change variable letting $x = eV/(mc^2)$, then (5) becomes

$$\frac{j e \lambda^2}{2mc^3 \varepsilon_o} = 1/4 \left[\int_0^{\frac{eV}{mc^2}} \frac{dx}{(x^2 + 2x)^{1/4}}\right]^2 \qquad (6)$$

Equation (6) has been numerically integrated using a Runge-Kutta algorithm to produce its solution shown in Figure 2. We see that for low voltage ($eV \leq mc^2$) the result approaches the classical Child-Langmuir Law:

$$j = \frac{4\varepsilon_o}{9} \sqrt{\frac{2e}{m}} \frac{V^{3/2}}{\lambda^2} . \qquad (7)$$

However, for the ultra-relativistic case ($eV \geq 100\ mc^2$) the result is a linear current-voltage relation:

$$j = 2c\varepsilon_o \frac{V}{\lambda^2} . \qquad (8)$$

(We thank J. W. Poukey for pointing out that the relativistic Child-Langmuir Law was first derived by Boers and Kelleher[9]). The solution of (6) can be expressed conveniently for our purposes as

$$V = (\text{const.}) (\lambda^2 j)^W, \quad 2/3 \leq W \leq 1 \qquad (9)$$

For electrons, equation (8) applies for $eV \gtrsim 51$ MeV.

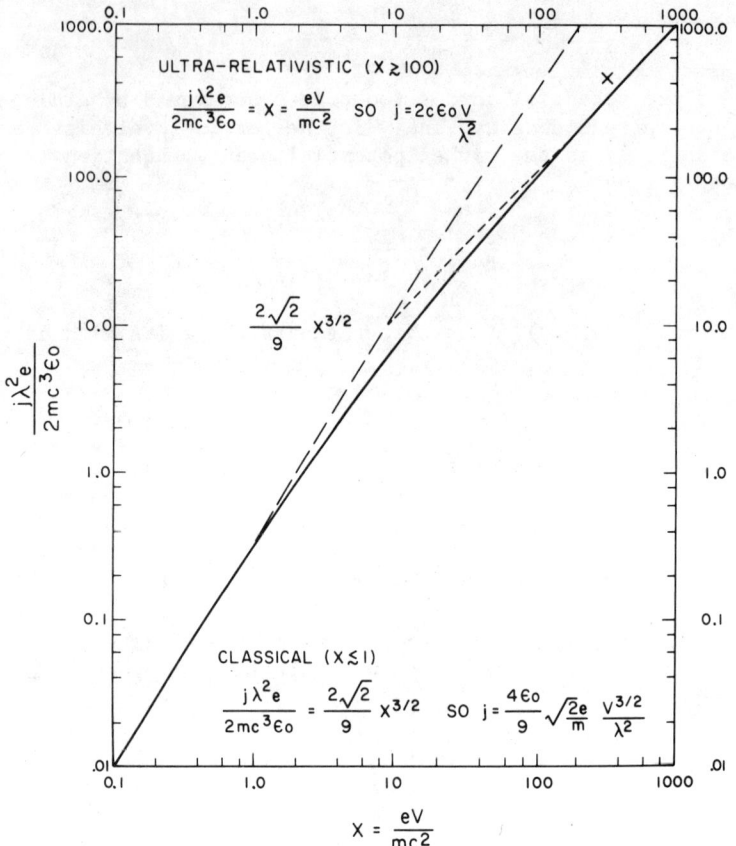

Fig. 2. The solid line shows the solution of equation (6) for the relativistic Child-Langmuir Law. At low voltages the solution asymptotically approaches the longer dashed line which is the classical result $j \sim V^{3/2}$. At very high voltage the solution approaches the shorter dashed line where $j \sim V$.

In the preceding analysis we have not treated a second species because of the complexity of the general case. We note two limits however. In the classical case, the ions are much more massive than the electrons so that the electrons carry most of the current and little is lost by neglecting the ions. In general, when written in terms of $\gamma = 1/\sqrt{1 - (v/c)^2}$ equation (4) becomes $\gamma = eV/(m_0 c^2) + 1$ so that

$$m = m_0 \gamma = \frac{eV}{c^2} + m_0 . \qquad (10)$$

Therefore, for $eV \gg m_0 c^2$ both ions and electrons are accelerated up to approximately the same mass since both are approaching the limiting drift velocity $v \sim c$. And therefore, in the ulta-relativistic case we expect ions to perform essentially as electrons, each carrying half the current. In intermediate cases electrons will carry somewhat more current than ions and the effect of ion inclusion will decrease w (eq. (9)) somewhat.

Ion inclusion in discussed quantitatively by Poukey[10].

SPONTANEOUS GROWTH OF THE DOUBLE LAYER

As indicated in the BASIC ASSUMPTIONS section, the generation of separator-aligned currents is a normal part of solar activity and this current is the site of a hyperbolic pinch with attached slow-mode shocks. Should the electric field substantially exceed the critical runaway field a double layer should form. Two points need noting however; first, small amplitude double layers need not be dangerous and we may expect these to form in addition to the flare itself. Second, a larger amplitude double layer can be stabilized by a particle flux from upstream of the hyperbolic pinch which effectively shorts out the double layer. This condition cannot last however, and an eventual catastrophe is inevitable. As a consequence of the shock transition the ratio n/T (density/temperature), which is approximately proportional to the runaway field threshold, is a decreasing function of inflow velocity. The situation is particularly sensitive near the top of the arch where the initial density is least.

BBK noted that the flare "conversion goes through at least two steps; magnetic to kinetic at the site of instability, and then kinetic to thermal at some remote site." The double layer is the site of the magnetic to kinetic conversion and the thermalization takes place outside the double layer. Smith and Priest[11] correctly noted that plasma turbulence would quench a current instability through thermalization (negative feedback) and this is the process which occurs outside the double layer. They failed to notice, however, that the thermalization could not occur for many Debye lengths and that the direct electric field acceleration (due to the voltage induced by the current cutoff) would occur over this small region. Buneman[12] had computed the effect of electric field acceleration (from a double layer we shall assume) into plasma (at the end of the double layer),: "The effect stops 'runaway' in about 100 plasma periods after which there is 'heating' by 'collective collisions' instead." His computation was certainly inadequate for our purposes as it was nonrelativistic; however, it illustrates the general effect. (We note that a rather similar point of view coupling double layers with the Buneman instability has recently been presented by Belova et al.[13]).

Normally, the plasma pressure at the edges of a double layer is balanced by the pressure of the electrostatic energy density within it but in the case of a very strong double layer the additional force due to the external beam-plasma interaction causes the double layer to expand[14]. Under the particular present circumstance of very large circuit inductance, the current density cannot change rapidly (Eq. 1) and consequently, the voltage drop (9) must increase. This increases the beam-plasma force still further, and so on. The result is a catastrophic growth of V, an effect of positive feedback as shown in the next section.

THE FLARE TRANSIENT

If the total current I flows in a sheet of width 2L, thickness 2ℓ, then

$$I = 4L\ell j \qquad (11)$$

so that from (1), (8), and (11)

$$R_{eff} = V/I = \lambda^2/(8c\varepsilon_o L\ell) \qquad (12)$$

From the feedback process we might expect λ to increase exponentially, if so,

$$R_{eff} = R_o \exp(\Gamma t), \qquad (13)$$

with $R_o = \lambda_o^2/(8c\varepsilon_o L\ell)$. This exponential behavior of R_{eff} was used to produce the solar flare circuit flash phase solution shown in Figure 3 (taken from BBK). However, the cause is now double layer growth.

The dimensionless scales of Figure 3 should be multiplied by scale factors: multiply I by 10^{12} amperes; P by 10^{20} watts; V by 10^8 volts; R by 10^{-4} ohms. 10^{32} ergs of energy is released during the 100 second flash phase (hatched area under the P-curve).

From Figure 3 we see that maximum voltage occurs for $R_{eff} \sim 0.3$ ohms at which point the double layer will be 900 km thick. Also from Figure 3, the current I is nearly constant for $\Delta t \sim 200$ seconds. Thus, the number of particles accelerated during the flare is

$$N = 1/e \int I \, dt \simeq I_o \Delta t/e \qquad (14)$$

or $N \sim 10^{33}$ particles. And for the high voltages developed here, about half will be electrons, half ions. Since the current drops during the flash phase, the magnetic field B never exceeds its initial value of 23 Gauss and no reverse currents are needed. The release of 10^{32} ergs at the ends of the double layer results in a temperature of about 2×10^7 degrees which explains the thermal origin of the flare x-rays.

The protons are accelerated to the opposite end of the layer from electrons and perform heating there producing both x-rays and neutrons from quasithermonuclear spallation reactions. The neutrons are captured by protons at the photosphere to produce nuclear gamma-ray emission. We expect flares producing large amounts of high energy protons directed at earth to produce few photospheric nuclear

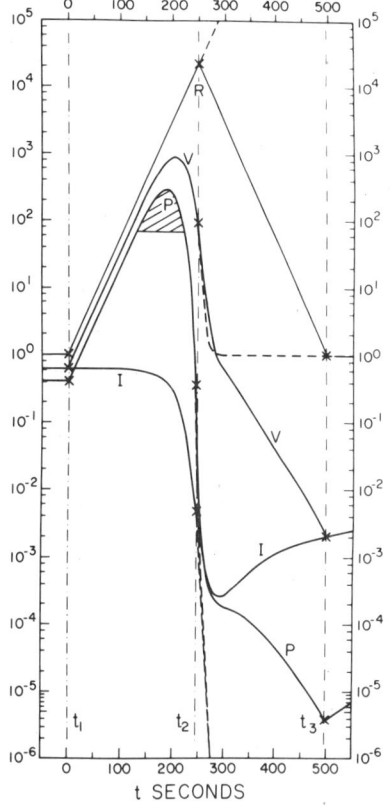

Fig. 3. The current, voltage, power, and the effective resistance are presented for the case of a large solar flare. The effective resistance comes from the ratio V/I in the ultra-relativistic Child-Langmuir law (eq. 13). The other parameters follow self-consistently from the circuit solutions presented by Baum, Bratenahl, and Kamin[4]. An electric double layer forms during the period where the P, V, and R curves are elevated above their baseline (interflare storage phase) values. The flash phase occurs at the hatched portion of the Power curve wherein 90% of the energy is released. The vertical scale is dimensionless but conversion factors appear in the text.

reactions and vice-versa.

The double layer growth will cut off the separator current by diverting it downstream ("sheet rupture")[15]. The effect of tearing the current will be to form two flare ribbons which will propagate away from one another. There will be a low level of filamentation in the sheet, both before and after rupture, which Colgate[6] feels explains the energetic particle spectrum. In our model the particles leave the double layer with a monoenergetic spectrum, energy eV(t), and the collective interactions will steepen the spectrum.

ACTIVE REGION PARAMETERS

BBK produced the following estimates: length scale $h \sim 10^5$ km, inductance $L' \sim 10$ H, current $I \sim 10^{12}$ amperes, $R \sim 10^{-4} \Omega$, preflare separator voltage $V_s \sim 3.7 \times 10^7$ volts. We assume that at the threshold of instability the electric field crosses the Dreicer runaway point E_D and plasma enters the hyperbolic pinch at (ion) mach one ($v = v_{\theta i}$). Then the magnetic field is

$$B = E_D/v_{\theta i} \sim n/T^{3/2} \qquad (15)$$

If the current flows in a separatrix-attached sheet[16] of area $L\ell$ then amperes law gives

$$L = \frac{\mu_o I}{2B} = \frac{\mu_o I v_{\theta i}}{2E_D} \sim T^{\frac{3}{2}}/n \qquad (16)$$

and in differential form, amperes law is

$$\ell = \frac{B}{\mu_o j} = \frac{B}{\mu_o nev_{\theta e}} = \frac{E_D}{\mu_o nev_{\theta i} v_{\theta e}} \sim 1/T^2. \qquad (17)$$

Here we assume that at flare threshold the double layer has just started to form as an electrostatic shock at (electron) mach one ($v_d \sim v_{\theta e}$). It is now apparent that ℓ will be unrealistically small unless $T \sim 10^4 °K$ and B will be unrealistically small unless $n \sim 10^{12} cm^{-3}$ which are typical of values in a quiescent prominence where the thermal instability has produced a condensation. These preflare values are radically altered during the flare itself where the double layer evacuation process lowers the density locally to $n \sim 10^8 cm^{-3}$ and the heating at the ends of the layer locally raise the temperature to $T \sim 10^7 °K$.

REFERENCES

1. H. Alfven and P. Carlquist, Solar Phys. **1**, 220 (1967).
2. A. Bratenahl and P. J. Baum, Solar Phys. **47**, 345 (1976).
3. A. Bratenahl and P. J. Baum, Geophys. J. Roy. Astr. Soc. **46**, 259 (1976).
4. P. J. Baum, A. Bratenahl and G. Kamin, Astrophys. J. **226**, 286 (1978). Called "BBK".
5. D. S. Spicer, Sol. Phys. **53**, 305 (1977).
6. S. A. Colgate, Astrophys. J. **221**, 1068 (1978).
7. D. S. Spicer, Sol. Phys. **70**, 149 (1981).
8. P. G. Bergmann, Introduction to the Theory of Relativity, Prentice-Hall, N.J., p. 136 (1942).
9. J. E. Boers and D. Kelleher, J. Appl. Phys. **40**, 2409 (1969).
10. J. W. Poukey, J. Vac. Sci. Tech. **12**, 1214 (1975).
11. D. F. Smith and E. R. Priest, Sol. Phys. **53**, 305 (1977).
12. O. Buneman, Phys. Rev. **115**, 503 (1959).
13. N. G. Belova, A. A. Galeev, R. Z. Sagdeev and Yu. S. Sigov, JETP Lett. **31**, 551 (1980).
14. L. P. Block, Cosmic Electrodynamics **3**, 349 (1972).
15. Syrovatskii, S. I. and Somov, B. V., in Solar and Interplanetary Dynamics, eds. M. Dryer and E. Tandberg-Hanssen, IAU, D. Reidel Pub. Co., Boston, p. 425 (1980).
16. P. J. Baum and A. Bratenahl, Sol. Phys. **67**, 245 (1980).

Sec. IV Instrumental Concepts

THE BURST AND TRANSIENT SOURCE EXPERIMENT
FOR THE GAMMA-RAY OBSERVATORY

G. J. Fishman, C. A. Meegan, and T. A. Parnell
Space Sciences Laboratory
NASA/Marshall Space Flight Center
Huntsville, Alabama

and

R. B. Wilson
Physics Department
University of Alabama in Huntsville
Huntsville, Alabama   35807

ABSTRACT

The Burst and Transient Source Experiment (BATSE) is a highly sensitive, all-sky monitor for the Gamma-Ray Observatory (GRO). The eight scintillation detector modules of BATSE are positioned around the GRO spacecraft to provide a uniform view of the entire sky above the horizon. Each module is 20 in. in diameter, 0.5 in. thick and operates in the energy range from 40 to 600 keV. Rough burst locations can be derived from the relative counting rates of the detectors; longer-lived (> 100 min) transient sources can be detected and located by Earth occultations. The data system contains mass memory to allow storage of large amounts of burst data in short time intervals. Data are sorted onboard in various ways into programmable energy, time and detector domains. Details of the design and capabilities of the BATSE are presented.

INTRODUCTION

The Gamma-Ray Observatory (GRO) is the next major observatory for high-energy astrophysics, planned by NASA for launch by the Space Shuttle in 1987. This paper describes the Burst and Transient Source Experiment (BATSE) for the GRO. The free-flying observatory is planned to operate for 2 years in a low-altitude, low-inclination orbit. The four large experiments selected for flight will perform a multitude of investigations at energies above ~ 100 keV which include observations of all known astrophysical sources and production mechanisms of gamma rays. Figure 1 shows a concept of the GRO spacecraft with BATSE and the other GRO instruments: EGRET (Energetic Gamma-Ray Experiment Telescope), OSSE (Oriented Scintillation Spectrometer Experiment), and COMPTEL (Imaging Compton Telescope).
Even though detector components of the other three experiments on the GRO could detect gamma-ray bursts, only the BATSE detectors are designed and optimized to provide highly sensitive

full-sky (above the Earth's horizon) observations of burst and transient sources. A microprocessor-based data system allows commandable on-board burst trigger criteria to be stored and the processing and storage of large amounts of burst data in various formats with different energy and time resolutions. The large area of the detector array will permit studies of very fast time structure within a burst, spectral changes during a burst, and the observation of weak bursts at a sensitivity not possible with the present generation of gamma-ray burst detectors.

The following briefly describes the scientific objectives of the BATSE: (1) To detect and study the spectral and temporal structure of gamma-ray bursts down to $\sim 10^{-8}$ erg/cm^2; (2) To provide rapid, rough locations, via a single spacecraft measurement, of hundreds of gamma-ray bursts during the GRO mission not detectable by other spacecraft; (3) To provide accurate Earth-based timing data for stronger bursts for use in arc-second locations by an interplanetary burst timing network; (4) To provide rapid detection and location of longer-lived hard X-ray and gamma-ray transient sources by Earth occultation measurements; and (5) To monitor the stronger, known hard X-ray sources for variability.

DETECTORS

The eight detector modules of the BATSE array are positioned around the periphery of the GRO spacecraft, as shown in Figure 1. The eight faces of the array are parallel to the eight sides of a regular octahedron. This provides uniform sky coverage and ensures that every detectable burst will be viewed by four detectors. Gamma-ray bursts may be located by measuring the ratios of the counting rates of the detectors observing a burst. Three detectors are required to make a location determination; a fourth detector provides redundancy and a measure of the uncertainty in the location determination.

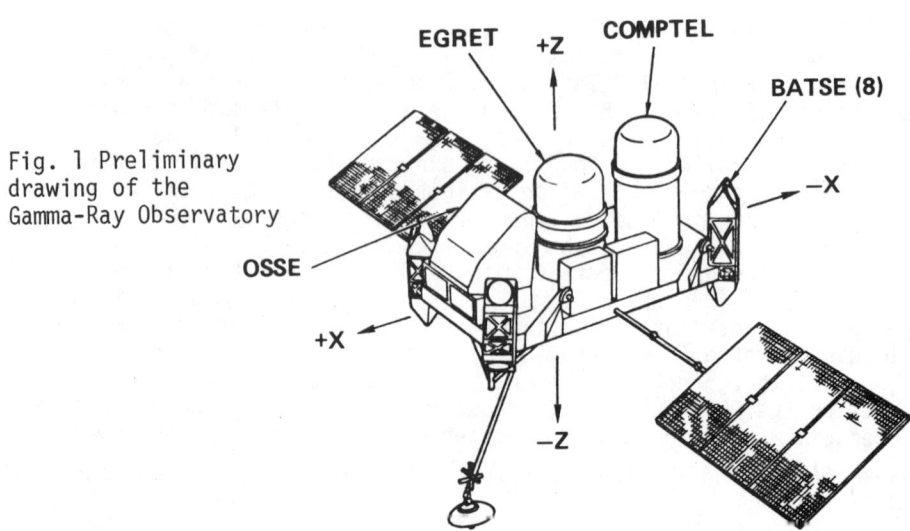

Fig. 1 Preliminary drawing of the Gamma-Ray Observatory

The front faces of the detectors have a clear 2π steradian field-of-view except for several spacecraft components which are relatively transparent at 100 keV. The large diameter-to-thickness ratio of the scintillation crystal gives the detector an angular response which is nearly a cosine function. The front face of each detector will have a thin aluminum absorber to reduce the response to low-energy photons. This will prevent deadtime and pulse pile-up from the diffuse X-ray background and strong X-ray sources. It will also provide a well-determined low-energy threshold for the detectors. The rearward 2π steradian of each detector module has a passive shield, as described later.

The primary detector is a NaI(Tℓ) crystal 50.8 cm (20 in.) in diameter and 1.27 cm (0.5 in.) thick. This detector crystal is similar in design and size to those used in gamma-ray imaging cameras for medical diagnostic applications. It is planned to suspend the crystal and its window around the edge by a mechanical isolator. Scintillation light is collected by three 12.7 cm (5 in.) photomultiplier tubes (PMT's). The light collection technique is different from that usually used with crystal scintillation detectors. Instead of directly coupling the PMT to the crystal window, a large light-integrating collector housing is used (Figure 2). Measurements have shown that for a detector of

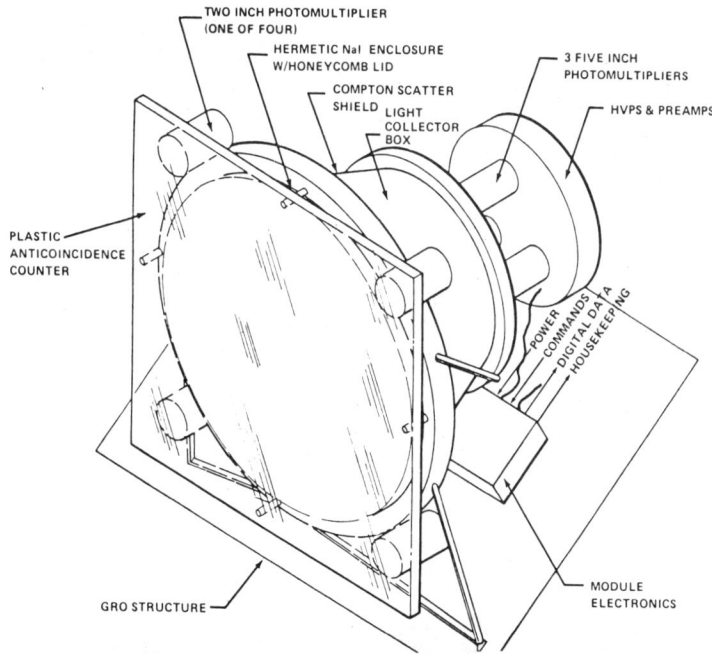

Fig. 2 Conceptual drawing of a BATSE module.

this type, the transfer variance provides a larger contribution to the detector resolution than does the variance of the total light collected due to statistical fluctuation at energies above ~ 100 keV.[1] Measurements have also been performed to optimize the geometry of the light collector housing for maximum resolution.[1] All internal surfaces of the housing are painted with a thick, barium sulfate white coating. The measured resolution in the laboratory, using a broad-beam Cd^{109} source, is 32 percent FWHM at 88 keV for a detector with a similar geometry. Anticipated improvements in the crystal packaging are expected to improve the energy resolution for the flight detectors. The photopeak at 22 keV is well separated from the noise.

The front of each module is covered by a 0.25 in. thick plastic scintillator detector. Pulses from this detector are normally used to anticoincidence a large part of the charged-particle-related background; however, pulses can be used in coincidence with the NaI(Tℓ) detector for calibration, using background single-charged, minimum ionizing particles. The sides and rear of each module will be covered by ~ 1 mm of a high-Z metal (probably a lead-tin combination) to reduce the number of Comptonscattered photons from a burst that reach the detector from the back.

In each module, detector pulses are processed in parallel by: (1) a high-speed, four-energy-channel discriminator capable of rates up to $\sim 10^6$ s^{-1} and (2) a pulse-amplitude-to-pulse-duration converter integrated circuit (LeCroy MQT-200) which has a resolution of greater than $10^3:1$. The former pulse data are used in the highest time resolution data types during a burst. The latter pulse data are used in all other data types since it contains considerably more energy information.

DATA SYSTEM

Pulses from all eight detector modules are routed to a central electronics unit where they undergo considerable on-board processing in order to reduce the high counting rates into the allocated 1500 bits-per-second data rate. This section briefly describes the data flow and various data types.

The pulse height data are initially converted into one of 128 energy channels. These 128 channels are quasi-logarithmic, providing an overall dynamic range of over 600:1 for the 128 channels. Several of the data types utilize 16 energy channels which are derived from the 128 channels through a programmable look-up table, which may be changed by command.

The gamma-ray pulses are sorted in parallel into different "data types." Each data type has a particular end use or scientific objective. The basic characteristics of the BATSE data types are given in Table 1. The sorting of gamma-ray pulses into data types is accomplished "off-line" by specially designed circuits. The formating of data and certain data parameters (e.g., time constants) are under the control of the central processing unit (CPU), which utilizes a Texas Instruments SPB-9900 microprocessor.

Table 1

BATSE DATA TYPES

	TYPE	DETECTORS	ENERGY CHANNELS	TIME RESOLUTION	PRIMARY USES
NORMAL	CONTINUOUS	8 SEPARATE	16	2 OR 4 SEC	GENERAL MONITORING; BURST LOCATION; DETECTION OF TRANSIENTS BY EARTH OCCULTATION
	SPECTRAL	8 SEPARATE	128	100 – 500 SEC *	ENERGY CALIBRATION; BACKGROUND OBSERVATIONS
	PULSAR	1 TO 4, COMBINED	16	64 PHASE BINS (SEE TEXT)	PULSAR PROFILES, CALIBRATION
BURST DATA	MEDIUM RESOLUTION	{ 8 SEPARATE[1] 1 TO 4, COMBINED[2] }	16	10 – 50 mSEC *	BURST SPECTRAL EVOLUTION ON SHORT TIME SCALES
	SPECTRAL	4 SEPARATE **	128	2 SEC *	BURST SPECTRAL EVOLUTION
	TIME–TO–SPILL	1 TO 4, COMBINED **	4	~ 10 μ SEC (SEE TEXT)	HIGH TEMPORAL RESOLUTION DURING BURST
	TIME–TAGGED	{ 8 SEPARATE[1] 4 SEPARATE[2], ** }	4	2 μ SEC	HIGHEST TIME RESOLUTION DURING INITIAL PHASE OF BURST; DETECTION OF RAPID PERIODICITIES
	PRE–BURST	8 SEPARATE	4	64 mSEC	WEAK PRECURSOR EMISSION PRIOR TO TRIGGER

* TYPICAL, UNDER CPU CONTROL
** FROM DETECTORS WITH HIGHEST RATES
1) PRIOR TO BURST TRIGGER
2) AFTER BURST TRIGGER

The continuous data type contains basic, 16-channel data from all eight detectors. During a burst or when other data types are not being transmitted, it is read out every 2 s. When other data are interleaved with the continuous data, it is read out every 4 s.

The spectral data consist of 128-channel spectra, described previously, from all eight detectors. During a burst it can be read into memory as often as every 2 s for 64 s. At other times, it will be used for calibration with the light emitting diode (LED) calibration system described later and to accumulate background gamma-ray spectra. The accumulation time is under CPU control. The pulsar data type is used to collect spectra from known periodic sources, primarily for calibration purposes. Data are sorted into 64 phase bins at a precise period for a specified number of sweeps. The period resolution is 2_μ s. Data are taken from a single detector or combined from up to four detectors in 16 energy channels.

Other data types are triggered by the onset of a burst and are accumulated rapidly into a solid-state mass memory ($\sim$ 2 M bits) for later playback into the allocated telemetry rate. The time-to-spill data type records the time to accumulate a fixed number of counts (programmable to 16, 64, or 256). In this way the time resolution improves (up to ~ 10 μs) during the intense portions of a burst. The highest time resolution data are derived at the onset of a burst, when each gamma-ray is time-tagged to the nearest 2 μs. In addition, 2 s of pre-trigger burst data are recorded with 64 ms resolution.

The burst trigger signal is derived by the CPU by monitoring the discriminator rates from all eight detectors on time scales of 64, 256 and 1024 ms. An increase in the counting rates from two or more detectors above a preselected trigger level is required before a burst signal is generated. Additional spectral or directional criteria may be added to the burst trigger algorithm to eliminate false triggers from other sources such as solar flares. Balloon flight experience is expected to help determine the best algorithm in order to increase the burst trigger sensitivity and minimize false triggers.

The BATSE burst trigger signal will be routed to the three other instruments on the GRO. Each of these instruments has wide-field detector components which can provide additional data on gamma-ray bursts. If a burst source happens to be in the primary field-of-view of the other experiments, sensitive burst data can be derived over the entire energy range of the GRO.

EXPECTED PERFORMANCE AND RESULTS

The sensitivity of the BATSE detector array to typical gamma-ray bursts will be $\sim 10^{-8}\sqrt{t}$ erg/cm^2, where t is the burst duration. However, this sensitivity is dependent upon the burst spectrum and time profile; significantly increased sensitivity is expected for bursts with soft spectra or rapidly varying emission. The large area of the detectors will produce counting rates approaching 10^6/s during the peaks of perhaps ~ 20 bursts per year. These high rates will allow statistically significant intensity variations on a time scale of tens of microseconds or spectral variations on a time scale of milliseconds to be observed. These measurements will put severe constraints on emission and cooling models of burst sources and on the models of source geometry and beaming characteristics. The time-tagged data will allow periodic emission to be detected from strong bursts on time scales as short as tens of microseconds. Such periodic emission may be expected from burst models involving oscillations of a neutron star.[2]

The use of multiple, anisotropic detectors on a single spacecraft to determine burst locations has been successfully accomplished by the KONUS experiment on the Venera 11 and Venera 12 spacecraft. The location accuracy, for that experiment has been stated as being from 1 to 5 degrees.[3] In principle, the BATSE/GRO experiment could achieve location accuracies better than 1 degree for bursts with intensities greater than 10^{-6} erg/cm^2 if the measurements were limited only by counting statistics. In reality, it is expected that systematic uncertainties will dominate the statistical uncertainties in deriving locations for the stronger bursts. The primary sources of systematic error are: (1) unequal and/or uncertain response of different detector modules, (2) scattering of burst photons by spacecraft components and the Earth's atmosphere into detectors, and (3) uncertainties in detector alignment and aspect.

The absolute energy response of each detector will be determined using background spectral lines and features. The relative response of different detectors will be determined by an LED/fiber optics calibration system. Light pulses from a single, central LED will be routed to all eight detector modules by thin fiber optics cables. The gain of all eight detectors can thus be compared to each other. The amplitude and rate of LED pulses will be controled by the central processing unit. An attempt will be made to minimize detector variations due to temperature fluctuations and magnetic field variations.

Compton-scattered burst radiation will be a major source of location uncertainty. As mentioned earlier, most of the rear surface of each module will be protected by a high-Z graded shield. However, residual radiation penetrating the shield, coming through unshielded areas or scattering from components in the forward 2π steradian field-of-view will contribute to the indirect burst signal. This cannot easily be modeled prior to flight, therefore, a post-integration spacecraft radiation scattering angular survey will be made using radioactive sources in order to accurately estimate these contributions. The scattering of burst radiation from the Earth's atmosphere could, in principle, be modeled since the geometry and composition of the scattering medium is relatively simple and well known.

The alignment of the detectors with respect to each other, the spacecraft, and inertial space will each be determined to 0.1 degree through preflight alignment and calibration and the GRO aspect system. There are several additional opportunities for on-orbit calibration of the detectors responses. Solar flares will provide a source of hard X-rays from a known location, although the majority of flares have very soft spectra. The Crab pulsar, NP0532, is a steady, strong source of hard X-rays that is detectable in the on-board pulsar data mode.[4] Strong bursts, locatable with an interplanetary timing network, as described later, will provide location calibration perhaps 20 to 30 times per year.

If the corrections and calibrations are successfully applied, hundreds of bursts during the 2-year mission will be located from 1- to 10- degree accuracy, depending on their strength, spectrum, and profile. While not good enough for unambiguous optical, X-ray or radio identification, a celestial map with this many sources would show the galactic distribution of sources and/or the association of certain classes of bursts with nearby galaxies or clusters of galaxies. For example, if the March 5, 1979, gamma-ray transient source is assumed to be in the Large Magellanic Cloud (LMC), then this event could be seen as far away as the Virgo Cluster with the BATSE/GRO instrument. Events of this type may be detectable every few days, assuming a frequency of once per 10 years per galactic mass. Large numbers of these types of events would be associated with the Virgo Cluster or other rich, nearby clusters.

The BATSE experiment will become an important component in the interplanetary burst timing network in the late 1980's. The original burst network, which successfully operated for several years, is now defunct.[5] During its operation, it demonstrated the capability for deriving arc-minute burst locations for strong bursts with rapid time variations. Although most locations showed empty optical and radio fields, the unique event of March March 5, 1979, has been identified with the supernova remnant N49 in the Large Magellanic Cloud.

In the time frame of the GRO, other instruments of the network will include detectors on the Solar Polar Mission and on Soviet-French interplanetary space probes. All detectors are expected to have absolute timing accuracies of 16 ms or better, providing arc-second locations for strong, rapid bursts. It is hoped that mission operations would allow burst locations to be derived on a much shorter time scale than has previously been possible so that short-lived accompanying optical, X-ray or radio emission could be detected.

Sources with an intensity greater than ~ 0.3 Crab will be detectable as they are occulted by, and re-emerge from behind, the Earth's atmosphere near the horizon. The change in counting rate will be strongest in the detectors that are most nearly facing the approaching and receding horizon. Increased sensitivity may be achieved by summing data from successive orbits. The transition time for a source to go from 10 percent attenuation to 90 percent attenuation at 100 keV is ~ 8 s for a source in the orbital plane. For sources out of the orbital plane, the transition time is longer; sources within 19 degrees of orbital poles will not be occulted. This includes only ~ 6 percent of the sky. The unocculted region continuously changes, as the orbital plane precesses ~ 7 degrees per day in the inertial frame.

The time of occultation will provide an accurate measure of the source location since the spacecraft location and the horizon location will be well known. The location accuracy is largely independent of the spacecraft or module attitude but will depend on the intensity of a source. For a source with the spectrum and intensity of the Crab, a location accuracy of ~ 0.1 degree is expected.

The detection of long-lived transient sources with intensities greater than ~ 0.3 Crab implies that all "steady" sources above this intensity will be continuously monitored. This will include ∼10 sources in the energy range 26 to 122 keV as determined from the recent Ariel-5 survey,[6] including the confused region near the galactic center. This continuous monitoring may itself provide valuable data on the long time scale variations of these stronger sources. It may also be possible to deconvolve a sky map of selected regions in the BATSE energy range by utilizing large segments of data and orbital precession.

It is difficult to say how many of these long-lived transient sources may be detected during the mission since a wide-field sky monitor in the BATSE energy range has never been flown. A rough estimate of the number of observable galactic sources similar to A0535+26 may be made. That source is seen to flare about twice per year at a level of ∼2xCrab. Assuming a distance to A0535+26 of 1 kpc[7], BATSE would be able to detect about seven of these sources in the galactic plane. Active galaxies such as Cen A also undergo active periods in the hard X-ray and gamma-ray regions. The early detection of these sources would be valuable for detailed study by more sensitive instruments at other wavelengths on GRO, Spacelab, the X-Ray Timing Explorer, as well as ground-based instruments.

REFERENCES

1. C. A. Meegan, "Performance of Photomultiplier Tubes and Sodium Iodide Detector Systems," NASA TM-82406, March 1981.

2. R. Ramaty, S. Bonazzola, T. L. Cline, D. Kazanas, P. Meszaros and R. E. Lingenfelder, Nature, $\underline{287}$, 122 (1980).

3. E. P. Mazets, et al., "Cosmic Gamma-Ray Burst Spectroscopy," A. F. Ioffe Physical-Technical Institute, Leningrad, Preprint No. 719 (1981).

4. R. B. Wilson, G. J. Fishman and C. A. Meegan, these proceedings.

5. T. L. Cline, et al., Ap. J., $\underline{246}$, L133 (1981).

6. M. J. Coe, et al., "High Energy X-Ray Observations from Ariel-5," M.N.R.A.S., in press (1981).

7. L. Maraschi, et al., Nature, $\underline{263}$, 34 (1976).

MODULATED MULTIPLE SLIT CAMERA FOR IMPROVED LOCALIZATION OF GAMMA RAY BURSTS*

Paul Gorenstein
Harvard/Smithsonian Center for Astrophysics
Cambridge, MA 01238

ABSTRACT

The addition of a modulation collimator technique to a one-dimensional coded aperture imaging system known as the Multiple Slit Camera (MSC) can improve the location precision for gamma ray bursts and slower hard X-ray transients to about a tenth of a square arcminute. Wires are wound coaxially around the random slit collimator and imaging proportional counter of one half of a MSC. The ratio of counts between the modulated and unmodulated halves provides an array of possible fine positions. The ambiguity is reduced to one or a few fine positions by a coarse position from another MSC orthogonal to the first. The addition of the modulation collimator reduces the system sensitivity only slightly. The fine position capability operates over about two-thirds of the nearly all-sky field of view of the MSC. This system offers excellent sensitivity and provides gamma burst positions from a single spacecraft that are as precise as the best of those from the long baseline network of interplanetary spacecraft.

INTRODUCTION

Precise localization remains one of the principal objectives for investigations of gamma ray burst events. Accurate positions that permit the search for low intensity persistent optical, radio, and X-ray counterparts will allow the burst sources to be studied by astronomers in other wavelength ranges. These studies may result in information that is critical for understanding the physical conditions that lead to the gamma ray burst phenomena. A few precise positions have been obtained from the interplanetary network of gamma ray detectors that has been operating in recent years. The most intense and unusual event detected to date has been precisely located by the network and identified with the supernova N49 in the Large Magellanic Cloud (Cline et al. 1980, Evans et al. 1980).[1,2] Because of its interesting consequences, this identification has restimulated a great deal of interest in the study of gamma ray bursts. The implications of the N49 association have been indirectly one of the principal reasons why this workshop was organized. The search for counterparts to other bursts with good positions is still in progress; communications presented at recent meetings suggest that possible radio and an X-ray counterparts are being investigated.

It is generally agreed that any future dedicated mission to study gamma ray bursts must include a means of obtaining precise locations. In this paper, an instrument is described that has the capacity of obtaining positions from a single spacecraft in near

*Work supported by NASA grant NSG-5138.

Earth orbit that are as precise as the best of those derived from the long baseline network. In addition, the instrument has excellent sensitivity, a large field of view, and operates equally well on transients of any time scale including steady sources. The instrument is a new modification of the coded aperture multiple slit camera system described by Gorenstein, Helmken, and Gursky (1976)[3]. That device already had position determination capability of the order of an arcminute. This now becomes the coarse position for the new system. The new modification consists of adding a cylindrical modulation collimator to one-half of the detector. An array of fine positions is obtained whose precision is of the order of 10"-20". These include the correct fine position, as well as a large number of extraneous ones. The superposition of the coarse position with the fine position array eliminates all or nearly all of the extraneous ones. The modulation collimation action has no effect upon the multiple slit camera action and reduces the sensitivity of the system only slightly.

THE MULTIPLE SLIT CAMERA

In designing a system for detecting gamma ray bursts, it is important to recognize that the best ratio of signal to noise, i.e. source counts to background, exists in the hard X-ray region, 20-100 keV. The typical burst photon number spectrum is: $dN/dE \sim E^{-1} \text{Exp}(-E (keV)/500)$. Even though the spectrum may continue down to the 2-10 keV X-ray region, the contribution of the cosmic X-ray sources to background is more severe than at higher energies. Galactic X-ray sources typical spectra are thermal bremsstrahlung with kT in the range 2-20 keV. The diffuse X-ray background itself can be characterized by a kT of 40 keV. In the 200-1000 keV or gamma ray range, there are fewer burst photons, and they are difficult to collimate. Thus, while the transients have been historically labeled "gamma ray bursts" and while they do indeed have considerable intensity and interesting spectra in the gamma ray region, it appears that the hard X-ray band should be the one to be addressed for purposes of detection and position determination.

A coded aperture system with good efficiency in the hard X-ray band was suggested by Gorenstein, Helmken, and Gursky (1976)[3]. A cylindrical random slit collimator that is 35% open forms one dimensional images in the hard X-ray band. Four parts of Figure 1 illustrate how it operates. Figure 1a is a simplified version of the multiple pinhole camera suggested by Dicke in 1968. In this case, the holes are cross sections of slits that extend in a direction perpendicular to the figure. Two openings are shown. Each is assumed to admit one photon from the source direction and another from a random direction. A position sensitive detector determines the X coordinate of each photon. A method for determining the source direction is illustrated in Figure 1b. One plots the number of photons, $N(\theta)$, versus direction, θ, by assigning every possible arrival direction equal a priori probability. Thus, there are two counts associated with θ_1, the

455

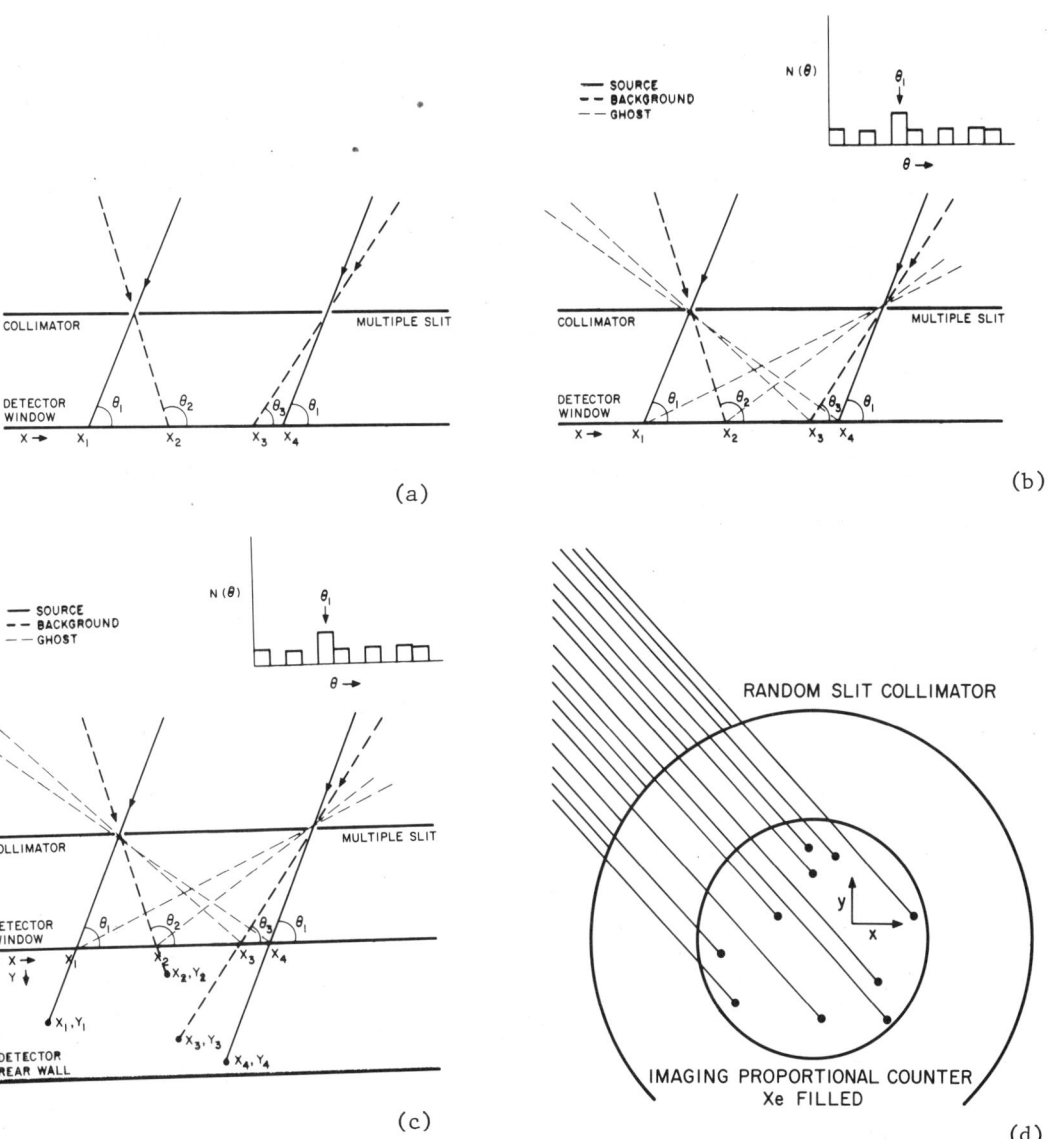

Figure 1. Explanation of Multiple Slit Camera. Part (a) is cross section of a random slit camera with two slit openings. Each slit admits one source and one background photon. Their directions are determined by a position sensitive detector. In (b) all possible arrival directions are plotted, including 4 "GHOST" directions obtained by associating a photon with the incorrect slit. The true source direction is the only one with two counts. In Figure 1(c) this method is applied to penetrating hard X-rays in a two-dimensional position sensitive proportional counter. A cylindrical geometry that provides a broad field of view is shown in 1(d).

true source direction. The two background photons are not from the same direction. In addition, there are four "ghost" directions which correspond to assigning a photon to an opening through which it did not pass. Given the random nature of the background and ghost directions, only the true source direction, θ_1, has two counts. An alternative method of determining the source direction is to cross correlate the photon positions in the detector with the locations of the slit openings in the collimator. The peak of the correlation function corresponds to the source direction.

Figure 1c illustrates how this method can be applied effectively in the hard X-ray band where X-rays can penetrate to considerable depth in a position sensitive proportional counter. We consider a two dimensional position sensitive proportional counter, such as the Imaging Proportional Counter (IPC) of the Einstein Observatory. Positions at which photons are absorbed are measured in the (X,Y) plane of the figure. By measuring Y as well as X, the information on the angular direction is preserved no matter how deeply into the detector the photon penetrates. Using xenon gas and making the pressure-depth factor some 30 atmosphere centimeters, it is possible to achieve a high average quantum efficiency in the 20-100 keV range. Primary electrons produced by the photoelectric absorption of the X-rays drift along an electric field perpendicular to the (X,Y) plane. Their position is read out when the electrons arrive at the anode-cathode planes. Drift regions on both sides of the anode-cathode planes constitute the active volume of the detector. To achieve a broad field of view, the system is made in a cylindrical geometry as shown in Figure 1d. A source can be incident along any azimuthal direction. It can have a significant directional component in Z (perpendicular to the (X,Y) plane) without affecting the position determination. Each detector produces a measurement of the position in one dimension, i.e. along one half of a great circle. Two units oriented orthogonally to each other provide the precise two dimensional celestial position. In practice, the field of view of a two unit detector system is less than all sky. Attachment to the spacecraft will occlude some of the field of view. When the source direction has a very large Z component, the projected area of the system is small. A four unit system will provide all sky coverage with some redundancy.

In the years since this system was first described, significant progress has been achieved in demonstrating the feasibility of imaging proportional counters. Good performance of a sealed, pressurized xenon IPC with an extended drift region was described by Gorenstein et al. (1979).[4] Doi and Fujii (1978)[5] have described a system for background rejection by risetime discrimination in a pressurized xenon proportional counter. A very effective but simple method of position readout for a xenon IPC has been described by Gilvin, Mathieson, and Smith (1981)[6] and Sims, Thomas, and Turner (1981).[7] The long term behavior of the first position sensitive proportional counter in orbit, the Einstein Observatory/HEAO-2 IPC, has been very favorable with rather low background rates (Gorenstein, Harnden, and Fabricant 1981).[8]

Typical parameters for a Multiple Slit Camera are the following:

```
radius of collimator:              60 cm
length of collimator:              100 cm
radius of drift field region:      10 cm
length of drift field region:      25 cm (each of two sides)
width of slits:                    1 mm
spatial resolution of detector:    1 mm
```

When the position of a burst is analyzed: $\Delta\theta_1$, the full width at half maximum of the peak, will be equal to:

$$\Delta\theta_1 = 2\text{mm}/600\text{mm} = 10'$$

The location precision, or the statistical error, in the determination of the centroid of the peak is approximately equal to:

$$10'/(N\sigma)_1 - 2$$

where $(N\sigma)_1$ is the number of standard deviations in the difference between the source counts in the peak and background. For most of the bursts detected, the number of background counts, B, is much larger than the number of source counts, S.

$$(N\sigma)_1 \approx S/\sqrt{B}$$

It is reasonable to be able to obtain locations down to one-tenth of the FWHM or about 1'. For events of higher statistical significance, systematic errors will probably be dominant. In the new arrangement to be described below, the error box derived from the Multiple Slit Camera becomes the "coarse position".

THE MODULATION FEATURE

It has been demonstrated that the counterparts of gamma ray bursts at optical, radio, and X-ray wavelengths are faint. Thus, it is desirable to obtain as high a location precision as possible to search for the counterparts. The coupling of a modulation collimator to the Multiple Slit Camera results in a method for refining the location precision by about a factor of three in each dimension. The modulation collimator was developed for X-ray astronomy investigations by M. Oda, now at the Institute for Space and Aeronautical Sciences, Tokyo. It was used successfully in a rocket experiment for determining the location of Sco X-1 (Gursky et al. 1966)[9] which lead to the identification of its optical counterpart (Sandage et al. 1966).[10] Various modulation collimators, including scanning and rotating versions, have produced important results from satellites, such as SAS-3, HEAO-1, and Hakucho.

The most elementary form of the modulation collimator consists of two parallel grid planes of wires where the center to center spacing is equal to twice the wire diameter. Figure 2 illustrates the transmission of the two grids when radiation is exactly perpendicular to the planes and when it is inclined from the perpendicular by an angle equal to the wire diameter divided by the distance between the two grids. As the angle varies, the transmission modulates between 50% and 0 as illustrated in Figure 3. (At large angles from the normal, the maximum transmission is less than 50% because the roundness of the wires reduces the projected area.)

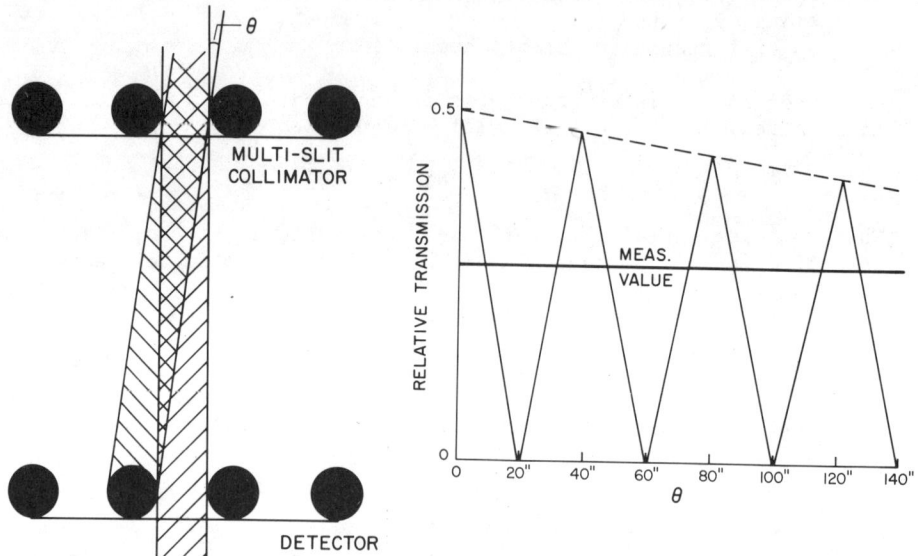

Figure 2. Two grid modulation collimator. One wire is wound around multiple slit collimator, the other around the drift region of the detector. Normally, incident radiation is 50% transmitted. Radiation inclined from the normal is totally extinguished.

Figure 3. Transmission of modulation collimator as a function of θ, the angle to the normal. Measurement of the transmission determines a quasi-periodic series of values for θ.

The modulation collimator technique can be applied to the Multiple Slit Camera by winding a wire around the outside of the collimator and around the drift field volume of the detector. Wires cross the slits at right angles. The wires are applied to only one half of the Multiple Slit Camera or one of the two drift field volumes. This arrangement is shown in Figure 4. Two separate sets of anode-cathode planes could be used to distinguish the halves. In practice, signals in two halves of a single detector are distinguishable from each other. The cathode, which is closer to the drift region where the hard X-ray is absorbed, experiences an ion pulse at some time after both cathodes receive the induced signals which are used to determine the (X,Y) positions. Thus, there is a single bit of information in the Z direction indicating whether the photon was absorbed at a positive or negative value of

Z. (The cost of saving a set of anode-cathode planes is accepting some "crosstalk" which is discussed below.) A cross-sectional view is shown in Figure 5. Radiation incident in a direction perpendicular to the axis of the cylinder will be transmitted through the modulated half of the detector at 50% of the rate through the unmodulated half. As the incident angle varies, the ratio of the number of counts of the two halves of the detector modulates according to the manner illustrated in Figure 3. Thus, measurement of the count ratio limits the angle to a series of possible values. The series is quasi-periodic with two possible values occurring within a period. In practice, the modulation is not perfectly triangular because the wires are not actually parallel planes. A cross-sectional view of the (X,Y) plane would show that the opposing wires are actually concentric circles of different radii. Thus, the distance between wires varies slightly across the area intercepted by source direction. It is minimum along the common radial direction of the detector and multiple slit collimator and maximum at the extremes where the source photons are tangent to the detector edge. Keeping in mind that the detector is position sensitive in this plane, we can exclude those photons that are absorbed too close to the edge of the detector and keep the triangular response within acceptable bounds.

The series of angles represent a set of possible source positions all perpendicular to the unambiguous great circle of position that is determined by the multiple slit collimator. They represent a "fine" position for another Multiple Slit Camera that is orthogonal to the first one. Because the modulation wires are orthogonal to the slits, they have no effect upon the position determination action of the Multiple Slit Camera. It is shown below that they reduce its sensitivity by only a very small amount.

The wires would be 4 mil (.1mm) diameter tungsten, a standard stock item. At 40 keV, the photoelectric absorption of this material is two mean paths through the center. An approximate picture of the effect of the wires is that photons in the 20-40 keV range or about half of the total detected from a typical burst, are modulated. Those in the range 50-100 keV are essentially unaffected by the wires. Tungsten fluorescence radiation is not a problem; it is 8-10 keV and is absorbed in the walls of the detector. As the two halves of each detector are identical (except for the modulating wires) systematic errors in the count ratio due to differences in efficiency should be small. The typical parameters for the modulation collimator are the following:

    wire diameter: .1mm
    center to center wire spacing: .2mm
    separation between the two windings: 60cm - 10cm = 50cm

One half of the modulation period, $\Delta\theta_2$, is equal to

$$\Delta\theta_2 = .1mm/500mm = 40"$$

The fine position is the phase determined by the ratio of the counts between the modulated and unmodulated halves of the detector.

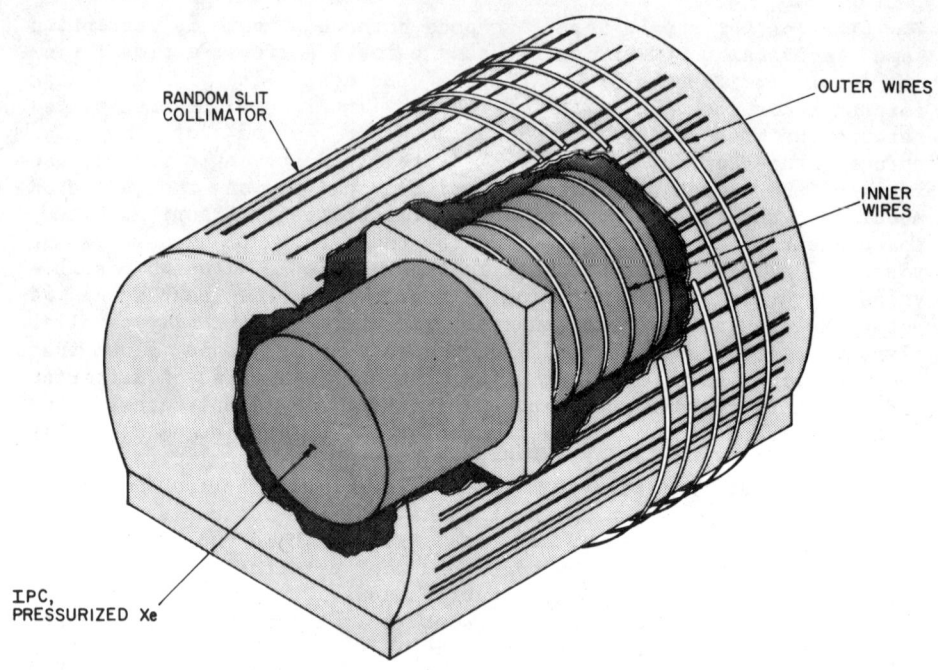

Figure 4. Multiple Slit Camera with modulation windings on one side is shown.

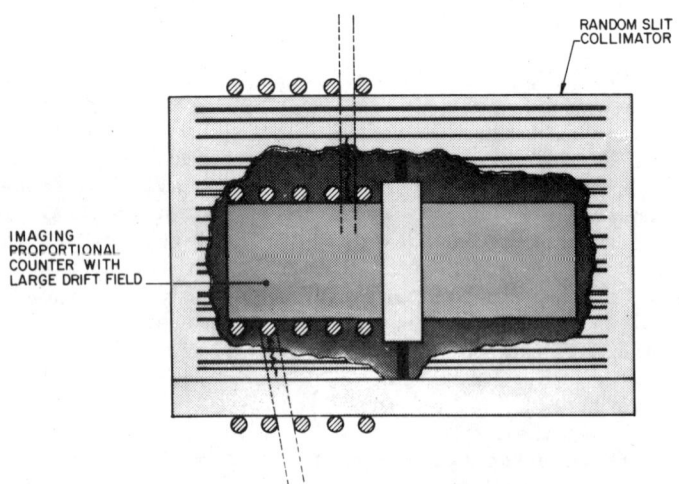

Figure 5. Cross sectional view of Modulated Multiple Slit Camera. Dotted lines indicate directions where the modulated flux is 50% and 0% of that of the unmodulated flux.

The fine location precision is then some fraction of the 40" period which depends on the strength of the burst. Within one modulation period, two phases will give the same ratio, the positive and negative slope of the triangle. Thus, there are two possible fine positions within a modulation period.

Assuming that N counts from a source are incident upon an unmodulated detector, one can calculate the number that are available for the coarse position measurement and the number for the fine position.

$$
\text{N unmodulated detector} \rightarrow \begin{cases} N/2 \text{ unmodulated half} \\ N/2 \text{ modulated half} \rightarrow \begin{cases} N/4 \quad E>50 \text{ keV, unmodulated wires transparent} \\ N/4 \quad E<40 \text{ keV, modulated} \rightarrow 25\% \text{ transmission average} \rightarrow N/16 \end{cases} \end{cases}
$$

The total number of counts detected is equal to $N/2 + N/4 + N/16 = 13N/16$. All of these are available for the measurement of coarse position. $N/4$ of the counts are subject to modulation. The number that appears in the 20-40 keV band of the modulated half of the detector will vary between 0 and $N/8$, with an average of $N/16$. Deriving the phase angle from the count ratio does require knowledge of the burst's spectrum. The IPC provides some energy resolution through the pulse amplitude of its signal. The energy resolution is typically 10% for a 25 keV signal, but there is a complication to the energy resolution because of the escaping fluorescence radiation (29.5 keV) when photons exceed the 34.6 keV K edge of xenon. This makes it difficult to achieve a clean a priori separation of the 20-40 keV and > 50 keV bands on the basis of raw pulse height information alone. Knowledge of the source's spectrum is needed to interpret the count ratio in terms of the phase angle. Sufficient spectral information is obtained from the analysis of the pulse height distribution of the IPC itself which considers escaping fluorescence radiation. Alternatively, the Modulated Multiple Slit Camera system is likely to be present in an instrument complement that includes a detector with good energy resolution.

The modulation collimator action is relatively free from systematic errors. The same anode can be used for both halves of the detector. In that case, differences in gain cannot affect the count ratio. Separate detectors might be preferable to completely eliminate "cross talk" that occurs for sources incident at low angles. In that case, a photon might enter the collimator through the unmodulated half of the detector and be absorbed in the other. This can be accounted for in the analysis as the coarse position is highly accurate for this purpose. Nevertheless, if two anodes (with a partition) are used, it is a simple matter to equalize their gain to the level required. It is conceivable that along any given source direction that one half of the detector could be shaded more than the other or that there is back scattering with peculiar structure from either the spacecraft or the atmosphere. (This is an intrinsic problem for all detector systems that measure locations by count ratios.) However, it should be remembered that the source position is already known to a high degree of accuracy from the multiple slit coarse position. Systematic differences in shading or back scattering geometry are unlikely to vary on a scale of an arcminute. The continual observation of sources with highly accurately known positions, such as the Crab Nebula, Cyg X-1, or Cyg X-3, serves as a calibration of the systematic differences. Observation of these sources at several orientations (which is guaranteed to occur in a mission of long duration) will calibrate out all of the systematic differences in geometry between the two halves of the detector system. The field of view of the modulation collimator is about two thirds that of the Multiple Slit Camera. The roundness of the wires limits the field of view to about ± 60° inclination from the axis of the cylinder.

EXPECTED PERFORMANCE

Detailed simulations of the detector performance have not been carried out. However, it is possible to describe the performance in an approximate manner on the basis of count estimates. We consider a gamma ray burst with a spectral shape similar to the Apollo 16 event (Metzger et al. 1974),[11] but at 1/30 of the intensity. On the Vela scale its integrated intensity would be 10^{-6} ergs/cm^2. About 60 bursts at least as intense as that occur on the average each year in a 2π region of sky. We assume that the flux comes in a total time interval of seven seconds. The parameter of the system are:

        detector radius: 10cm
        depth of drift region: 25cm
          (each half)
        multiple slit collimator: 0.35
          transmission
        pressure of xenon: 1.5 atm

Thus, the effective area of the system (for normal incidence to the cylinder axis) is equal to 2 x 20cm x 25cm x 0.35 = 350cm^2.

For each detector, we expect N = 3400 source counts (without modulation collimator) and the number of background counts, B, in 7 seconds is 18,200. As the background is principally diffuse X-rays, we can assume it is attenuated by the collimator in a manner similar to the source. The number of source and background counts that enter into the coarse position measurement are:

13/16N = 2760

13/16B = 15,000 (7 sec)

$(N\sigma)_1$ = 2760/$\sqrt{17,760}$ = 20

Location Precision = 10′/(20-2) = 0.63′
(coarse position)

Systematic errors may not allow division of the 10′ cell by more than a factor of 10, limiting the coarse position precision to 1′ in practice.

For the fine position measurement, we calculate the ratio of the counts in the modulated half (S_1) of the detector to that of the unmodulated half (S_2). We take the modulated side to be at its average transmission of 25%.

modulated half (20-40 keV)	unmodulated half (20-40 keV)
1/16N = 212	1/4N = 850
1/16B = 1138 (7 sec)	1/4B = 4700
S_1 = 212 ± $\sqrt{1350}$	S_2 = 850 ± $\sqrt{5550}$

$R = S_1/S_2$, which determines the phase within the 40″ modulation period. There are two phases which correspond to this ratio within a period.

$$\Delta R/R = \sqrt{\left(\frac{37}{212}\right)^2 + \left(\frac{75}{850}\right)^2} = 0.194$$

Location Precision = 40″ (0.194) ≈ 8″ (2 positions)
(fine position)

The superposition of the coarse position with the array of fine positions (as shown in Fig. 6) produces a final result consisting of two error boxes; each is 16″ x 16″ giving a total area of 0.15 (arcmin)^2. We obtain this precision or better for 40 bursts per year.

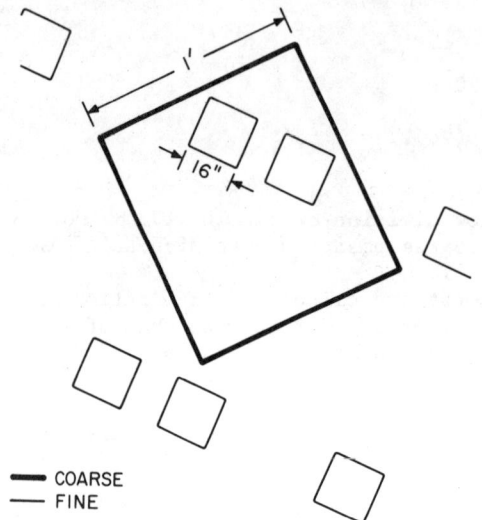

Figure 6. Position determination by Modulated Multiple Slit Camera. The "coarse" position is derived from the coded aperture imaging action of the random multiple slits and the two dimensional position sensitive proportional counter. The "fine" position array is derived from the modulation ratio.

SUMMARY AND CONCLUSIONS

The discussion in this paper attempts to demonstrate that the Modulated Multiple Slit Camera system can provide celestial positions that are as fully accurate as the best positions obtainable from a long baseline network of detectors aboard interplanetary spacecraft. The technique offers the following advantages in a program of gamma ray burst investigations.

(1) It is completely self-contained in a single spacecraft.
(2) It can be carried out in near Earth orbit aboard a simple satellite or platform.
(3) The coded aperture approach yields true one-dimensional iamges which give it immunity from bursts of diffuse radiation associated with the Earth's magnetosphere.
(4) The sensitivity is very high; many bursts per year can be detected and localized to good precision. Faint bursts may include different types of objects, as well as more distant examples of those already detected.
(5) The field of view is very broad; essentially all sky for "coarse" positions of ~1´ and most of the sky for the fine positions.
(6) The system functions for transients in all time domains. Steady sources serve useful calibration functions. Detection and localization are decoupled from the requirement for fast time variations in the source. We can investigate and localize slow transients as well as rapid bursts.
(7) Broad temporal coverage is provided. It is possible to look for precursor activity, repeat bursts or transients, or measure

the steady flux from any region inte sky at any time. Thus, this technique has a good duty cycle because useful measurements can be carried out between bursts or transients.

Because the system has such a broad field of view, the pointing requirements are not stringent. Solar or Earth pointing is adequate. We require only that either the celestial altitude not change significantly during a transient or that it change in a manner that is known accurately. However, in order to make full use of all of the capabilities of the Modulated Multiple Slit Camera system, ample telemetry will be needed.

Given that the Modulated Multiple Slit Camera system has the capabilities listed above, it would be, along with spectrometers, an essential component of any future instrument complement that is developed for a comprehensive study of gamma ray bursts and transients. The instrumentation can be aboard a single spacecraft or platform in near Earth orbit.

REFERENCES

1. T.L. Cline, et al., Ap.J. (Letters) 237, L1 (1980).
2. W.D. Evans, et al., Ap.J. (Letters) 237, L7 (1980).
3. P. Gorenstein, H. Helmken, and H. Gursky, Astrophys. and Space Sci. 42, 89 (1976).
4. P. Gorenstein, D. Perlman, D. Parsignault, and R. Burns, I.E.E.E. Trans. on Nuc. Sci. NS-26, 502 (1979).
5. K. Doi and M. Fujii, Nuc. Instr. and Meth. 155, 305 (1978).
6. P.J. Gilvin, E. Mathieson, and G.C. Smith, I.E.E.E. Trans. on Nuc. Sci. NS-28, 835 (1981).
7. M.R. Sims, H.D. Thomas, and M.J.L. Turner, I.E.E.E. Trans. on Nuc. Sci. NS-28, 825 (1981).
8. P. Gorenstein, F.R. Harnden, Jr., and D.G. Fabricant, I.E.E.E. Trans. on Nuc. Sci. NS-28, 869 (1981).
9. H. Gursky, R. Giacconi, P. Gorenstein, J.R. Waters, M. Oda, H. Bradt, G. Garmire, and B.V. Sreekantan, Ap.J. (Letters) 146, L311 (1966).
10. A.R. Sandage, et al., Ap.J. (Letters) 146, L317 (1966).
11. A.E. Metzger, R.H. Parker, D. Gilman, L.E. Peterson, and J.I. Trombka, Ap.J. (Letters) 494, L19 (1974).
12. S.E. Woosley and R.K. Wallace, Lick Observatory preprint (1981).

APPENDIX
Measurement of X-ray Afterglow With Multiple Slit Camera System

The thermonuclear model for γ-ray bursts by Woosley and Wallace (1981)[12] predicts that a γ-ray burst will be followed by X-ray emission for a period of about 100 seconds. The X-ray emission may be characterized by interesting temporal structure. The total integrated flux in the X-ray region is expected to be 10^{-2} that of the gamma ray burst itself. Burst observations by the Vela satellites reported by Klebesadel at this workshop suggest that an X-ray afterglow is indeed present. The best method for studying the X-ray afterglow would seem to be to point an imaging telescope or a narrow field of view X-ray detector directly at the burst direction. To achieve high sensitivity, it is important to exclude diffuse background and the strong galactic sources. Logistically, it is difficult to determine a burst direction, analyze aspect information, and point a spacecraft or a scanning platform towards the burst all within a few seconds. The Multiple Slit Camera system offers an alternative, a simple and reliable means of pointing a reasonably large area X-ray detector at a burst. It utilizes the same instrument that detects the burst and thus does not require another detector to be in the instrument package. The only limitation is that the X-rays should be above 5 keV to be transmitted through the detector window.

The xenon IPC used in the multiple slit camera has a high intrinsic efficiency for X-rays above 5 keV. In fact, for the detection of gamma ray bursts, the X-ray efficiency <20 keV is deliberately suppressed by absorbing walls to enhance signal to noise by reducing background from cosmic X-ray sources. If instead the walls themselves were transmissive above 5 keV, an external, thin, aluminum cylinder concentric with the drift field volume could be used to absorb X-rays. If the cylinder were open along part of its circumferential area, the system could acquire X-ray sensitivity along any direction by a simple rotation. Figure 7 illustrates this method. In the normal condition, the open portion of the aluminum absorber is pointed downward (Fig. 7a). When a burst occurs and its azimuthal angular direction is determined by the Multiple Slot Camera, the aluminum cylinder is rotated and pointed such that the opening is centered on the burst direction. The open area, which is about 60° of the cylinder, contains an "egg-crate" collimator which limits the field of view to about 1°. The field of view in the other direction could either be left at ~140° or for greater sensitivity it would be limited to 10° by movable vanes. These vanes are conditioned by the other detector that measures positions perpendicular to the first. While instantaneous determination of position and movement are still necessary, these operations are now comparatively simple. The source direction is a coordinate that is totally internal to the instrument; no aspect or other information is required. An on-board micro-processor can easily compute the azimuthal angle. Pointing at the source requires only the simplest and most reliable of all possible mechanical movements, the rotation of a light weight, dynamically balanced cylinder about its own axis (and the rotation of light vanes within the collimator along a perpendicular direction). This can be accomplished in less than a second. The short delay between detecting the burst and rotating the X-ray collimator to point at it makes the sensitivity of this detector effectively equal to that of a much larger area instrument which takes a longer time to acquire the source. Figure 7b shows the collimator rotated towards the burst direction. X-rays from the afterglow do not penetrate as deeply into the detector volume as the burst photons, but the two dimensional position sensitivity is maintained. The multiple slit collimator action is still operative during the afterglow so that there is angular resolution within the field of view to measure background simultaneously, thus providing a good determination of the afterglow intensity as a function of time.

Assuming that 1% of the burst energy appears as an afterglow of 100 seconds duration in the 5-10 keV band, we can estimate the detection sensitivity. For the sample event of 10^{-6} ergs/cm^2 considered earlier, there will be 10^{-8} ergs/cm^2 in the afterglow, or about 1 photon/cm^2. The burst will be detected in at least two detectors. Taking the transmissions of the modulation collimator (on one side) and the multiple slit collimator into account, the total effective area for radiation normal to the axis is equal to:

$$2 \times 20 \text{ cm} \times 25 \text{ cm} \times 0.35 (1 + 1/4) = 438 \text{ cm}^2.$$

Assuming a field of view of 10 square degrees, the 5-10 keV background (mostly diffuse X-rays) is 8×10^{-3} photons/cm^2-sec.

Total number of counts in X-ray afterglow = 438 x 1 = 438
Total number of background counts = 438 x 8 x 10^{-3} x 100 = 350

For purposes of detection, the significance of the afterglow is $438/\sqrt{350}$ or 23σ. This is sufficient to allow study of temporal structure. As described earlier, the sample event considered is not particularly intense. There are 60 per year/2π that are more intense. In fact, the most productive study of the X-ray afterglow will come from the many more intense events that occur each year.

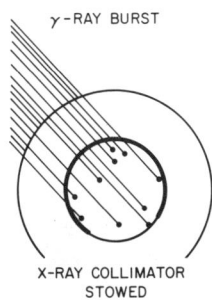
X-RAY COLLIMATOR STOWED

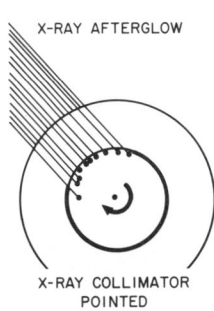
X-RAY COLLIMATOR POINTED

Figure 7. Detection of X-ray afterglow. Quiescent X-ray sensitivity is suppressed by a rotatable cylinder surrounding the drift field volumes. The cylinder contains a collimated open area that is normally pointed downwards. When a burst direction is determined, the open area is rotated to observe the X-ray afterglow.

A THIRD-GENERATION SMALL SPECTROSCOPY EXPERIMENT
FOR HARD TRANSIENT EVENTS

R.W. Klebesadel, W.D. Evans, and J.G. Laros
Los Alamos Scientific Laboratory

G.H. Nakano, D.W. Datlowe, and W.L. Imhof
Lockheed Palo Alto Research Laboratory

H.S. Hudson
University of California, San Diego

ABSTRACT

We describe an experiment for monitoring high-energy transient events (X-ray and γ-ray bursts from both solar and stellar sources) as proposed for the OPEN (Origin of Plasmas in the Earth's Magnetosphere) mission. This experiment contains Si(Li) detectors, a high-purity germanium detector, and a bismuth germanate scintillation counter for high-energy response. Cooling for the solid-state detectors is provided by a passive radiator. The instrument gives broad spectral response with high energy resolution; its mass of 15 kg and telemetry requirement of 200 bps impose only modest demands upon spacecraft resources. The use of a passive cooler does also place a constraint upon the orbit of the satellite; deep space would be the preferred location but other orbits may also be suitable. The ideas embodied in the design may be of interest in the design of other spectroscopy measurements from deep space, as for example for participants in a future triangulation network for transient events.

INTRODUCTION

The earliest observations of cosmic gamma-ray bursts[1,2] were made accidentally with instrumentation designed for other purposes. Confirmatory observations with other "first generation" instrumentation quickly followed, and in the intervening years a second generation of dedicated burst instruments has come into existence, with the interplanetary network of small detectors making possible extremely accurate source locations in some cases.

At the same time instrumentation for the observation of solar flares has undergone a steady development, with emphasis in the recent Solar Maximum Mission upon broad spectral coverage as a means of maximizing the information derived from a given transient event. Important solar measurements still are missing, especially high-resolution spectroscopy in the hard X-ray and γ-ray bands.

We therefore have proposed a lightweight package to accomplish both the solar and cosmic observations, the STEP (Spectroscopy of Transient Energetic Photons) instrument. The description given here refers to its configuration for the OPEN mission (Origin of Plasmas in the Earth's Neighborhood), but we feel that some of the design properties would be useful in instruments optimized for other opportunities. In view of the systematic effects that may limit the precision of timing intercomparisons among different burst detectors, it is most important that a future third generation of transient monitors incorporate modern detector schemes and have sufficient attention paid to problems of calibration to reduce these systematic effects.

EXPERIMENT LAYOUT

A schematic drawing of the proposed configuration of the STEP experiment is given in Figure 1. The location of the satellite at the L1 Lagrangian point of the Earth-sun system conveniently permits the use of a passive radiator rather than a cryogenic cooler; this has many advantages, including simplicity, reliability, low cost, long life, and ease of detector temperature manipulation. Temperatures on the order ~70 K are required for the solid-state detectors described below, and can be achieved with a two-stage device as illustrated in Figure 1. The high-energy detectors are located on the axis of symmetry of the radiators, since some absorbing material can be tolerated. The low-energy detectors, however, must face the sky through a minimal entrance window. The STEP design therefore places them in four locations at the periphery of the first-stage radiator, where the spinning motion of the spacecraft causes them to scan past the sun each rotation. The sky coverage is not completely uniform, especially at low energies, but design modifications for a different opportunity could improve this situation.

A different layout embodying some of the same design principles might be more desirable for a different spacecraft. In particular a passive radiator may not be appropriate for all missions, but it might be consistent, for example, with a near-Earth orbit of a satellite stabilized by gravity gradient. However, it should be emphasized that a deep-space orbit, such as the "halo" orbit of ISEE-3 at the Lagrangian point, has many advantages over a near-Earth orbit, especially in the area of background effects for high-energy observations as well as in providing a good baseline for timing determination of source locations.

DETECTOR COMPLEMENT

The STEP experiment observes in the 2 keV - 20 MeV spectral range with an array of three detector types: Si(Li) solid-state detectors for low energies, a high-purity germanium detector for high spectral resolution for hard X-rays and γ-rays, and a bismuth germanate scintillator for high sensitivity at high energies. We briefly review the reasons for choosing these detector types. Interestingly, even though the detectors selected have very desirable properties for the observation of transient events, only germanium[3,4] has had any application to such observations. The STEP layout includes a cup-shaped plastic anticoincidence shield to reject the cosmic-ray background; this is especially significant at higher energies. Figure 2 shows the spectral resolution of these detectors in comparison with previously used types.

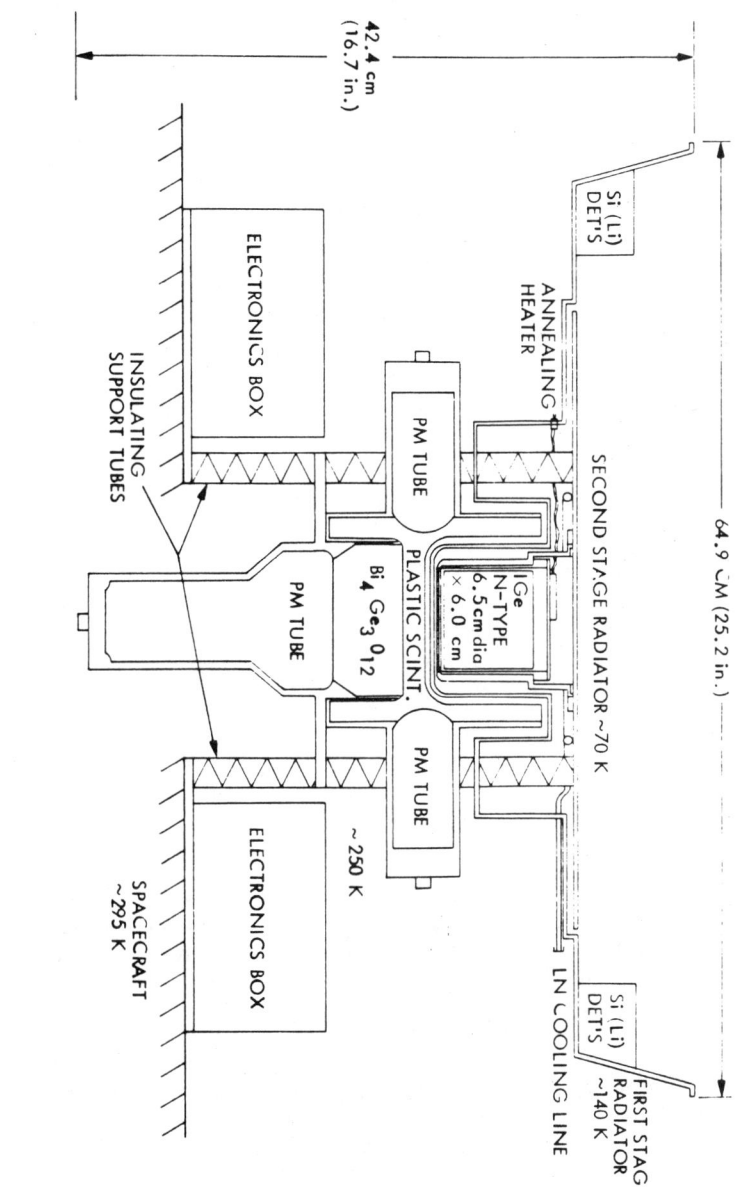

1. Schematic layout of the instrument. The axis of symmetry is aligned with the spin axis, so that the side-mounted Si(Li) detectors (in four places) scan across the solar direction, in the configuration of the Interplanetary Physics Laboratory spacecraft.

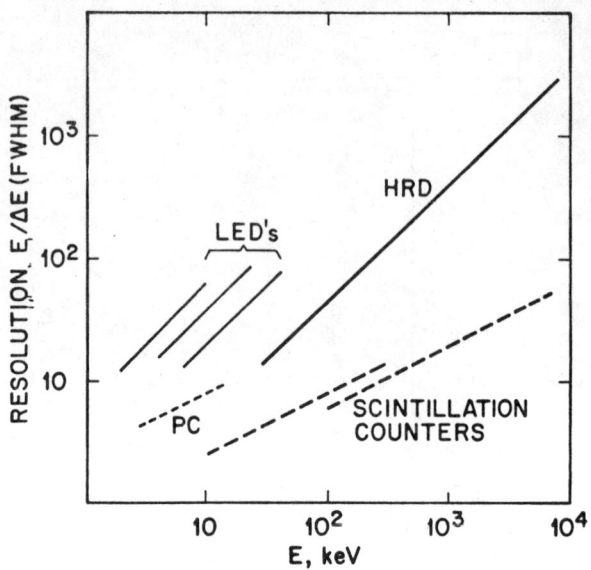

2. Comparison of the spectral resolution for the detector array (LED: silicon detectors; HRD: high-purity germanium detector) described here, in comparison with typical proportional-counter (PC) and scintillation-counter performance.

The Si(li) detectors give high resolution at low energies, but with the disadvantage of small area. STEP contains a total area of 20 cm^2 in Si(Li), with an energy resolution of about 0.5 keV (FWHM) over the 2 - 30 keV range. Smaller detector elements give higher spectral resolution, but with correspondingly limited sensitivity. For the STEP experiment, an array of three detector areas with different entrance-window thickness was optimized for different regions of the parameter space of temperature and emission measure for solar flares. A different optimization - the minimum window thickness - would be more appropriate for a purely cosmic X-ray observation.

The advantages of high-purity germanium are clear, in that it can achieve spectral resolution on the order of 1 keV in the hard X-ray and γ-ray spectral regions. The STEP experiment includes one large-volume detector (~180 cm^3) of n-type germanium for high-resolution observations of bright transients. The passive cooling scheme employed by STEP is particularly convenient for operating germanium detectors over long durations, since re-annealing, if needed, can be accomplished with a small heater and a high thermal impedance between the detector unit and the cryostat.

Finally, the bismuth germanate detector combines high density (ρ = 7.13 gm/cm^3) with high Z - the crystal is 40% by weight of bismuth[5,6] Thus a small detector can have excellent photopeak efficiency for high-energy γ-radiation. We illustrate in Figure 3 the photopeak efficiency of the STEP detectors, and in Figure 4 the

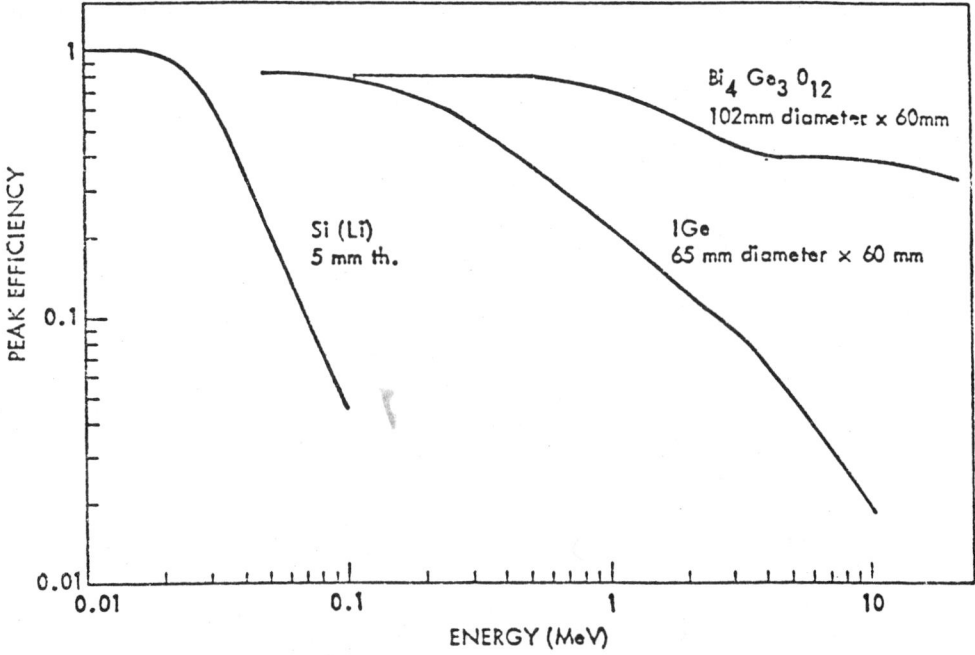

3. Peak photon efficiencies for the three principal detectors as a function of energy, as derived from a detailed Monte Carlo simulation of the response at high energies.

sensitivity to unresolved emission lines of the two high-energy detectors, based upon conservative estimates of the background spectra. For shorter integration times, the noise in the signal may predominate, so we have displayed the unresolved-line sensitivity for integration times from 1 to 1000 sec. The basic equation[7] governing the sensitivity is:

$$F_{min} = \frac{S^2}{eAT} \left[\frac{1}{2} + \sqrt{\frac{1}{4} + \frac{ABET}{S^2}} \right] \quad (1)$$

where F_{min} is the minimum detectable flux (ph/cm²sec), eA the photopeak efficiency (cm²), AB the differential background counting rate (counts/sec per MeV), T the integration time (sec), E the energy resolution (MeV), and S the statistical significance of the desired result (S = 5 for 5 σ). It should be noted that the actual line sensitivity is also strongly affected by the continuum spectrum of the source, and that the higher resolution of the germanium detector may give it a relative advantage in this respect.

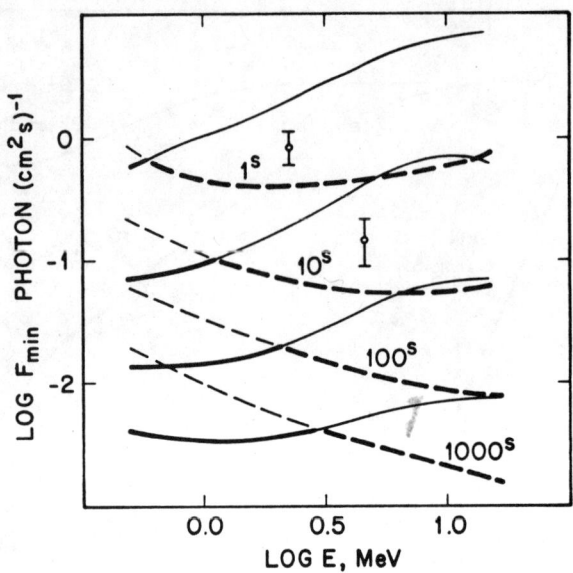

4. Sensitivity of the high-energy detectors for unresolved γ-ray lines, as a function of energy and integration time. The points represent the line intensities of the 1978 July 11 solar γray burst; the sensitivities are based upon equation (1) with no allowance for source continuum. Dashed lines give the germanium sensitivity, solid lines the bismuth germanate sensitivity.

CONCLUSIONS

The instrument described briefly here contains a relatively powerful array of detectors in a light (15 kg) package; with on-board storage (~3.5 megabits) and some on-board processing, a continuous telemetry bit rate of about 200 bps can easily provide both monitoring functions and complete event descriptions. None of the components of the package represents difficult or untested technology, so the costs should be relatively low. The energy resolution, both for X-rays and γ-rays, greatly exceeds that of existing data. Even more important for theoretical interpretation, the instrument by itself provides a very broad spectral range, possibly more than three decades in a favorable event. We encourage the development of instrument packages with these properties, and suggest that extensive laboratory work be carried out to establish the detector properties thoroughly before launch. Finally, the ideal observation with detector packages of this type would consist of multiple spacecraft in solar orbit.

REFERENCES

1. Klebesadel, R.W., Strong, I.B., and Olson, R.A., Astrophys. J. (Lett.) 182, L85 (1973).
2. Cline, T.L., Desai, U.D., Klebesadel, R.W., and Strong, I.B., Astroph J. (Lett.) 185, L1 (1973).
3. Imhof, W.L., Nakano, G.H., Johnson, R.G., Kilner, J.R., Reagan, J.B., Klebesadel, R.W., and Strong, I.B., Astrophys. J. 191, L7 (1974
4. Teegarden, B.J., and Cline, T.L., Astrophys. J. (Lett.) 236, L67 (1980); erratum 240, L175 (1980).
5. Cho, Z.H., and Faroukhi, M.R., J. Nucl. Medicine 18, 841 (1977)
6. Drake, D.M., Nilsson, L.R., and Faucett, J., Nucl. Inst. Measurements (to be published, 1981).
7. Peterson, L.E., Ann. Revs. Astron. Astrophys. 13, 423 (1975).

PROPOSED HARD X-RAY IMAGING AND GAMMA RAY BURST STUDIES FOR XTE

J. E. Grindlay and S. S. Murray
Harvard-Smithsonian Center for Astrophysics

ABSTRACT

A new hard x-ray detector concept is proposed for XTE which would enable very high sensitivity studies of persistent hard x-ray (~15-300 KeV) sources and gamma ray bursts (~50-550 KeV). Coded aperture imaging is employed so that ~2′ source locations can be derived within a 3 field of view. Gamma bursts could be located initially to within ~2° and x-ray/hard x-ray spectra and timing, as well as precise locations (~2′), derived for possible burst afterglow emission.

INTRODUCTION

A high sensitivity study of cosmic x-ray sources at energies of ~20-200 KeV and above has not yet been done; the promise of the band remains unfulfilled. The EINSTEIN Observatory has demonstrated (for soft x-ray astronomy) the incredible wealth of new results obtained when large sensitivity increases are achieved. Such an increased sensitivity is possible with the Energetic X-ray Imaging and Timing Experiment (EXITE) described here. The EXITE detector will allow a factor of up to ~30 increase in sensitivity over the HEAO-A4 results (see Figure 1) for detection of persistent hard x-ray sources. EXITE would also enable significant new studies of gamma ray bursts to be made.

The scientific objectives (in abbreviated form) of EXITE are: (1) to measure Comptonization tails and cyclotron features in source spectra to determine physical conditions at the source; (2) to measure spectral-temporal variability at high energies and study accretion flows; (3) to locate and identify hard x-ray sources in complex fields; (4) to measure high energy spectra of quasars and locate serendipitous sources; and (5) to measure the size spectrum (log N - log S), broad energy spectrum (including line emission) and source locations for gamma ray bursts.

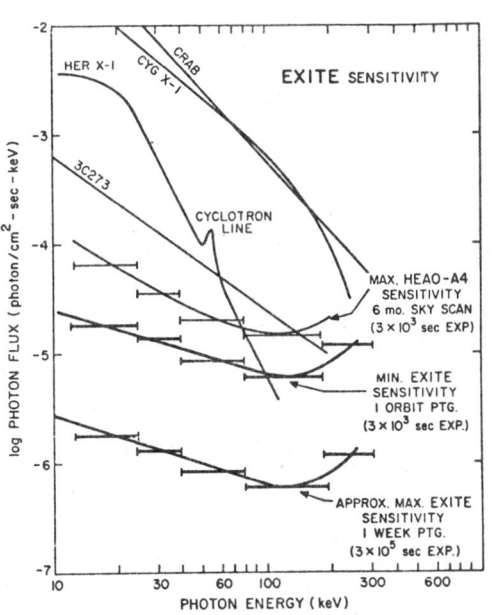

Figure 1: EXITE Sensitivities for 5σ Detection

0094-243X/82/770477-11$3.00 American Institute of Physics

GENERAL DESCRIPTION OF EXITE

EXITE represents a fundamentally new concept in hard x-ray astronomy. It applies coded aperture imaging to the hard x-ray band (~15-300 KeV) and at once realizes several key advantages previously only possible at low x-ray energies: (1) background is measured simultaneously with the object flux; there is no need to scan or off-set point; (2) a source can be resolved in complex fields; and (3) precise (~2-5 arcmin) positions can be derived for the first time at hard x-ray energies (~100 KeV) by centroiding between resolution elements of the reconstructed image.

Coded aperture imaging systems are based on the multiple pinhole camera[1] where N overlapping images of a source distribution are formed on a detector viewing the source through N pinholes in an aperture mask. Although Dicke suggested using a random hole pattern, images reconstructed using a "random" pattern (never truly random) contain an additional source of systematic noise. To overcome this we suggested (in a 1976 Spacelab 2 proposal) using a "pseudo-noise" (PN) array since these have the optimum property[2] that their autocorrelation function (ACF) is identically constant off axis and a δ-function on axis. PN arrays were independently pointed out by Gunsen and Polychronopulos[3] and have been discussed extensively by Fenimore[4] (and references therein) who call them uniformly redundant arrays (URA). The (systematic) noise-free imaging possible with a URA mask is not compromised by a physical realization of such a pattern where the hole size is smaller than the hole spacing. Furthermore, the URA is cyclic and has the very desirable "window property" that a complete pattern is available to surround every element of an extended mask.

The EXITE detector makes optimum use of these desirable properties of the URA in that each single detector (a 15 cm diameter CsI scintillator crystal) views the sky through a URA mask composed of a number (~50) of contiguous cycles of a basic URA pattern (each of dimension 13 x 11; see Figure 2). A single detector, which is read out as one device, then effectively (in the analysis) produces ~50 independent and parallel 13 x 11 pixel images which are then co-added. This allows a very much faster analysis (by a factor of 50) of the total image than if the entire detector area viewed the entire URA mask as in a "conventional" coded aperture camera. The mosaicing of the detector and mask also allows the optimum packing of the rectangular URA pattern into the round detector elements required (because of the need for a diode image intensifier tube to amplify the scintillator optical output). Finally, the URA mask is made self-supporting by placing the (1.5 mm square) holes on a 1.7 mm grid. This is turned to additional advantage by using the 1.5 mm holes as the detector collimator. Collimation (2.9° FWHM) is achieved by

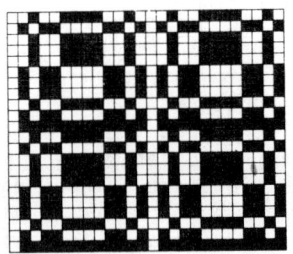

Figure 2: A 2 x 2 Cycle Portion of the Extended URA Coded Aperture Collimator

making the mask effectively very thick (3 cm) using spaced tungsten laminates (see Figure 3).

DESIGN PRINCIPLES AND OPERATION OF EXITE

The principles of operation of the EXITE detector are illustrated in Figure 4. The URA mask projects a collimated image of the sky onto a CsI scintillator. The light produced in the CsI scintillator by an incident x-ray is amplified by the intensifier while preserving the spatial distribution of brightness. The intensified image is decreased in scale to match the input of the MAMA detector through a reducing fiber-optic (12:1). The light losses in this reduction are compensated by the gain of the intensifier so that the total number of photons incident on the photon-counting imaging detector is the same as initially collected from the scintillator. This preserves the energy resolution capability of the system which is basically limited by the counting statistics of the light from the CsI. The imaging optical detector required would be a resistive anode[5] crossed grid[6] or multi-anode[7] type system. The Multi-Anode-Microchannel-Array (MAMA) detector[7] is especially well suited for EXITE and, therefore, included in the design here.

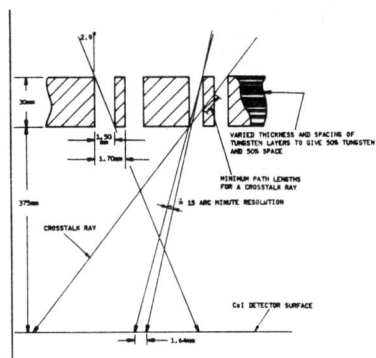

Figure 3: Laminated Tungsten Coded Aperture Collimator

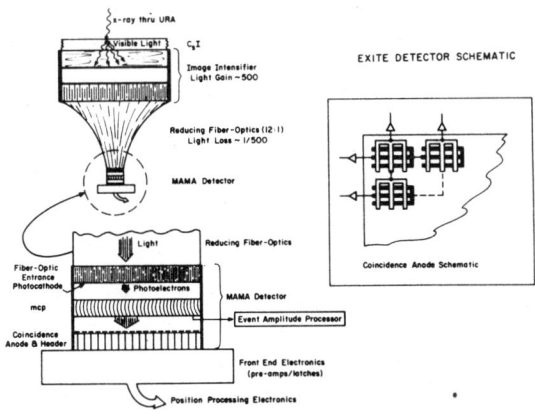

Figure 4: Schematic of Position-Sensitive Scintillation Detection with Optical Imaging Readout

The anodes of the MAMA detector are 140 μm x 140 μm which corresponds to the hole spacing in the URA mask (1.7 mm) when the reduction factor of 12:1 is considered. Light reaching the MAMA from a single x-ray results in the production of a few tens to several hundreds of photoelectrons depending on incident energy. These will be spread out over several of the 140 μm pixels due to the thickness of the CsI crystal and the front window of the image intensifier. Thus, several of the anode lines along each axis will detect the event. A centroid estimate of the event position is then determined for each axis by the processing electronics. Since the photoelectrons are spread out on a large scale relative to the MCP channel size (~15 μm), each electron will be independently amplified in the microchannel plate (MCP) array (with gain ~10^6) and the total charge coming out of the plate will be proportional to the incident number of photoelectrons and thus the primary x-ray energy. This will be measured using a separate amplifier at the output of the MCP to provide energy information (64 channel PHA). Single electron events due to thermal generation either in the MAMA or the image intensifier can easily be discriminated against using the total charge output from the MCP since a 20 KeV x-ray is expected to produce at least 20 photoelectrons.

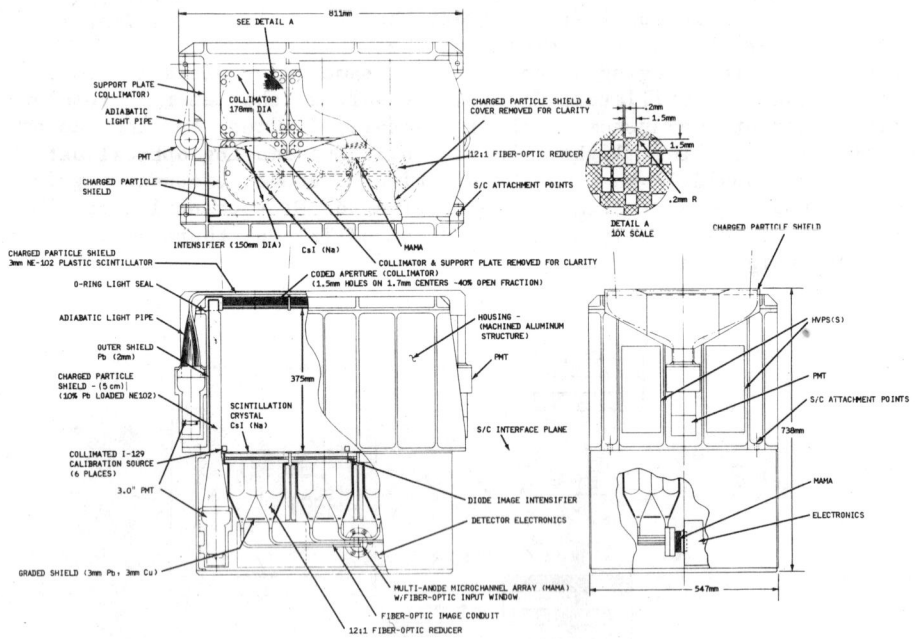

Figure 5: EXITE System Layout

EXITE is a modular detector concept. As proposed for XTE, two independent MAMA detectors (each viewing three 15 cm diameter scintillator x-ray detectors) are included for maximum area and the desired redundancy against single point failures. Each detector, however, is actually composed of three independent and parallel scintillators (for maximum usage of the 25 mm circular photocathode area of the MAMAs), and each single 15 cm diameter scintillator operates (in the coded aperture imaging) as a mosaic of adjacent but independent image-forming subdetector elements. All detector subelements (each is 2.2 cm x 1.9 cm) view the same 2.9° (FWHM) field of view on the sky. Thus, each of the two MAMA detector systems is itself actually multiredundant while at the same time easy to test, calibrate, and read out as a single device.

The EXITE design for XTE has a total active detection area of A_{eff} = 438 cm^2 (= geometric area times 0.4 transmission of the collimating coded aperture mask). The detector employs both active and passive shielding to reject and reduce background. The system is shown in Figure 5 above and occupies an area of 81 cm x 54 cm and height (above the spacecraft interface) of 40 cm. It weighs 237.9 kg and requires 3.3 Kbps (max.) of telemetry and 39 watts of power.

ANTICIPATED EXITE PERFORMANCE

A full simulation of the EXITE detector and optical imaging characteristics has been carried out. Comparison of the EXITE detector system with that[8] flown on HEAO-A4 indicates the total background count rate in the ~15-300 KeV band should be ~35 counts sec^{-1} outside the SAA. Model source spectra were allowed to be incident (on axis) on the full system for a given integration time and were "detected" with the complete system efficiencies and a gaussian blurring (σ = 0.5 pixel) in the centroid determination.

Images of Cyg X-1 for a short (10 sec) and long (500 sec) exposure and of 3C273 for a long (5 x 10^4 sec) exposure are shown in Figure 6 as they should appear from EXITE. These images contain the full 15-300 KeV bandwith of the instrument; similar images could be shown for any band. The full 13 x 11 pixel (2.9°) image is shown; each pixel is 15 arcmin. The sources are detectable in a ~1 pixel radius about the center due to blurring in the MAMA detector readout;

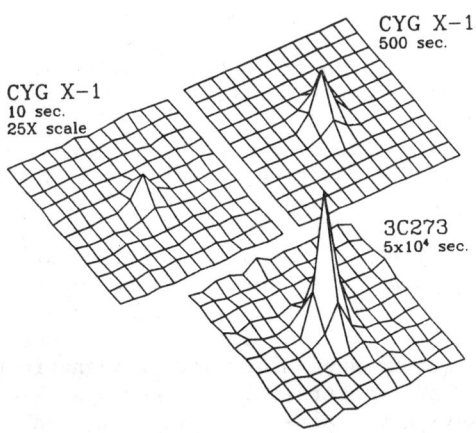

Figure 6: Images (15-300 KeV) from Simulated EXITE Data

centroiding is then possible and (for the long exposures) would yield positions with 2 arcmin accuracy. Additional sources in the field which were up to 86 times fainter than Cyg X-1 and ~42 times fainter than 3C273 would be detectable (and located to ~5 arcmin) in the two "long" exposure images, respectively. The total detected count rate of Cyg X-1 by EXITE is 110 counts sec^{-1} while 3C273 is 2.8 counts sec^{-1}. Since Cyg X-1 is the brightest known (steady) source in the ~20-300 KeV energy band, the 100 counts sec^{-1} allowed for in the telemetry has been selected.

The spectra of counts detected in both long exposures are shown in Figure 7 as the plotted points (with actually "observed" 1σ error bars). Also plotted as lines are the assumed input (power law) spectra for both sources. The spectral indices could be derived from the data for either source with uncertainties $\Delta\alpha/\alpha$ of only a few percent.

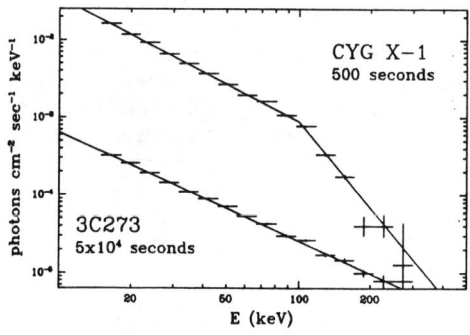

Figure 7: Spectra Derived from Simulated Images

GAMMA RAY BURST STUDIES WITH EXITE ON XTE

The EXITE detector on XTE would enable detailed studies of gamma ray bursts to be made. For the first time, both high sensitivity x-ray (~1-30 KeV) as well as hard x-ray (~15-500 KeV) spectra and timing of gamma bursts could be obtained by the large area proportional counter (LAPC) assumed present on XTE and EXITE instruments, respectively. Since recent (Soviet) experimental evidence[9] for cyclotron and positron annihilation lines from bursts as well as the ~8 sec pulsations from the spectacular March 5, 1979, burst[10] point toward a neutron star origin, gamma bursts are more than ever appropriate for study from XTE. Nuclear flash models for gamma bursts can account for many of the essential features[11] of bursts and--because of the requirement for strong magnetic fields--would predict an x-ray (~20 KeV) afterglow for perhaps ~10^3 sec. This afterglow emission may have already been detected by the Vela satellite x-ray detectors[12] but more sensitive x-ray and hard x-ray observations are very much needed. This could be conducted on XTE with the EXITE detector, which would both initially detect gamma bursts and measure their spectrum (~50-500 KeV) and approximate source location (~2°) as well as precisely locate (~2´) and study the afterglow emission. No other gamma burst detector systems now being considered could accomplish such broad band (x-ray through gamma ray) studies (spectra and single-station burst location capability) of gamma bursts.

BURST DETECTOR SYSTEM

The burst system consists of six identical CsI(Na) scintillation detectors (see Figure 8) mounted externally to the EXITE instrument (or even elsewhere on the spacecraft) in such a way as to provide a constant complete sky survey of x-ray/gamma-ray radiation in the ~50-550 KeV energy range. Each detector assembly is composed of a 15.2 cm square by 0.64 cm thick CsI(Na) scintillator, an adiabatic twisted light pipe, a short 38 mm diameter photomultiplier tube (PMT), and associated electronics. Approximately half of the sky is visible to groups of three adjacent detectors. Burst arrival direction cosines may be computed from the relative rates in these various groups of three detectors. Moderately intense bursts ($\geq 1 \times 10^{-5}$ erg cm^{-2}) can thus be located to within ~2° and a real-time spacecraft slew performed (for events sufficiently close to the XTE pointing direction) to obtain a precise (~2 arcmin) location with the primary EXITE imaging detector system.

For the detectors mounted at 45° to the EXITE shields, a graded shield of lead/copper has been placed on the back side of each scintillator to limit the scintillators to forward viewing only in the gamma-ray energy range below 500 KeV. The window over the front side of all six scintillators is about 1.6 mm thick aluminum. This thickness provides electron shielding to 900 KeV but at the same time only attenuates the x-ray/gamma-ray intensity at 50 KeV by 15 percent (but the high background from lower energy discrete sources and diffuse background is attenuated). Alternatively, a plastic scintillator (~1 cm NE-102) could be mounted on top of the CsI crystals and viewed by the same light pipe and PMT. Particle events could be easily discriminated against by a pulse decay-time discriminator circuit on each detector. Particle-induced burst events are otherwise rejected (in the analysis) by the large area anti-coincidence shields (lead-doped plastic scintillator) on which the burst detector scintillators are mounted (see Figure 8). Rate data from these shields during a burst should also allow the burst radiation scattered from the spacecraft and Earth´s atmosphere (thus producing incident particles) to be isolated and the systematic errors in gamma burst arrival directions to be corrected.

BURST DETECTOR DATA HANDLING SYSTEM

The burst data consists of both high time resolution (0.1 sec) count rates in a fixed band (nominally ~50-300 KeV, but commandable and with a veto for phosphorescence events) and lower time resolution (2 sec) energy spectra (32 channel PHA--~50-550 KeV and 16 bits per channel, or sufficient for even a ~10^{-3} erg cm^{-2} burst) for each

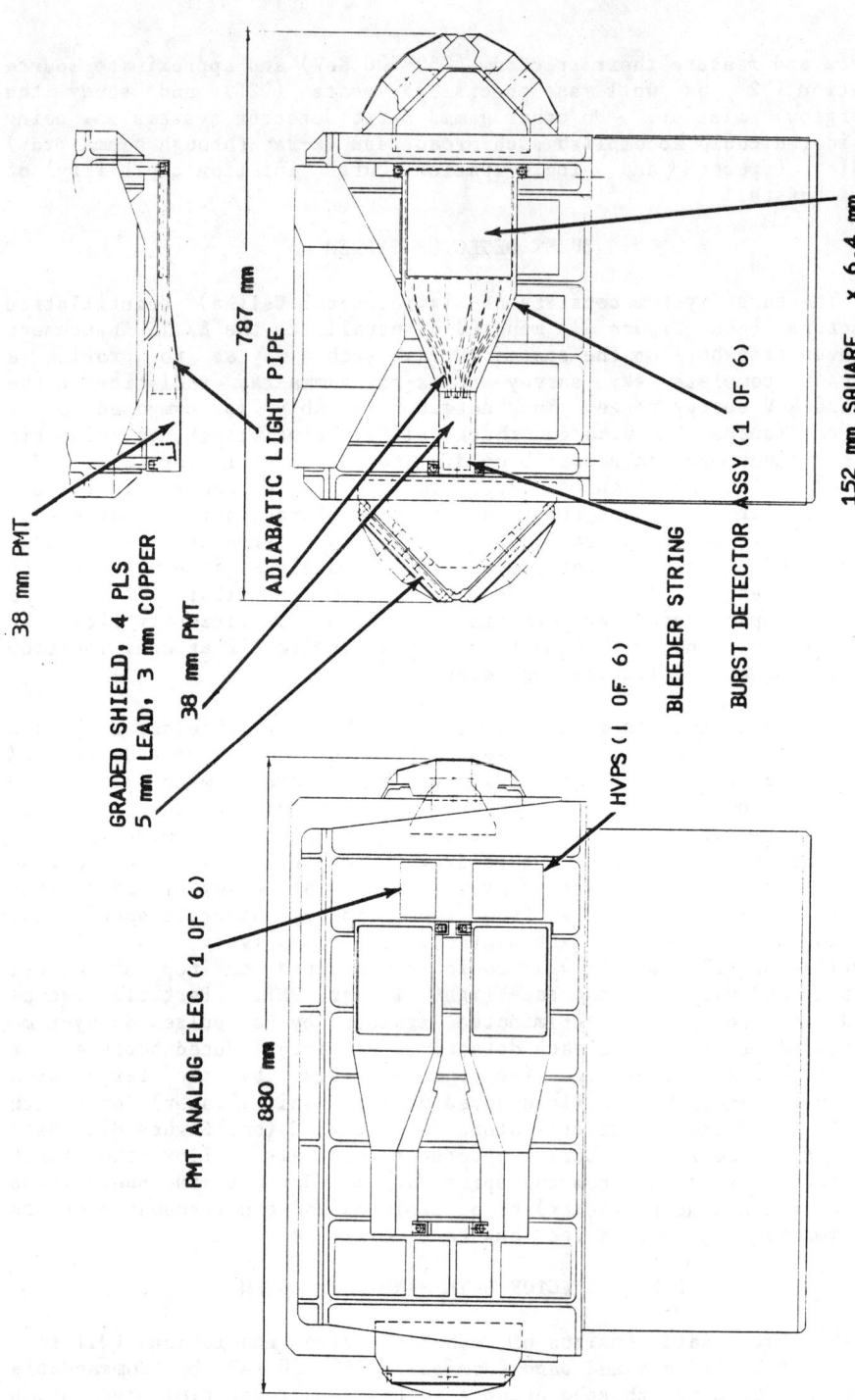

Figure 8: Burst Detector Add On Option

detector. Each detector event rate contained in the 50 to 300 KeV window is counted into 16 bit counters once every 0.1 seconds. Each counter's content is stored sequentially in memory which has the capability (~0.5 Mbit) to hold three minutes worth of data. The counter data are also routed to a µp subsystem where they are reformatted into 0.3-second samples for the continuous on-going normal telemetry data stream. In addition, the µp continually monitors these six rates via a burst mode algorithm to detect bursts (required in 2 detectors). If a burst is detected, a burst flag is set. The flag has 0.1 second time resolution. A sixteen level command buffer is provided for inflight control of the burst level criterion. As an option, the µp can also use the rate data to calculate the direction cosines of the burst location. These direction cosine values are routed through a special spacecraft interface port for possible use in automatically slewing the spacecraft to point at the burst source.

When a burst is detected, the burst flag is set but the memory continues to accept/spill data for an additional two minutes. At the end of this 2-minute period, the memory contents are frozen and remain frozen until the burst flag is cleared by command. Thus, three minutes of time/energy spectrum burst data, one minute before the flag and two minutes after the flag, are preserved. At the end of the 2-minute period, these memory data are automatically sent to the data stream by usurping the telemetry space normally taken by the prime science (imaging) data and housekeeping data. Approximately 164 seconds are required to transfer all the data to the ground. After the intial dump, additional memory data dumps can be made as desired (by command) until the burst flag is reset.

The total EXITE telemetry is organized in two formats. The normal format contains 0.3 second rate data, coded aperture prime science data, and housekeeping data. In the event of a burst (or if the burst flag is set by command) a burst mode format is set. In this format, the rate data continues as before with 0.3 second samples of the six burst detectors and five CPDs, but the memory dump usurps the 3.398 Kb/sec prime data plus housekeeping data rate long enough to be automatically transmitted at the end of flag plus two minutes, or upon command. When no memory dump is needed, the system reverts back to sending prime imaging data and housekeeping data as in the normal mode.

The weight addition for the Burst Detection System (including added electronics and HVPSs) is estimated at 33.4 kg. The power estimate for the burst detector system electronics is 12 watts.

BURST DETECTOR SENSITIVITY AND PERFORMANCE

The background detected by the burst system is due to Earth albedo gamma rays and neutrons, spacecraft albedo, activation of the scintillators, diffuse and discrete source background and phosphorescence from charged particle interactions in the scintillators. Above the ~50 KeV threshold, phosphorescence is negligible and the largest background contribution for these wide-angle detectors is from atmospheric and spacecraft albedo. Using results from Ling[13] and Trombka et al.[14] for these components, respectively, and allowing for a factor of ~2 uncertainty, we estimate a total background spectrum $dN/dE \simeq 2.6\ E^{-0.9}$ cts/cm^2 sec KeV in the ~50-300 KeV band. This gives a total average background count rate $B \simeq 1700$ cts/sec in each of the six detectors. For orbits transiting the SAA (approximately one-third of the orbits), this average rate will increase substantially (factor of ~2-3) due to long-lived (~0.5 hour) activation decays. An increased effective background rate, incorporating the SAA duty cycle, has been used in deriving the sensitivities below.

Using the typical gamma burst spectra derived by Cline and Desai[15] the total signal count rate expected in each detector is $S \simeq 3.7 \times 10^4$ cts/sec for a burst with total energy 10^{-4} erg cm^{-2} and assumed ~1 sec (peak) duration. Thus, the threshold for (5σ) burst detection (in three detectors) should be for energies $>4 \times 10^{-7}$ erg cm^{-2}. This is the approximate threshold for coarse ~20° positions and log N - log S studies. To achieve ~2° burst positions (and high S/N spectra to search for line features), a signal-to-noise ratio of >30 is required in each of three detectors. This should occur for bursts larger than ~1.0×10^{-5} erg cm^{-2}, which are expected (from present log N-log S results) at a rate of ~40 per year within the ~2π steradians for which accurate locations can be derived. (Note that although bursts are detectable with approximately all-sky coverage, accurate locations are possible for about half the sky.) Thus, even if XTE slews could only be carried out within a ~1000 sec afterglow period for burst directions within ~45° of the pointing position, ~10 bursts per year would be accurately located by EXITE and observed with high sensitivity by EXITE and the LAPC. We note that estimates by Woosley[11] for the afterglow spectrum ($L_x \simeq 10^{35}$ erg s^{-1}, $kT \simeq 20$ KeV) would predict EXITE sensitivities of ~25σ per second for a typical burst source at ~500 pc distance. Thus, high signal-to-noise spectra and timing observations should be possible.

Finally, the EXITE burst detection system could detect (5σ) persistent hard x-ray sources with strengths ~1/3 Crab and a Crab-like spectrum in ~300 sec. Thus, hard sources can be monitored continuously and an all-sky monitor capability provided for XTE.

REFERENCES

1. R. Dicke, Ap. J. (Letters), 153, L101 (1968).
2. D. Calabro and J. Wolf, Inform Control, 11, 537 (1968).
3. J. Gunsen and B. Polychronopulos, MNRAS, 177, 485 (1976).
4. E. Fenimore, Applied Optics, 19, 2465 (1980).
5. C. Martin, P. Jelinsky, M. Lampton, R. Malina, and H. Anger, Rev. Sci. Instr., in press (1981).
6. E. Kellogg, S. S. Murray, and D. Bardas, IEEE Trans. Nuc. Sci., NS-26, 403 (1979).
7. J. Timothy, G. Mount, and R. Bybee, SPIE-Optics, 183, 169 (1979).
8. J. Matteson, AIAA 16th Aerospace Sc. Meeting (1978).
9. E. P. Mazets, et al., FTI (Leningrad), Preprint No. 599, 618 (1979).
10. T. L. Cline, et al., Ap. J. (Letters), 237, L1 (1980).
11. S. Woosley, et al., Preprint (1981).
12. J. Terrell, Talk at Texas Symposium, Baltimore, MD (1980).
13. J. C. Ling, J. Geophys. Res., 80, 3241 (1975).
14. J. Trombka, et al., Ap. J., 181, 737 (1973).
15. T. Cline and U. Desai, Ap. J. (Letters), 196, L43 (1975).

ION CHAMBER GAMMA BURST DETECTOR

Stirling A. Colgate
Universty of California
Los Alamos National Laboratory
P.O. Box 1663
Los Alamos, NM 87545

ABSTRACT

A gamma ray burst detector of x-ray photons 2 to 10 keV is designed to maximize area, 100 m^2, and sensitivity, 10^{-10} ergs cm^{-2} s$^{-\frac{1}{2}}$, modest directionality, 2×10^{-4} sr, and minimize thickness, 3 mg cm^{-2}, as a plastic space balloon ion chamber. If the log N - log S curve for gamma bursts extends as the -3/2 power, the sensitivity is limited by gamma-burst peak overlap in time so that the question of the size spectrum and isotropy is maximally tested. Supernova type I prompt x-ray bursts of $\cong$ 3-ms duration should be detected at a rate of several per day from supernova at a distance greater than 100 Mpc.

INTRODUCTION
ASTROPHYSICS

There are several reasons to attempt to extend the sensitivity of detection of gamma bursts to lower levels. The first is the tantalizing question of the origin of gamma bursts. The astrophysical circumstance is reflected in the log N - log S curve (N observed number, S observed burst size) coupled with the angular distribution function. A galactic distribution should show a galactic structure anisotropy at a burst level corresponding to a departure from a linear 3/2 power of the log N, log S curve. There is almost universal expectation that gamma bursts originate from neutron stars; in which case we expect a galactic distribution which reflects the higher velocity and hence higher galactic latitude associated with neutron stars. On the other hand, most likely, only neutron stars of low velocity such as those that would retain either a companion star for mass accretion or planets for comet or asteroid accretion are responsible. It is a tantalizing question of astrophysics to resolve these questions.

Precise locations of the site of the largest gamma bursts give the maximum probability of "seeing" an associated object because regardless of the distribution function at the origin, the largest bursts will be most likely the closest; this despite the controversy of the location of the so-far singular event (March 5) in the Magellanic Cloud. Furthermore, the obtaining of detailed spectra is biased towards large and presumably close events. It is an unsafe but a practical hypothesis to assume that the spectral properties are only weakly dependent on burst size. This assumes that the distribution n of an event source size, SS, at the origin is more abruptly limited, $P(n) < (SS)^{-3/2}$, than the spatial distribution for the isotropic part of the angular distribution. On the other hand, if the number of events at a given source increases rapidly for decreasing event size as for instance asteroid or comet collisions where $P(n) \propto (SS)^{-2}$ to $^{-3}$, then we would expect to see, at very low flux level, primarily a subset of nearby isotropic sources. If the temporal and spectroscopic characteristics vary as a function of source size, as would be expected, then detectability may also be affected, changing the log N - log S curve. In general, a smaller event at the source is likely to be softer in spectra and shorter in time. Hence, a burst detector that has the maximum sensitivity for soft photons and short time has the highest probability of seeing events that depart from the log N versus log S curve because of source characteristics. In addition, the maximum sensitivity detector regardless of spectral or temporal bias has the maximum probability of detecting departures from log N - log S and isotropy.

SUPERNOVA

There are numerous predictions of x-ray bursts from supernova that depend upon the then favored mode of supernova origin as well as presupernova structure. The early detection of SN by any means would greatly increase the probability of understanding SN and being able to use SN as a standard candle to measure the size scale of the universe.[1]

The size of the expected x-ray burst depends upon the heat radiated from the outer layers of the SN just after the explosion shock reaches the surface. The competition between expansion cooling and radiation diffusion determines the x-ray pulse. The extended envelope models of type II SN by Falk and Arnette[2] give a long time, (radius/ velocity $\cong$ several hours) very soft ($<h\nu> \cong 100$ eV) x-ray burst with significant total energy, approximately 10^{48} ergs, or a flux of 10^{44} ergs sec^{-1}. Because of the softness, the fraction of photons above typical x-ray detection threshold (1 to 2 keV) is greatly reduced leading to a low probability of detection. Recently two SN type II's have been observed in ultraviolet spectra (Benvenute 1981) and confirming the very extended envelope model, i.e., a pre-SN stellar wind (larger in one case than in the other) as well as a late time (several weeks to a month) x-ray emission in the second case. The early x-ray pulse for these models is expected to be small because of x-ray absorption blanketing by the tenuous external stellar wind.

Supernova type II on the other hand are now known not to have the very extended envelope, $R \cong 10^{14}$ cm, because the recent early (near maximum light) ultraviolet measurements on SN[3] show an ultraviolet flux that is $\lesssim 1\%$ of the expected Planck spectrum extrapolated from the visible spectrum at the derived temperature of 15,000° K. Since the lack of hydrogen in the spectra is so well confirmed, it is reasonable to expect an initial compact structure typical of a white dwarf. In this case, the surface layer will be shocked to a relativistic energy and the radiation pulse will be short because of size and velocity. This pulse was calculated in the extreme limit of a small size presupernova, radius = 2×10^8 cm and large ejected relativistic mass giving a kinetic energy of 10^{49} ergs.[4] The radiated energy is then a very small fraction of the kinetic energy, 10^{-6} or 10^{43} ergs per logarithmic decade of photon energy. This very small fraction of radiation energy is caused by the same phenomena that restricts gamma bursts to circumstances requiring a strong magnetic field, i.e., the rapid surface blowoff, expansion, and cooling before radiation can diffuse out of the previously shocked hot surface layers. This radiation pulse extends from the highest gamma energies of 10^{12} eV down to the visible and contains roughly 10^{43} ergs per logarithmic decade. The characteristic time of each energy interval is inversely proportional to the energy such that the 3 keV photon band is emitted in $\cong$ 3 ms. If the presupernova star has a larger envelope, the prompt radiation pulse will be more energetic, $\propto R^2$, and last longer, $\propto R$, and so increases detectability.

The detection of the radiation pulse would serve two important astrophysical objectives: it would be the strongest confirmation of the so-far theoretical estimate of the relativistic ejecta and hence possible origin of cosmic rays and, secondly, it would allow the early localization of an expected optical SN type I. A localized optical search should then find an SN type I and the observation of the early SN light curve becomes a distinct possibility. The early light curve becomes critical in understanding the SN phenomenon itself as well as using SN type I as a standard candle for measuring the scale of the universe.[1]

Weak gamma bursts as well as SN bursts lead to very small radiation pulses; current detectors see down to several × 10^{-7} ergs cm^{-2} per pulse, or one 100 keV photon cm^{-2} per pulse. Hence to gain adequate statistics one needs a large area and a sensitivity to a low photon energy. (Present spectra would indicate comparable total energy per logarithmic photon energy internal for both gamma bursts as well as for theoretical SN early bursts.)

DETECTOR CONSIDERATIONS

One then asks what would a detector look like that is optimized for both these qualities. For maximum area and low photon energy one wants the thinnest window material and lowest weight per unit area. The thinnest detector window material that is easily manageable and available in large quantities is Mylar in a thickness range of ¼ mil, i.e., 2.5×10^{-4} in. thick. Only a gas ionization detector can be made as thin as the window. A gas detector also requires high strength in the window material which is satisfied by Mylar. Mylar has an x-ray photon absorption coefficient such that ¼ mil (10^{-3} g cm^{-2}) corresponds to an energy threshold (e-fold attenuation) of 2 keV. At 3 keV the transmission is roughly 75%. In turn a gas filling of A, Ne, Kr, and Xe, and thickness of 10^{-3} g cm^{-2} will result in 75% absorption for photons of 3 to 6 keV. We, therefore, expect 50% photon absorption in the gas of the thinnest detector made of standard ¼-mil Mylar and an equal mass (one Mylar layer) of noble gas filling. The critical question is whether the Mylar strength is adequate to contain this gas pressure. Novick[5] successfully built a Mylar-scrim inflatable proportional counter detector as long ago as 1965.

We assume that the thermal environment cannot be significantly altered from earth-solar equilibrium or 300° K. Hence, the gas pressure for a thickness $\Delta = 10^{-3}$ g cm^{-2} and a spacing d will be

$$P_{gas} = \frac{\Delta}{d\rho_{air}} \frac{A_{air}}{A_{gas}} \text{ atmospheres} .$$

The stress τ in stringers of thickness w = Mylar thickness = 6.2×10^{-4} cm spaced periodically equal to the spacing between window layers, d, becomes

$$\tau = \frac{P_{gas} d}{w} = \frac{\Delta}{w \rho_{air}} \frac{A_{air}}{A_{gas}} \text{ atmospheres} .$$

If we use a heavy gas mixture such that $A_{gas}/A_{air} = 3$, then

$$\tau = 1/3/(w \rho_{air}) = 1.3 \times 10^3 \text{ atmospheres}$$

$$= 20,000 \text{ psi} .$$

This is a near maximum working stress for Mylar whose yield stress is 60 to 100×10^3 psi and so some compromise in gas filling may be necessary. However, scrim threads can greatly increase the strength.

One notes that the stress limitation is independent of the spacing between layers d because we have assumed a spacing d of the stringers equal to that of the spacing between window layers, figure 1. We can, therefore, make our detector with an arbitrary aspect ratio of L/d, i.e., a thin plane of large area L^2.

GEOMETRY

We therefore envision three mutually orthogonal large planar detectors, gas-filled balloons, similar to air mattress construction made of aluminized Mylar. The three planes will have different projected areas (as well as detectability) as a function of viewing angle of an incoming radiation pulse and so give direction, limited by the aspect ratio, d/L, and statistical variation within the signal on each plane.

A very large area would be $L^2 = 100$ m^2 or 10^6 cm^2 and L = 10 meters. Then statistical variations will limit the useful directionality of L/d < 100 or d $\leq$ 10 cm. On the other hand, the spacing d should be as large as possible to ease construction complexity.

We then obtain a balloon-like air mattress 10-cm thick, 10×10 meters in dimension and filled with 1/30 atmosphere of a noble gas mixture.

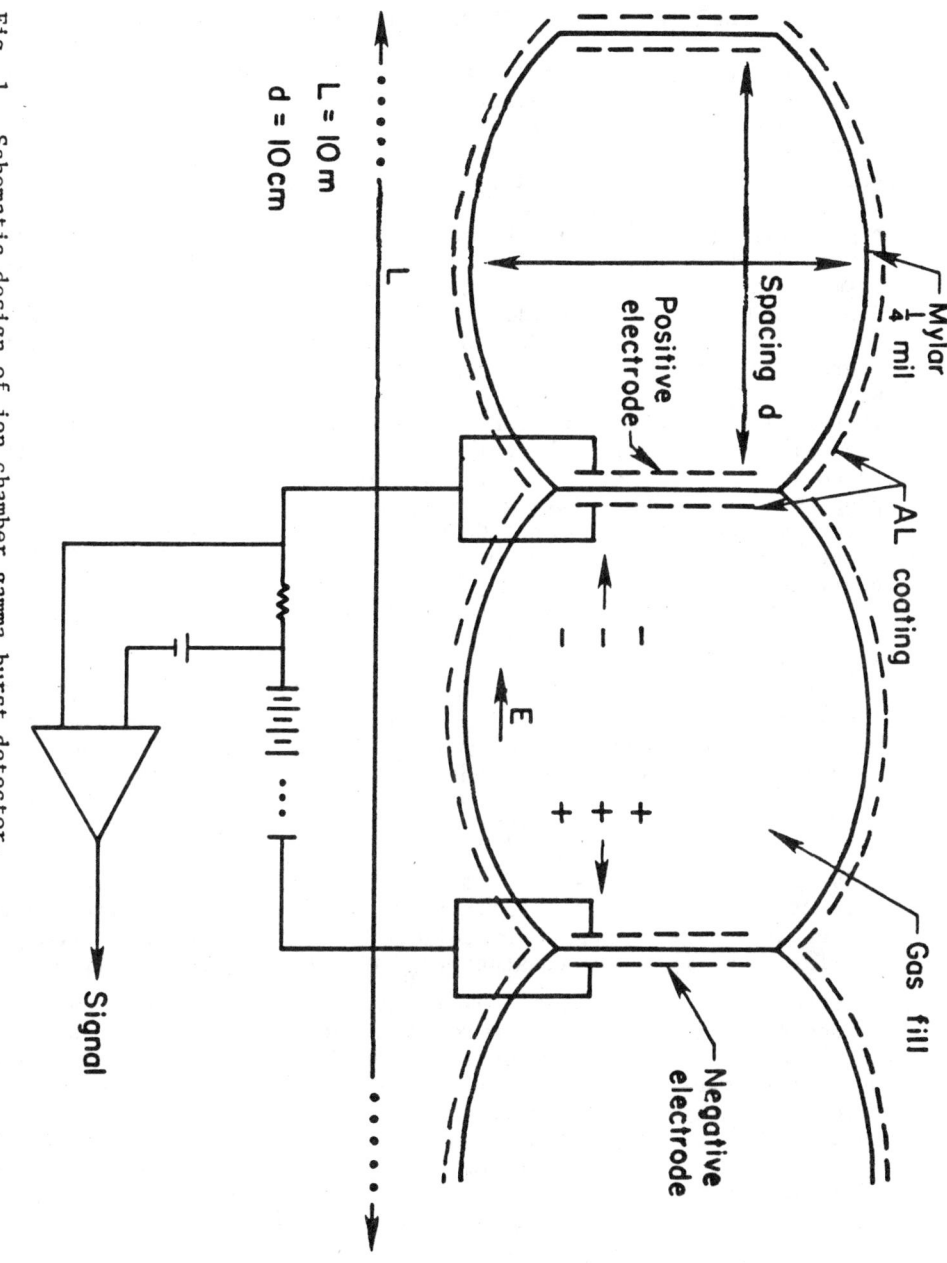

Fig. 1. Schematic design of ion chamber gamma burst detector.

COLLECTION MODE, ION CHAMBER, OR PROPORTIONAL COUNTER

The large area implies a large background counting rate because of cosmic rays and x-ray background. We assume an orbit that avoids the radiation belts and we also assume that the detector is most useful when the earth shields the solar x-ray flux, i.e., ½ the observing time. This strategy has the serendipitous advantage of maximizing the gas filling because of reduced temperature on the night side and also offers the possibility of optical identification of SN, if detected. At 2 keV the isotropic x-ray flux is just the cosmic ray flux of 1 sterad cm^{-2} s^{-1}.^{6} Discrete sources, such as Sco-X1 and the Crab will double this x-ray background and so will correspond to just the CR background or a total of 2 counts sterad^{-1} cm^{-2} s^{-1}. Furthermore, a CR proton at minimum ionization will deposit the same energy (dE/dx) w = 2 keV as an x-ray photon. Therefore, the pulse rate for 2π steradians, i.e., half shielded by the earth and assuming the earth shine above 2 keV is less than the space flux, then results in 10^7 pulses s^{-1} per detector plane. A proportional counter taking into account collection time could probably discriminate against CR's, but for only a reduction of 2 in background. This is hardly worth the complexity of 10^7 s^{-1} event analyses.

Instead, we integrate the ionization signal for a characteristic time Δt that is determined by collection time, bit rate, storage, transmission, etc.

The collection time of an ionization pulse is the drift time in the chamber. The ions will contribute the same as the electrons unless proportional counter gain is needed. If we exclude the complexity of proportional counter gain, then the collection time becomes $\Delta t_{col} = d/v_d = d\, P_{gas}/2E$ where E is the electron field in volts cm^{-1}, P_{gas} in atmospheres, and d in cm. Then for P_{gas} = 1/30 atmosphere and E = 150 V cm^{-1}, i.e., ¼ of the breakdown field, then Δt ≅ 1 ms. This is adequate time resolution but shorter could be an advantage. A shorter collection time is possible using a smaller spacing, d, and constant (P_{gas}/E), but the complexity of the structure becomes greater. Hence Δt = 1 ms and d = 10 cm is a reasonable compromise. The background rate of 10^7 s^{-1} then becomes 10^4 events per time resolution interval of 10^{-3} s. The 5σ detection pulse presumably observable in each of the other 2 planes for 50% detection efficiency becomes 2 × 500 photons = 10^3 photons or 10^{-3} photons cm^{-2} or 3 × 10^{-12} ergs cm^{-1} per pulse.

For a typical gamma burst where the resolution time might be several seconds, the 5σ detection level becomes × 30 larger, or 10^{-10} ergs cm^{-2}. These limits are to be compared to current detectors, the Mazets experiment of 10^{-7} ergs cm^{-2} and the proposed BASTE experiment of 5 × 10^{-8} ergs cm^{-2} for a several second burst. The electrical charge collected per millisecond signal of 5σ n ≅ 3 × 10^4 ions, or well above simple amplifier noise.

EXPECTED BURST RATE

If the log N - log S curve is extended down to this burst level, we expect a frequency $\dot{\varepsilon}$ of one half the basic rate $\dot{\varepsilon}_{01}$ event per year of 10^{-5} ergs cm^{-2} × (gain)$^{3/2}$ = $\frac{1}{2}(10^{-5}/10^{-10})^{3/2}$ = 1.5 × 10^7 per year or 0.5 s^{-1}. This exceeds the effective burst time of several seconds and so only bursts of $\cong 2 \times 10^{-10}$ ergs would be time resolved.

Supernova on the other hand perhaps give rise to pulses of 10^{43} ergs in 3×10^{-3} s at 3 keV. At a distance of 100 Mpc, the pulse size becomes

$$S_{100} = \frac{10^{43}}{4\pi(3 \times 10^{26})^2} = 10^{-11} \text{ ergs cm}^{-2}$$

or 10σ. The rate of events assuming 1 per 200 years per standard galaxy and 0.05 standard galaxies per Mpc^3, is 10^3 per 4π sr or 500 per year per half sky. This is an event rate of several per day, or one per semishielded night-time intervals per orbit day. The maximum optical brightness will be about 16th to 17th magnitude with roughly 100 candidate galaxies in the resolution field of 2×10^{-4} sr.

HEAVY COSMIC RAYS

A significant source of bkg will be very heavy cosmic rays - iron nuclei with a flux of 5.57×10^{-2} m^{-2} sterard^{-1} s^{-1} above the earth's atmosphere corresponding to a low orbit.[7] The ionization signal proportional to Z^2 is 640 times larger. This signal is 6 times larger than our statistical bkg of 100 photons. This rate is 35 s^{-1} per plane detector (2π sr) and so becomes a serious background. On the other hand each plane detector must be divided into several sections for redundancy insurance against a gas leak or electrical failure. Hence a relatively simple signal processing can remove heavy CR nuclei pulses by coincidence analysis. This requires a processing rate of 12 bits per millisecond per detector or 10^5 baud or well within microprocessor capabilities. The storage or transmission requirement is roughly 1 event per second which is also modest.

LAUNCH MODE

This experiment is adaptable to the shuttle mode of launch and retrieval resulting in several days in orbit.

In summary, a large area, thin wall, and small mass (10 kg + electronics) ionization chamber gamma burst detector is designed that can extend the log N - log S curve down to a presumed temporal confusion limit and also detect an expected prompt type I supernova in galaxies at 100 Mpc at a rate of one per day.

I am indebted to many questions from the audience at the Supernova Workshop, Cambridge, England, June 26-July 10, 1981, and the Gamma Burst Conference, La Jolla, California, August 4-8, 1981. This work was supported by the DOE and the Astronomy Section of the NSF.

REFERENCES

1. S. A. Colgate, Ap. J., 232, 404-408 (1979).
2. S. W. Falk, and W. D. Arnett, Ap. J. Suppl., 33, No. 4, 515 (1977).
3. P. Benvenuti, N. Panagia, and W. Wamsteker, private communication, NATO-ASI Conference on Supernovae Cambridge, England, (June 29-July 11, 1981).
4. S. A. Colgate and A. G. Petschek, Supernovae and Supernova Remnants, Memorie Della Societá Astronomica Italiana, 49, No. 2-3, 417, Aprile-Settembre, 1978. ISSN:0037-8720 (1978).
5. R. Novick, private communication (1981).
6. L. E. Peterson, Ann. Rev. of Astron. & Astro. Phys. 13, 423 (1975).
7. W. R. Webber, 17th ICCR, OG1, 1.6, p. 80 (1981).

SAC/rep:54I

Sect. V. PANEL DISCUSSION AND RECOMMENDATIONS

The last morning of the meeting was spent in a panel discussion that involved the active participation of the audience. Panel members were Doyle Evans, Paul Gorenstein, Don Lamb, Reuven Ramaty, Ian Strong (moderator), Bonnard Teegarden, and Stan Woosley. Strong focused the panel's attention on two fundamental questions: 1) "What progress has been made during the last several years toward understanding the nature of high energy transient phenomena?" and 2) "What, as best as can be determined at the present time, should be our priorities for future experiments?"

There was consensus that considerable progress had indeed been achieved recently in our understanding of several types of high energy transient phenomena: X-ray bursts, gamma-ray bursts and solar gamma-ray transients. Key developments catalogued by the panel and the audience included:

* Discovery (although still controversial in some cases) in the spectra of a number of gamma-ray bursts of line features attributed to redshifted pair annihilation, cyclotron emission and scattering or absorption in a strong magnetic field, and possible nuclear deexcitation. Most of the recent data come from instruments on Soviet spacecraft.

* Convergence of theoretical models upon a __magnetized__ neutron star ($B > 10^{12}$ gauss) as the likely site for gamma-ray burst production. This convergence has been motivated by the spectroscopic observations and the historical dilemma of producing gamma-ray emission on a neutron star that at the Eddington limit has a characteristic temperature of only 2 keV. Details of the energizing process that leads to hot magnetically confined plasma remain controversial, but may involve a thermonuclear explosion, sporadic accretion or structural changes.

* Location of the spectacular gamma-ray burst of March 5, 1979 coincident with a supernova remnant in the Large Magellanic Cloud. The rapid rise time, much softer than average spectrum, periodically pulsing tail, enormously super-Eddington luminosity, and recurrence properties of this burst are special, although not necessarily unique.

* Precise locations (0.05 to 14 arc min^2 error boxes) for roughly half a dozen gamma-ray burst sources. Although no clear systematics have yet emerged, several interesting objects have turned up in searches at radio, optical and X-ray wavelengths. Continued studies of this kind may lead to the discovery of prototypical gamma-ray burst sources that may then be monitored at various wavelengths and over a long time period.

* Considerable progress in understanding of X-ray bursts, implying that these phenomena originate from accreting neutron stars (demonstrated by black body estimates of the emission area) and suggesting a relationship between X-ray bursts and gamma-ray bursts. There are at least two classes of X-ray bursts, one apparently involving thermonuclear runaway and the other involving sporadic accretion. There appear to be several classes of gamma-ray bursts and these may also arise from different mechanisms.

* Observations of a rich variety of nuclear gamma-ray lines from solar flares, providing new information on the acceleration of particles in flares, the structure of the energetic particle interaction region and the physics of flares in general. For example, these observations show that protons and nuclei are accelerated to more than 10 MeV/nucleon in less than a few seconds, and comparisons with interplanetary measurements show that gamma-ray emission results primarily from the interaction of particles that remain trapped at the sun.

Most of the panel's time was spent in considering experimental priorities for the future. There was considerable discussion regarding the relative merits of various observational programs, e.g. spectroscopy vis-á-vis positions, and some discussion of the likely costs and technology involved in carrying out these objectives.

The following conclusions concerning future experimental priorities were agreed upon by the panel, with the full participation of the audience, and reflect the consensus of the workshop.

Missions already planned for this decade, principally the Gamma-Ray Observatory (GRO) and the X-ray Timing Explorer (XTE) will greatly extend our knowledge of gamma-ray bursts, perhaps even leading to a qualitative understanding of the underlying mechanisms. In particular, GRO will be capable of broad spectral analysis above 100 keV, positions accurate to 1 degree (with much greater accuracy by timing, if it can be used in conjunction with burst detectors on other spacecraft) and the detection of gamma-ray bursts on at least a daily basis (flux limit of 6×10^{-8} erg cm^{-2}). Any remaining questions concerning the Galactic or extra-Galactic origin of most gamma-ray bursts should be resolved; further spectral study may help distinguish different classes of bursts, and if other spaceprobes are available for a timing network more accurate positions can be obtained for examination at other wavelengths. XTE, while suffering somewhat for the study of gamma-ray bursts from its restricted field of view, may provide much needed spectral information in the lower energy (< 100 keV) range for some bursts, depending on the final instrumental configuration.

Even with the powerful capabilities of GRO, however, there are questions of fundamental importance to our eventual understanding of gamma-ray bursts that will not be adequately addressed.

Chief among these is higher resolution spectroscopy, capable not only of confirming the existence of emission and absorption features but of determining the detailed shape of the apparent cyclotron and annihilation features and of possible higher energy emission lines, such as those reported in the November 19, 1978 burst. Such studies of line features and their temporal evolution would yield unique information concerning thermodynamic conditions, redshifts, magnetic field strengths, and composition at the burst sites. Such information is essential if gamma-ray bursts are to be really studied, not just detected. It is also essential that studies be carried out at softer energies, roughly 1 keV to 100 keV, as well as in the 100 keV to 10 MeV range traditionally associated with gamma-rays. For example, most of the emission from events like that of March 5, 1979 is in this lower energy range as are the cyclotron features and an expected X-ray afterglow at a temperature of $\sim$ 2 keV. X-ray transients have already been detected by the Los Alamos group shortly before and after a gamma-ray burst, and are presumed to originate from the burst source.

Similar spectroscopic observations of gamma-ray lines in solar flares are also needed during the next solar maximum to better understand the rich array of emission lines observed with SMM.

Accurate source locations are also essential if gamma-ray bursts are to cease being astrophysical curiosities and enter the mainstream of astronomical research. We hope that within the decade intensive study of existing burst position error boxes at other wavelengths will have identified prototypical gamma-ray burst sources, and that these sources will be under routine surveillance. But we cannot be certain that this will be the case, particularly if the deep space timing network is not re-established to provide additional precise locations for study. So the ability to precisely determine gamma-ray burst sources from a single spacecraft may still be important.

Thus there exists a need for an Explorer class mission devoted to the study of high energy transient phenomena. The primary objective of such a mission would be the study of gamma-ray bursts, but solar flare and X-ray burst observations would be important secondary goals that could be studied using the same complement of experiments. The primary experimental capabilities should be:

1) Spectroscopy over a broad energy range ($\sim$ 1 keV to $\sim$ 10 MeV) with sufficient energy resolution to resolve the shape of the emission and absorption lines;

2) Source location with sufficient accuracy for follow-up observations at other wavelengths; and

3) Temporal resolution and throughput sufficient to study the evolution of the spectrum.

A number of experimentalists present felt that all these goals could be addressed concurrently using a single satellite of the Explorer class. This would be an Explorer in the traditional sense; that is, oriented toward an in-depth study of a specific class of phenomena and built at modest cost.

We recommend that NASA form a scientific working group to study the design of such a spacecraft.

No. 26	High-Energy Physics and Nuclear Structure - 1975 (Santa Fe and Los Alamos)	75-26411	0-88318-125-8
No. 27	Topics in Statistical Mechanics and Biophysics: A Memorial to Julius L. Jackson (Wayne State University, 1975)	75-36309	0-88318-126-6
No. 28	Physics and Our World: A Symposium in Honor of Victor F. Weisskopf (M.I.T., 1974)	76-7207	0-88318-127-4
No. 29	Magnetism and Magnetic Materials - 1975 (21st Annual Conference, Philadelphia)	76-10931	0-88318-128-2
No. 30	Particle Searches and Discoveries - 1976 (Vanderbilt Conference)	76-19949	0-88318-129-0
No. 31	Structure and Excitations of Amorphous Solids (Williamsburg, VA., 1976)	76-22279	0-88318-130-4
No. 32	Materials Technology - 1975 (APS New York Meeting)	76-27967	0-88318-131-2
No. 33	Meson-Nuclear Physics - 1976 (Carnegie-Mellon Conference)	76-26811	0-88318-132-0
No. 34	Magnetism and Magnetic Materials - 1976 (Joint MMM-Intermag Conference, Pittsburgh)	76-47106	0-88318-133-9
No. 35	High Energy Physics with Polarized Beams and Targets (Argonne, 1976)	76-50181	0-88318-134-7
No. 36	Momentum Wave Functions - 1976 (Indiana University)	77-82145	0-88318-135-5
No. 37	Weak Interaction Physics - 1977 (Indiana University)	77-83344	0-88318-136-3
No. 38	Workshop on New Directions in Mossbauer Spectroscopy (Argonne, 1977)	77-90635	0-88318-137-1
No. 39	Physics Careers, Employment and Education (Penn State, 1977)	77-94053	0-88318-138-X
No. 40	Electrical Transport and Optical Properties of Inhomogeneous Media (Ohio State University, 1977)	78-54319	0-88318-139-8
No. 41	Nucleon-Nucleon Interactions - 1977 (Vancouver)	78-54249	0-88318-140-1
No. 42	Higher Energy Polarized Proton Beams (Ann Arbor, 1977)	78-55682	0-88318-141-X
No. 43	Particles and Fields - 1977 (APS/DPF, Argonne)	78-55683	0-88318-142-8
No. 44	Future Trends in Superconductive Electronics (Charlottesville, 1978)	77-9240	0-88318-143-6
No. 45	New Results in High Energy Physics - 1978 (Vanderbilt Conference)	78-67196	0-88318-144-4
No. 46	Topics in Nonlinear Dynamics (La Jolla Institute)	78-057870	0-88318-145-2
No. 47	Clustering Aspects of Nuclear Structure and Nuclear Reactions (Winnepeg, 1978)	78-64942	0-88318-146-0
No. 48	Current Trends in the Theory of Fields (Tallahassee, 1978)	78-72948	0-88318-147-9
No. 49	Cosmic Rays and Particle Physics - 1978 (Bartol Conference)	79-50489	0-88318-148-7

AIP Conference Proceedings

No. 50	Laser-Solid Interactions and Laser Processing - 1978 (Boston)	79-51564	0-88318-149-5
No. 51	High Energy Physics with Polarized Beams and Polarized Targets (Argonne, 1978)	79-64565	0-88318-150-9
No. 52	Long-Distance Neutrino Detection - 1978 (C.L. Cowan Memorial Symposium)	79-52078	0-88318-151-7
No. 53	Modulated Structures - 1979 (Kailua Kona, Hawaii)	79-53846	0-88318-152-5
No. 54	Meson-Nuclear Physics - 1979 (Houston)	79-53978	0-88318-153-3
No. 55	Quantum Chromodynamics (La Jolla, 1978)	79-54969	0-88318-154-1
No. 56	Particle Acceleration Mechanisms in Astrophysics (La Jolla, 1979)	79-55844	0-88318-155-X
No. 57	Nonlinear Dynamics and the Beam-Beam Interaction (Brookhaven, 1979)	79-57341	0-88318-156-8
No. 58	Inhomogeneous Superconductors - 1979 (Berkeley Springs, W.V.)	79-57620	0-88318-157-6
No. 59	Particles and Fields - 1979 (APS/DPF Montreal)	80-66631	0-88318-158-4
No. 60	History of the ZGS (Argonne, 1979)	80-67694	0-88318-159-2
No. 61	Aspects of the Kinetics and Dynamics of Surface Reactions (La Jolla Institute, 1979)	80-68004	0-88318-160-6
No. 62	High Energy e^+e^- Interactions (Vanderbilt, 1980)	80-53377	0-88318-161-4
No. 63	Supernovae Spectra (La Jolla, 1980)	80-70019	0-88318-162-2
No. 64	Laboratory EXAFS Facilities - 1980 (Univ. of Washington)	80-70579	0-88318-163-0
No. 65	Optics in Four Dimensions - 1980 (ICO, Ensenada)	80-70771	0-88318-164-9
No. 66	Physics in the Automotive Industry - 1980 (APS/AAPT Topical Conference)	80-70987	0-88318-165-7
No. 67	Experimental Meson Spectroscopy - 1980 (Sixth International Conference, Brookhaven)	80-71123	0-88318-166-5
No. 68	High Energy Physics - 1980 (XX International Conference, Madison)	81-65032	0-88318-167-3
No. 69	Polarization Phenomena in Nuclear Physics - 1980 (Fifth International Symposium, Santa Fe)	81-65107	0-88318-168-1
No. 70	Chemistry and Physics of Coal Utilization - 1980 (APS, Morgantown)	81-65106	0-88318-169-X
No. 71	Group Theory and its Applications in Physics - 1980 (Latin American School of Physics, Mexico City)	81-66132	0-88318-170-3
No. 72	Weak Interactions as a Probe of Unification (Virginia Polytechnic Institute - 1980)	81-67184	0-88318-171-1
No. 73	Tetrahedrally Bonded Amorphous Semiconductors (Carefree, Arizona, 1981)	81-67419	0-88318-172-X
No. 74	Perturbative Quantum Chromodynamics (Tallahassee, 1981)	81-70372	0-88318-173-8
No. 75	Low Energy X-ray Diagnostics-1981 (Monterey)	81-69841	0-88318-174-6
No. 76	Nonlinear Properties of Internal Waves (La Jolla Institute, 1981)	81-71062	0-88318-175-4
No. 77	Gamma Ray Transients and Related Astrophysical Phenomena (La Jolla Institute, 1981)	81-71543	0-88318-176-2